Wireless Sensor Networks:
A Systems Perspective

Wireless Sensor Networks: A Systems Perspective

Editor: Phoebe Hill

MURPHY & MOORE
www.murphy-moorepublishing.com

Published by Murphy & Moore Publishing
1 Rockefeller Plaza,
New York City, NY 10020, USA
www.murphy-moorepublishing.com

Wireless Sensor Networks: A Systems Perspective
Edited by Phoebe Hill

International Standard Book Number: 978-1-63987-572-6 (Hardback)

Cataloging-in-Publication Data

Wireless sensor networks : a systems perspective / edited by Phoebe Hill.
    p. cm.
Includes bibliographical references and index.
ISBN 978-1-63987-572-6
1. Wireless sensor networks. 2. Wireless communication systems.
3. Sensor networks. I. Hill, Phoebe.
TK7872.D48 W57 2022
681.2--dc23

Contents

Preface

Every book is initially just a concept; it takes months of research and hard work to give it the final shape in which the readers receive it. In its early stages, this book also went through rigorous reviewing. The notable contributions made by experts from across the globe were first molded into patterned chapters and then arranged in a sensibly sequential manner to bring out the best results.

Wireless sensor network can be described as a self-configured and infrastructureless wireless network that is used to monitor physical or environmental conditions, such as temperature, sound, vibration, pressure, motion, or pollutants. It also conveys its data through the network to the main location or sink where the data can be observed and analyzed. A sink or base station is an interface between the users and the network. The required details from the network can be retrieved by putting queries in the sink. Wireless sensor networks are applied in various domains such as the military, area monitoring, transportation, health applications, environmental sensing, structural monitoring, industrial monitoring, and the agricultural sector. This book elucidates the concepts and innovative models around prospective developments with respect to wireless sensor networks. The ever-growing need for advanced technology is the reason that has fueled the research in WSNs in recent times. This book will provide comprehensive knowledge to the readers.

It has been my immense pleasure to be a part of this project and to contribute my years of learning in such a meaningful form. I would like to take this opportunity to thank all the people who have been associated with the completion of this book at any step.

Editor

The sustainable development information management of Winter Olympics based on Internet-based wireless sensor networks

Maomao Zhang* and Feng Zhai

Abstract

Environment is the foundation for human survival. Sustainable development is the main direction of social progress and the common choice of human progress. The development of the Winter Olympic Games is closely related to the progress of mankind and society. Therefore, it is particularly important to study the sustainable development of the Winter Olympic Games. Based on this, an Internet-based wireless sensor for Winter Olympic environment information acquisition scheme is proposed. The hardware and sensor modules of the wireless sensor are designed, and the adaptive weighting algorithm is used to fuse several wireless sensor nodes. By testing and analyzing the communication ability of a single wireless sensor node and the error rate under different distance, the data is collected for the environment information of a Winter Olympic sports field. The results of the data collection are very close to the actual environmental parameters. It shows that the wireless sensor network system based on the Internet has strong stability and reliability in data collection. It can provide effective data support for the information management research of sustainable development of Winter Olympic Games.

Keywords: Internet, Wireless network, Sensors, Winter Olympic Games, Sustainable development

1 Preface

As the environmental problems become more and more serious, after "movement" and "culture," "environmental protection" has become the third main theme of the Olympic movement. Its importance and concern are increasing [1]. In 2010, the Turin Olympic Games put forward a series of green motions, such as reducing greenhouse gas emissions, reducing the water consumption to the lowest, and promoting the construction of environment-friendly hotels. It indicates that the Green Olympics will reach a new peak [2]. At the 42nd Congress of the United Nations, the World Commission on environment and development formally proposed the concept of sustainable development: sustainable development is a development that meets the needs of contemporary people and does not constitute a harm to the ability of future generations to meet their needs [3]. In recent 30 years, sustainable development has become the universal values and codes of conduct widely recognized by the international community. The sustainable development of the Olympic movement refers to the active

and effective process of the Olympic movement, taking the Olympic Games, the Paralympics, and the Youth Olympic Games as the platform and taking sports, culture, and education as the means to promote the active and effective transformation of the people, the environment, the city, and the society [4, 5]. Sustainable development is the mainstream direction of the society and the common choice of human progress. It is also the best goal of the Olympic Games to make unremitting efforts. The sustainable development of the Winter Olympic Games follows this concept [6].

With environmental sustainability becoming a hot topic, environmental monitoring technology has also encountered unprecedented challenges, mainly reflected in the growing demand for dynamic data acquisition, real-time tracking, visual management, and dynamic analysis and forecasting of monitoring data [7]. The technology of remote monitoring of environment is integrated with sensing technology, communication technology, and computer technology. It effectively completes the functions of collection, storage, remote transmission, and real-time processing of various environmental parameters. It has changed the past backward situation

* Correspondence: ka92695@163.com
China University of Mining and Technology, Xuzhou 221000, Jiangsu, China

only by manpower, improved the work efficiency, and increased the transmission distance. It plays a very important role in disaster prevention and mitigation, environmental condition prediction, disaster prediction, and so on [8, 9]. Remote monitoring of the environment has its specific application background, generally in unpopulated uninhabited areas (deserts, mountains, jungles, high temperatures, high pressure, high altitude, high danger) in a remote and harsh environment. This makes it difficult to complete data transmission by installing cables. In this case, the use of Internet wireless communication is a good choice [10]. In recent years, because of the rapid development of computer technology, network technology, and modern electronic technology, wireless communication technology has been widely used in many fields, such as medical treatment, automation control, and remote monitoring [11–13]. In addition, the wireless sensor network based on the Internet can monitor, perceive, and collect the information of various environmental information monitoring objects in the distributed area of the network in real time, and process these information to the users who need these information [14].

2 Design of wireless sensor network monitoring system based on the Internet

2.1 The role of wireless sensor in the sustainable development of Winter Olympic Games

The sustainable development is the inevitable choice of the human society. In the journey towards the modern industrial society, the expansion of the city, the rapid expansion of the population, and the one-sided pursuit of economic indicators have broken the harmony with nature. The difficulties of air pollution, ecological vulnerability, resource exhaustion, and endangered species are difficult to be ignored, and social problems such as the spread of lifestyle diseases and mental subhealth cannot be underestimated [15]. These problems are intertwined, intricate, global spread, and far-reaching influence, and pose a serious threat to the survival of mankind [16]. Therefore, the path of sustainable development is an inevitable choice for human progress and social benign operation. Since the beginning of its founding, the Olympic Movement has been steadfastly committed to "building a peaceful and beautiful world." Therefore, the issue of sustainable development has naturally attracted the attention and active enthusiasm of the Olympic people headed by the International Olympic Committee [17]. Figure 1 shows the importance of environmental sustainability for the Winter Olympics. Therefore, it is very necessary to use the Internet-based wireless sensor network to monitor the environmental information of the Olympic Winter Games to promote the sustainable development of the environment.

Internet-based wireless sensor technology and wireless communication capabilities between nodes provide a broad application prospect for wireless sensor networks. As a ubiquitous sensing technology, wireless sensor networks are widely used in military, industrial control, intelligent buildings, medical care, material tracking, and smart agriculture [18]. In addition, the wireless sensor network also shows great vitality in the field of environmental monitoring and has become more and more important. It will continue to provide a large number of

Fig. 1 The importance of environmental sustainability to the Winter Olympics

continuous and comprehensive environmental information in the macro and micro fields, and contribute to the sustainable development of the environment [19]. How to use the low-cost information collection equipment to efficiently realize the Winter Olympic environment information collection and obtain the key environmental information and knowledge of the Winter Olympic Games. The environmental information includes air temperature, humidity, solar radiation illumination, and concentration. In the collection process, synchronization and real-time and distribution characteristics should be taken into account, as well as the possible problems of noise and abnormal values. These challenges bring realistic challenges to the realization of environmental sustainable management information [20].

2.2 Hardware design scheme of a wireless sensor node based on the Internet

A typical wireless sensor node is composed of a sensor module, microcontroller module, radio frequency communication module, and power supply module. The sensor module gets the perception and acquisition of the parameters of the target monitoring object in the monitoring area through various sensors and transmissions to the microcontroller module through the corresponding I/O interface. The microcontroller module is the core part of the wireless sensor node, which completes the acquisition and preprocessing of the sensor data, and then encapsulates these data into wireless data packets, and transfers them to the radio frequency communication module. The radio frequency communication module mainly realizes the sending and receiving of wireless data packets. According to the situation, it may also need to forward other wireless packets. The energy supply module provides energy for the various functional units of wireless sensor nodes. In many wireless sensor network applications, it is often battery powered. Therefore, low power design is of great importance for the low power characteristics of wireless sensor nodes and the life cycle of the entire network.

In the process of scheme design, the research, analysis, and comparison of micro controller and wireless transceiver are carried out, and the MSP430F5438 and CC2520 of TI company are used as the micro controller chip and wireless transceiver chip in the wireless sensor nodes of this project. Using the sensor satisfying the performance index, the design block diagram of the whole sensor node is shown in Fig. 2.

2.3 Design of sensor module based on the Internet

In order to allow environmental experts to study the relationship between environmental micro molecules and sustainable development in the environment monitoring area of Winter Olympic Games, the main micro-environmental indicators that the nodes need to monitor include environmental temperature, humidity, light intensity, and CO_2 concentration. For different microenvironment factors, it is necessary to select corresponding sensors for monitoring, considering the factors such as energy consumption, measurement range, accuracy, cost, and volume.

The air temperature and humidity sensor are SHT10P, and the power supply voltage is 2.4–5.5 V. The ambient temperature measurement range is – 40–123.8 °C, accuracy ± 0.5 °C (0 °C). The ambient humidity measurement range is 0–100%, the accuracy is + 3%RH (0 °C), and the output is a I2C interface of the digital communication interface, so there is no need for sensing circuit design.

Light intensity has an important influence on the Winter Olympic Games venue. Therefore, it is necessary to collect this data. In this design, the light intensity sensor uses S1087, the principle of which is a light-sensitive diode. For different light intensity, S1087 passes through the current of different sizes. In the application circuit design, it is directly connected to the A/DC input of the microcontroller and parallel to a 100-k resistor.

Carbon dioxide concentration is also one of the important factors in the Winter Olympic Games. In this design, the carbon dioxide concentration sensor selects COZIR, its power consumption is only 3.5 mW, the peak current is 33 mA, the average current is less than 1.1 mA, and the range of power supply voltage is 3.3–5 V, with outer zone temperature and humidity compensation range 0–2000 ppm.

2.4 Data fusion modeling of wireless sensor TT & C and Internet based on the adaptive weighting algorithm

The Winter Olympic sports field monitoring system based on the Internet wireless sensor network is a

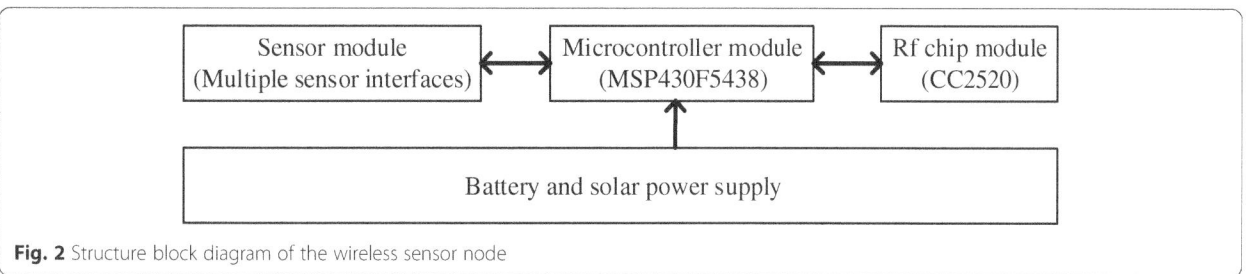

Fig. 2 Structure block diagram of the wireless sensor node

network system integrated with monitoring, control, and wireless communication. There are many nodes in the network, and the nodes are densely distributed and have some redundancy. Due to the complex environment and energy limitations of the sports field, nodes are more prone to failure. In addition, there is a monitoring requirement within a single Winter Olympic sports field, which makes it inappropriate to use each sensor node to transmit data to the coordinator of the field wireless sensor network in the process of information collection and transmission in the network.

The main idea of data fusion is to use certain means, methods, and tools to combine data from different nodes to improve the accuracy of data or to reduce the data transmission of network data. Like other applications of the wireless sensor network, data fusion plays an important role in the wireless sensor network of Winter Olympic Games. There are n sensors to measure a measured object, and there are different weighting factors for different sensors. Under the optimal condition of the minimum mean square error, according to the measured values obtained by each sensor, the optimal weighting factor corresponding to each sensor is found in an adaptive way, so that the value of the fusion $X^\wedge$ is optimal.

The variance of n sensors is $\sigma_1^2, \sigma_2^2, \cdots, \sigma_n^2$, respectively. The estimated true value is X, and the measured values of each sensor are $X_1, X_2, \cdots, X_n$. They are independent of each other and are unbiased estimates of X. The weighting factors of each sensor are $W_1, W_2, \cdots, W_n$, respectively. The $X^\wedge$ values and weighting factors of the fusion meet the following two formulas.

$$\hat{X} = \sum_{p=1}^{n} W_p X_p, \quad \sum_{p=1}^{n} W_p = 1 \tag{1}$$

The total mean square error is:

$$\sigma^2 = E\left[\left(X-\hat{X}\right)^2\right] = E\left[\sum_{p=1}^{n} W_p^2(X-X_p)^2 + 2\sum_{p=1,q=1}^{n} W_p W_q(X-X_p)(X-X_q)\right] \tag{2}$$

They are independent of each other and are unbiased estimates, so $E[(X-\overline{X}_p)(X-X_q)] = 0$. So σ^2 can be written as:

$$\sigma^2 = E\left[\sum_{p=1}^{n} W_p^2(X-X_p)^2\right] = \sum_{p=1}^{n} W_p^2 \sigma_p^2 \tag{3}$$

From formula (3), it shows that the total mean square error σ^2 is a multivariate two-degree function of each weighting factor, so σ^2 must have a minimum value. The minimum value is calculated by weighting factor W_1, $W_2, \cdots, W_n$, which satisfies formula (1).

According to the extreme value theory of multivariate function, the weighted factor corresponding to the minimum mean square error can be obtained.

$$W_p' = 1/\sigma_p^2 \sum_{i=1}^{n} \frac{1}{\sigma_i^2} \quad (p = 1, 2, \cdots, n) \tag{4}$$

The minimum mean square error at this time is as follows:

$$\sigma_{\min}^2 = 1/\sum_{p=1}^{n} \frac{1}{\sigma_p^2} \tag{5}$$

The above is estimated according to the measured values of each sensor at a certain time. When the estimated value X is a constant, it can be estimated according to the mean of the historical data of each sensor. Set:

$$\overline{X}_p(k) = \frac{1}{k} \sum_{i=1}^{k} X_P(i) \tag{6}$$

The estimated value at this time is:

$$\hat{\overline{X}} = \sum_{p=1}^{n} W_p \overline{X}_P(k) \tag{7}$$

The total mean square error is:

$$\overline{\sigma}^2 = E\left[\left(X-\hat{\overline{X}}\right)^2\right] \tag{8}$$

The same reason can be obtained:

$$\overline{\sigma}^2 = \frac{1}{k} \sum_{p=1}^{n} W_p^2 \sigma_p^2 \tag{9}$$

Obviously, the optimal weighted factor W_p' corresponding to the $\overline{\sigma}^2$ at the earliest time still satisfies formula (4), and the corresponding minimum mean square error is:

$$\overline{\sigma}_{\min}^2 = 1/\left(k \sum_{p=1}^{n} \frac{1}{\sigma_p^2}\right) = \sigma_{\min}^2/k \tag{10}$$

It can be seen from formula (10) that $\overline{\sigma}_{\min}^2$ O must be less than $\sigma_{\min}^2$, and $\overline{\sigma}_{\min}^2$ will decrease with the increase of k.

According to the actual needs of sustainable development, the Winter Olympic venues are refreshed with environmental parameters per minute. At the same time, according to the principle of adaptive weighting algorithm, in order to eliminate the influence of errors and improve the measurement accuracy, the wireless sensor node sampling every 10 s and collecting 6 measured values per minute do arithmetic average, and then calculate the variance and send it to the cluster head, the Winter Olympics motion field controller. The controller sends the data of multiple wireless sensor nodes to the controller and sends it to the coordinator. The application model of adaptive weighted fusion algorithm in sports field wireless sensor network is shown in Fig. 3.

Among them, $\overline{X}_p(5)$ represents the average value of the p wireless sensor's sampling time five times per minute, and σ_p^2 represents its variance. W'_{np} represents the optimal weighting factor of the wireless sensor node p calculated by the controller of the Winter Olympics moving site n in accordance with the three mean and variance provided by all wireless sensor nodes within its internal wireless sensor nodes. $\overset{\wedge}{\overline{X}}_n$ represents the estimated value of n in Winter Olympic Games based on adaptive weighting algorithm. Taking the temperature as an example, the steps of the application of the algorithm are as follows:

(1) The wireless sensor node p collects temperature data every 10 s and obtains five measurements every 1 min. The average value of $\overline{X}_p(5)$ is calculated by formula (5). Get the variance O, then send these two values to the controller of the Winter Olympic Games venue.

(2) The Winter Olympics motion field controller receives the mean and variance of all the same type sensor nodes to remove the maximum and minimum values, and calculates the optimal weighting factor W'_{np} of the p temperature sensor of the remaining node according to formula (3). Then, according to formula (6), the fusion estimation value of temperature in Winter Olympic venues, $\overset{\wedge}{\overline{X}}_n$, can be calculated.

(3) The Winter Olympics sports field controller sends the fusion value to the liquid crystal display for local control and sends it to the coordinator of the Winter Olympics sports field group wireless sensor monitoring network.

The fusion values of humidity and light parameters can also be obtained by the same method. The above application models and implementation steps fully take into account that the wireless sensor nodes have certain computing power, which improves the sampling period and improves the accuracy. After data fusion is processed by the controller, the energy consumption is reduced. At the same time, the whole process is also consistent with the specifications for the design of the venue control system of the Winter Olympic Games.

3 Experimental
3.1 Analysis and test of communication range
In order to evaluate the communication performance of wireless sensor nodes in the Winter Olympics application environment, the communication range of nodes is measured in a Winter Olympic sports field environment, so as to provide the necessary reference information for the deployment of our actual application system. The following test strategies are adopted: using a coordinator and a terminal node to act as a source and a sink for a communication process respectively. A set of data is sent to the coordinator every 5 s from the terminal node. After the coordinator receives the data, it is sent to the PC through the UART communication interface, and then the PC is displayed in the serial debug assistant window to display the received data. During the distance test, the location of the fixed coordinator is unchanged, and then it moves slowly after a certain distance. It is temporarily stable and the observation data is correctly received. If the communication is normal, then the mobile terminal node will be unable to achieve stable communication until the critical boundary of the communication area is determined. Then, the terminal

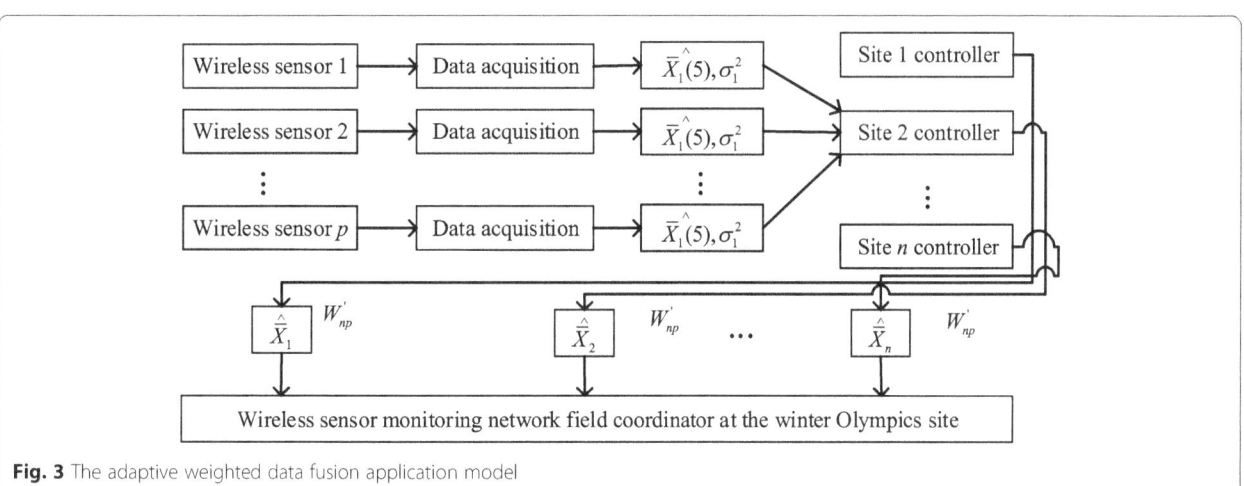

Fig. 3 The adaptive weighted data fusion application model

node is placed back, and the distance between the steady and stable communication is found, the distance between the coordinator and the terminal node is measured, and the distance is defined as the communication range. The communication range of wireless sensor nodes is measured in the Winter Olympic venues.

The communication range of wireless sensor nodes is tested at the Winter Olympic venues. The temperature is 6 °C. Wireless sensor nodes are powered by three dry cells. Different transmission power is used to measure the communication range without transmitting power, as shown in Table 1.

From Table 1, it shows that the communication range of wireless sensor nodes is closely related to the transmitting power: the larger the transmitting power, the larger the communication range will be. But at the same time, the power consumption of wireless sensor nodes will also become large. Therefore, according to the actual application needs, the appropriate radio frequency transmission power must be determined. For example, in this project, the power of the wireless transmitting signal is set to 0 dBm to meet the needs of the application, so the transmission power of the wireless signal is 0 dBm, which can meet the needs of the project.

3.2 Calculation of bit error rate and communication distance

By calculating the bit error rate (PER) at different distances, the effective communication distance of the device can be estimated in the actual environment. The test was carried out in the sports field of the Winter Olympics. The test process is as follows:

Test equipment: Winter Olympic venue controller and wireless sensor node, all use SMA antenna.

Test method: the Winter Games playground controller acts as the main node, and the wireless sensor acts as a slave node and connects to the Winter Olympic sports site network to send a short data frame to the Olympic motion field controller at a rate of 1000 frames/s. The controller of the Winter Olympic venue calculates the number of data frames within 10 s and calculates the BER. When the bit error rate is very small, it shows that the wireless communication is reliable. The closer the bit error rate is to 100%, the worse the signal quality.

When the bit error rate is 100%, it is impossible to receive data and communicate effectively.

Test place: a Winter Olympic venue, covering an area of 5000 m².

Test results: as shown in Fig. 4, when two devices are within 160 m, PER is very small, almost near zero, and can be reliably communicated. When the distance between the equipment is greater than 200 m, the PER begins to rise in a straight line, and after more than 260 m, the PER is 100%. The Winter Olympic sports field controller cannot receive the data of the wireless sensor nodes. It can be seen that when the distance between devices is within 200 m, it can maintain normal communication.

4 Results and discussion

4.1 Collection and test of environment temperature and humidity in Winter Olympic Games venues

After completing the deployment of the system, the monitoring function of the whole system will be tested and analyzed, mainly including the collection and analysis of various parameters. Three wireless sensor nodes in the system are used to collect the environmental parameters of the Winter Olympic Games in the environment, and transmit the data to the coordinator through the wireless transmission of the wireless sensor network on the ground floor. Then, the coordinator is transmitted to the Internet through GPRS and finally arrives at the data center. The monitoring data can be displayed in the monitoring system of the data center.

The temperature and humidity sensor SHT10P assembled by the wireless sensor nodes in the system is used to collect the temperature and humidity parameters of the environment and is presented through the curves, as shown in Figs. 5 and 6, respectively.

In Figs. 5 and 6, different colors represent data curves of different sensor nodes. It can be seen from Figs. 5 and 6 that the data collected by these sensor nodes are basically the same and the trend of change is the same. The data collected by the system are displayed with an intuitive data curve, which is close to the temperature and humidity values measured by the standard instrument and displays the temperature change tendency intuitively. By comparing the temperature and humidity data curves, it shows that the higher the temperature is, the

Table 1 The range of communication under different wireless transmission powers

Testing environment	Transmitted power (dBm)	Communication distance (m)
Temperature 6 °C	− 20	69
The height of wireless sensor node 0.5 m	0	220
	10	358

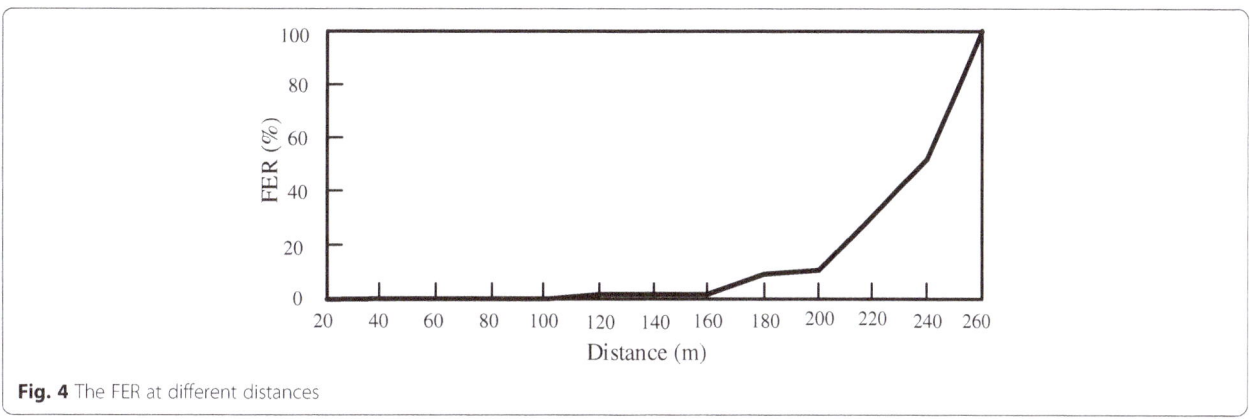

Fig. 4 The FER at different distances

lower the corresponding humidity is. Therefore, the data obtained from the Winter Olympic sports site environment information collection system based on the Internet wireless sensor network can provide an important reference for environmental experts to study the sustainable development of the Winter Olympic Games.

4.2 Collecting and testing the environmental light intensity of Winter Olympic Games venues

The light intensity sensor S1087, which is assembled by the wireless sensor nodes in the system, is used to collect the light intensity information of the Winter Olympic sports field environment and is shown through the curve, as shown in Fig. 7.

As shown in Fig. 7, the curves of different colors in the data graph represent the data curves collected by different wireless sensor nodes. It can be seen from the diagram that the size of their respective data is roughly the same, and the change curves of light intensity are basically the same. It can be used as a reference for the follow-up study of the sustainable development of the Winter Olympic Games based on the environmental expert knowledge system.

4.3 Collecting and testing carbon dioxide concentration in Winter Olympic Games venues

The carbon dioxide concentration sensor COZIR is used to control the sensor by the WSN node in the system and to collect the carbon dioxide concentration information in the Winter Olympic sports field environment and shows it through the curve, as shown in Fig. 8.

In Fig. 8, the curves of different colors in the data graph represent the data curves collected by different wireless sensor nodes. It can be seen from the figure that the data collected by them are approximately of the same size, and the curve of carbon dioxide concentration is basically the same. Compared with the light intensity curve of the crop growth environment, it is found that when the intensity of light intensity is very high, the corresponding carbon dioxide concentration is the lowest, which is also consistent with the laws of the natural environment.

The continuous monitoring system is tested and evaluated for the deployed system, and the system works stably. From the test results, we can see that the designed system meets the requirements of the Winter Olympic Games environmental monitoring system for system

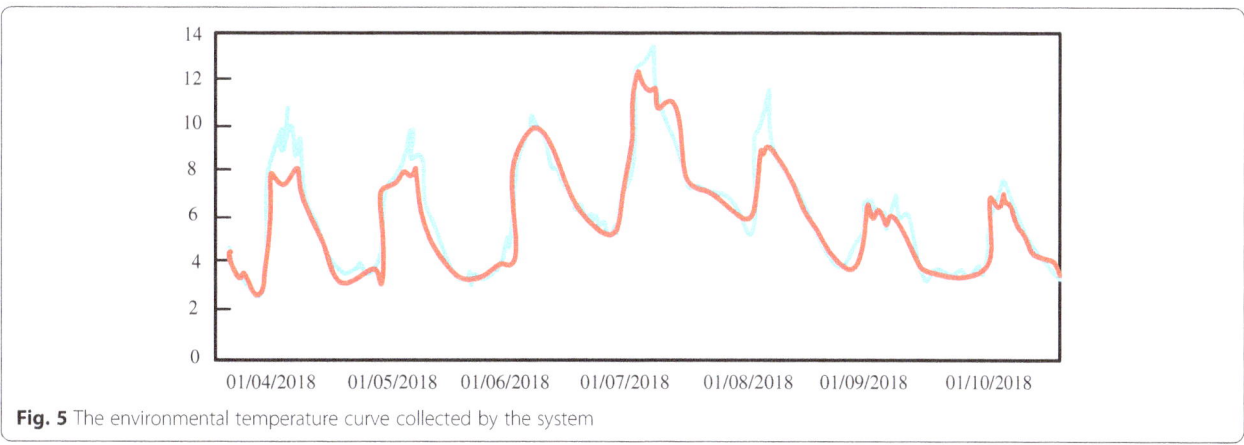

Fig. 5 The environmental temperature curve collected by the system

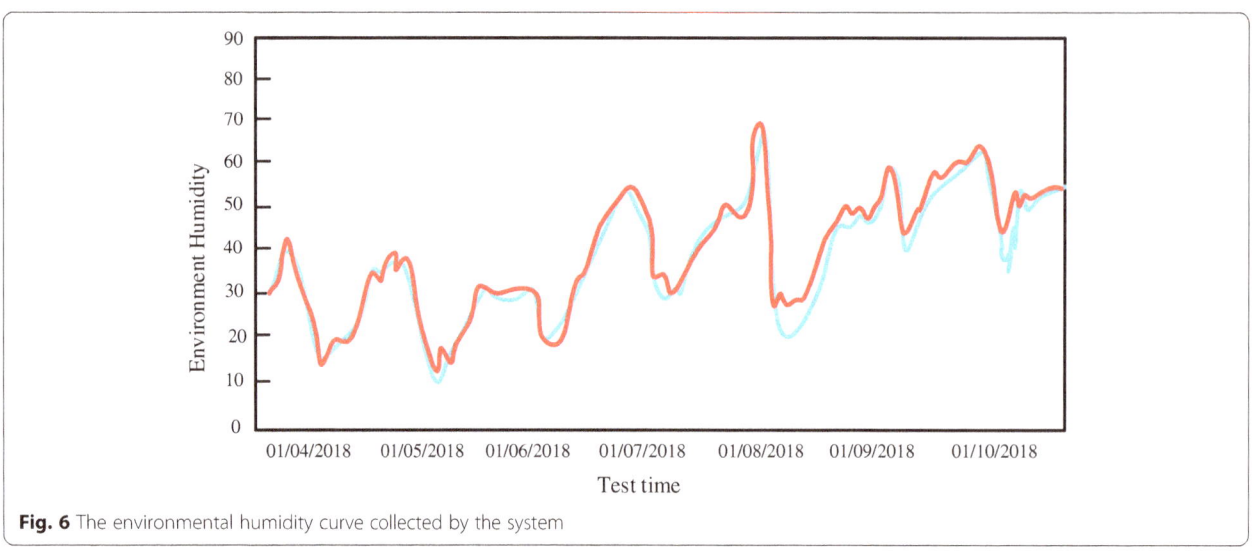

Fig. 6 The environmental humidity curve collected by the system

functions and corresponding technical indicators. Therefore, the data collected by the WSN-based environment information collection system is consistent with the actual environmental parameters and can be used as an important data reference for the follow-up Winter Olympic environment analysis system. This provides an important reference for the scientific management and behavior decision of the sustainable development of the sports field environment of the Winter Olympic Games, so as to promote the greater significance of the Olympic Games' sports venues to achieve more uses.

5 Conclusion

The path of sustainable development is an inevitable choice for human progress and social benign operation. Since the beginning of the Olympic movement, the

Olympic movement has committed itself to "building a peaceful and beautiful world," so the issue of sustainable development has naturally aroused the high attention and active pursuit of the International Olympic Committee-headed Olympians. Based on this, a wireless sensor network based on the Internet is proposed to collect the environmental information of the Winter Olympic Games, which provides data support for the environmental experts to study the sustainable development of the Winter Olympic Games. The hardware and sensor modules of the wireless sensor network based on the Internet are designed, and the adaptive weighting algorithm is used to fuse several wireless sensor nodes. By testing and analyzing the communication ability of a single wireless sensor node and the error rate under different distances, the results show that the designed wireless

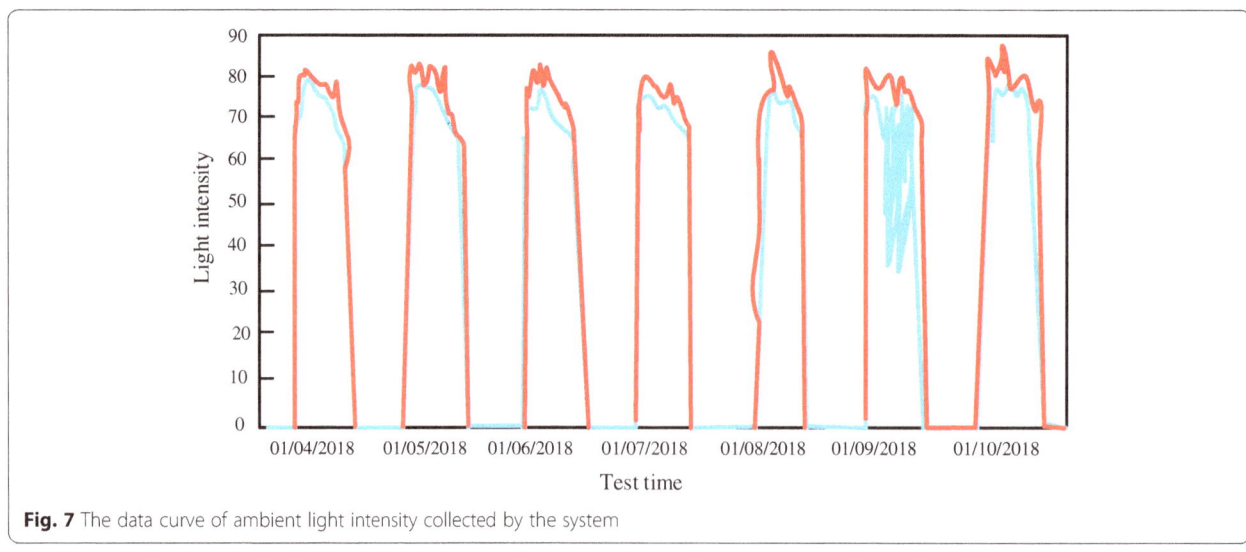

Fig. 7 The data curve of ambient light intensity collected by the system

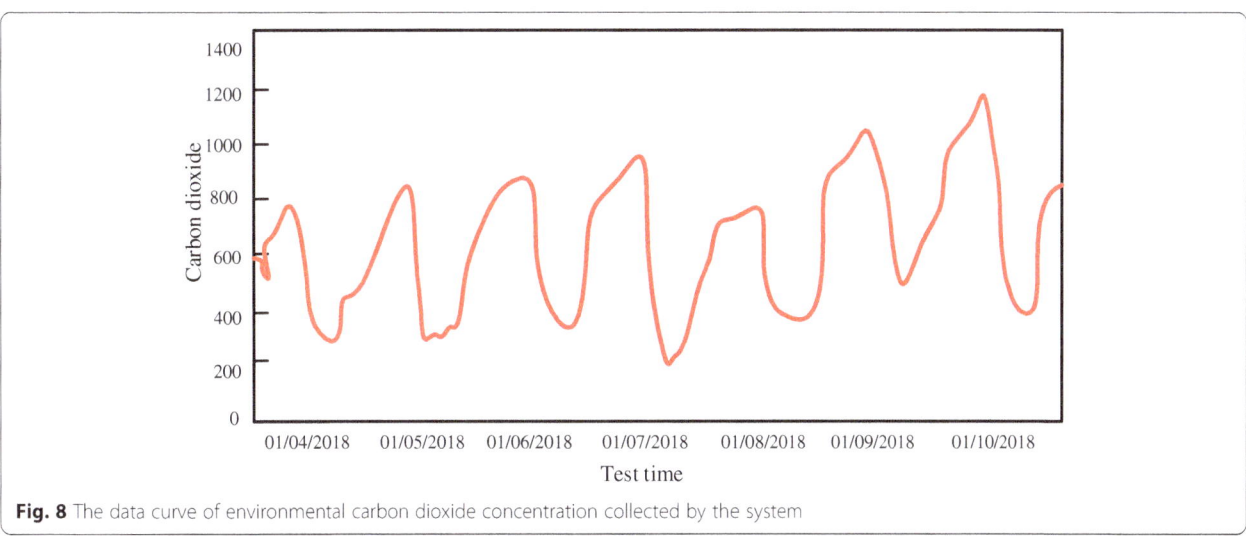

Fig. 8 The data curve of environmental carbon dioxide concentration collected by the system

sensor has the characteristics of long working distance and high stability. Finally, through the field data collection of the sports site environment information of a Winter Olympic Games, the results show that the data collected are in good agreement with the actual environmental parameters. Therefore, it shows that the wireless sensor network system based on the Internet can provide effective data support for the information management research of the sustainable development of the Winter Olympic Games and can provide some reference for the future research of the wireless sensor network system based on the Internet.

Abbreviations
dBm: Deadly Boss Mods; GPRS: General packet radio service; PC: Personal computer; PER: Packet error rate; RH: Relative humidity; SMA: Stone mastic asphalt; UART: Universal asynchronous receiver/transmitter; WSN: Wireless sensor network

Funding
The study was supported by "Project of Philosophy and Social Science Research of Jiangsu Province, China (Grant No. 6R179628";The study was supported by "Humanities and Social Science Foundation of Ministry of Education of China"(Grant No. 1R179487).

Authors' contributions
MZ has made great contributions to the direction of the Internet wireless sensor network. FZ contributed a lot to the wireless sensor network and to the sustainability of the Winter Olympics. Both authors read and approved the final manuscript.

Authors' information
Maomao Zhang, Doctor of Education, Associate Professor. Graduated from the Beijing Sport University in 2009.Worked in China university of Mining and Technology.Her research interests include Sport culture and social sport. Feng Zhai, Master of Education, Professor. Graduated from the Beijing Sport University in 2003.Worked in China university of Mining and Technology. His research interests include Sport Training and social Sport education.

Competing interests
The authors declare that they have no competing interests.

References
1. R. Amos, H. Robertson, The sustainable development of the London 2012 Olympic Park: a real controversy 11- to 15-year-old students' perspectives right from the scene. Pituitary **3**(3), 153–158 (2012)
2. D. Cyranoski, Chinese biologists lead outcry over Winter Olympics ski site. Nature **524**(7565), 278 (2015)
3. H. Preuss, The contribution of the FIFA World Cup and the Olympic Games to green economy. Sustainability **5**(8), 3581–3600 (2013)
4. J.R. Gold, M.M. Gold, "Bring it under the legacy umbrella": Olympic host cities and the changing fortunes of the sustainability agenda. Sustainability **5**(8), 3526–3542 (2013)
5. J. Mailhot, S. Bélair, M. Charron, et al., Environment Canada's experimental numerical weather prediction systems for the Vancouver 2010 Winter Olympic and Paralympic Games. Bull. Am. Meteorol. Soc. **91**(8), 69–82 (2010)
6. K. Mike, M. Gordon, Aquatics Centre, London 2012 Olympic and Paralympics Games. Bautechnik **89**(10), 701–711 (2012)
7. R. Phillips, The hunt for the gray wolf: a case study in recovering top-predator management policy in Washington state. Endocrinology **137**(10), 4363–4371 (2013)
8. Z. Du, E. Dai, Environmental ethics and regional sustainable development. J. Geogr. Sci. **22**(1), 86–92 (2012)
9. A. Kallioras, N. Ruzinski, Special issue: sustainable development of energy, water and environment systems. Water Resour. Manag. **25**(12), 2917–2918 (2011)
10. H. Bilal, Eucalyptus in social forestry and sustainable development-District Malakand Pakistan. Desalination **280**(s 1–3), 183–190 (2014)
11. S. Mandžuka, XIVth Winter Road Congress - reconciling road safety and sustainable development in a context of climate change and economic constraints. Promet-traffic & Transportation **26**(2), 187–188 (2014)
12. M. Keshtgari, A. Deljoo, A wireless sensor network solution for precision agriculture based on Zigbee technology. Wirel. Sens. Netw. **4**(1), 25–30 (2012)
13. DD Chaudhary, SP Nayse, LM Waghmare, Application of wireless sensor networks for greenhouse parameter control in precision agriculture[J]. International Journal of Wireless & Mobile Networks (IJWMN) **3**(1), 140–149 (2011)
14. A. Tzounis, N. Katsoulas, K.P. Ferentinos, et al., Development of a WSN for greenhouse microclimate distribution monitoring. Annals "Valahia" University of Targoviste - Agriculture **10**(1), 7–13 (2016)
15. J. Hou, Y. Gao, Greenhouse wireless sensor network monitoring system design based on solar energy. Wireless Communication Technology **2**, 475–479 (2010)
16. J.I.Z. Chen, H.G. Yue, W.B. Wu, et al., A novel apparatus for surveillance of green energy system based on WSSs. Engineering **05**(1), 135–140 (2013)
17. C. Bertelle, M. Alobaidy, A. Ayesh, et al., Intelligent land-use management and sustainable development: from interacting wireless sensors networks to spatial emergence for decision making. Ecography **24**(5), 555–568 (2010)

Improved energy-balanced algorithm for underwater wireless sensor network based on depth threshold and energy level partition

Pan Feng[1†], Danyang Qin[1*†] (iD), Ping Ji[1], Min Zhao[1], Ruolin Guo[1] and Teklu Merhawit Berhane[2]

Abstract

Considering the insufficient global energy consumption optimization of the existing routing algorithms for Underwater Wireless Sensor Network (UWSN), a new algorithm, named improved energy-balanced routing (IEBR), is designed in this paper for UWSN. The algorithm includes two stages: routing establishment and data transmission. During the first stage, a mathematical model is constructed for transmission distance to find the neighbors at the optimal distances and the underwater network links are established. In addition, IEBR will select relays based on the depth of the neighbors, minimize the hops in a link based on the depth threshold, and solve the problem of data transmission loop. During the second stage, the links built in the first stage are dynamically changed based on the energy level (EL) differences between the neighboring nodes in the links, so as to achieve energy balance of the entire network and extend the network lifetime significantly. Simulation results show that compared with other typical energy-balanced routing algorithms, IEBR presents superior performance in network lifetime, transmission loss, and data throughput.

Keywords: UWSN, Energy-balanced routing, Network lifetime, Depth threshold, EL partition

1 Introduction

Wireless communication and information technology have been developed to the fifth generation (5G) [1–5], which enabled the realization of various applications based on radio signals [6–10], including satellite systems [11–13], however, they could not be used in an underwater environment. Wireless sensor network (WSN) is wildly used in an underwater environment to collect and transmit data. Underwater WSN (UWSN) can realize wide-area information transmission by underwater sensors, which has certain application value in underwater target detection, underwater Internet of Things construction, marine data collection, disaster prevention, and underwater sonar communication [14]. However, the energy cost by data transmission and the difficulty

in battery-replacement in an underwater environment require an efficient and energy-balanced routing protocol to extend the underwater network lifetime. Many existing protocols for UWSN [15–18] might reduce energy consumption, but most of them only consider the problem of local energy consumption.

At present, the energy routing research of UWSN mainly considers consuming energy efficiently. Wahid and Kim [19] proposed a depth-based routing protocol (DBR) by selecting a transponder node based on depth and residual energy, named energy-efficient DBR (EEDBR). Cao et al. [20] studied the balanced transmission mechanism (BTM) for UWSN in the view of energy pattern, in which each node selects a transmission pattern based on its energy level (EL). Li et al. [21] proposed a relative distance-based forwarding (RDBF) protocol. In another work [22], a routing algorithm with efficient energy consumption was proposed based on the sensors' distance and the residual energy. Mahmood et al. [23] extended DBR and EEDBR, improving the network lifetime. Shen

*Correspondence: qindanyang@hlju.edu.cn
[†]Pan Feng and Danyang Qin contributed equally to this work.
[1]Heilongjiang University, Harbin, China
Full list of author information is available at the end of the article

et al. [24] proposed a new energy-efficient centroid-based routing protocol (EECRP) to improve the energy performance of the network, which requires a long lifetime round and base stations located in the network. Azam et al. [25] proposed a balanced load distribution (BLOAD) in order to avoid energy holes caused by energy consumption imbalance, prolonging the stability period and lifetime of UWSN. Javaid et al. [26] proposed two UWSN routing protocols. The first protocol used adaptive hop-by-hop vector-based forwarding (AVN-AHH-VBF) to avoid a void node. The second protocol was cooperation-based AVN-AHH-VBF (CoAVN-AHH-VBF). Ali et al. [27] proposed two protocols: forward layered multipath power control-one (FLMPC-One), and FLMPC Two, reducing the energy consumption and achieving reliability by eluding energy holes. Bengheni et al. [28] proposed an energy management scheme which enhanced energy harvesting. Yousaf et al. [29] proposed a joint rate and power allocation policy (JRPAP), which balanced fairness, throughput, and energy consumption. Yang et al. [30] proposed a hybrid TDMA/CSMA protocol in the MAC layer to improve network energy efficiency and throughput.

For any energy-balanced routing algorithm, the transmitting power of the sensors is the greatest of all working states, which is about 100 times higher than the receiving power [31, 32]. So improving the transmitting energy efficiency is important to improve the data throughput and lifetime of the network. Improved energy-balanced routing (IEBR) adopts the frames of two classical UWSN protocols, BTM and data-aggregating ring (DAR) [33], which will be descripted in Section 2, and it modifies their routing and data transmission mechanisms based on the actual needs of UWSN. IEBR will focus on the global optimaization which can hardly be achieved by the existing energy balance algorithms. Simulation results show that compared with other typical energy-balanced routing algorithms, IEBR processes superior performances in network lifetime, transmission loss, and data throughput.

2 UWSN energy-balanced routing analysis

2.1 BTM and UDAR routing models

The energy balance problem is always an important research field of UWSN. In the routing algorithms for UWSN energy balance, BTM and DAR have good energy balance performance, and their frames are widely adopted to construct the routing models. Figure 1 illustrates the data transmission mechanisms in the models, where the above is BTM and the following is underwater DAR (UDAR).

BTM is a routing protocol based on hybrid data transmission, which includes two algorithms. Firstly, a tree, whose nodes are sensors and directed edges are links between sensors, is established through efficient routing algorithm (ERA) to determine the transmission route of

the data packets, e.g. the multi-hop routes in Fig. 1. Then, the data packets will be transmitted according to the data balanced transmission (DBT) algorithm. BTM divides the initial energy of each node into m ELs. During multi-hop transmission (MT), the closer a sensor to the sink, the more energy it takes to convey the increasing traffic. When the EL of a node decreases from m to $m - 1$, it will broadcast notice packets to all its predecessors. If the EL of the predecessor is higher than that of its successor, the direct transmission (DT) will be adopted to deliver the packets to the sink. For example, in Fig. 1, when EL_B is lower than EL_A, the node A will transmit data to the sink directly. Otherwise, MT will be used, as the multi-hop transmission link from the node A to the sink in Fig. 1. By the way, the transmission pattern can be constantly changed, achieving continuous transmission with balanced energy consumption. However, such mode conversion makes BTM only suitable for small-scale networks, because there are too many direct transmission links in BTM.

UDAR is the derivative model of DAR in underwater environments and it divides all nodes in UWSN into different sets based on the hops to the sink, which is called as hop grade (HG). HG_i is a set of the nodes with hop grade i. HG_i is a circular area in space, which is called as ring sector. As shown in Fig. 1, the nodes in UWSN are divided into HG_1, HG_2, HG_3, and HG_4. The nodes in some HG will collect the data from the nodes in other HGs by MT in a different period and then directly transmit the data to the sink so as to avoid the rapid energy exhaustion of the nodes lying close to the sink. In Fig. 1, the nodes in HG_4 are responsible for collecting data of nodes in other HGs by MT (the green arrows), and then the nodes in HG_4 transmit them to the sink by DT (the blue arrows) along with their own data. It should be noted that the nodes in HG_1 directly transmit data to the sink. UDAR achieves energy balance among different nodes, but it may cause the problem of data transmission loop.

2.2 UWSN energy consumption model

Since the coverage area of the sensor is a circle, given the network radius R, the node density ρ, the maximum number of hops H, and the width of each ring sector w, the total number of nodes can be defined as Eq. (1):

$$N = \rho \pi R^2 = \rho \pi (Hw)^2 \qquad (1)$$

Then, the number of the nodes in HG_1 is shown in Eq. (2). The area of HG_1 is a circle, and the other ring sectors are ring, so they are called as ring sectors. sensors in HG_1, HG_H or other ring sectors have given the network radius R, so they are discussed separately.

$$N_1 = \rho \pi w^2 \qquad (2)$$

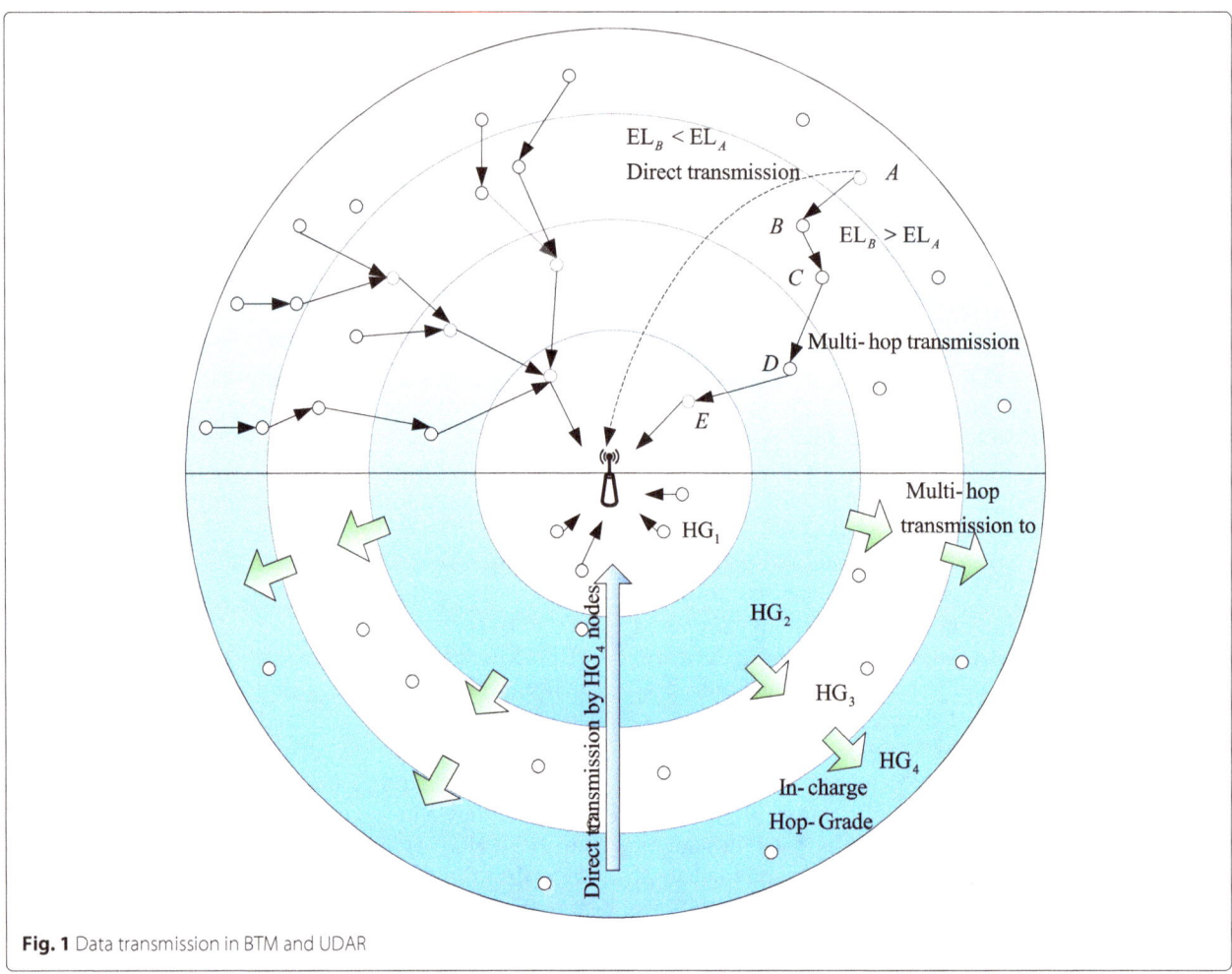

Fig. 1 Data transmission in BTM and UDAR

Assuming that every node in the UWSN send a data packet at first, according to the different ring sectors, all nodes can be divided into three groups to calculate their energy consumption respectively as follows:

(1) The whole energy consumed by nodes in HG_1 is shown as follows:

$$E_1 = E_{rx}(\rho\pi H^2 w^2 - \rho\pi w^2) + E_{tx}\rho\pi H^2 w^2$$
$$= \rho\pi w^2(E_{rx}(H^2 - 1) + E_{tx}H^2) \tag{3}$$

where E_{rx} and E_{tx} is energy consumption of receiving and transmitting a data packet respectively. The first term in Eq. (3) is the energy consumption of receiving data from nodes in other HGs, and the second term represents the energy consumption of transmitting data to the sink.

(2) The nodes in HG_k $(1 < k < H)$ need to receive data from the neighbors in HG_{k+1} and send the data along with their own data to the neighbors in HG_{k-1}. The energy consumption is shown in Eq. (4).

$$E_k = E_{rx}(\rho\pi H^2 w^2 - \rho\pi(kw)^2) \tag{4}$$
$$+ E_{tx}(\rho\pi H^2 w^2 - \rho\pi((k-1)w)^2)$$
$$= \rho\pi w^2(E_{rx}(H^2 - k^2) + E_{tx}(H^2 - (k-1)^2))$$

(3) The nodes in HG_H only need to send their own data to the nodes in HG_{H-1}, without accepting data from other nodes. Their energy consumption is shown in Eq. (5).

$$E_H = \rho\pi w^2 E_{tx}(H^2 - (H-1)^2) \tag{5}$$

The practical conditions, including the propagation delay and the working frequency range, should also be considered in setting up the underwater energy model [34] to calculate specific energy consumption. The attenuation of the signal with transmission distance d and the frequency f in underwater acoustic channel is defined in Eq. (6).

$$A(d,f) = A_0 d^k v^d \tag{6}$$

where A_0 is the normalized coefficient, k is the spreading factor, and v is the absorption coefficient, which is based on the signal frequency expressed in kilohertz. The value of k relies on the geometrical shape of the propagation. k is 2 in spherical spreading and is 1 in cylindrical spreading. In addition, v is defined by Eq. (7), where the parameter α is related to the signal frequency f [35]. α can be calculated by Eq. (8) with f exceeding 100 Hz and by Eq. (9) with f lying below. If f is lower, α is shown in Eq. (9). Given f

(kHz) and d (km) , the transmitting power consumption P_t can be obtained by Eq. (10).

$$v = 10^{\alpha(f)/10} \qquad (7)$$

$$10\log\alpha(f) = 0.002 + \frac{0.11f^2}{1+f^2} + \frac{44f^2}{4100+f^2} + 2.75 \times 10^{-4}f^2 + 0.003 \qquad (8)$$

$$10\log\alpha(f) = 0.002 + \frac{0.11f^2}{1+f^2} + 0.011f^2 \qquad (9)$$

$$P_t = \frac{P_0}{A(d,f)} \qquad (10)$$

It can be seen from Eq. (10) that factual transmitting power is $P_t A(d,f)$ if the transmitting power is P_t. The transmitting and receiving energy consumption are shown in Eq. (11) and Eq. (12) respectively, where x is the size of transmitted data. r is a constant depending on the receiver.

$$E_{tx}(x,d) = P_t A(d,f)xt \qquad (11)$$

$$E_{rx}(x) = rxt \qquad (12)$$

3 Improved energy-balanced routing algorithm

Since existing UWSN energy-balanced routing models such as BTM and UDAR have the problem of insufficient global energy balance and transmission loop, IEBR selects the relay nodes according to the distance and depth at the same time, so as to minimize the hops and eliminate the transmission loop in every link. Then, the EL model in BTM will be adopted to establish dynamic links, achieving balanced energy consumption in the same ring sectors and prolonging the network lifetime. In addition, IEBR will also use cross-sector data transmission to achieve energy balance in different ring sectors.

3.1 Network node deployment

Considering the complexity of underwater environments and the inconsistency of the propagating energy consumption, the UWSN topology with ring sector structure in UDAR has been constructed as shown in Fig. 2. The coverage of UWSN is divided into spaced ring sectors Sr_1, Sr_2,..., Sr_n from the inside to the out. Given R, which is the network radius, and O_t, which is the optimal communication distance threshold of sensors, the maximum number of ring sectors is R/O_t . If the transmitting distance exceeds the given O_t, the signal quality will drop sharply to be regarded as unavailable. This threshold has a close corresponding relation with the radius R as shown in Fig. 3. The whole area of UWSN is a concentric circle, where the sink is in the center and other sensors are in the target area randomly and uniformly. Assuming UWSN satisfy the following conditions:

- All sensor nodes have limited battery power

- The sensor nodes use the positioning methods (received signal strength indication (RSSI) [36] and MoteTrack position recognition scheme [37]) for position sensing in a given underwater environment to make the location known to every sensor
- There are enough data to send for the sensors
- The data reporting mechanism is periodic
- The sink is static and above the water

3.2 Energy-balanced routing (EBR) construction
3.2.1 Relay selection based on optimal distance threshold
After all sensors are deployed as above, if node i not in Sr_1 has data to send, it will select the neighbor at the optimal distance as the relay. Node i will broadcast the location of both itself and the sink s to the neighbors, and every neighbor receiving the data will calculate the parameter N_j according to Eq. (13):

$$N_j = \alpha|d(i,j) - O_t| + (1-\alpha)d(j,s) \qquad (13)$$

where $d(i,j)$ is the distance between i and j, α is a system parameter and there is α = 0.5. α gives the same weight to the two distances (distance from node i to relay j and distance from relay j to sink), so the system will consider the effect of two distances on data transmission equally and choose the most appropriate relay based on distance.

The node with the smallest N_j will be the relay, which ensures that the relay node is located at the optimal distance from node i to sink. Figure 4 shows tree typical structures and a list of relays, through which the routing information can be obtained.

When a relay j is obtained, the parameter N_j will be stored in the routing table of its predecessor i. Then j will inform its successor the fact of j's being selected as the relay and the node number in the routing table along with N_j. To reduce the node power consumption, the algorithm will allow each node forwarding packets from at most two neighbors.

The process of selecting a relay node with N_j is shown in Fig. 5. Firstly, node i identifies all its neighbors by sending query packets, such as the neighbors q, k, and m in Fig. 5. Then node i will select a relay from its neighbors based on N_j. The algorithm will calculate the distance between node i and its neighbor (black arrow) and the distance between the neighbor and the sink (blue arrow) . In Fig. 5, node i selects neighbor j as the relay. After that, the node j will select its relay x. This process will continue until a complete link from i to the sink has been established. The specific routing algorithm can be seen in Table 1.

3.2.2 Data transmission mode based on EL consumption
EBR may get a set of nodes based on the value of N_j and establish a link from node i to the sink among these

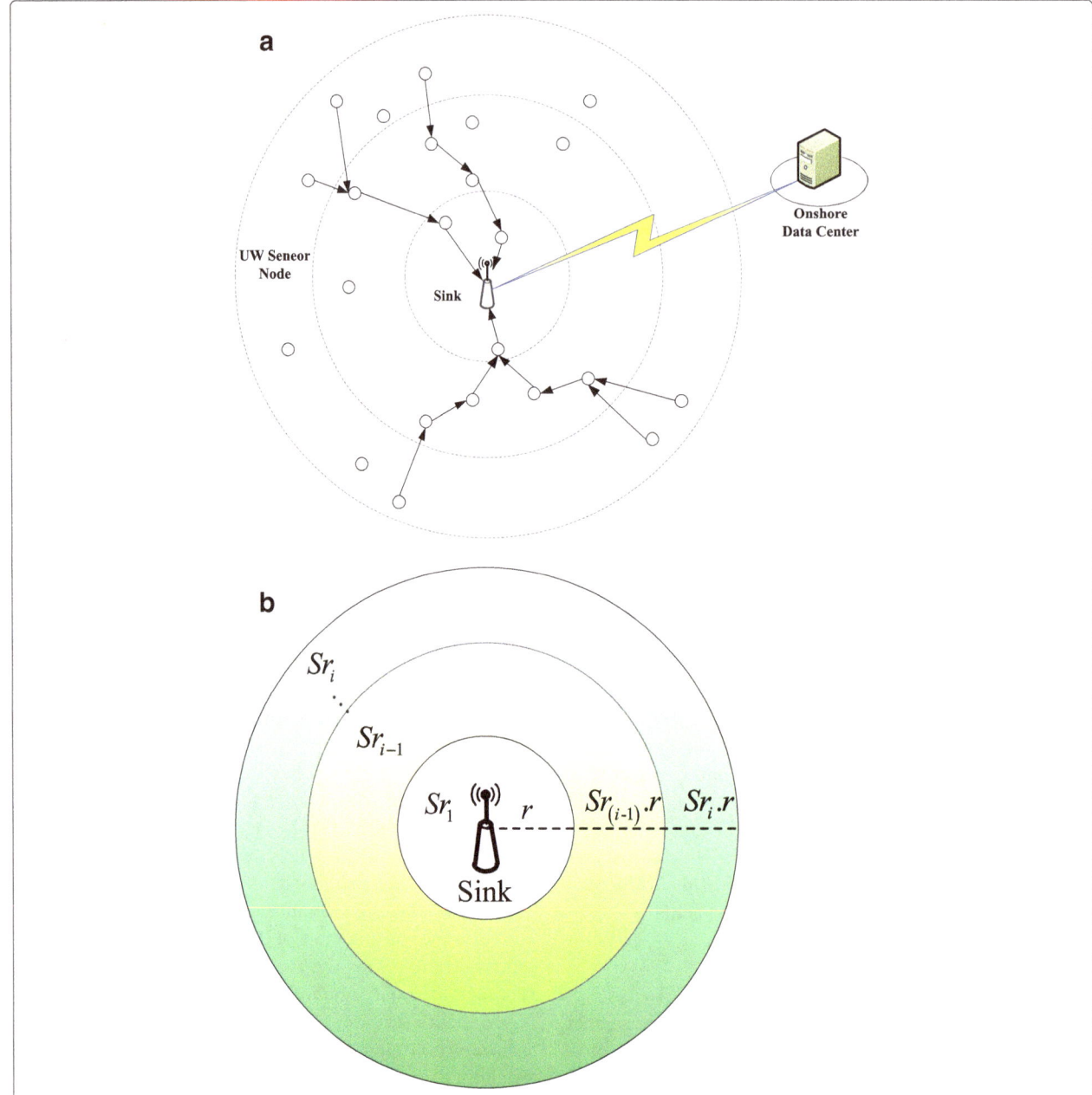

Fig. 2 The structure of UWSN. **a** Network topology. **b** Ring sector. Sink is on the center to gather data from other nodes and transmits them to data center. All nodes are located in ring sections (Sr) randomly and the radius of each section is r

nodes to transmit data. The initial energy E_0 of each node is divided into L ELs, the Unit EL (UEL) is defined as Eq. (14).

$$UEL = \frac{E_0}{L} \qquad (14)$$

The energy consumption of nodes i and j is calculated by Eq. (15), including the energy consumption of sensing, receiving, and transmitting. x is the data size and d is the transmitting distance.

$$E_t^{i,j}(x, d) = E_{sen}^i(x) + E_{rx}^j(x) + E_{tx}^i(x, d) \qquad (15)$$

During the data transmission, all the sensors in different ring sectors have the same initial ELs. If there is UEL=β, EL of node j in Sr_i and node k in Sr_{i-1} are shown in Eqs. (16) and (17), respectively.

$$E_{EL}^j = \frac{E_0}{\beta} \qquad (16)$$

$$E_{EL}^k = \frac{E_0}{\beta} \qquad (17)$$

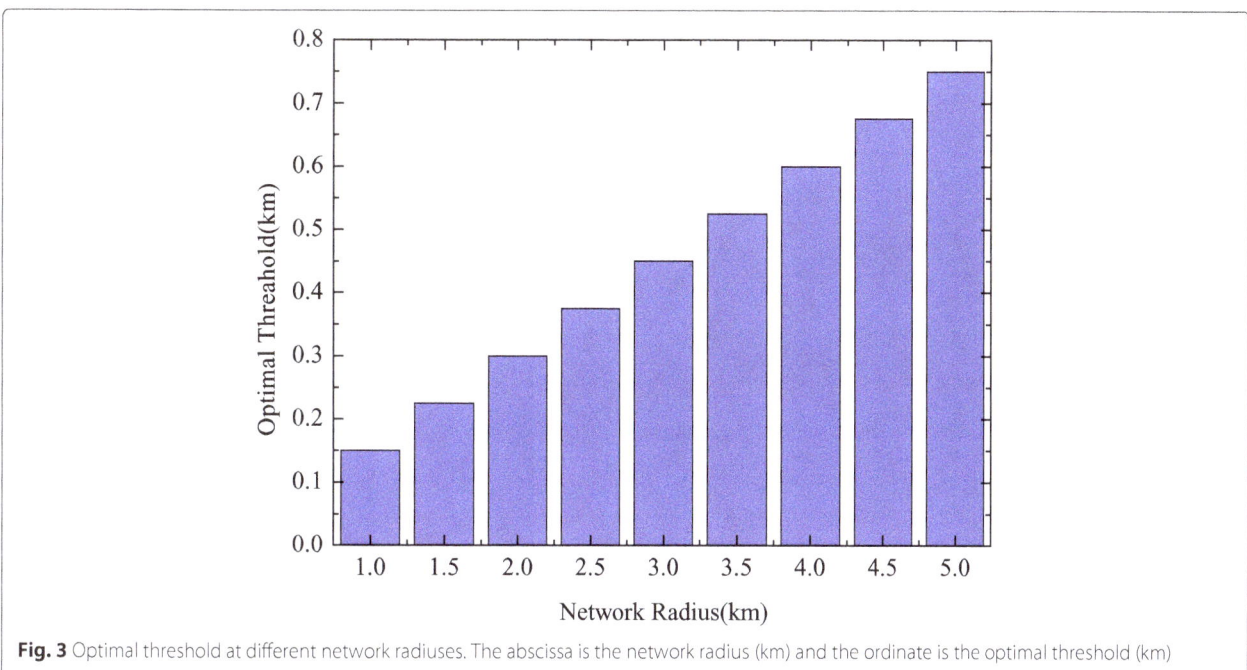

Fig. 3 Optimal threshold at different network radiuses. The abscissa is the network radius (km) and the ordinate is the optimal threshold (km)

As shown in Fig. 6, two conditions may occur during data transmission. In the right link, the EL of each node is equal to that of the successor, so the entire link topology remains the same. In the left link, the transmission load of i and j are different for they are located in different ring sections resulting in $EL_j < EL_i$. Then, node j will send a control packet to node i, and the link between them is cut off. Now, node j will only transmit its own data along the original link, and node i will have to build a new one. Each relay in the new link is the node with maximum EL in the neighbors. This operation can balance the energy consumption of all sensors in the same ring sectors. Related algorithms are shown in Table 2.

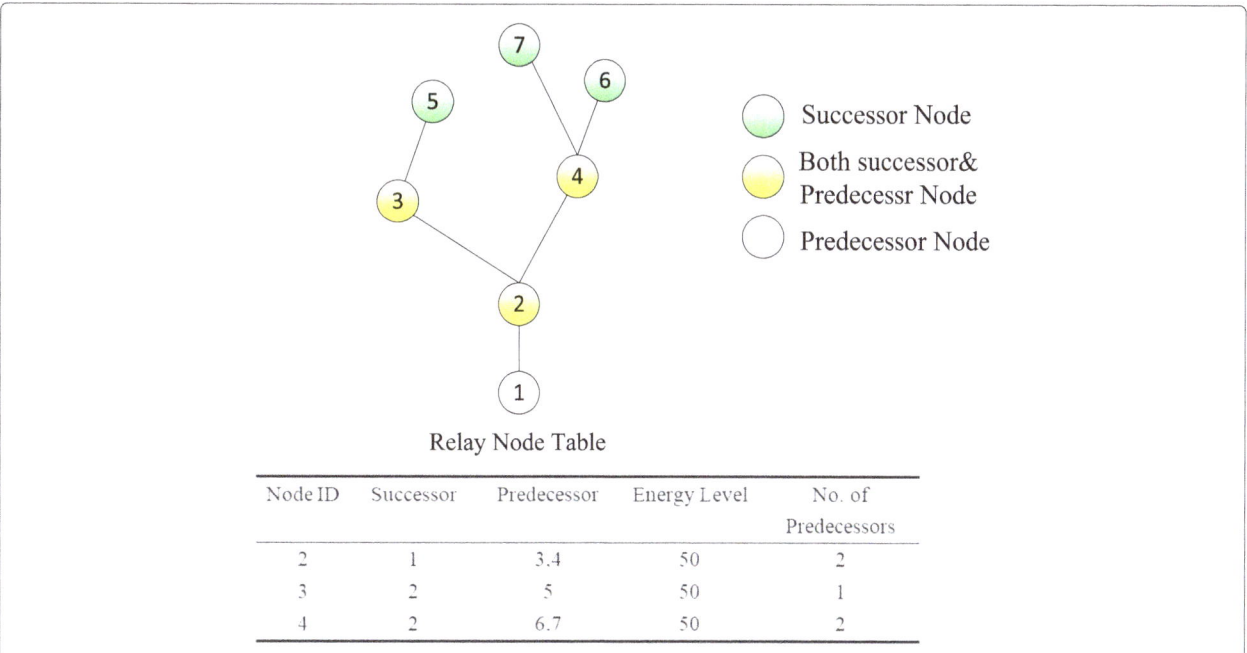

Relay Node Table

Node ID	Successor	Predecessor	Energy Level	No. of Predecessors
2	1	3,4	50	2
3	2	5	50	1
4	2	6,7	50	2

Fig. 4 Predecessor and successor with an exemplary relay node table. This is a routing tree, the vertex is the sensor node, the edge is the link between the nodes. The table below shows information about relay nodes

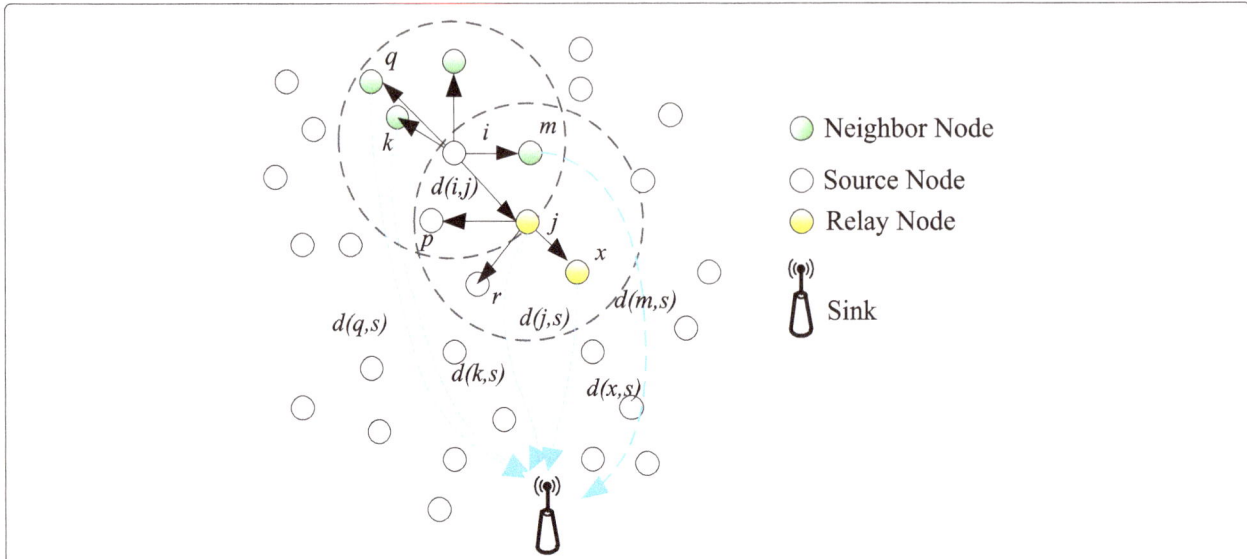

Fig. 5 Selection of relay nodes with N_j. Source node calculates the distance between and the neighbor; obtains the distance between the neighbor node and sink from the neighbor node, calculating N_j of all neighbors; and selects the relay node

3.3 Realization of improved EBR (IEBR)

3.3.1 Relay node selection model based on depth

To solve the problem of transmission loop, IEBR will take the depth threshold to limit the neighbor number, and the depth of a node depends on the ring sector where the node is located. The nodes in the same ring sector have the same depth. The closer a node lies to the sink, the smaller depth a node will have. A node will get the depth of its neighbors by broadcasting control packets when it has data to transmit. The sensor node will select the neighbors with smaller depth as the relay candidates. After that, the algorithm will select only one node with smallest N_j from all candidates as the relay. The data transmission of IEBR is also based on EL. A sensor will not reselect the relay until the EL of its successor falls below that of itself.

The sensor with greater depth will not be selected as the relay according to IEBR; however, traditional BTM and EBR leave the depth of the node out of account, resulting in the transmission loop, as $A \rightarrow B \rightarrow A$ in Fig. 7, so the data packets cannot be transmitted to the sink or additional hops are required, as $A \rightarrow B \rightarrow C$ in Fig. 7, which will increase the energy consumption and cause lifetime reduction. For the special condition with no neighbor or only one neighbor B existing, the sensor A will expand the communication range to contain more neighbors [38] so as to establish IEBR loop-free transmission. The routing establishment process of IEBR is shown in Table 3.

3.3.2 Cross-sector data transmission

To reduce the hops and the transmission loads, IEBR will search for the relays in every other ring sectors instead of in adjacent ring sectors, i.e., a node in Sr_i will look for the relay node in Sr_{i-2} instead of Sr_{i-1}.

Suppose that there are four nodes $A \in Sr_1$, $B \in Sr_2$, $C \in Sr_3$, and $D \in Sr_4$. A and B send data to C and D, respectively. The same volume of data being transmitted at the same distance will have the same energy consumption. C receives the data from A and will send the data as well as its own data, the UEL of C will drop faster than A, which will make A reselect the relay node. Then node C will head to send its own data, node A will begin to send

Table 1 Routing establishment of EBR

	Routing establishment: Relay Selection based Distance
1:	Initialization:
2:	TotalELs $= m$
3:	UEL $= E_0/m$
4:	$\alpha = 0.5$
5:	SelectRelayNode:
6:	SourceID $= i$
7:	NeighborID $= j$
8:	M is the number of neighbors
9:	for $j = 1 : M$ do
10:	$d(i,j) = d_{ij}$
11:	$d(j,s) = d_{j,s}$
12:	$N_j = \alpha |d_{ij} - O_t| + \alpha d_{j,s}$
13:	if $N_j =< N_{j-1}$ then
14:	$min(N_j) = N_j$
15:	RelayID$=j$
16:	end if
17:	end for

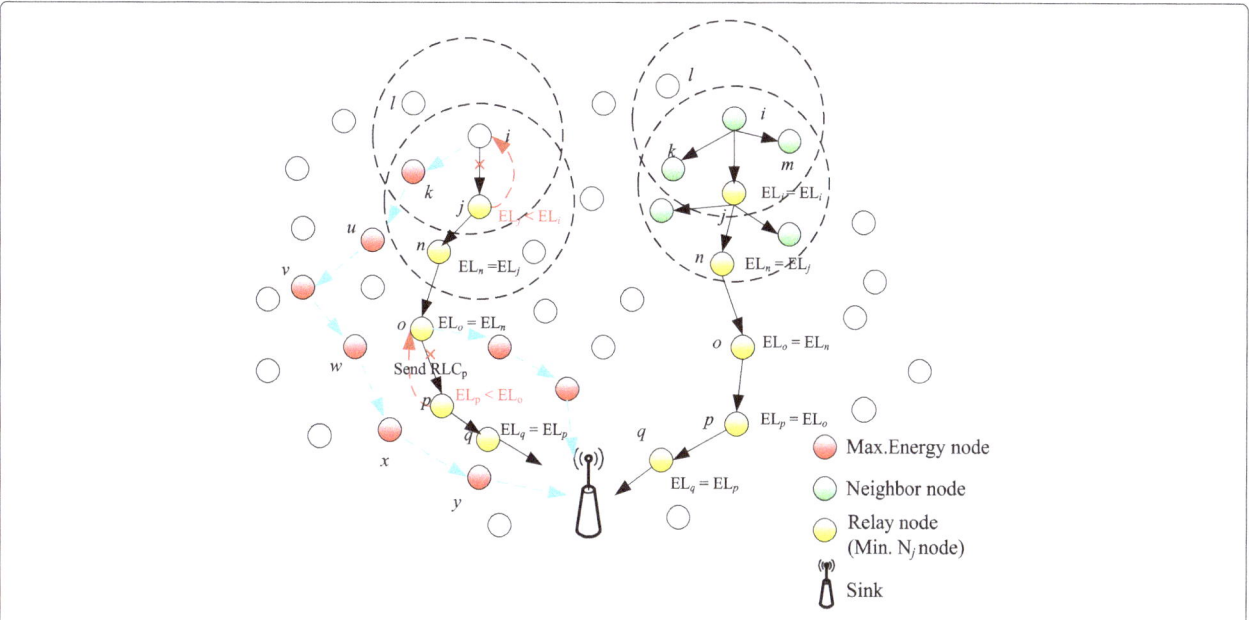

Fig. 6 Data transmission based on EL difference. The link between nodes are dynamic, if EL of successor is lower, the node will find a new relay, and a link become two links by this way

data to another node with a higher EL and the data load is shared by different nodes in this way. Finally, the energy consumption of all ring sectors can achieve balance.

Moreover, the node number in every ring sector is assumed to be the same fixed value in the mathematical mode for simplifying the calculation. In IEBR, it will vary according to the data load as well as the distance to the sink, and the nodes in a ring sector with higher energy consumption will be more, prolonging the lifetime of UWSN for longer lifetimes of these ring sectors.

Table 2 Data transmission of EBR

	Data transmission: Relay Selection based on Energy Level
1:	if $EL_j >= EL_i$ then
2:	continue
3:	else
4:	ELNoticePacketSend(j, i)
5:	NeighborFinding (i)
6:	LinkBuild.sourceID $= j$
7:	for neighborID $k = 1 : m$ do
8:	if $EL_k >= EL_i$ then
9:	node k be new relay node
10:	LinkBuild.soureID $= k$
11:	end if
12:	end for
13:	end if

3.3.3 Packet loss rate constraint of IEBR algorithm

The energy balance algorithm in UWSN will always cause packets lost increasingly so as to limit the practical application. Thus, a maximum throughput model is established in IEBR to reduce the packet loss rate along while achieving global energy balance. Linear programming is used in the paper to design the objective function Maximize $\sum_{t=1}^{t_{\max}} T_p(r)$, and it should satisfy the following constraints:

(i) $E_u \leq E_0, \quad \forall u \in N$;
(ii) $d_{u,v} \leq d_{\mathrm{opt}}, \quad \forall u, v \in N$;
(iii) $f_{u,v} \leq f_{\max}, \quad \forall u, v \in N$;
(iv) $d_{\min} \leq d_u \leq d_{\max}$;
(v) $P_l \geq P_g, \quad \forall u \in N$;
(vi) $\sum_{u=1}^{n} E(u) \cong \sum_{v=1}^{m} E(v), \quad \forall u, v \in N$;

The objective function will maximize the number of effective packets received by the sink during time $t_{\max}$. (i) is the energy constraint, and each sensor u's energy is E_0 at first. All sensors' energy should be consumed efficiently to extend the network lifetime and increase throughput. (ii) requires that the distance between two communication nodes u and v not exceed the optimal transmission distance d_{opt} to keep the packet loss rate from increasing. (iii) describes the constraint of data flow in physical link. $f_{\max}$ is the upper limit of data flow, and it can be defined as the maximum number of packets that can be transmitted per unit time when the size of each packet is fixed. The data flow from any node u to another node v should be less than $f_{\max}$ to ensure that packet loss rate is acceptable. (iv) indicates the upper limit and lower limit

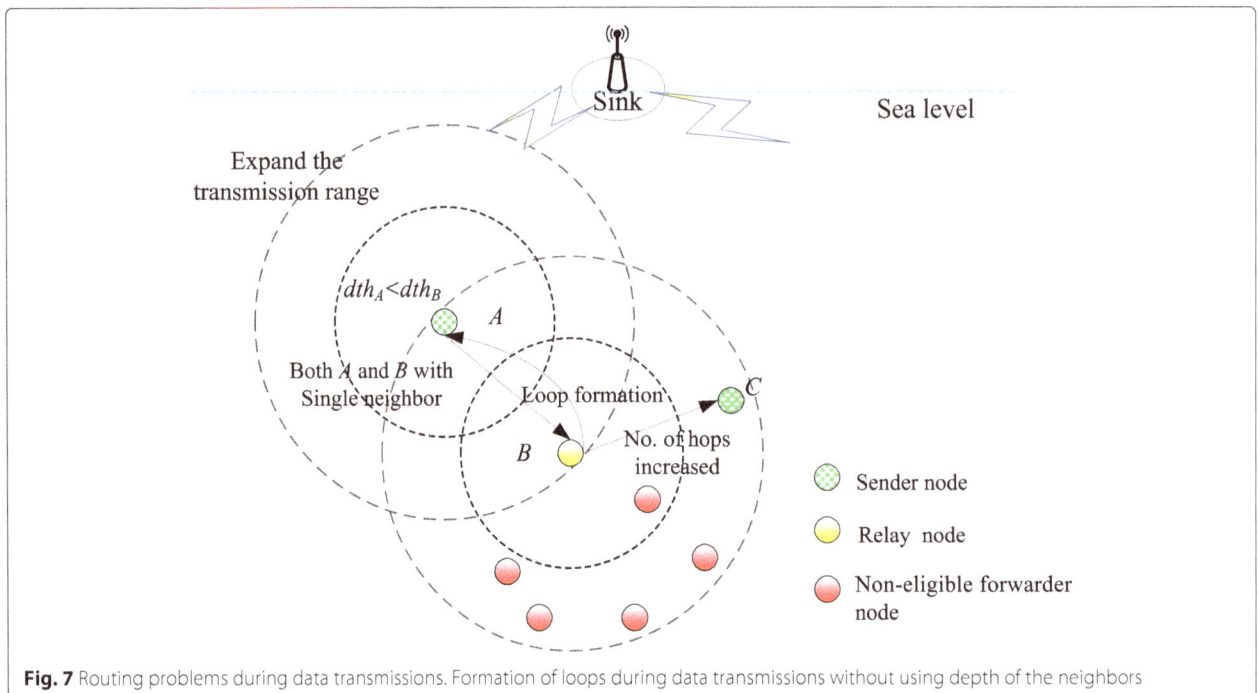

Fig. 7 Routing problems during data transmissions. Formation of loops during data transmissions without using depth of the neighbors

of the transmission distance, d_{max} and d_{min}. Transmitting data over a long distance by expanding the transmission range will result in a large amount of packets loss while reducing the transmission range will cause higher energy consumption, shorter network lifetime, and higher packet loss rate. IEBR is a reasonable trade-off in the view of the global performance. (v) indicates that the probability P_l of the current link state should be no less than P_g, which

is the minimum probability required for successful data transmission [39]. (vi) indicates that the energy consumption of every ring sector should be approximately equal. If the energy consumption is balanced, the network can achieve high throughput so as to prolong the effective lifetime.

4 Performance evaluation

The performance of the proposed IEBR will be verified by cross comparison with BTM and UDAR. In addition, EBR will be adopted as an independent algorithm to evaluate the impact of various elements and stages of the algorithm model.

For EBR, BTM, and UDAR, there are the same number of sensors lying in each ring sector of UWSN in the simulation. The initial energy of each sensor node is 300 J, and the transmission data packet size is 200 bits: 50 bits in the control field, 150 bits in the data field. Carrier sense multiple access with collision avoidance (CSMA/CA) is adopted under IEEE 802.15.4. IEBR model uses Linprog

Table 3 Routing establishment of IEBR

	Routing establishment: Relay selection based on depth
1:	Query packet $= Q_p$
2:	Depth threshold $= h_{th}$
3:	Depth of Node $i = h_i$
4:	Depth difference between node i and $j = h_{diff}(i, j)$
5:	RNT:Relay Node Table
6:	When neighbor j receiving Q_p
7:	if $h_i < h_j$ then
8:	if $h_{diff} > h_{th}$ then
9:	Algorithm 1
10:	if $N_j = min(N_j)$ then
11:	Add node ID in RNT
12:	else
13:	drop Q_p
14:	end if
15:	end if
16:	end if

Table 4 Simulation parameters

Parameters	Value
Network radius (R)	1–5 km, $\Delta = 0.5$ km
Number of nodes (N)	80
Initial energy (E_0)	300 J
Frequency (f)	20 kHz
Receiving constant (r)	0.2×10^{-4} J/bit

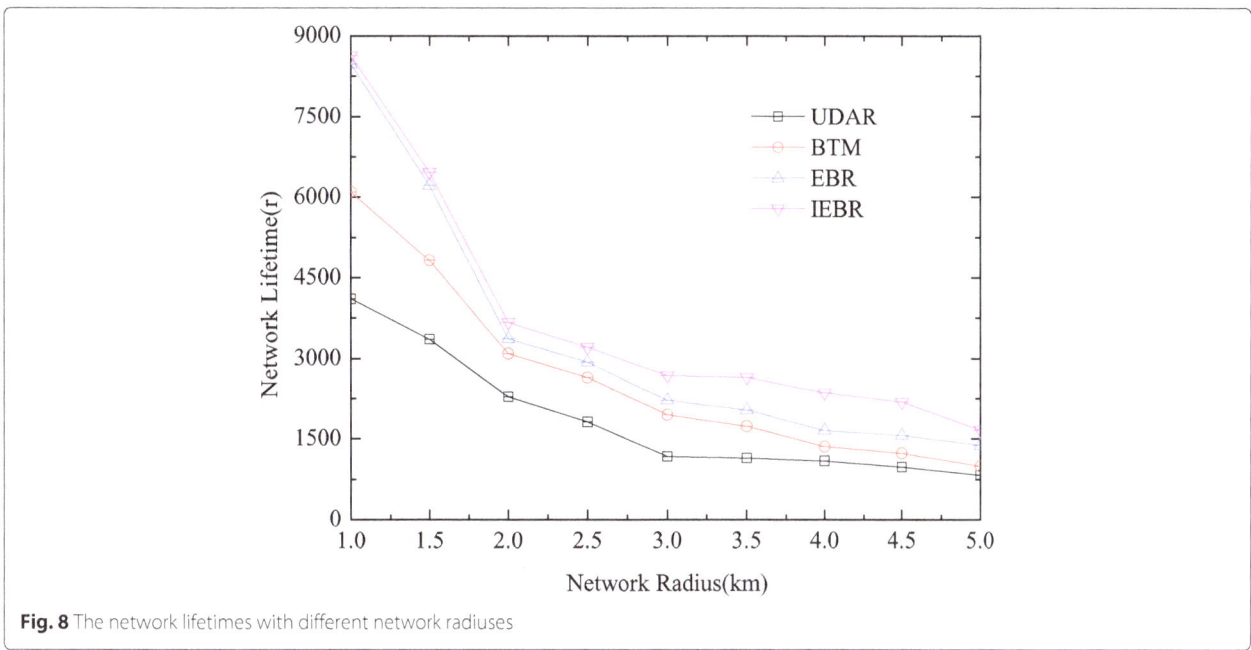

Fig. 8 The network lifetimes with different network radiuses

linear programming to achieve the throughput optimization. Network lifetime, effective throughput and transmission loss are used to evaluate the network performance. Network lifetime defined by BTM is evaluated by the maximum transmission rounds (r) that can be achieved. Effective throughput is the number of valid packets (p) received by the sink successfully. Some parameters are shown in Table 4.

4.1 Network lifetime with different network radiuses

The network lifetime comparison of the algorithms on different network scales is shown in Fig. 8. The curves show that the network lifetime of all algorithms will decrease with the increase of the network radius. When the radius is less than 2 km, the network lifetime falls obviously, but the decline curves of IEBR and EBR are an obvious flat to achieve better lifetime performance. Specifically, the

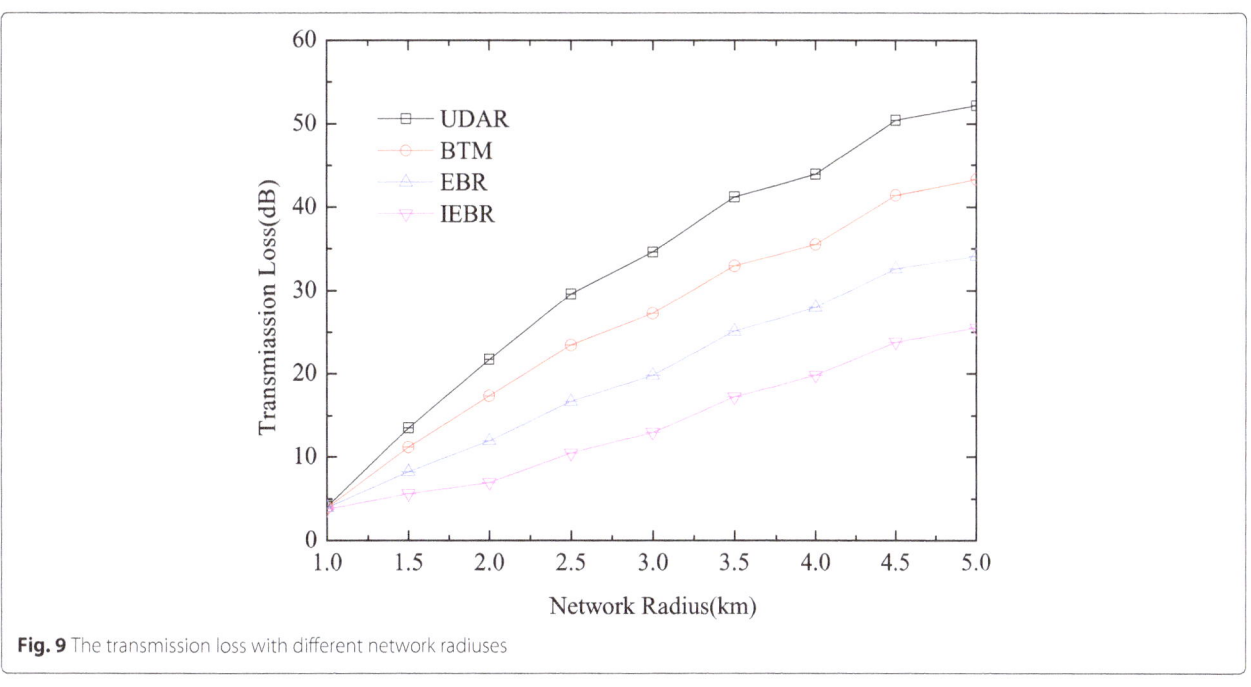

Fig. 9 The transmission loss with different network radiuses

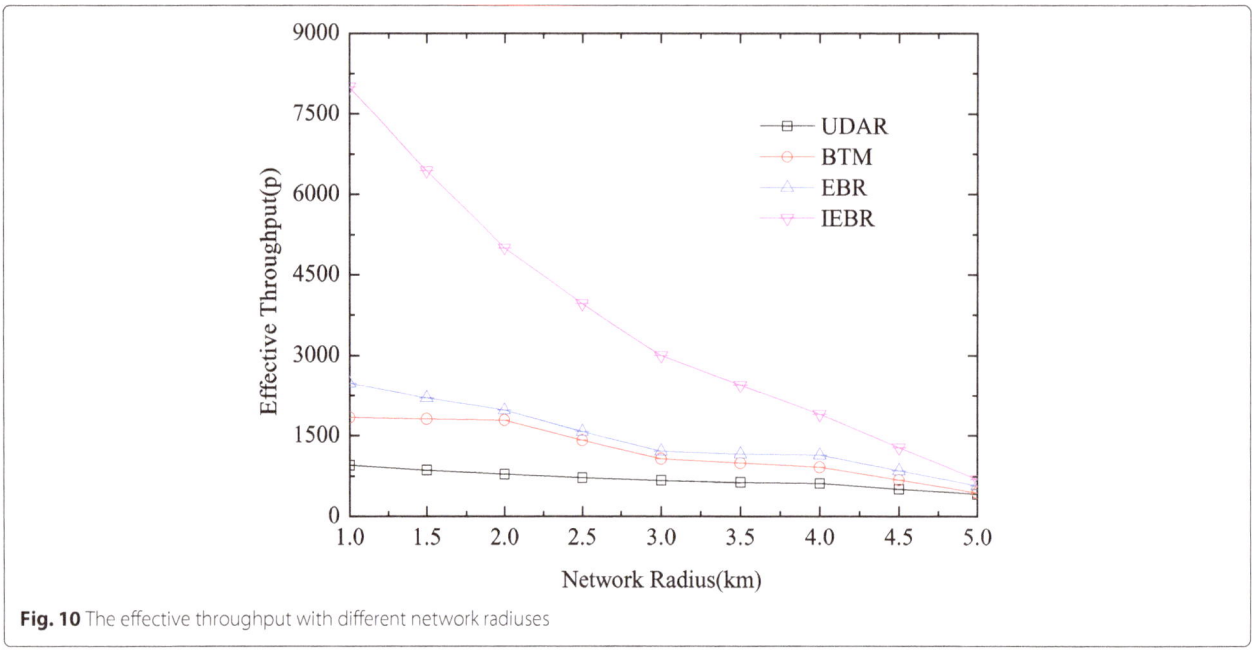

Fig. 10 The effective throughput with different network radiuses

network lifetime of IEBR is about 1.5 times and twice more than that of BTM and UDAR, respectively. When the radius of the network exceeds 3 km, the falling trend of the network lifetime tends to be flat. The network lifetime of IEBR is still higher than the other algorithms, about 1.5 times higher than EBR, and about twice higher than UDAR and BTM. Overall, IEBR performs better in the network lifetime on different network scales, which is mainly because IEBR is able to reduce the hops as well as the data load of the sensors near the sink.

4.2 Transmission loss with different network radiuses

The transmission loss caused by the balanced algorithms with different network sizes is shown in Fig. 9, which indicates that the transmission loss will increase with the network radius for all algorithms. The increasing trend of IEBR, however, is relatively flat. The transmission loss curves are roughly the same when the network radius is 1km. With the network radius increasing by every 1km, the transmission loss of IEBR and EBR will increase by about 5 dB and 7 dB, respectively, while that of BTM and UDAR will exceed 10 dB and 12 dB, respectively. The advantage of the IEBR algorithm is relatively obvious. When the network radius reaches 5 km, the transmission loss of UDAR is about 30 dB higher than IEBR, the increment is 20 dB for BTM, and is about 10 dB for EBR. It can be seen that IEBR will cause low transmission loss while processing a long network lifetime.

4.3 Effective throughput with different network radiuses

Considering the multi-hop forwarding characteristics of UWSN, the number of effective data packets received is used as the measure of throughput. The effective throughput comparison with different network radiuses is shown in Fig. 10. The effective throughput will decline with the network radius increasing. That is because the increase in the network radius will cause the decrease in node density with constant node number. And the increase in distance between two neighbors will cause the packet loss rate to increase. Specifically, when the radius is 1 km, the effective throughput of IEBR is approximately 8 times more than that of UDAR. After that, the effective throughput will drop sharply with the radius increasing, but it is still higher than that of the other algorithms. That means IEBR has an advantage in effective throughput, especially in the network with the limited radius, which is similar with the simulation results in network lifetime.

4.4 Network lifetime with different network radiuses and node numbers

In the previous simulation, the node number is a constant. The increase in the radius means the decrease in the node density. The comparison of network lifetime in different sizes and scales is shown in Fig. 11. There is a similar trend in the simulating lifetime curves of all balanced algorithms with the node number increasing from 80 to 160. The maximum network lifetime occurs when the number is about 120 and the network radius is 1 km, which is related to the construction of the network model. When the network radius increases to 3 km, IEBR shows a great sensitivity to the node number. When it increases (or decreases) by 10, the network lifetime will increase (or decrease) by about 200. With the node increasing from 80 to 120 in the simulation, IEBR will prolong the

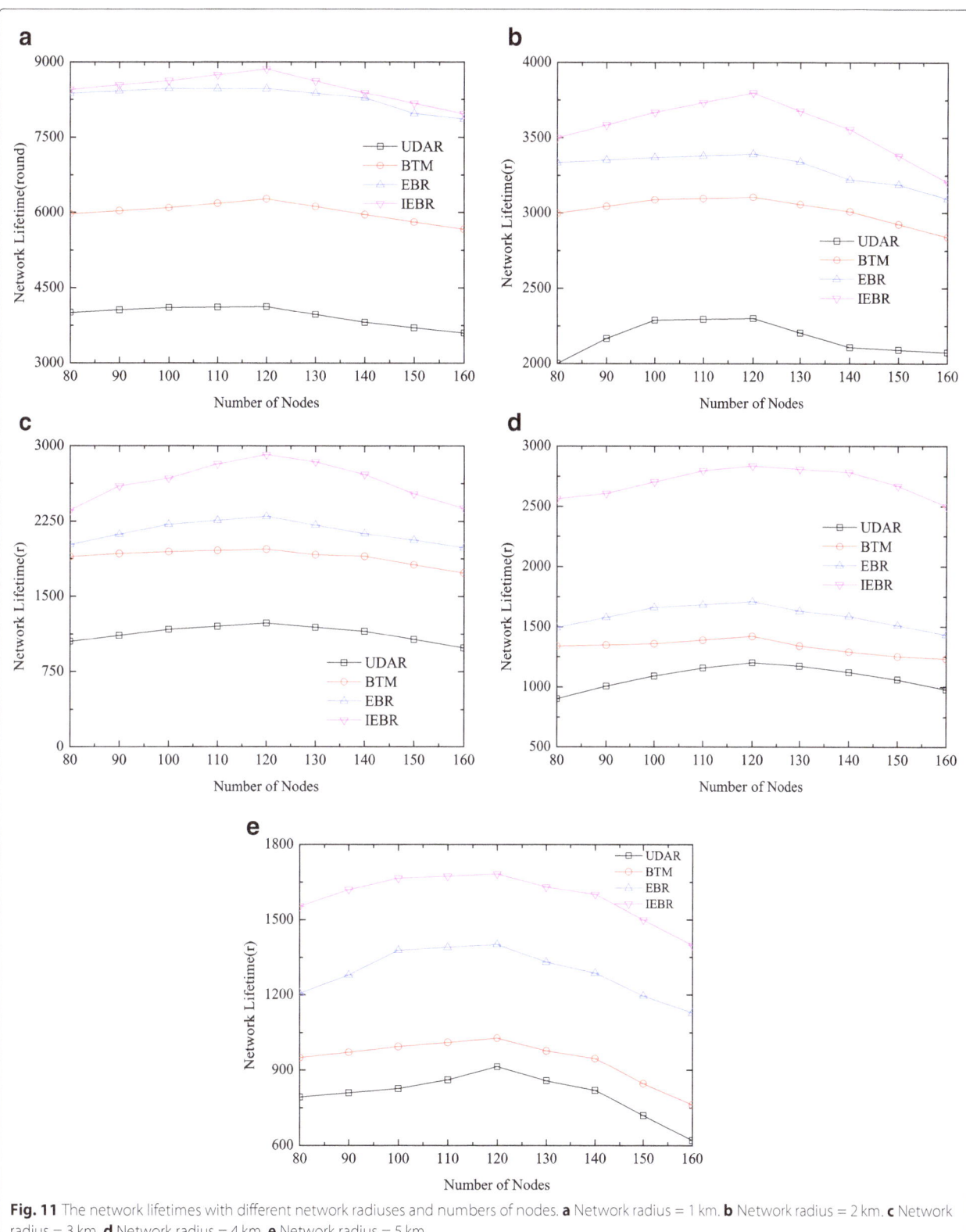

Fig. 11 The network lifetimes with different network radiuses and numbers of nodes. **a** Network radius = 1 km. **b** Network radius = 2 km. **c** Network radius = 3 km. **d** Network radius = 4 km. **e** Network radius = 5 km

network lifetime by about 25%. The comparisons show that IEBR has an advantage in network lifetime over other algorithms with different network radiuses and numbers of nodes.

4.5 Effective throughput with different network radiuses and node numbers

The comparison of the effective throughput with different network radiuses and numbers of nodes are shown

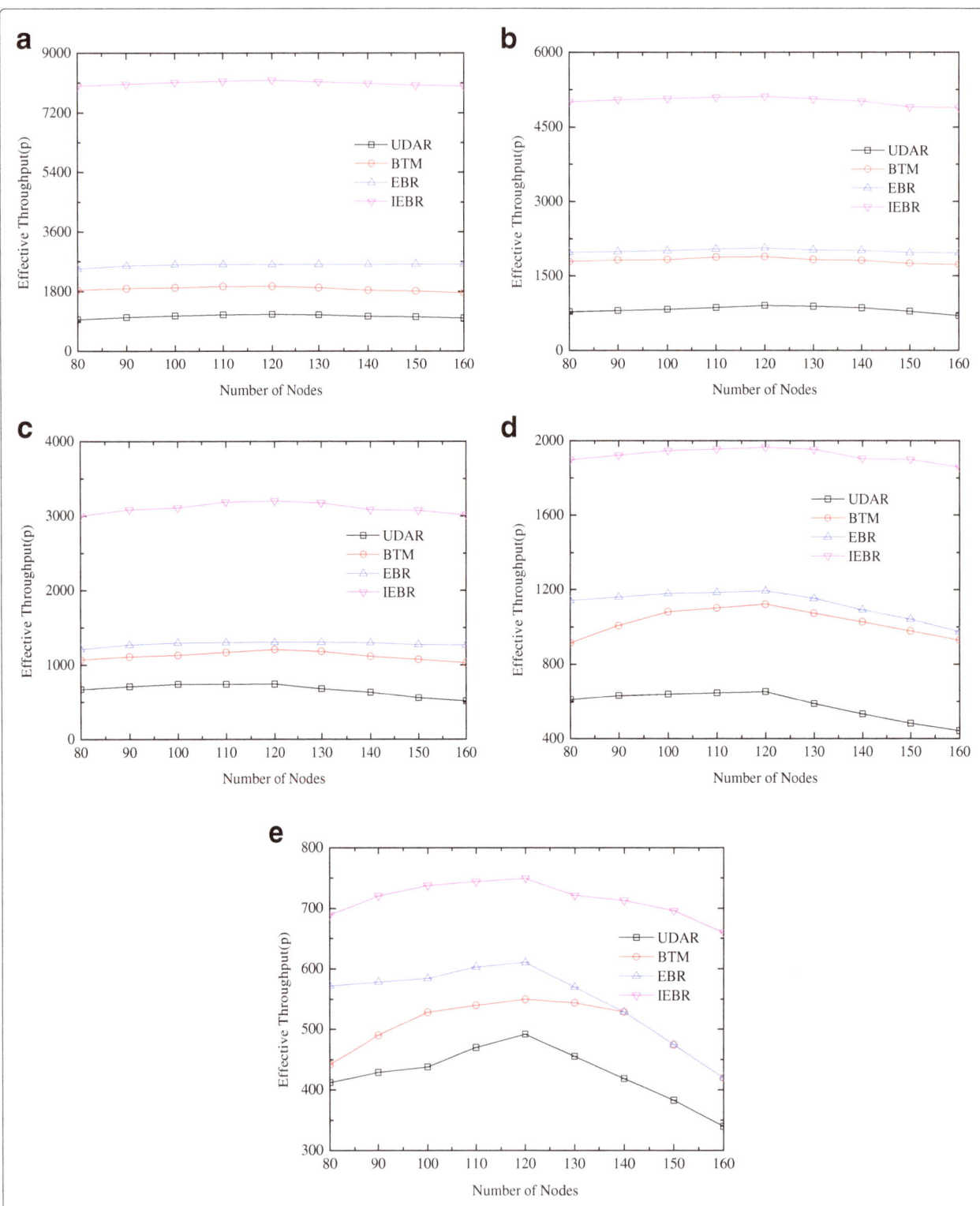

Fig. 12 The effective throughput with different network radiuses and numbers of nodes. **a** Network radius = 1 km. **b** Network radius = 2 km. **c** Network radius = 3 km. **d** Network radius = 4 km. **e** Network radius = 5 km

in Fig. 12. The effective throughput is less sensitive to the node density when the network radius is small. With the radius increasing, the effective throughput of the network will rise first and then gradually decline, which is similar to the lifetime curves. When the node number is about 120, the values of effective throughput are the largest with different network radiuses. The larger lifetime means more data transmission and reception, and the effective throughput could be larger, which is similar to the results in Fig. 11. Under different node density conditions, IEBR will have high effective throughput and relatively stable performance in a small-scale network.

5   Conclusion

To solve the problem of limited energy and short lifetime in UWSN, an improved energy balance routing (IEBR) algorithm is presented in this paper. A ring sector model is constructed, and the optimal relay node is selected by transmission distance and depth threshold to avoid the transmission loop. IEBR will select the optimal relay node among different ring sectors alternately, and the link structure will be adjusted dynamically according to the energy level difference. Simulation results show that IEBR has a longer network lifetime, larger effective throughput, and lower transmission loss than the existing typical algorithms in UWSN with different sizes and scales. Moreover, the research indicates that IEBR achieves global energy balance rather than the local balance as the existing algorithms do. Futural research will consider the spatial expansion of the energy ring sectors and the model of dynamic depth thresholds.

Abbreviations
BTM: Balanced transmission mechanism; DT: Direct transmission; EBR: Energy balance; EL: Energy level; HG: Hop grade; IEBR: Improved energy-balanced routing; MT: Multi-hop transmission; RSSI: Received signal strength indication; UDAR: Underwater data-aggregating ring; UEL: Union energy level; UWSN: Underwater Wireless sensor network

Authors' contributions
PF and DyQ conceptualized the idea and designed the experiments. PF contributed in writing and draft preparation and DyQ supervised the research. All authors read and approved the final manuscript.

Funding
The funding was supported by the National High Technology Research and Development Program of China (2012AA120802),National Natural Science Foundation of China (61771186), Postdoctoral Research of Heilongjiang Province (LBH-Q15121), and Undergraduate University Project of Young Scientist Creative Talent of Heilongjiang Province (UNPYSCT-2017125).

Competing interests
The authors declare that they have no competing interests.

Author details
[1]Heilongjiang University, Harbin, China. [2]Dire-dawa Institute of Technology, Dire Dawa, Ethiopia.

References
1. L. Xin, J. Min, Z. Xueyan, et al., A novel multi-channel Internet of Things based on dynamic spectrum sharing in 5G communication[J]. IEEE Internet of Things J. **6**, 1–1 (2018)
2. X. Liu, F. Li, Z. Na, Optimal resource allocation in simultaneous cooperative spectrum sensing and energy harvesting for multichannel cognitive radio[J]. IEEE Access. **5**, 1–1 (2017)
3. X. Liu, M. Jia, Z. Na, et al., Multi-modal cooperative spectrum sensing based on Dempster-Shafer fusion in 5G-based cognitive radio[J]. IEEE Access. **6**(99), 199–208 (2018)
4. Z. Na, Y. Wang, X. Li, et al., Subcarrier allocation based simultaneous wireless information and power transfer algorithm in 5G cooperative OFDM communication systems[J]. Phys. Commun. **29**, 164–170 (2018)
5. Z. Na, J. Lv, M. Zhang, et al., GFDM based wireless powered communication for cooperative relay system[J]. IEEE Access. **7**, 50971–50979 (2019)
6. M.A. Shaolin, Y. Wang, Design and implementation of an ultra-wideband high-accuracy ranging system[J]. J. Tianjin Norm. Univ.(Nat. Sci. Ed.) **37**(6), 55–57 (2017)
7. X. Zhang, Z. Zhang, Near-field plate applied in wireless power transmission system[J]. J. Tianjin Norm. Univ. (Nat. Sci. Ed.) **37**(6), 58–61 (2017)
8. J. Wu, C. Zeng, J. Sun, Research and application of wireless intelligent network monitoring smog system based on STM32F407[J]. J. Tianjin Norm. Univ. (Nat. Sci. Ed.) **37**(6), 62–66 (2017)
9. Z. Liu, J. Chen, Y. Tong, F. Duan, J.I. Maolin, Research and implementation of digital baseband signal transmission[J]. J. Tianjin Norm. Univ. (Nat. Sci. Ed.) **38**(1), 51–55 (2018)
10. B. Liang, J. Xie, J. Shi, W. Wang, Design and implementation of three-phase inverter for microgrid research[J]. J. Tianjin Norm. Univ. (Nat. Sci. Ed.) **38**(1), 59–63 (2018)
11. M. Jia, X. Gu, Q. Guo, W. Xiang, N. Zhang, Broadband hybrid satellite-terrestrial communication systems based on cognitive radio toward 5G. IEEE Wirel. Commun. **23**(6), 96–106 (2016)
12. M. Jia, X. Liu, X. Gu, Q. Guo, Joint cooperative spectrum sensing and channel selection optimization for satellite communication systems based on cognitive radio. Int. J. Satell. Commun. Netw. **35**(2), 139–150 (2017)
13. M. Jia, X. Liu, Z. Yin, Q. Guo, X. Gu, Joint cooperative spectrum sensing and spectrum opportunity for satellite cluster communication networks. Ad Hoc Netw. **58**(C), 231–238 (2016)
14. I.F. Akyildiz, D. Pompili, T. Melodia, Underwater acoustic sensor networks: research challenges[J]. Ad Hoc Netw. **3**(3), 257–279 (2005)
15. N.Z. Zenia, M.S. Kaiser, M.R. Ahmed, et al., in *Electrical Engineering and Information Communication Technology (ICEEICT), 2015 International Conference on*. An energy efficient and reliable cluster-based adaptive mac protocol for uwsn[C] (IEEE, Dhaka, 2015), pp. 1–7
16. A. Solayappan, M.BH. Frej, S.N. Rajan, in *Systems, Applications and Technology Conference (LISAT), 2017 IEEE Long Island*. Energy efficient routing protocols and efficient bandwidth techniques in Underwater Wireless Sensor Networks-a survey[C] (IEEE, Farmingdale, 2017), pp. 1–7
17. Z. Wan, S. Liu, W. Ni, et al., An energy-efficient multi-level adaptive clustering routing algorithm for underwater wireless sensor networks[J]. Clust Comput, 1–10 (2018)
18. S. Souiki, M. Hadjila, M. Feham, in *Programming and Systems (ISPS), 2015 12th International Symposium on*. Energy efficient routing for ,obile underwater wireless sensor networks[C] (IEEE, Algiers, 2015), pp. 1–6
19. A. Wahid, D. Kim, An energy efficient localization-free routing protocol for underwater wireless sensor networks[J]. Int. J. Distrib. Sensor Netw. **8**(4), 307246 (2012)
20. J. Cao, J. Dou, S. Dong, Balance transmission mechanism in underwater acoustic sensor networks[J]. Int. J. Distrib. Sensor Netw. **11**(3), 429340 (2015)
21. Z.L. Li, N.M. Yao, Q. Gao, Relative distance-based forwarding protocol for underwater wireless sensor networks[C]. Trans. Tech. Publ. Appl. Mech. Mater. **437**, 655–658 (2013)
22. A. Wahid, S. Lee, D. Kim, in *2011 IEEE-Spain OCEANS*. An energy-efficient routing protocol for UWSNs using physical distance and residual energy[C] (IEEE, Santander, 2011), pp. 1–6
23. S. Mahmood, H. Nasir, S. Tariq, et al., Forwarding nodes constraint based DBR (CDBR) and EEDBR (CEEDBR) in underwater WSNs[J]. Procedia Comput. Sci. **34**, 228–235 (2014)

24. J. Shen, A. Wang, C. Wang, et al., An efficient centroid-based routing protocol for energy management in WSN-assisted IoT[J]. IEEE Access. **5**, 9–18479 (1846)

25. I. Azam, N. Javaid, A. Ahmad, et al., Balanced load distribution with energy hole avoidance in underwater WSNs[J]. IEEE Access. **5**, 15206–15221 (2017)

26. N. Javaid, T. Hafeez, Z. Wadud, et al., Establishing a cooperation-based and void node avoiding energy-efficient underwater WSN for a Cloud[J]. IEEE Access. **5**, 11582–11593 (2017)

27. B. Ali, N. Javaid, A.R. Hameed, et al., in *Wireless Communications and Mobile Computing Conference (IWCMC), 2017 13th International*. Energy hole avoidance based routing for underwater WSNs[C] (IEEE, Valencia, 2017), pp. 1654–1659

28. A. Bengheni, F. Didi, I. Bambrik, EEM-EHWSN: Enhanced energy management scheme in energy harvesting wireless sensor networks[J]. Wirel. Netw. **25**(6), 3029–3046 (2019)

29. R. Yousaf, R. Ahmad, W. Ahmed, et al., A unified approach of energy and data cooperation in energy harvesting WSNs[J]. Sci. China Inf. Sci. **61**(8), 082303 (2018)

30. X. Yang, L. Wang, J. Xie, et al., Energy efficiency TDMA/CSMA hybrid protocol with power control for WSN[J]. Wirel. Commun. Mob. Comput. **2018** (2018)

31. I. Harris AF, M. Stojanovic, M. Zorzi, in *Proceedings of the 1st ACM international workshop on Underwater networks*. When underwater acoustic nodes should sleep with one eye open: idle-time power management in underwater sensor networks[C] (ACM, Los Angeles, 2006), pp. 105–108

32. A.A. Syed, W. Ye, J. Heidemann, in *The 27th Conference on Computer Communications. IEEE. INFOCOM 2008*. T-Lohi: A new class of MAC protocols for underwater acoustic sensor networks[C] (IEEE, Phoenix, 2008), pp. 231–235

33. Y. Bi, N. Li, L. Sun, DAR: An energy-balanced data-gathering scheme for wireless sensor networks[J]. Comput. Commun. **30**(14-15), 2812–2825 (2007)

34. J. Poncela, M.C. Aguayo, P. Otero, Wireless underwater communications[J]. Wirel. Pers. Commun. **64**(3), 547–560 (2012)

35. M. Stojanovic, On the relationship between capacity and distance in an underwater acoustic communication channel[J]. ACM SIGMOBILE Mob. Comput. Commun. Rev. **11**(4), 34–43 (2007)

36. K.M. Kwak, J. Kim, Development of 3-dimensional sensor nodes using electro-magnetic waves for underwater localization[J]. J. Inst. Control Robot. Syst. **19**(2), 107–112 (2013)

37. K. Lorincz, M. Welsh, in *International Symposium on Location-and Context-Awareness*. Motetrack: A robust, decentralized approach to rf-based location tracking[C] (Springer, Berlin, 2005), pp. 63–82

38. M. Zorzi, P. Casari, N. Baldo, et al., Energy-efficient routing schemes for underwater acoustic networks[J]. IEEE J. Sel. Areas Commun. **26**(9), 1754–1766 (2008)

39. A. Ahmad, N. Javaid, Z.A. Khan, et al., $(ACH)^2$: Routing Scheme to Maximize Lifetime and Throughput of Wireless Sensor Networks[J]. IEEE Sensors J. **14**(10), 3516–3532 (2014)

Energy-efficient and secure mobile node reauthentication scheme for mobile wireless sensor networks

BoSung Kim[*] ⓘ and JooSeok Song

Abstract

Mobile wireless sensor networks (MWSNs) are a relatively new type of WSN where the sensor nodes are mobile. Compared to static WSNs, MWSNs provide many advantages, but their mobility introduces the problem of frequent reauthentication. Several mobile node reauthentication schemes based on symmetric key cryptography have been proposed to efficiently handle frequent reauthentication. However, due to security weaknesses such as unconditional forwarding or low-compromise resilience, these schemes do not satisfy the security requirements of MWSNs. In this paper, we propose an energy-efficient and secure mobile node reauthentication scheme (ESMR) for MWSNs that satisfies the security requirements of MWSNs by addressing the security weaknesses of previous studies. ESMR prevents unconditional forwarding by allowing a foreign cluster head to authenticate mobile nodes, providing high-compromise resilience because it limits the use of cryptographic keys for different purposes. Security analysis shows that ESMR meets security requirements and can prevent relevant security attacks. Performance evaluation shows that ESMR is suitable for multi-hop communication environment, where the number of hops between the mobile node and cluster head is two or more. Specifically, ESMR requires less than 6% increased total energy consumption and 3% increased reauthentication latency compared with previous studies; hence, it introduces negligible performance overhead. Considering both performance and security aspects, ESMR also can be applied to single-hop communication environment.

Keywords: Wireless sensor networks, Mobile wireless sensor networks, Mobile sensor node, Security, Authentication, Key agreement, Multi-hop communication

1 Introduction

With the development of the internet of things (IoT), wireless sensor networks (WSNs) are attracting significant attention as a fundamental technology in the IoT. WSNs consist of many sensor nodes that collect data from their surrounding environments and send them to a base station in a multi-hop fashion. Mobile WSNs (MWSNs) are a new type of WSN in which sensor nodes are mobile. By supporting mobility, they can achieve better network performance than traditional static WSNs. Recent studies [1, 2] have shown that MWSNs not only extend network lifetime, but also improve connectivity and coverage. MWSNs also support more types of applications

than static WSNs in the following fields: patient monitoring [3], animal monitoring and tracking [4], and object monitoring [5].

Although MWSNs provide many advantages compared to static WSNs, the underlying sensor mobility introduces the problem of frequent reauthentication. Many MWSN applications, such as battlefield surveillance, habitat monitoring, and healthcare, require secure communications, and authentication is an essential first step for secure communication. In MWSNs, sensor nodes continuously change their positions, causing frequent communication link and network topology changes [6, 7]. Consequently, authentication is repeatedly required to establish secure communications whenever the communication link changes. Such frequent reauthentication can cause significant energy consumption for resource-constrained sensor nodes, and it is important to efficiently handle

*Correspondence: bokor@yonsei.ac.kr
Department of Computer Science, Yonsei University, 50 Yonsei-ro, Seodaemun-Gu, Seoul, Republic of Korea

communication and computation overheads. Although there have been many studies to support mobility in existing internet services, they are not applicable to MWSNs due to resource constraints of the sensor node [8]. Therefore, lightweight security mechanisms that explicitly consider sensor mobility are required to handle frequent reauthentication in MWSNs.

Several mobile node reauthentication schemes for MWSNs have been proposed to efficiently handle frequent reauthentication based on symmetric key cryptography [9–11], which can provide lightweight reauthentication mechanisms suitable for resource-constrained sensor nodes. However, these schemes generally have significant security weaknesses.

In most applications, the sensor node is placed in a location easily accessible by an adversary; hence, the adversary is more likely to be able capture the sensor node and extract cryptographic secrets. Therefore, any mobile node reauthentication scheme should provide high-compromise resilience, where the compromised node reveals no information about links it is not directly involved with. However, Han et al.'s schemes [9, 10] do not provide high-compromise resilience. If even a single-cluster head is captured, other nodes that do not share pairwise keys with the compromised cluster head can be affected, i.e., an adversary can compromise pairwise keys of other nodes that are not directly involved with the compromised cluster head.

Adversaries can also eavesdrop on the radio frequencies of MWSNs due to the open nature of wireless communication, and alter or spoof messages to launch denial of service (DoS) attacks or drain receiver resources. Authentication to provide the assurance of identity of the communicating node is essential to prevent such attacks. However, Jiang et al.'s scheme [11] allows a foreign cluster head to unconditionally forward mobile nodes' reauthentication requests to the home cluster head without preliminary verification, which could be used to launch DoS attacks on the home cluster head.

In this paper, we propose an energy-efficient and secure mobile node reauthentication scheme (ESMR) for MWSNs. ESMR focuses on satisfying the security requirements of MWSNs by addressing the security weaknesses of the existing mobile node reauthentication schemes [9–11]. ESMR uses two additional keys (key derivation key (KDK) and authentication key (AK)) in addition to the pairwise key to prevent unconditional forwarding. ESMR also provides high -compromise resilience because it limits the use of the KDK and AK for different purposes. Each cluster head has its own KDK that it shares as a group key with neighboring cluster heads. During initial authentication, the home cluster head generates an AK for a mobile node for the next authentication by using its KDK. When the mobile node moves to a new location and initiates the

reauthentication procedure with a foreign cluster head, the foreign cluster head can authenticate the mobile node by using the AK. Consequently, unconditional forwarding is prevented. The KDK is only used to generate AKs, and AKs are only used for authenticating mobile nodes. Thus, the compromised cluster head does not affect other nodes that do not share pairwise keys with the compromised cluster head. Security analysis shows that by addressing the security weaknesses of the existing schemes, ESMR meets security requirements of MWSNs and can prevent relevant security attacks. The main contributions of this work can be described as follows:

- We show the security weaknesses of existing mobile node reauthentication schemes based on symmetric key cryptography [9–11].
- We propose ESMR, which satisfies the security requirements of MWSNs by addressing the security weaknesses of existing mobile node reauthentication schemes.
- The security of ESMR is formally verified by using the automated validation of internet security protocols and applications (AVISPA) tool [12].
- Simulations are conducted using OMNeT++ with the INET framework 3.6.3 to evaluate the performance of ESMR in terms of energy consumption and reauthentication latency.

The remainder of this paper is organized as follows. Section 2 discusses related works and their problems. Section 3 briefly reviews existing mobile node reauthentication schemes [9–11] and discusses identified security weaknesses. Section 4 provides an overview of ESMR, and Section 5 provides the details. Section 6 analyzes the security of ESMR. Section 7 evaluates the performance of ESMR. Finally, Section 8 summarizes and concludes the paper.

2 Related works

Mobile node reauthentication schemes for MWSNs can be classified into two types according to their encryption techniques: symmetric key cryptography schemes [9–11, 13–15] and public key cryptography schemes [16–19]. Symmetric key cryptography schemes can be further classified into two types: post-deployment key establishment schemes [9–11, 13] and hybrid schemes that use both random key pre-distribution and post-deployment key establishment mechanisms [14, 15].

2.1 Symmetric key cryptography schemes

Han et al. proposed two ticket-based schemes [9, 10]. In [9], each cluster head has its own ticket generation key (TGK) that is used to generate tickets, which is shared as a group key with neighboring cluster heads. When a mobile node moves to a new location and initiates the reauthentication procedure, the mobile node and foreign cluster head authenticate each other and establish a pairwise key using the ticket. However, this scheme may not work properly in situations where sensor nodes are irregularly distributed and mobile nodes move between non-neighboring cluster heads.

Therefore, an improved ticket-based scheme was proposed [10] that utilizes the neighboring cluster head list (NCL) for reauthentication. Two non-neighboring cluster heads compare their NCLs to find a common neighboring cluster head. Then reauthentication can be performed through the common neighboring cluster head even if two cluster heads are not neighbors.

The main disadvantage of these schemes is that communication overhead is concentrated in mobile nodes, because the mobile nodes directly transfer the information required for reauthentication, e.g., tickets and NCLs. The schemes also do not provide high-compromise resilience, since each cluster head shares its TGK as a group key with neighboring cluster heads. Therefore, multiple TGKs are exposed if even a single-cluster head is captured. An adversary can then obtain secret information included in all generated tickets using the exposed TGKs and compromise multiple nodes' communication security using secret information obtained from the tickets.

In [13], Bilal and Kang proposed a ticket-based authentication suite that supports multiple secure connections. The authentication suite consists of two mobile node reauthentication protocol, SRP1, and SRP2, which support multiple secure connections between the mobile node and multiple cluster heads. In SRP1 and SRP2, the mobile node and foreign cluster head can authenticate each other directly using the ticket. However, SRP1 and SRP2 have the same disadvantage as [9] and [10], in which the communication overhead is concentrated in the mobile node. Moreover, since SRP2 requires the involvement of the base station, there is a problem that the reauthentication latency is long.

To reduce the communication overhead of mobile nodes, Jiang et al. [11] proposed a mobile node reauthentication scheme without using tickets. In the scheme, a foreign cluster head forwards the reauthentication request of a mobile node to the home cluster head. The home cluster head then authenticates the request on behalf of the foreign cluster head. Since the information required for reauthentication is exchanged between the foreign and home cluster head, the communication overhead in mobile nodes is reduced. However, the scheme has a security weakness called unconditional forwarding. Since there are no shared secrets between the foreign cluster head and mobile node, foreign cluster heads cannot authenticate reauthentication requests of mobile nodes and unconditionally forward reauthentication requests to

the home cluster head. This unconditional forwarding can be lead to DoS attacks on the home cluster head.

The previous schemes use only the post-deployment key establishment mechanism, whereas hybrid schemes [14, 15] use both random key pre-distribution and post-deployment key establishment mechanisms. In these schemes, by default, two nodes authenticate each other and establish a pairwise key based on existing random key pre-distribution schemes [20, 21]. For random key pre-distribution schemes, a set of keys are randomly chosen from a large key pool and pre-stored in each sensor node. During the key discovery phase, two nodes exchange pre-stored key identifiers to find a common key. They use the post-deployment key establishment scheme if two nodes do not share a common key to establish a pairwise key with the help of the base station.

The main disadvantage of hybrid scheme is that they require a minimum network density for the random key pre-distribution mechanism. Although two nodes that do not share a common key can establish a pairwise key using the post-deployment key scheme, multi-hop communication with the base station incurs longer communication delay and consumes more energy as hop distance between the mobile node and base station increases.

2.2 Public key cryptography schemes

In [16], Gandino et al. proposed an authentication and key establishment scheme based on public key cryptography for mobile and static WSNs. In the scheme, authentication tables are used to reduce communication and computational overhead due to the verification of digital certificate. The authentication table stores information necessary for a node to authenticate the other nodes in the network and is distributed to each node before deployment. In the key establishment phase, two nodes can authenticate each other's public keys directly using the authentication table instead of the digital certificate. However, there is a problem that the sensor node generates a pairwise key by performing public key encryption and decryption operations which cause a high-computation overhead. Especially, the computation overhead of the mobile node becomes more severe because the mobile node generates a new pairwise key every time reauthentication is performed.

With the development of elliptic curve cryptography (ECC) optimization techniques, several ECC-based mobile node reauthentication schemes have been proposed [17–19]. Zhang et al. [17] used the elliptic curve digital signature algorithm and elliptic curve Diffie-Hellman key agreement to dynamically generate pairwise keys. However, the scheme is not suitable for large-scale WSNs because of the overhead required for certificate management.

Seo et al. [18] proposed an ECC-based scheme without certificates to overcome this limitation using pairing-free certificateless hybrid signcryption scheme (CL-HSC) to dynamically provide both mobile node authentication and key agreement. The properties of the CL-HSC ensure that a pairwise key can be generated without requiring expensive pairing operations or certificate exchange. However, the scheme still requires expensive ECC point multiplications for mobile nodes to generate long-term pairwise keys, i.e., mobile nodes must perform multiple expensive ECC multiplications repeatedly whenever they move and are connected to a new cluster head.

Omar et al. [19] proposed a trusted third party based mobile node reauthentication scheme using ECC. In the scheme, when a mobile node wants to join a new cluster, it sends a join request message, including its public key, to a foreign cluster head. Upon receiving the message, the foreign cluster head requests the base station, which is a trusted third party, to authenticate the mobile node. Since the base station performs verification of the public key, there is no computation overhead of the sensor node due to the expensive ECC multiplication. However, the longer the hops distance between the cluster head and the base station, the longer the re-authentication latency becomes.

3 Analysis of post-deployment schemes

This section briefly reviews existing schemes [9–11] and highlights their security weaknesses.

Han et al.'s schemes [9, 10] use tickets for mobile node reauthentication, and have the same security weakness because they have almost the same reauthentication process. We focus on the mobile node reauthentication process based on [9]. Each cluster head has its own ticket generation key (TGK) that it shares with neighboring cluster heads. In the initial authentication, the home cluster head CH_A generates ticket T for the next reauthentication as follows using its ticket generation key TGK_{CH_A}:

$$
\begin{aligned}
T &= (t, w) \\
t &= E(K_{TGK_{CH_A}}, TS||R_1||K_{MN-CH_A}) \\
w &= MAC(K_{TGK_{CH_A}}, ID_{MN}||t)
\end{aligned}
\tag{1}
$$

where R_1 is a random number generated by the mobile node MN and K_{MN-CH_A} is a pairwise key between MN and CH_A.

When MN moves to a new location and receives the HELLO message from the foreign cluster head CH_B, it launches the reauthentication process by sending following reauthentication request to CH_B:

$$
\begin{aligned}
MN &\to CH_B : ID_{MN}||ID_{CH_B}||t||w||v_1 \\
v_1 &= MAC(K_{M-CH_A}, ID_{MN}||ID_{CH_B}||t||w||v_0)
\end{aligned}
\tag{2}
$$

Since CH_B is a neighboring cluster head of CH_A, it also has the TGK of CH_A. Therefore, CH_B can verify w, and obtain R_1 and K_{MN-CH_A} by decrypting t. CH_B then authenticates MN by verifying v_1 using K_{MN-CH_A} and

generates a pairwise key with MN as follows:

$$
K_{MN-CH_B} = KDF(R_1||R_0)
\tag{3}
$$

where R_0 is a random number generated by CH_B.

CH_B finally generates a new ticket T', in the same way that CH_A did in (1) using its TGK and sends following message to MN:

$$
\begin{aligned}
CH_B &\to MN : ID_{CH_B}||ID_{MN}||u_3||v_3 \\
u_3 &= E(K_{MN-CH_A}, R_0||v_2||T') \\
v_2 &= H(K_{MN-CH_B}||R_0) \\
v_3 &= MAC(K_{MN-CH_A}, ID_{CH_B}||ID_{MN}||u_3)
\end{aligned}
\tag{4}
$$

Upon receiving the message from CH_B, MN verifies v_3 and obtains R_0 using K_{MN-CH_A}. MN then generates K_{MN-CH_B} in the same way that CH_B did in (3) and finally authenticates CH_B by verifying v_2 using K_{MN-CH_B}.

Assume that cluster head CH_C is another neighboring cluster head of CH_A, but CH_C and CH_B are not neighbors. If an adversary captures CH_C and extracts TGK_{CH_A} from it, the adversary can obtain R_1 and K_{MN-CH_A} by decrypting t included in T using TGK_{CH_A}, as shown in (1). The adversary can then decrypt u_3, included in the message in (4), using K_{MN-CH_A} to obtain R_0. The adversary can finally compromise K_{MN-CH_B} by directly generating K_{MN-CH_B} using R_1 and R_0, as shown in (1). Consequently, the adversary can compromise communication security between MN and CH_B not directly involved with the compromised cluster head. This problem is more serious because each cluster head has multiple TGKs, including TGKs of the neighboring cluster head and its own TGK. Thus, an adversary can compromise communication security for a number of nodes by compromising a single-cluster head.

On the other hand, Jiang et al.'s scheme [11] does not use tickets to reduce communication overhead of the mobile node. In the scheme, reauthentication process is proceeded with the help of the home cluster head, CH_A, rather than using a ticket. When MN wants to launch the reauthentication process, it sends following reauthentication request to the foreign cluster head CH_B:

$$
\begin{aligned}
MN &\to CH_B : ID_{MN}||ID_{CH_A}||t_1||MAC_1 \\
MAC_1 &= MAC(K_{MN-CH_A}, ID_{MN}||t_1||H(I))
\end{aligned}
\tag{5}
$$

where ID_{CH_A} is the identity of CH_A, t_1 is a timestamp, K_{MN-CH_A} is a pairwise key between MN and CH_A, and $H(I)$ is a hashed random number I shared between MN and CH_A.

Upon receiving the reauthentication request from MN, CH_B checks ID_{CH_A} included in the request and finds that the home cluster head of MN is CH_A. CH_B then forwards the reauthentication request of MN to CH_A:

$$
\begin{aligned}
CH_B &\to CH_A : ID_{MN}||t_2||t_1||MAC_1||MAC_2 \\
MAC_2 &= MAC(K_{CH_B-CH_A}, ID_{MN}||t_2||t_1||MAC_1)
\end{aligned}
\tag{6}
$$

where t_2 is a timestamp and $K_{CH_B-CH_A}$ is a pairwise key between CH_B and CH_A. After receiving the message from CH_B, CH_A verifies MAC_1 using K_{MN-CH_A} and $H(I)$ and authenticates MN on behalf of CH_B.

In the above scheme, CH_B cannot verify validity of the reauthentication request of MN in (5), because there is no shared secret between CH_B and MN. Consequently, CH_B unconditionally forwards the reauthentication request of MN as shown in (6). This unconditional forwarding allows DoS attacks on CH_A through neighboring cluster heads of CH_A, because an adversary can easily alter or spoof eavesdropped messages in MWSNs.

4   Overview of ESMR

In this paper, we propose an energy-efficient and secure mobile node reauthentication scheme (ESMR) for MWSNs. ESMR supports mutual authentication and key establishment between mobile nodes and cluster heads. ESMR is based on Jiang et al.'s scheme [11], which has the lowest computation and communication overheads for mobile nodes among existing schemes, and falls into the post-deployment key establishment scheme category. Figure 1 presents an overview of ESMR.

Each cluster head in ESMR has its own key derivation key (KDK) to generate the authentication keys (AKs) for mobile nodes. Prior to network operation, each cluster head shares its KDK as a group key with neighboring cluster heads (phase 0). During initial authentication (phase 1), the home cluster head CH_A, that a mobile node MN first connects after deployment, generates an AK for the mobile node by using its KDK for the next authentication. MN initiates the reauthentication procedure, when it moves to a new location and connects to a foreign cluster head CH_B. During reauthentication (phase 2), CH_B generates an AK using the KDK of CH_A and authenticates MN.

ESMR addresses the security weaknesses of schemes proposed by Han et al. [9, 10] and Jiang et al. [11]. First, ESMR prevents unconditional forwarding by allowing foreign cluster heads to authenticate mobile nodes. Second, ESMR provides high-compromise resilience by limiting the use of the KDK and AK for different purposes: the KDK is only used to generate AKs, and AKs are only used for authenticating mobile nodes. Table 1 provides the major notations used in this paper.

4.1   Network model

Following mobility-aware medium access control protocols for WSNs [22, 23], we consider a heterogeneous sensor network, consisting of a base station, cluster heads, and sensor nodes, as shown in Fig. 2. There are N_1 and N_2 cluster heads and sensor nodes, respectively, with where $N_1 << N_2$, and hence the total number of nodes in the network = $N_1 + N_2$.

Cluster heads are high-end sensor nodes with more resources in terms of computational power, storage, and battery life than the sensor nodes. The communication range of a cluster head is also larger than that of a sensor node. Cluster heads compose a stationary backbone network and periodically broadcast lightweight beacon messages to inform nodes of their presence. Considering the wide communication range of cluster heads, we assume that a foreign cluster head is a neighboring cluster head to the home cluster head.

Sensor nodes act as cluster members and can be stationary or mobile. A sensor node initiates the reauthentication procedure with a foreign cluster head when it moves to a new location and receives a beacon message from the foreign cluster head. Since the communication range of the sensor node is smaller than that of a cluster head, stationary nodes relay mobile node messages to cluster heads.

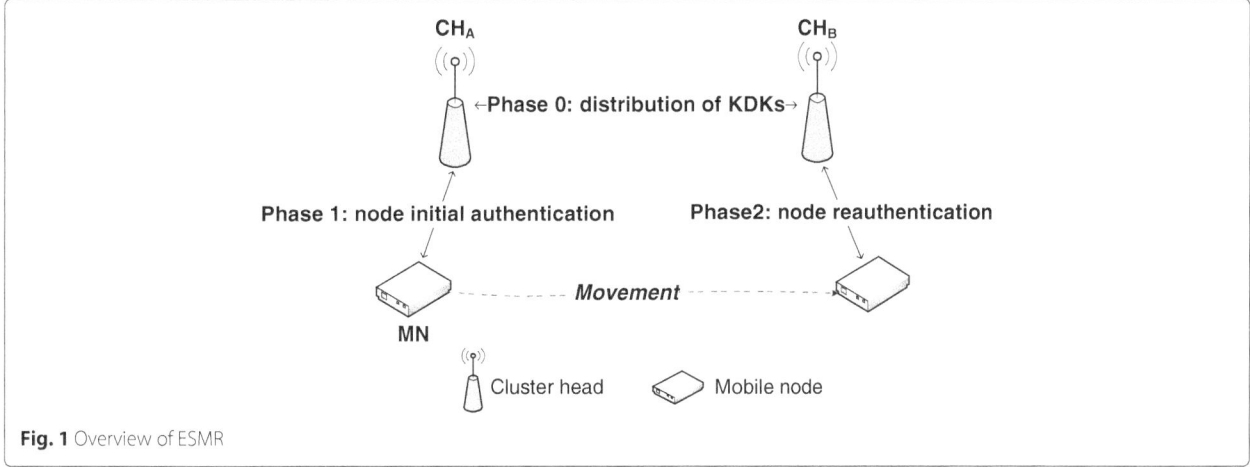

Fig. 1 Overview of ESMR

Table 1 Notations

Symbol	Description
ID_A	Identity of node A
TS_i	Timestamp
K_{A-B}	Pairwise key between node A and node B
KDK_A	KDK of node A
AK_A	AK of node A
$E(K, m)$	Encrypt message m with key K
$MAC(K, m)$	Message authentication code of message m with key K
$\|\|$	Concatenation
$H(\cdot)$	Hash function
f	Pseudorandom function
$\oplus$	Exclusive-OR operation

4.2 Adversary model

We assume that mobile nodes and cluster heads can be attacked passively or actively. Because of the open nature of wireless communication channels, an adversary can easily perform passive attacks, such as eavesdropping and traffic analysis, to gather information without being detected. In active attacks, an adversary may inject, intercept, or replay messages to disrupt network functionality or degrade network performance. We also assume that mobile nodes and cluster heads can be captured by an adversary. Because of the unattended nature of WSNs, nodes can also be physically captured by an adversary. Once a node is captured, all its stored secret information can be revealed to an adversary. An adversary can then utilize this secret information to perform the impersonation attack or compromise communication security. General security mechanisms, such as authentication and encryption, cannot prevent such insider attacks. However, secure reauthentication schemes should minimize insider attacks.

5 Details of ESMR

ESMR consists of three phases: the distribution of KDKs, node initial authentication, and node reauthentication. For simplicity of explanation, we describe ESMR based on Fig. 1.

5.1 Phase 0: distribution of KDKs

Each cluster head has its own KDK to generate AKs for mobile nodes. Prior to network operation, each cluster head shares its KDK as a group key with neighboring cluster heads. We assume that each cluster head has established pairwise keys with neighboring cluster heads after deployment. This can be accomplished with the help of the base station in a similar way to Han et al.'s scheme [9]. Each cluster head then uses the pairwise key to securely distribute its KDK to neighboring cluster heads. For example, when CH_A wants to share KDK_{CH_A} with CH_B, it computes e_0 and v_0, and then sends the following message to CH_B with current timestamp TS_0:

$$CH_A \to CH_B : ID_{CH_A}||ID_{CH_B}||TS_0||e_0||v_0$$
$$e_0 = E(K_{CH_A-CH_B}, KDK_{CH_A})$$
$$v_0 = MAC(K_{CH_A-CH_B}, ID_{CH_A}||ID_{CH_B}||TS_0||e_0)$$

The reason for sharing the KDK as a group key is that it is difficult to predict the movement of mobile node. To use a pairwise key between two cluster heads to generate AKs, we have to accurately predict the network cluster head the mobile node will connect to after movement. However, existing movement estimation techniques are inaccurate or require additional special hardware [24]. For example, although calculating an angle of arrival incurs no additional costs, is error prone, and energy inefficient. On the other hand, although global positioning systems can measure precise and absolute coordinates, they require additional expensive hardware.

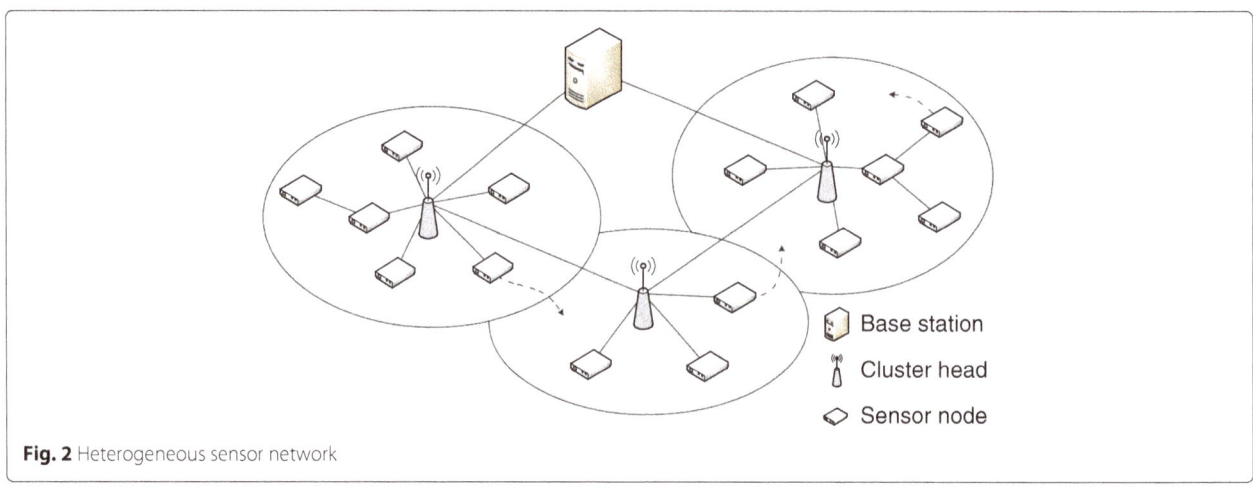

Fig. 2 Heterogeneous sensor network

5.2 Phase 1: node initial authentication

Initial authentication is performed when MN joins the network for the first time after deployment. Let CH_A be the home cluster head that MN first connects to after deployment. We assume that MN has been authenticated by CH_A with the help of the base station in a similar way to Han et al.'s schemes [9, 10] or Jiang et al.'s scheme [11]. After being authenticated by CH_A, MN shares information with CH_A, including a hashed value $H(I)$ of a random number I, a random number N_{MN}, and a pairwise key K_{CH_A-MN}. CH_A also generates AK_{MN} using its own KDK for reauthentication and passes it to MN:

$$AK_{MN} = f(KDK_{CH_A}, ID_{MN})$$

5.3 Phase 2: node reauthentication

Each cluster head CH_i periodically broadcasts beacon messages with current timestamp TS_1 to inform nodes of its presence:

$$CH_i \to * : ID_{CH_i} || TS_1$$

When MN moves to a new location and receives a beacon message from a foreign cluster head CH_B, it initiates the reauthentication procedure with CH_B, as shown in Fig. 3:

(1) MN computes v_1 and v_2, and sends the message "1" with current timestamp TS_2 to CH_B to rejoin the network:

$$MN \to CH_B : ID_{MN} || ID_{CH_A} || TS_2 || v_1 || v_2$$
$$v_1 = MAC(K_{MN-CH_A}, ID_{MN} || ID_{CH_A} || H(I))$$
$$v_2 = MAC(AK_{MN}, ID_{MN} || ID_{CH_A} || TS_2 || TS_1 || v_1)$$

(2) Upon receiving the message "1," CH_B generates AK_{MN} using KDK_{CH_A} and ID_{MN}:

$$AK_{MN} = f(KDK_{CH_A}, ID_{MN})$$

CH_B then verifies v_2 using AK_{MN} and authenticates MN. CH_B also checks whether or not TS_2 exceeds the specified time limit. If the result is valid, CH_B

computes v_3 and sends the message "2" with current timestamp TS_3 to CH_A:

$$CH_B \to CH_A : ID_{MN} || TS_3 || v_1 || v_3$$
$$v_3 = MAC(K_{CH_A-CH_B}, ID_{MN} || TS_3 || v_1)$$

Since AK_{MN} is only used to authenticate MN, the foreign cluster head CH_B must ask the home cluster head CH_A for the information required to generate a pairwise key K_{MN-CH_B} by sending the message "2."

(3) After receiving the message "2," CH_A verifies v_3 and v_1 using $K_{CH_A-CH_B}$ and K_{MN-CH_A}, respectively, and checks whether or not TS_3 exceeds the specified time limit. If the result is valid, CH_A computes e_1, which includes the information required to generate a pairwise key K_{MN-CH_B}, and v_4. CH_A then sends the message "3" with current timestamp TS_4 to CH_B:

$$CH_A \to CH_B : TS_4 || e_1 || v_4$$
$$e_1 = E(K_{CH_A-CH_B}, H(I) || N_{MN})$$
$$v_4 = MAC(K_{CH_A-CH_B}, TS_4 || e_1)$$

(4) Upon receiving the message "3," CH_B verifies v_4 using $K_{CH_A-CH_B}$ and checks whether or not TS_4 exceeds the specified time limit. If the result is valid, CH_B obtains $H(I)$ and N_{MN} by decrypting e_1. CH_B then generates a random number N_{CH_B} and computes pairwise key K_{MN-CH_B}:

$$K_{MN-CH_B} = H(H(I) || N_{MN} || N_{CH_B})$$

For the next reauthentication, CH_B also generates a new AK AK'_{MN}:

$$AK'_{MN} = f(KDK_{CH_B}, ID_{MN})$$

CH_B then computes h_1, e_2, and v_5, and sends message "4" with current timestamp TS_5 to MN:

$$CH_B \to MN : TS_5 || h_1 || e_2 || v_5$$
$$h_1 = H(N_{MN}) \oplus N_{CH_B}$$
$$e_2 = E(K_{MN-CH_B}, AK'_{MN})$$
$$v_5 = MAC(K_{MN-CH_B}, TS_5 || h_1 || e_2)$$

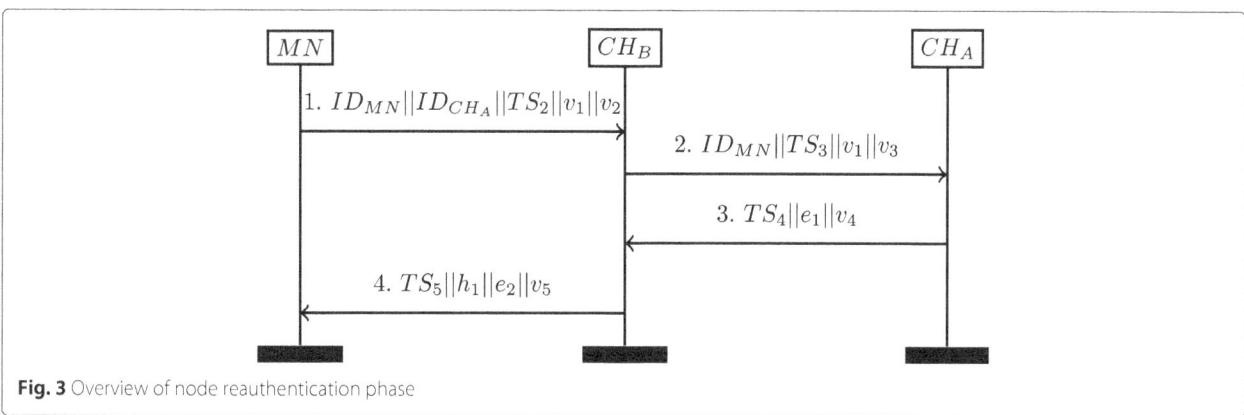

Fig. 3 Overview of node reauthentication phase

(5) When MN receives the message "4" from CH_B, MN obtains N_{CH_B} from h_1 and computes a pairwise key K_{MN-CH_B} in the same way as CH_B. MN then verifies v_5 using this key and checks whether or not TS_5 exceeds the specified time limit. If the result is valid, MN obtains AK'_{MN} by decrypting e_2.

After completing the reauthentication phase, CH_B updates $H(I)$ and N_{MN} as $H(I')$ and N'_{MN}, respectively, for the next reauthentication, and sends them to MN during the communication process.

6   Security analysis

In this section, the security of ESMR is both formally and informally analyzed. First, we formally verified ESMR by modeling it using AVISPA tool [12]. Second, we conducted an informal security analysis of ESMR and confirmed that it meets security requirements and can prevent relevant security attacks.

6.1   Formal verification using AVISPA

AVISPA is a push-button tool that is widely used by many academic researchers to automatically validate various kinds of security protocols. The architecture of AVISPA is illustrated in Fig. 4.

In AVISPA, a security protocol is specified using a role-based formal language called a high-level protocol specification language (HLPSL). The HLPSL2IF translator translates the HLPSL specification into an intermediate format (IF). The IF specification is then validated by any of four back-end tools: OFMC, CL-AtSe, SATMC, and TA4SP under the Dolev-Yao intruder model [25]. Using these back-end tools, we can validate two kinds of security goals: secrecy and authentication. The secrecy goal is used to validate the confidentiality of information. In AVISPA, the secrecy goal is modeled using the goal predicate

secret(T, id, $\{A, B\}$), which indicates that the value of term T is a secret shared only between agents A and B. The label id is used to identify the goal. The authentication goal is used to check whether or not two participants agree on a certain value in the current session. In AVISPA, the authentication goal is modeled using the goal predicates witness(B, A, id, T) and request(A, B, id, T) (for strong authentication) or wrequest(A, B, id, T) (for weak authentication). These predicates indicate that agent A authenticates agent B on some information T. The label id is used to identify the goal. The difference between strong and weak authentications is that weak authentication precludes replay attacks, but strong authentication does not. In this section, we briefly describe how we modeled ESMR in HLPSL and present the formal verification results of ESMR.

6.1.1   HLPSL specification of ESMR

Among the three phases of ESMR, the reauthentication phase, which is the main target phase of this study, was modeled and verified. We modeled a mobile node MN as *role_MN*, home cluster head CH_A as *role_CH_A*, and foreign cluster head CH_B as *role_CH_B*. Code block 1 presents the roles we modeled in the HLPSL. For the keyed message authentication code, we utilized a hash function by adding the symmetric key as one of the inputs. For example, v_1 is modeled as $MAC1' := MAC(M.H(Im).Kma)$, where Kma is the pairwise key K_{MN-CH_A} between MN and CH_A, and $MAC(\cdot)$ is a hash function called MAC.

Listing 1 HLPSL specification for roles

```
role role_MN(M, A, B: agent, H, MAC: hash_func,
    Kma: symmetric_key, AKm: message, Nm, Im:
    text, SND, RCV: channel(dy)) played_by M
    def=
local
State: nat, Tm, Tb1, Tb3: text, Nb: text,
```

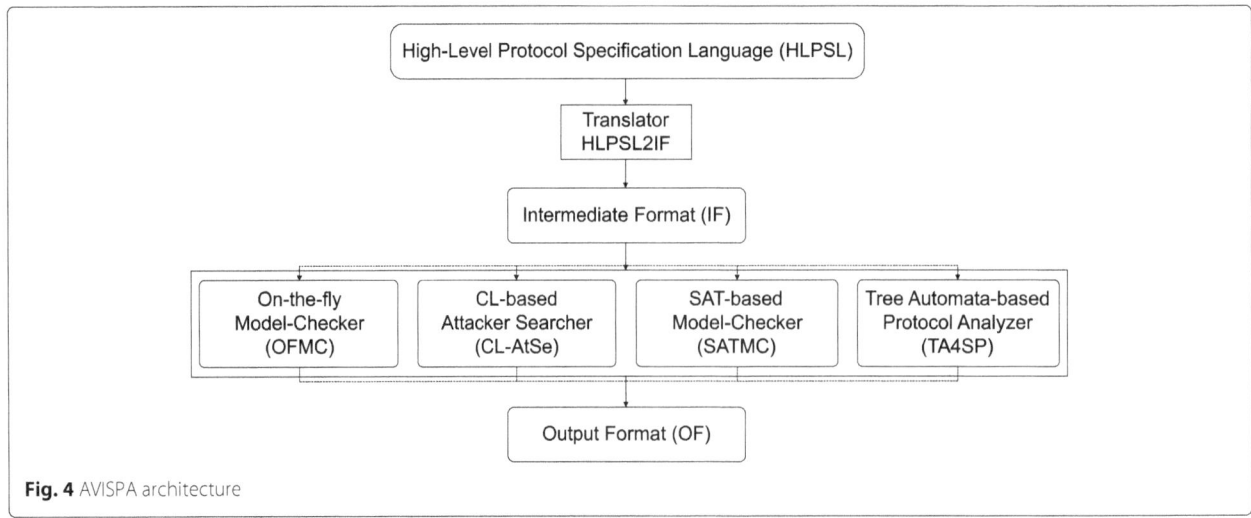

Fig. 4 AVISPA architecture

```
Kmb,AKm2: message ,  MAC1,MAC2,MAC5: message

init
State := 0
transition
1.  State =0/\RCV(B.Tb1')  =|>
State ':=1/\Tm':= new () /\MAC1':=MAC(M.H(Im).
     Kma)  /\MAC2':=MAC(M.A.Tm'.Tb1 .MAC(M.H(
     Im).Kma).AKm) /\SND(M.A.Tm'.MAC1'.MAC2
     ') /\ witness (M,B,b_m_v2 ,MAC2') /\ witness
     (M,A,a_m_v1 ,MAC1')

4.  State =1
/\RCV(Tb3'.xor (H(Nm),Nb').{AKm2'}_Kmb'.MAC
     (Tb3'.Tm.xor(H(Nm),Nb').{AKm2'}_Kmb'.H
     (Im).Kmb'))  =|>
State ':=2/\Kmb':=H(H(Im).Nm.Nb') /\request(
     M,B,m_b_kmb,Kmb')  end  role

role  role_CH_{A}(M,A,B:agent , H,MAC:
     hash_func , Kma,Kga,Kgb,Kab:
     symmetric_key , Nm,Im:text ,  SND,RCV:
     channel(dy)) played_by A def=
local
State:nat , Ta,Tb1:text , MAC1,MAC4: message

init
state := 0
transition
2.  State =0/\RCV(M. Tb2'.MAC(M.H(Im).Kma).
     MAC(M.Tb2'.MAC(M.H(Im).Kma).Kab))  =|>
State ':=1/\Ta':= new () /\MAC4':=MAC(Ta'.{H(
     Im).Nm}_Kab.Kab) /\SND(Ta'.{H(Im).Nm}
     _Kab.MAC4') /\MAC1':=MAC(M.H(Im).Kma) /\
     wrequest(A,M,a_m_v1 ,MAC1')  end  role

role  role_CH_{B}(M,A,B:agent , H,MAC:
     hash_func , Kga,Kgb,Kab:symmetric_key ,
     SND,RCV: channel(dy)) played_by B def=
local
State:nat , Tm,Ta,Tb1,Tb2,Tb3:text ,
Im,Nm,Nb:text , Kma:symmetric_key ,
Kmb,AKm2: message , MAC2,MAC3,MAC5: message

const
sec_kmb,sec_akm2:protocol_id

init
State := 0

transition
1.  State =0/\RCV(start)  =|>
/\Tb1':= new () /\SND(B.Tb0')

3.  State =1/\RCV(M.A.Tm'.MAC(M.H(Im').Kma')
     .MAC(M.A.Tm'.Tb0 .MAC(M.H(Im').Kma').H(
     Kga ,M)))  =|>
State ':=2/\Tb1':= new () /\MAC3':=MAC(M.Tb1'.
     MAC(M.H(Im').Kma').Kab) /\SND(M.Tb1'.
     MAC(M.H(Im').Kma').MAC3')  /\ MAC2':=
     MAC(M.A.Tm'.Tb1'.MAC(M.H(Im').Kma').H(
     Kga ,M)) /\ request(B,M,b_m_v2,MAC2')

5.  State =2/\RCV(Ta'.{H(Im).Nm'}_Kab.MAC(Ta
     '.{H(Im).Nm'}_Kab.Kab))  =|>
State ':=3/\Nb':= new ()
/\Kmb':=H(H(Im).Nm'.Nb') /\AKm2':=H(Kgb,M)
     /\Tb3':= new () /\ MAC5':=MAC(Tb3'.Tm.
     xor (H(Nm'),Nb').{AKm2'}_Kmb'.H(Im).Kmb
     ')
```

```
/\SND(Tb3'.xor (H(Nm'),Nb').{AKm2'}_Kmb'.
     MAC5') /\ witness (B,M,m_b_kmb,Kmb') /\
     secret (Kmb',sec_kmb,{M,B}) /\ secret (
     AKm2',sec_akm2,{M,A,B})  end  role
```

In the HLPSL specification, three authentication and two secrecy goals are defined. For authentication goals, the mutual authentication between MN and CH_B, and the authentication of CH_A on MN are defined as follows:

(1) Upon receiving the message "1" from MN, CH_B authenticates MN through v_2. We use the label b_m_v2 to identify the authentication goal. To verify the authentication goal, we add the witness and request predicates for the label b_m_v2 to the roles of MN and CH_B, respectively.

(2) Upon receiving the message "2" from CH_B, CH_A authenticates MN through v_1. We use the label a_m_v1 to identify the authentication goal. To verify the authentication goal, we add the witness and wrequest predicates for the label a_m_v1 to the roles of MN and CH_A, respectively. Because CH_B checks the freshness of MN's message and prevents the replay attack, CH_A only needs to perform weak authentication on MN.

(3) Upon receiving the message "5" from CH_B, MN authenticates CH_B using K_{MN-CH_B}. We use the label m_b_kmb to identify the goal. To verify the authentication goal, we add the witness and request predicates for the label m_b_kmb to the roles of CH_B and MN, respectively.

For secrecy goals, the secrecy of the pairwise key K_{MN-CH_B} and the secrecy of the authentication key AK'_{MN} are defined as follows:

(1) The pairwise key K_{MN-CH_B} should be only known to MN and CH_B. We use the label sec_kmb to identify the secrecy goal. To verify the secrecy of K_{MN-CH_B}, we add the secret predicate for the label sec_kmb to the role CH_B, where K_{MN-CH_B} is used.

(2) The authentication key AK'_{MN}, which is newly generated by CH_B, should only be known to MN, CH_A, and CH_B. We use the label sec_akm2 to identify the secrecy goal. To verify the secrecy goal, we add the secret predicate for the label sec_akm2 to the role B, where AK'_{MN} is generated.

Code block 2 presents the session and environment we modeled in the HLPSL. The environment section contains intruder knowledge and a composition of sessions. Because of the complexity of our model, we only defined two parallel sessions. Finally, we defined the goal facts to verify the three authentication goals and two secrecy goals outlined above.

Listing 2 HLPSL specification for session and environment

```
role session (M,A,B: agent, Kga,Kgb,Kma,Kab:
    symmetric_key, Im,Nm: text, H,MAC:
    hash_func) def=
local
AKm: message,
SND1,RCV1,SND2,RCV2,SND3,RCV3: channel(dy)

init
AKm:=H( Kga ,M)

composition
role_MN (M,A,B,H,MAC,Kma,AKm,Nm,Im,SND1,
    RCV1) /\
role_CH_{A}(M,A,B,H,MAC,Kma,Kga,Kgb,Kab,Nm
    ,Im,SND2,RCV2) /\
role_CH_{B}(M,A,B,H,MAC,Kga,Kgb,Kab,SND3,
    RCV3) end role

role environment() def=
const
node,cha,chb: agent, im,nm: text,
kga,kgb,kma,kab: symmetric_key,
hfunc,mac: hash_func,
b_m_v2,a_m_v1,m_b_kmb: protocol_id

intruder_knowledge = {node,cha,chb,hfunc,
    mac}

composition
session (node,cha,chb,kga,kgb,kma,kab,im,nm
    ,hfunc,mac) /\
session (node,cha,chb,kga,kgb,kma,kab,im,nm
    ,hfunc,mac)

end role

goal
 authentication_on b_m_v2
 weak_authentication_on a_m_v1
 authentication_on m_b_kmb
 secrecy_of sec_kmb
 secrecy_of sec_akm2

end goal

environment()
```

6.1.2 Formal verification results

Figure 5 presents the formal verification results of our model. In ESMR, an exclusive-OR (XOR) operation is used. Among the four back-end tools, only OFMC and CL-AtSe support algebraic properties of operators, such as XOR operators and exponential operators. Therefore, two back-end tools, namely, OFMC and CL-AtSe, were used to verify our model. Figure 5a, b presents the formal verification results under OFMC and CL-AtSe, respectively. We confirmed that ESMR is safe under OFMC and CL-AtSe. Specifically, ESMR securely provides mutual authentication and pairwise key establishment between MN and CH_B, while preventing the replay attack. ESMR also prevents unconditional forwarding, which can used to launch DoS attacks on CH_A, because CH_B can authenticate MN.

6.2 Informal security analysis

We informally analyzed the security of ESMR in terms of satisfying security requirements and preventing relevant security attacks under the adversary model described in Section 4.2. Table 2 compares the security of ESMR with schemes proposed by Han et al. [9, 10] and Jiang et al. [11].

Mutual authentication ESMR supports mutual authentication between mobile nodes and cluster heads. In ESMR, CH_B authenticates mobile node MN by verifying v_2 using AK_{MN}. Since CH_B is one of the neighbors of CH_A, it has KDK_{CH_A} and can compute AK_{MN}. When CH_A receives the message "2" from CH_B, it also authenticates MN by verifying v_1 using K_{MN-CH_A} and $H(I)$. If one of neighbors of CH_A is captured, KDK_A is exposed to the adversary. The adversary can then utilize the exposed KDK_A to make an illegal node to bypass the authentication of CH_B. However, because CH_A authenticates the illegal node again using K_{MN-CH_A} and $H(I)$, the illegal node cannot join the network even if it bypasses the authentication of CH_B. MN authenticates CH_B using N_{MN} and $H(I)$. When MN receives the message "4" from CH_B, it first obtains N_{CH_B} from h_1 and computes K_{MN-CH_B} using $H(I)$, N_{MN}, and N_{CH_B}. MN then verifies v_5 using K_{MN-CH_B} and $H(I)$. Thus, ESMR satisfies the mutual authentication.

Key freshness In ESMR, a new pairwise key is generated whenever MN moves and is connected to a new cluster head. MN and CH_B generate a pairwise key K_{MN-CH_B} using $H(I)$, N_{MN}, and N_{CH_B}. Since $H(I)$ and N_{MN} are updated after reauthentication is completed and N_{CH_B} is freshly generated for each session, a new pairwise key is generated for each session. Thus, ESMR satisfies key freshness.

Replay attack prevention In ESMR, all messages contain timestamps to prevent replay attacks. Thus, the network is assumed to be loosely synchronized. If the timestamp of a received message exceeds a specified time limit, it is determined to be a potential replay attack and the message is dropped. Since a new pairwise key is generated each session, this effectively prevents the replay attack.

Outsider DoS attack prevention DoS attacks can be launched by inside adversary and outside adversary. The inside adversary can launch DoS attacks by capturing the mobile node or cluster head, and replicating them. Since the replicated nodes have cryptographic materials such as secret keys, general security mechanisms such as authentication and encryption cannot prevent and detect insider DoS attacks. Thus, all the four schemes are vulnerable against insider DoS attacks.

Authentication is the first step to detect and prevent DoS attacks from outside adversary. However, in Jiang et

```
% OFMC                                    SUMMARY
% Version of 2006/02/13                    SAFE
SUMMARY                                   DETAILS
 SAFE
DETAILS                                   BOUNDED_NUMBER_OF_SESSIONS
                                           TYPED_MODEL
BOUNDED_NUMBER_OF_SESSIONS                PROTOCOL
PROTOCOL
                                          /home/span/span/testsuite/results/test.if
/home/span/span/testsuite/results/test.if GOAL
GOAL                                        As Specified
 as_specified                            BAKCEND
BAKCEND                                     CL-AtSe
 OFMC                                     STATISTICS
COMMENTS                                    Analysed  : 10262 states
STATISTICS                                  Reachable : 8136 states
 parseTime:0.00s                           Translation: 0.02 seconds
 searchTime: 1.19s
 visitedNodes: 452 nodes                   Computation: 61.22 seconds
 depth: 12 piles
```

(a) OFMC   (b) CL-AtSe

Fig. 5 Formal verification results. **a** OFMC **b** CL-AtSe

al.'s scheme [11], since there are no shared secrets between foreign cluster heads and mobile nodes, a foreign cluster head cannot verify the reauthentication request of a mobile node and unconditionally forwards it to the home cluster head. This leads to DoS attacks on the home cluster head. In contrast, ESMR can prevent unconditional forwarding, thus preventing potential DoS attacks based on unconditional forwarding. In ESMR, the message "1," sent by MN to CH_B, contains the message authentication code v_2 generated using AK_{MN}. Because CH_B is one of neighbors of CH_A, it can compute AK_{MN} using KDK_{CH_A} and ID_{MN}. CH_B then verifies v_2 using AK_{MN} and authenticates MN. Thus, ESMR prevent DoS attacks based on unconditional forwarding by allowing the foreign cluster head to authenticate the mobile node directly.

Compromise resilience It refers that even if a node is captured by an adversary, the compromised node reveals no information about links it is not directly involved with. Compromise resilience is high if an adversary cannot deduce the cryptographic secrets of any other nodes

not directly involved with the compromised node. However, if even a single-cluster head is captured in Han et al.'s schemes [9, 10], multiple TGKs are exposed to the adversary. The adversary can then obtain the secret information included in all tickets generated using the exposed TGKs and compromise the communication security for multiple nodes. Although ESMR uses the KDK as a group key in a similar way to the TGK in Han et al.'s schemes, the KDK is only used to generate AK, which are only used by CH_B to authenticate mobile nodes. Therefore, even if a cluster head is captured and multiple KDKs are exposed to the adversary, an adversary cannot obtain any pairwise keys other than those between the compromised cluster head and mobile node. Thus, ESMR provides higher-compromise resilience than Han et al.'s schemes.

Forward security It refers that even if a node is captured and its current secrets are leaked, an adversary cannot decrypt any data collected and encrypted before the compromise. In ESMR, the pairwise key between MN and CH_B is generated using freshly generated random numbers for each reauthentication; hence, pairwise keys are independent of each other. In other words, even if an adversary obtains the current pairwise key by compromising MN, the adversary cannot derive the previous pairwise key. When a cluster head is captured, multiple KDKs are exposed to the adversary. However, the KDK is only used to generate AKs used for mobile node reauthentication. Therefore, even if an adversary obtains the KDK by compromising the cluster head, the adversary cannot derive any pairwise keys. Thus, ESMR satisfies forward security.

Table 2 Security comparison between ESMR and related schemes

	[9]	[10]	[11]	ESMR
Mutual authentication	Yes	Yes	Yes	Yes
Key freshness	Yes	Yes	Yes	Yes
Replay attack prevention	Yes	Yes	Yes	Yes
Outsider DoS attack prevention	Yes	Yes	No	Yes
Compromise resilience	Low	Low	High	High
Forward security	No	No	Yes	Yes

In contrast, Han et al.'s schemes [9, 10] do not satisfy the forward security. They employ ticket for reauthentication, containing the previous pairwise key encrypted with the TGK. Therefore, if an adversary obtains the TGK by capturing the cluster head, the adversary can obtain the previous pairwise key from the ticket and decrypt any previous messages encrypted using that key.

7 Performance evaluation

We evaluated the performance of ESMR by comparing it with schemes proposed by Han et al. [9] and Jiang et al. [11] in terms of energy consumption and reauthentication latency. We also analyzed the performance of ESMR under DoS attacks based on unconditional forwarding by comparing it with Jiang et al.'s scheme [11] in terms of reauthentication latency and packet delivery ratio.

7.1 Evaluation methodology

The performance of the three schemes have been evaluated by means of simulation experiments. Simulations were performed by using the OMNeT++ simulator with INET framework version 3.6.3 to measure energy consumption and reauthentication latency. For the simulation, we considered the situation where a mobile node has moved to a new location and initiates the reauthentication procedures with a foreign cluster head.

The simulation sets up IEEE 802.15.4 using the IEEE 802.15.4 narrow band network interface card module provided by the INET framework and Tmote sky datasheet [26]. The data transmission speed was set to 250 kbps and the receiver sensitivity was set to -95 dBm. The transmission power of the mobile node and cluster head were set to -10 dBm and 0 dBm, respectively. The mobile node and cluster head had communication ranges of 30 m and 100 m, respectively, measured by the INET framework. The simulation used static routing, the simplest form of routing. Message size was calculated based on the following base parameter settings: ID of 2 bytes, MAC of 4 bytes, timestamp of 8 bytes, random number of 8 bytes, and key size of 16 bytes. It is also assumed that encryption does not change the message size.

7.2 Energy consumption analysis result and discussion

Because all the three schemes use symmetric key cryptography, there is no significant difference in energy consumption due to computations. Therefore, only energy consumption arising from communication was measured and compared. Since communication is generally the major energy consumption, this comparison is sufficient to investigate energy efficiencies of the three schemes. Based on the Tmote sky datasheet [26], the receiving current consumptions for the mobile node and cluster head were all set to 19.7 mA and the transmitting current consumptions for them were set to 11.2 and 17.4 mA,

respectively. The simulation was conducted for five scenarios where the number of hops (n) between the mobile node and cluster head varies from one to five.

Figure 6 presents comparisons of the total energy consumption for a single reauthentication based on the number of hops (n) between the mobile node and cluster head. The total energy consumption is calculated as the sum of the energy consumed by all nodes participating in the reauthentication process.

When $n = 1$, ESMR has the greatest total energy consumption among the three schemes. ESMR has the same reauthentication process as Jiang et al.'s scheme, but has a larger total message size. Consequently, ESMR incurs 3% more total energy consumption than Jiang et al.'s scheme. In Han et al.'s scheme, the mobile node and cluster head authenticate each other and establish a pairwise key using a ticket without the help of other nodes. In contrast, in ESMR, the information required for reauthentication is exchanged between the foreign cluster head and home cluster head. Consequently, ESMR incurs 16% more total energy consumption than Han et al.'s scheme.

However, when $n \geq 2$, Han et al.'s scheme has the greatest total energy consumption among the three schemes. The message size that the mobile node sends to the cluster head is larger than for Jiang et al.'s scheme and ESMR by at least 30 bytes. Hence, the energy consumption of Han et al.'s stationary node, which relays the mobile node's message to the cluster head, is larger than for Jiang et al.'s scheme and ESMR. Consequently, Han et al.'s scheme has greater total energy consumption than Jiang et al.'s scheme and ESMR. ESMR also incurs up to 6% more total energy consumption than Jiang et al.'s scheme when $n \geq 2$, because ESMR has the same reauthentication process as Jiang et al.'s scheme, but a larger total message size.

Thus, ESMR is suitable for multi-hop communication environment in terms of energy consumption, where the number of hops between the mobile node and cluster

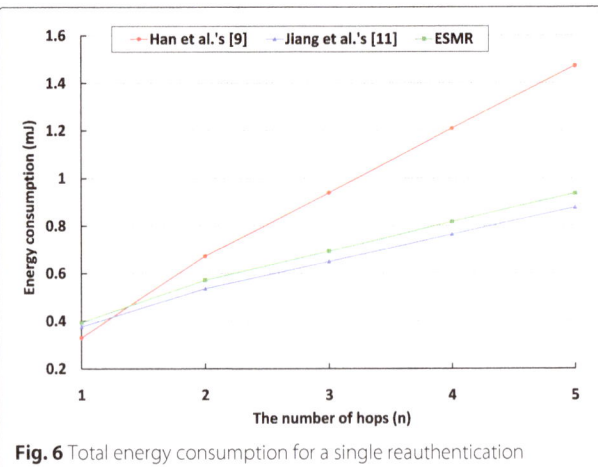

Fig. 6 Total energy consumption for a single reauthentication

head is two or more. Specifically, ESMR has only up to 6% increase in total energy consumption compared with existing schemes, which is negligible.

7.3 Reauthentication latency analysis result and discussion

Since the CC2420 featured by Tmote sky provides hardware support for AES-128, the computational delays caused by encryption and decryption are very small and were not considered. For example, the time required to encrypt 16 bytes is 449.203 μs for Tmote sky using CC2420 hardware encryption [27]. Therefore, only reauthentication latency as a result of communication was measured. Five scenarios were considered for the simulation where the number of hops (n) between the mobile node and cluster head varies from one to five. Simulations were conducted 100 times per scenario for each scheme.

Figure 7 compares reauthentication latency when $n = 1$. Average reauthentication latencies of Han et al.'s scheme, Jiang et al.'s scheme, and ESMR were 10.4 ms, 11.8 ms, and 12.2 ms, respectively. ESMR has the longest average reauthentication latency since it requires the same number of messages as Jiang et al.'s scheme, but has larger total message size. Thus, ESMR has 3% longer average reauthentication latency than Jiang et al.'s scheme. ESMR also requires communication between cluster heads to exchange the information required for reauthentication. In contrast, Han et al.'s scheme does not require communication between cluster heads because the mobile node and cluster head authenticate each other and establish pairwise keys using the ticket. Consequently, ESMR has 16% longer average reauthentication latency than Han et al.'s scheme. However, the actual difference in average reauthentication latency between ESMR and Han et al.'s scheme is very small at 1.8 ms. Moreover, since average reauthentication latency for all the three schemes is less than 13 ms, the three schemes are sufficiently fast.

Figure 8 compares reauthentication latency when $n = 5$. Average reauthentication latencies of Han et al.'s scheme, Jiang et al.'s scheme, and ESMR were 39.6 ms, 25.4 ms,

and 26.2 ms, respectively. Han et al.'s scheme has the longest reauthentication latency since the message size that mobile nodes send to cluster heads is larger than for Jiang et al.'s scheme and ESMR. Han et al.'s scheme also requires an additional message for reauthentication compared with Jiang et al.'s scheme and ESMR; hence, reauthentication latency of a stationary node that relays a mobile node's message to a cluster head is longer than for Jiang et al.'s scheme and ESMR. ESMR has 3% longer average reauthentication latency than Jiang et al.'s scheme because although ESMR requires the same number of messages for reauthentication, it has a larger total message size than Jiang et al.'s scheme.

Figure 9 compares average reauthentication latency based on the number of hops (n) between the mobile node and cluster head. Han et al.'s scheme has the shortest reauthentication latency when $n = 1$, but the longest when $n \geq 2$. ESMR has slightly longer reauthentication latency than Jiang et al.'s scheme regardless of n due to the larger message size. However, the actual difference is very small which is less than 0.9 ms. Thus, Jiang et al.'s scheme and ESMR have similar reauthentication latency regardless of n.

Thus, in terms of reauthentication latency, ESMR is suitable for both single-hop and multi-hop communication environments. Specifically, when $n = 1$, ESMR has a maximum increase in the average reauthentication latency of 1.8 ms compared with existing schemes, but it is fast enough because the actual average reauthentication latency is less than 13 ms, like the existing schemes.

7.4 Performance analysis under DoS attacks based on unconditional forwarding

We considered a scenario in which outside adversary sends a huge amount of spoofed or altered mobile nodes' reauthentication requests to neighboring cluster heads of a home cluster head in order to attempt DoS attacks based on unconditional forwarding on the home cluster head. We also assumed that each adversary node sends the messages only to one of neighboring cluster heads of the home

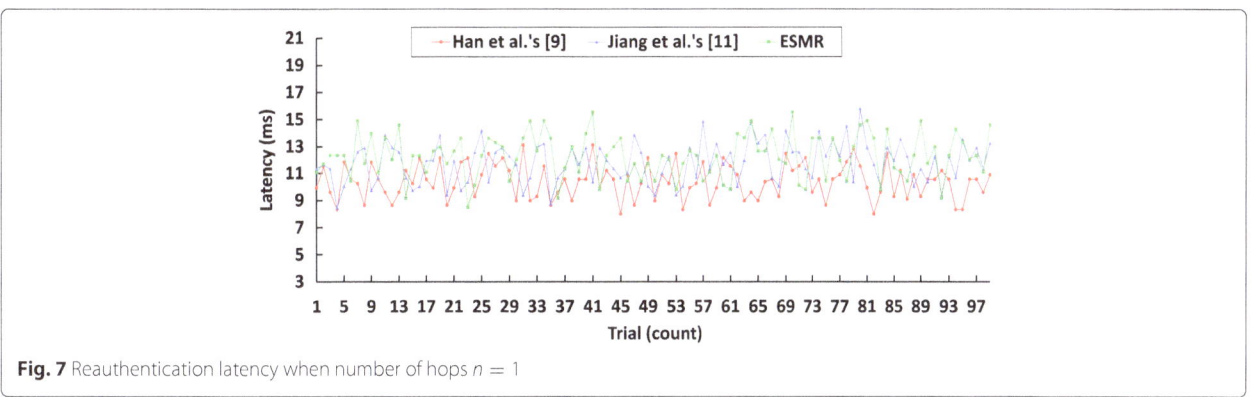

Fig. 7 Reauthentication latency when number of hops $n = 1$

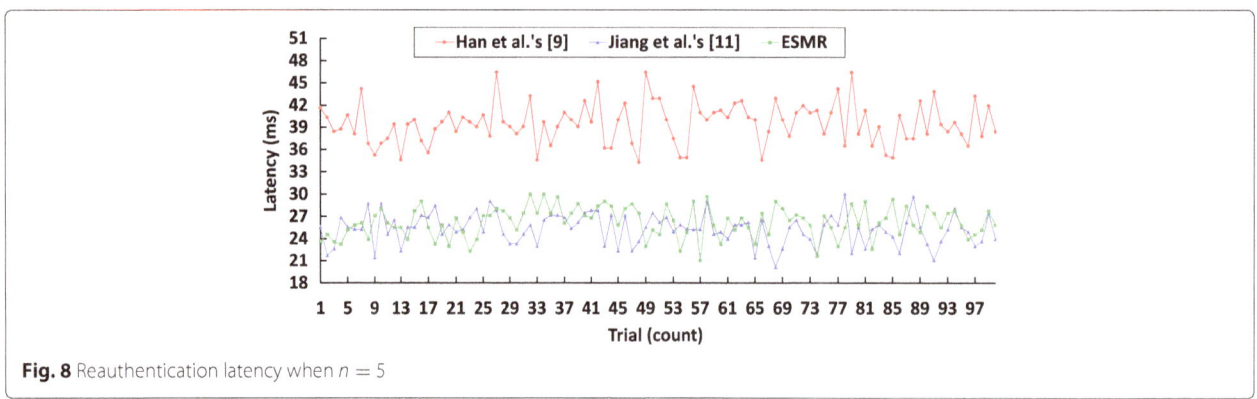

Fig. 8 Reauthentication latency when $n = 5$

cluster head except for the foreign cluster head. Hop distance between the mobile node and home cluster head was set to 1, and message transmission interval of adversary nodes was set to 10 ms. Three cases were considered for the simulation where the number of adversary nodes (k) varies from one to three. Simulations were conducted 500 times per case for Jiang et al.'s scheme [11] and ESMR.

Figure 10 compares packet delivery ratio under DoS attacks based on unconditional forwarding according to the number of adversary nodes (k). The packet delivery ratio of ESMR was 100% regardless of k. On the other hand, the packet delivery ratio of Jiang et al.'s scheme decreases from 100 to 11% as k increases from 0 to 3.

Figure 11 compares average reauthentication latency under DoS attacks based on unconditional forwarding according to the number of adversary nodes (k). The average reauthentication latency of ESMR was approximately 12 ms regardless of k. On the other hand, the average reauthentication latency of Jiang et al.'s scheme increases from 11.8 ms to 24 ms as k increases from 0 to 3. Although the packet delivery ratio of Jiang et al.'s scheme was 100% when $k = 1$, the average reauthentication latency was 20.2 ms which is 28.5% longer than for ESMR.

Thus, ESMR can effectively prevent DoS attacks based on unconditional forwarding on the home cluster head because it can prevent unconditional forwarding itself by allowing the neighboring cluster heads to verify validity of the spoofed or altered reauthentication requests, i.e., the neighboring cluster heads do not forward invalid reauthentication requests to the home cluster head unlike Jiang et al.'s scheme.

8 Conclusion

Several mobile node reauthentication schemes based on symmetric key cryptography have been proposed to efficiently handle frequent reauthentication [9–11]. However, we found that Han et al.'s schemes [9, 10] do not provide high-compromise resilience and Jiang et al.'s scheme [11] have a problem of unconditional forwarding which can be used to launch DoS attack on the home cluster head.

In this paper, we proposed the energy-efficient and secure mobile node reauthentication (ESMR) for MWSNs, which satisfies the security requirements of MWSNs by addressing the security weaknesses of the existing schemes [9–11]. Security analysis verified that ESMR meets the security requirements of MWSNs and

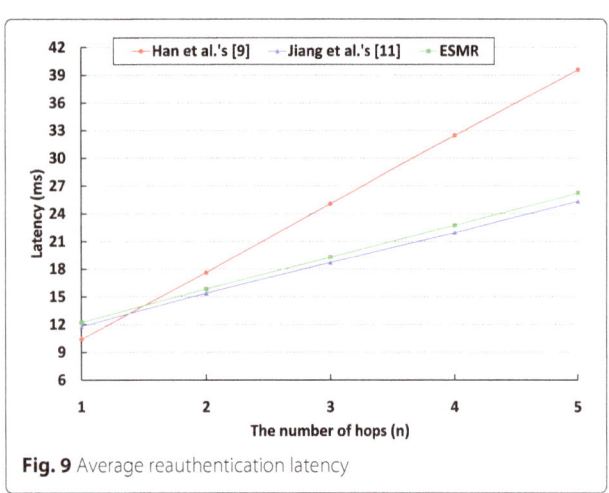

Fig. 9 Average reauthentication latency

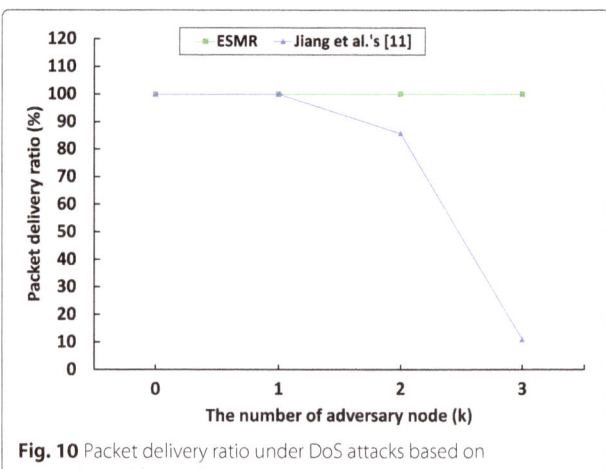

Fig. 10 Packet delivery ratio under DoS attacks based on unconditional forwarding

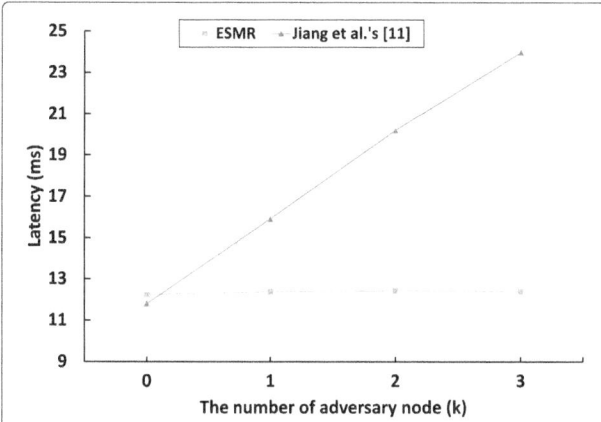

Fig. 11 Average reauthentication latency under DoS attacks based on unconditional forwarding

can prevent relevant security attacks. ESMR is suitable for multi-hop communication environment, but ESMR can be applied to single-hop communication environment. ESMR requires less than 16% increased total energy consumption when the number of hops between the mobile node and cluster head is one, but ESMR is fast enough because the average reauthentication latency is less than 13 ms, like the existing schemes. Moreover, ESMR meets the security requirements of MWSNs by addressing the security weaknesses of the existing scheme. Thus, ESMR can provide secure and fast mobile node reauthentication with slightly increased total energy consumption for single-hop communication environment.

Abbreviations
AES: Advanced encryption standard; AK: Authentication key; AVISPA: Automated validation of internet security protocols and applications; CL-AtSe: CL-based attacker searcher; CL-HSC: Certificateless hybrid signcryption scheme; DoS: Denial of service; ECC: Elliptic curve cryptography; HLPSL: High-level protocol specification language; IF: Intermediate form; IoT: Internet of things; KDK: Key derivation key; MWSNs: Mobile wireless sensor networks; NCL: Neighboring cluster head list; OFMC: On-the-fly model-checker; OF: Output format; SATMC: SAT-based model-checker; TA4SP: Tree automata-based protocol analyzer; TGK: Ticket generation key; WSNs: Wireless sensor networks; XOR: Exclusive-OR

Acknowledgements
This research was supported by Basic Science Research Program through the National Research Foundation of Korea(NRF) funded by the Ministry of Education (NRF-2019R1I1A1A01062743).

Funding
This research was supported by Basic Science Research Program through the National Research Foundation of Korea(NRF) funded by the Ministry of Education (NRF-2019R1I1A1A01062743).

Authors' contributions
BK proposed the main concept, conducted formal verification and simulations, analyzed results, and wrote the manuscript. JS contributed to revising the manuscript and fine-tuning the proposed scheme. Both authors read and approved the final manuscript.

Competing interests
The authors declare that they have no competing interests.

References
1. X. Wang, S. Han, Y. Wu, X. Wang, Coverage and energy consumption control in mobile heterogeneous wireless sensor networks. IEEE Trans. Autom. Control. **58**(4), 975–988 (2013)
2. Y. Yang, M. I. Fonoage, M. Cardei, Improving network lifetime with mobile wireless sensor networks. Comput. Commun. **33**(4), 409–419 (2010)
3. O. Chipara, C. Lu, T. C. Bailey, G.-C. Roman, in *Proceedings of the 8th ACM Conference on Embedded Networked Sensor Systems. SenSys '10.* Reliable clinical monitoring using wireless sensor networks: Experiences in a step-down hospital unit (ACM, New York, 2010), pp. 155–168
4. S. Ehsan, K. Bradford, M. Brugger, B. Hamdaoui, Y. Kovchegov, D. Johnson, M. Louhaichi, Design and analysis of delay-tolerant sensor networks for monitoring and tracking free-roaming animals. IEEE Trans. Wirel. Commun. **11**(3), 1220–1227 (2012)
5. M. Li, Y. Liu, Underground coal mine monitoring with wireless sensor networks. ACM Trans. Sens. Netw. (TOSN). **5**(2), 10 (2009)
6. C. Zhu, L. Shu, T. Hara, L. Wang, S. Nishio, L. T. Yang, A survey on communication and data management issues in mobile sensor networks. Wirel. Commun. Mob. Com. **14**(1), 19–36 (2014)
7. A. Ghosal, S. Halder, in *Cooperative Robots and Sensor Networks 2015.* Security in mobile wireless sensor networks: Attacks and defenses (Springer International Publishing, Cham, 2015), pp. 185–205
8. A. Achour, L. Deru, J. C. Deprez, Mobility management for wireless sensor networks a state-of-the-art. Procedia Comput. Sci. **52**, 1101–1107 (2015)
9. K. Han, K. Kim, T. Shon, Untraceable mobile node authentication in wsn. Sensors. **10**(5), 4410–4429 (2010)
10. K. Han, T. Shon, K. Kim, Efficient mobile sensor authentication in smart home and wpan. IEEE Trans. Consum. Electron. **56**(2), 591–596 (2010)
11. S. Jiang, J. Zhang, J. Miao, C. Zhou, A privacy-preserving reauthentication scheme for mobile wireless sensor networks. Int. J. Distrib. Sens. Netw. **9**(5), 913782 (2013)
12. A. Armando, D. Basin, Y. Boichut, Y. Chevalier, L. Compagna, J. Cuellar, P. H. Drielsma, P. C. Heám, O. Kouchnarenko, J. Mantovani, S. Mödersheim, D. von Oheimb, M. Rusinowitch, J. Santiago, M. Turuani, L. Viganò, L. Vigneron, in *Computer Aided Verification.* The avispa tool for the automated validation of internet security protocols and applications (Springer, Berlin, Heidelberg, 2005), pp. 281–285
13. M. Bilal, S.-G. Kang, An authentication protocol for future sensor networks. Sensors. **17**(5), 979 (2017)
14. Y. Qiu, J. Zhou, J. Baek, J. Lopez, Authentication and key establishment in dynamic wireless sensor networks. Sensors. **10**(4), 3718–3731 (2010)
15. S. H. Erfani, H. H. S. Javadi, A. M. Rahmani, A dynamic key management scheme for dynamic wireless sensor networks. Secur. Commun. Netw. **8**(6), 1040–1049 (2015)
16. F. Gandino, C. Celozzi, M. Rebaudengo, A key management scheme for mobile wireless sensor networks. Appl. Sci. **7**(5), 490 (2017)
17. X. Zhang, J. He, Q. Wei, Eddk: Energy-efficient distributed deterministic key management for wireless sensor networks. EURASIP J. Wirel. Commun. Netw. **2011**, 1–11 (2011)
18. S.-H. Seo, J. Won, S. Sultana, E. Bertino, Effective key management in dynamic wireless sensor networks. IEEE Trans. Inf. Forensic. Secur. **10**(2), 371–383 (2015)
19. M. Omar, I. Belalouache, S. Amrane, B. Abbache, Efficient and energy-aware key management framework for dynamic sensor networks. Comput. Electr. Eng. **72**, 990–1005 (2018)
20. H. Chan, A. Perrig, D. Song, in *2003 Symposium on Security and Privacy.* Random key predistribution schemes for sensor networks (IEEE, Berkeley, 2003), pp. 197–213
21. L. Eschenauer, V. D. Gligor, in *Proceedings of the 9th ACM Conference on Computer and Communications Security. CCS '02.* A key-management scheme for distributed sensor networks (ACM, New York, 2002), pp. 41–47
22. A. Gonga, O. Landsiedel, M. Johansson, in *2011 International Conference on Distributed Computing in Sensor Systems and Workshops (DCOSS).* MobiSense: Power-efficient micro-mobility in wireless sensor networks (IEEE, Barcelona, 2011), pp. 1–8
23. M. Nabi, M. Blagojevic, M. Geilen, T. Basten, T. Hendriks, in *2010 7th Annual IEEE Communications Society Conference on Sensor, Mesh and Ad Hoc*

Communications and Networks (SECON). Mcmac: An optimized medium access control protocol for mobile clusters in wireless sensor networks (IEEE, Boston, 2010), pp. 1–9

24. Q. Dong, W. Dargie, A survey on mobility and mobility-aware mac protocols in wireless sensor networks. IEEE Commun. Surv. Tutor. **15**(1), 88–100 (2013)

25. D. Dolev, A. Yao, On the security of public key protocols. IEEE Trans. Inf. Theory. **29**(2), 198–208 (1983)

26. Tmote Sky Datasheet (2006). http://www.eecs.harvard.edu/~konrad/projects/shimmer/references/tmote-sky-datasheet.pdf. Accessed 23 Jan 2018

27. M. Healy, T. Newe, E. Lewis, in *Smart Sensors and Sensing Technology*. Analysis of hardware encryption versus software encryption on wireless sensor network motes (Springer, Berlin, Heidelberg, 2008), pp. 3–14

Energy-efficient filtering algorithm for a class of industrial sensor network systems with packet dropouts, time-varying delay, and multiplicative noises

Hui Li[1], Ming Lyu[1]* (iD), Baozhu Du[2], Jie Zhang[1] and Yuming Bo[1]

Abstract

In this paper, for the purpose of improving the energy efficiency of the industrial sensor networks, we investigated the event-based H_∞ filtering problem for a class of discrete-time nonlinear sensor network systems with time-varying delay, packet dropout, and multiplicative noises. Instead of traditional time-triggered communication mechanism, the event-triggered strategy is adopted in industrial sensor network, which could not only reduce the transmission frequency of the sensor measurement output, but also guarantee the prescribed filtering performance, if only the threshold in the event-triggered function is chosen suitably. The time-varying delay characteristic of systems is considered with the event-triggered strategy, which has seldom been studied due to the complexity of time-varying delay and event-triggered strategy. The most common network-induced phenomenon of packet dropout in industrial sensor network is described. The purpose is to design a filter satisfying exponentially stable and H_∞ indexes. The main result is that sufficient conditions are established, guaranteeing our proposed filter satisfying filtering performance constraints, and the parameters of filter could be got through the derived linear matrix inequality (LMI), if only it is feasible. At last, the filtering approach is demonstrated by a simulation.

Keywords: Energy efficiency, Event-triggered communication mechanism, Time-varying delay, Industrial sensor network system, H_∞ filtering, Multiplicative noises, Packet dropouts

1 Introduction

Over the past decades, the H_∞ filtering technique has attracted considerable research attention and fruitful results have appeared, see for example [1–13] and the references therein. This is mainly due to the following two reasons. Firstly, in a lot of practical engineering, it is hard to get the probabilistic information of disturbance and the H_∞ technique could well deal with this kind of noise signals. Secondly, no matter how precise the system model is, there is also some error between the physical plant and its model. And the robustness of the H_∞ filtering approach may tolerate such error in system model. From the above analysis, we could find that investigating the H_∞ filtering technique has not only theoretically importance but also engineering significance. As such, we will employ the H_∞ approach to design the filter for a class of sensor network systems.

It is well known that the limited network channel bandwidth and limited power are significant factors constraining the performance of industrial sensor network systems [14–19]. In traditional time-triggered communication mechanism, the signal of sensor is transmitted to the filter or controller at every time, which does not consider the limited bandwidth of communication channel and therefore increases the burden of industrial sensor network channel. To avoid the unnecessary frequent communication and save limited energy, an effective method is adopting event-triggered strategy [20–24], in which sensor measurement output is transmitted only when an

*Correspondence: lumtz@163.com
[1] School of Automation, Nanjing University of Science and Technology, Nanjing 210094, China
Full list of author information is available at the end of the article

event-triggered condition is satisfied. If only the event-triggered condition is suitably constructed, the transmission frequency of measurement will decrease while maintaining the prescribed filtering performance. During recent years, the event-triggered communication mechanism has been successfully applied to controller design for various engineer systems, such as networked systems [25, 26] and multi-agent systems [27–29]. Also, some results about event-based filter design have appeared, see for example [30–34]. However, when it comes to the industrial sensor network systems, considering the inevitable network-induced phenomena, the event-based filter design approach has not been adequately investigated and still has many problems needed to be solved. Therefore, the event-triggered communication mechanism will be adopted in the filtering problem for the proposed industrial sensor network systems.

Noting that, nonlinear control and filtering have attracted much interest [4, 35–41], due to the popular existence of nonlinearity in a lot of practical systems and its important effectiveness to systems. In [4], a sector-bounded approach is proposed to handle with a class of nonlinearities. It is pointed out that many plants may be modeled by systems with multiplicative noises and some characteristics of nonlinear systems can be closely approximately by models with multiplicative noises rather than by linearized models [42, 43]. Therefore, in this paper, the nonlinearity of addressed systems is described by a nonlinear function and state-multiplicative noises, which could better present the practical nonlinearity.

As a main source of system instability, time-delay widely exists in practical industrial sensor network systems and should be taken into the analysis process of systems. As such, the H_∞ filtering for various time-delay plants has attracted much interest, see [35, 44–46] and the reference therein. For example, the robust filter is designed for systems with packet dropout and constant delay in [44]. In [35], a delay-dependent H_∞ filtering method is proposed for delay systems whose postpone is time-varying. Very recently, in [30], the event-triggered strategy is adopted to address distributed H_∞ filtering problem for industrial sensor networks with time-invarying delay. Unfortunately, up to now, when event-triggered communication is adopted, the relative investigation about event-based H_∞ filter design problem has seldom taken time-varying delay into account. Therefore, we will investigate the event-based H_∞ filtering problem for industrial sensor networks whose postpone is time-varying.

Summarizing the above discussions, the event-based H_∞ filtering problem will be investigated for a class of nonlinear industrial sensor network systems with packet dropouts, multiplicative noises and time-varying delay. The main contributions are highlighted as follows:

1. During the design of filter for a class of discrete-time sensor network systems with time-varying delay, the event-triggered communication mechanism is adopted.

2. A comprehensive model of nonlinear sensor network systems is proposed which subjects to packet dropouts, multiplicative noises, and time-varying delay.

3. Sufficient conditions are built which could ensure proposed filter and corresponding event-based filtering algorithm is addressed.

Section 2 introduces the methods utilized for the energy-efficient filter. In Section 3, the delay sensor network with packet dropouts and multiplicative noises is introduced. The results and discussions are given in Section 4, where sufficient condition is derived for the H_∞ filter and the filtering method is addressed . A numerical example is given in Section 5. Finally, we conclude in Section 6.

2 Methods

In this paper, the energy-efficient filter is designed based on Lyapunov theory method and linear matrix inequality method. The simulation experiment is based on the LMI toolbox of MATLAB R2014a.

3 Problem formulation and preliminaries

Here, the following discrete nonlinear sensor network system with time-varying delay and multiplicative noise is considered:

$$\begin{cases} x(k+1) = \left[A + \sum_{i=1}^{\alpha} \tilde{w}_{i(k)} A_i\right] x(k) + \left[A_d + \sum_{j=1}^{\beta} \tilde{v}_j(k) A_{dj}\right] x(k-\tau(k)) \\ \qquad\qquad + f(x(k)) + Bw(k) \\ y(k) = Cx(k) + Dv(k) \\ z(k) = Lx(k) \\ x(l) = \varphi(l), \quad l = -d_M, -d_{M+1}, ..., 0, \end{cases}$$

$$(1)$$

where $x(k) \in \mathbb{R}^n$ represents the state vector, $y(k) \in \mathbb{R}^r$ is sensor output, $z(k) \in \mathbb{R}^m$ is the signal to be estimated, $w(k) \in \mathbb{R}^p$ and $v(k) \in \mathbb{R}^q$ are disturbance belonging to $l_2[0, \infty]$, $f(\cdot) : R^n \rightarrow R^n$ is nonlinear vector function, $\tilde{w}_i(k)(i = 1, 2, ..., \alpha)$ and $\tilde{v}_j(k)(i = 1, 2, ..., \beta)$ are zero mean Gaussian white noise with $\mathbb{E}\{\tilde{w}_i(k)\} = 0$, $\mathbb{E}\{\tilde{w}_i^2(k)\} = 1$, $\mathbb{E}\{\tilde{w}_i(k)\tilde{w}_j(k)\} = 0(i \neq j)$, $\mathbb{E}\{\tilde{v}_j(k)\} = 0$, $\mathbb{E}\{\tilde{v}_j^2(k)\} = 1$, $\mathbb{E}\{\tilde{v}_i(k)\tilde{v}_j(k)\} = 0(i \neq j)$, $\mathbb{E}\{\tilde{w}_i(k)\tilde{v}_j(k)\} = 0$. The time-varying delay $\tau(k) \in [d_m, d_M]$. A, A_i, A_d, A_{dj}, B, C, L, and D are known, real matrices with appropriate dimensions.

$f(x(k))$ is assumed to satisfy the following condition:

$$\| f(x(k)) \|^2 \leq \theta \| Gx(k) \|^2, \qquad (2)$$

where $\theta > 0$ is a known scalar and G is a known matrix.

Remark 1 *As an essential characteristic for many practical networked systems, time-delay should be considered, due to it is a main source of system instability. Although, for the purpose of decreasing the difficulty of filter design, in many filter design algorithm, time-delay is assumed to be constant. But, the fact is that time-delay is almost time-variant. Therefore, it is more practical significant to design filter for network systems with time-varying delay.*

Remark 2 *The addressed system (1) is a comprehensive model for industrial sensor network systems which includes the multiple noises, nonlinearity, and time-varying delay. As far as we know, due to the complexity of the addressed system (1), the relevant research results are few. This motivates our research interest.*

Different from traditional filter design, the event-triggered strategy is considered, which could reduce communication frequency. As such, a event generator function $g(\cdot, \cdot)$ is defined as follows:

$$g(\sigma(k), \delta) = \sigma^T(k)\sigma(k) - \delta^2 y^T(k)y(k), \qquad (3)$$

where $\sigma(k) = y(k_i) - y(k)$ with $y(k_i)$ being the measurement at the latest event time k_i and $y(k)$ is the current measurement. $\delta \in [0, 1]$ is the threshold. In practical engineering, δ can be determined on the basis of the filtering requirement. When a smaller filtering error is needed, δ is set to be smaller.

The current measurement $y(k)$ of the sensor is transmitted if only the following condition

$$g(\sigma(k), \delta) > 0 \qquad (4)$$

is met. Thus, the event-triggered sequence $0 \le k_0 \le k_1 \le \cdots \le k_i \le \cdots$ is determined iteratively by

$$k_{i+1} = \inf\{k \in N \mid k > k_i, f(\sigma(k), \delta) > 0\}. \qquad (5)$$

Remark 3 *The event-triggered strategy is adopted in the networked filter design for industrial sensor network. As is well known, in time-triggered communication mechanism, the measurement output of sensor is transmitted by network communication channel with limited bandwidth at every sampling time, even though the measurement output changes slightly in the next instant, which increases the burden of network channel and wastes a lot of source of industrial sensor network. However, in event-triggered communication mechanism, only when the designed condition is met, then measurement signal of sensor is transmitted . And a suitable threshold in the event generator function could not only reduce the measurement communication frequency but also make sure prescribed filtering performance.*

As is well known, the measurement of sensor transmitted by network may encounter packet dropouts. When the phenomenon of packet dropouts is considered, the real measurement obtained by filter can be depicted as

$$\tilde{y}(k_i) = \alpha(k_i)y(k_i). \qquad (6)$$

Here, stochastic variable $\alpha(k_i)$ is employed to govern the phenomenon of packet dropouts in industrial sensor network. It is assumed to be Bernoulli-distributed white sequence with

$$\text{Prob}\{\alpha(k) = 1\} = \mathbb{E}\{\alpha(k)\} = \bar{\alpha}, \text{Prob}\{\alpha(k) = 0\} = 1 - \bar{\alpha}.$$

For system (1), construct the following filter:

$$\begin{cases} x_f(k+1) = A_f x_f(k) + B_f \tilde{y}(k_i) \\ \quad z_f(k) = C_f x_f(k), \end{cases} \qquad (7)$$

where $x_f(k) \in \mathbb{R}^n$ is the estimate of the state $x(k)$, $z_f(k) \in \mathbb{R}^m$ represents the estimate of $z(k)$, and A_f, B_f, and C_f is the filter gain matrix to be designed.

By letting $\eta(k) = [x^T(k) \quad e^T(k)]^T$, $\tilde{z}(k) = z(k) - z_f(k)$, $e(k) = x(k) - x_f(k)$, $\bar{w} = [w^T(k) \quad v^T(k)]^T$, $h(\eta(k)) = [f^T(x(k)) f^T(x(k))]^T$, and $\tilde{\alpha}(k) = \alpha(k) - \bar{\alpha}$, we could get the augmented system:

$$\begin{cases} \eta(k+1) = \bar{A}\eta(k) + \tilde{\alpha}(k)\bar{A}_0\eta(k) + \sum_{i=1}^{\alpha} \tilde{w}_i(k)\bar{A}_i\eta(k) \\ \qquad + \bar{A}_d\eta(k - \tau(k)) + \sum_{j=1}^{\beta} \tilde{v}_j(k)\bar{A}_{dj}\eta(k - \tau(k)) + h(\eta(k)) \\ \qquad + \alpha(k_i)\bar{B}_f\sigma(k) + \bar{B}_1\bar{w}(k) + \tilde{\alpha}(k)\bar{B}_2\bar{w}(k) \\ \tilde{z}(k) = \bar{L}\eta(k), \end{cases} \qquad (8)$$

where,

$$\bar{A} = \begin{bmatrix} A & 0 \\ A - A_f - \bar{\alpha}B_f C & A_f \end{bmatrix}, \bar{A}_0 = \begin{bmatrix} 0 & 0 \\ -B_f C & 0 \end{bmatrix},$$

$$\bar{A}_i = \begin{bmatrix} A_i & 0 \\ A_i & 0 \end{bmatrix}, \bar{A}_d = \begin{bmatrix} A_d & 0 \\ A_d & 0 \end{bmatrix},$$

$$\bar{A}_{dj} = \begin{bmatrix} A_{dj} & 0 \\ A_{dj} & 0 \end{bmatrix}, \bar{B}_f = \begin{bmatrix} 0 \\ -B_f \end{bmatrix}, \bar{B}_1 = \begin{bmatrix} B & 0 \\ B & -\bar{\alpha}B_f D \end{bmatrix},$$

$$\bar{B}_2 = \begin{bmatrix} 0 & 0 \\ 0 & -B_f D \end{bmatrix},$$

$$\bar{L} = \begin{bmatrix} L - C_f & C_f \end{bmatrix}.$$

Definition 1 *[13]: The augmented system (8) with $\bar{w}(k) = 0$ is exponentially mean-square if there exist constant $\varepsilon > 0$ and $0 < \kappa < 1$ thus*

$$\mathbb{E}\left\{\| \eta(k) \|^2\right\} \le \varepsilon\kappa^k \max_{i \in [-d_m, 0]} \mathbb{E}\left\{\| \eta(i) \|^2\right\}, \quad k \in [0, \infty).$$

Our aim is to design a filter satisfying the following requirements: (Q1) the filtering error system (8) is exponentially mean-square stable, and (Q2) under the zero initial condition, for given scalar $\gamma > 0$, filtering error $\tilde{z}(k)$

satisfies

$$\sum_{k=0}^{\infty} \mathbb{E}\left\{\|\ \tilde{z}(k)\ \|^2\right\} < \gamma^2 \sum_{k=0}^{\infty} \mathbb{E}\left\{\|\ \bar{w}(k)\ \|^2\right\} \tag{9}$$

for all nonzero $\bar{w}(k)$.

4   Results and discussions
The main results and some discussions are presented in this section.

4.1   Analysis of H_∞ performance
First of all, we introduce the following lemma.

Lemma 1 *(Schur complement) Given constant matrices S_1, S_2, and S_3, where $S_1 = S_1^T$ and $0 < S_2 = S_2^T$, then $S_1 + S_3^T S_2^{-1} S_3 < 0$ if and only if*

$$\begin{bmatrix} S_1 & S_3^T \\ S_3 & -S_2 \end{bmatrix} < 0 \quad or \quad \begin{bmatrix} -S_2 & S_3 \\ S_3^T & S_1 \end{bmatrix} < 0. \tag{10}$$

Theorem 1 :*Consider the sensor network system(1) and let the filter parameters A_f, B_f, and C_f be given. Thus, the filtering error system(8) with $\bar{w}(k) = 0$ is exponentially stable in mean-square, if there exist positive definite matrixes $P > 0$, $Q > 0$ and positive constant scalars ε_1, satisfying*

$$\Phi_1 = \begin{bmatrix} \varphi_{11} + 2\varepsilon_1\theta\bar{G}^T\bar{G} & \bar{A}^T P\bar{A}_d & \bar{A}^T P & \bar{\alpha}\bar{A}^T P\bar{B}_f \\ +\delta^2\bar{C}^T C & & & +\bar{\alpha}(1-\bar{\alpha})\bar{A}_0^T P\bar{B}_f \\ * & \varphi_{22} & \bar{A}_d^T P & \bar{\alpha}\bar{A}_d^T P\bar{B}_f \\ * & * & P - \varepsilon_1 I & \bar{\alpha}P\bar{B}_f \\ * & * & * & \bar{\alpha}\bar{B}_f^T P\bar{B}_f - I \end{bmatrix} < 0, \tag{11}$$

where

$$\varphi_{11} = \bar{A}^T P\bar{A} + \bar{\alpha}(1-\bar{\alpha})\bar{A}_0^T P\bar{A}_0 + \sum_{i=1}^{\alpha} \bar{A}_i^T P\bar{A}_i$$
$$\quad - P + (d_M - d_m + 1)Q,$$

$$\varphi_{22} = \bar{A}_d^T P\bar{A}_d + \sum_{j=1}^{\beta} \bar{A}_{dj}^T P\bar{A}_{dj} - Q,$$

$$\bar{G} = \begin{bmatrix} G & 0 \end{bmatrix}, \ \bar{C} = \begin{bmatrix} C & 0 \end{bmatrix}.$$

Proof : Choose the following Lyapunov function

$$V(k) = V_1(k) + V_2(k) + V_3(k), \tag{12}$$

where

$$V_1(k) = \eta^T(k)P\eta(k), \ V_2(k) = \sum_{i=k-\tau(k)}^{k-1} \eta^T(i)Q\eta(i),$$

$$V_3(k) = \sum_{j=k-d_M+1}^{k-d_m} \sum_{i=j}^{k-1} \eta^T(i)Q\eta(i).$$

$\square$

Then, according to (8) with $\bar{w}(k) = 0$, there is

$$\mathbb{E}\{\Delta V_1(k)\}$$
$$=\mathbb{E}\{V_1(k+1) - V_1(k)\}$$
$$=\mathbb{E}\left\{\eta^T(k+1)P\eta(k+1) - \eta^T(k)P\eta(k)\right\}$$
$$=\mathbb{E}\left\{\left[\bar{A}\eta(k) + \tilde{\alpha}(k_i)\bar{A}_0\eta(k) + \sum_{i=1}^{\alpha} \tilde{w}_i(k)\bar{A}_i\eta(k)\right.\right.$$
$$\quad + \bar{A}_d\eta(k-\tau(k))$$
$$\quad \left. + \sum_{j=1}^{\beta} \tilde{v}_j(k)\bar{A}_{dj}\eta(k-\tau(k)) + h(\eta(k)) + \alpha(k_i)\bar{B}_f\sigma(k)\right]^T P$$
$$\quad \left[\bar{A}\eta(k) + \tilde{\alpha}(k_i)\bar{A}_0\eta(k) + \sum_{i=1}^{\alpha} \tilde{w}_i(k)\bar{A}_i\eta(k) + \bar{A}_d\eta(k-\tau(k))\right.$$
$$\quad \left. + \sum_{j=1}^{\beta} \tilde{v}_j(k)\bar{A}_{dj}\eta(k-\tau(k)) + h(\eta(k)) + \alpha(k_i)\bar{B}_f\sigma(k)\right]$$
$$\quad \left. - \eta^T(k)P\eta(k)\right\}$$

$$=\mathbb{E}\{\eta^T(k)\bar{A}^T P\bar{A}\eta(k) + 2\eta^T(k)\bar{A}^T P\bar{A}_d\eta(k-\tau(k))$$
$$\quad + 2\eta^T(k)\bar{A}^T Ph(\eta(k)) + 2\bar{\alpha}\eta^T(k)\bar{A}^T P\bar{B}_f\sigma(k)$$
$$\quad + \bar{\alpha}(1-\bar{\alpha})\eta^T(k)\bar{A}_0^T P\bar{A}_0\eta(k) + 2\bar{\alpha}(1-\bar{\alpha})\eta^T(k)\bar{A}_0^T P\bar{B}_f\sigma(k)$$
$$\quad + \sum_{i=1}^{\alpha} \eta^T(k)\bar{A}_i^T P\bar{A}_i\eta(k) + \eta^T(k-\tau(k))\bar{A}_d^T P\bar{A}_d\eta(k-\tau(k))$$
$$\quad + 2\eta^T(k-\tau(k))\bar{A}_d^T Ph(\eta(k)) + 2\bar{\alpha}\eta^T(k-\tau(k))\bar{A}_d^T P\bar{B}_f\sigma(k)$$
$$\quad + \sum_{j=1}^{\beta} \eta^T(k-\tau(k))\bar{A}_{dj}^T P\bar{A}_{dj}\eta(k-\tau(k)) + h^T(x(k))Ph(\eta(k))$$
$$\quad + 2\bar{\alpha}h^T(x(k))P\bar{B}_f\sigma(k) + \bar{\alpha}\sigma^T(k)\bar{B}_f^T P\bar{B}_f\sigma(k)$$
$$\quad - \eta^T(k)P\eta(k)\}. \tag{13}$$

Next, it can be derived that

$$\mathbb{E}\{\Delta V_2(k)\} = \mathbb{E}\{V_2(k+1) - V_2(k)\}$$

$$\leq \mathbb{E}\{\sum_{i=k+1-d_M}^{k-d_m} \eta^T(i)Q\eta(i) + \eta^T(k)Q\eta(k)$$

$$- \eta^T(k-\tau(k))Q\eta(k-\tau(k))\} \tag{14}$$

and

$$\mathbb{E}\{\Delta V_3(k)\} = \mathbb{E}\{V_3(k+1) - V_3(k)\}$$

$$= \mathbb{E}\{(d_M - d_m)\eta^T(k)Q\eta(k) - \sum_{i=k+1-d_M}^{k-d_m} \eta^T(i)Q\eta(i)\}. \tag{15}$$

Let

$$\zeta(k) = \begin{bmatrix} \eta^T(k) & \eta^T(k-\tau(k)) & h^T(x(k)) & \sigma^T(k) \end{bmatrix}^T.$$

It follows from (13)–(15) that

$$\mathbb{E}\{\Delta V(k)\} = \mathbb{E}\{V(k+1) - V(k)\}$$

$$= \sum_{i=1}^{3} \mathbb{E}\{\Delta V_i(k)\} \tag{16}$$

$$\leq \mathbb{E}\{\zeta^T(k)\tilde{\Phi}_1\zeta(k)\},$$

where

$$\tilde{\Phi}_1 = \begin{bmatrix} \bar{A}^T P\bar{A} + \bar{\alpha}(1-\bar{\alpha})\bar{A}_0^T P\bar{A}_0 \\ + \sum_{i=1}^{\alpha} \bar{A}_i^T P\bar{A}_i - P & \bar{A}^T P\bar{A}_d \\ + (d_M - d_m + 1)Q \\ * & \bar{A}_d^T P\bar{A}_d + \sum_{j=1}^{\beta} \bar{A}_{dj}^T P\bar{A}_{dj} - Q \\ * & * \\ * & * \end{bmatrix}$$

$$\begin{bmatrix} \bar{A}^T P & \bar{\alpha}\bar{A}^T P\bar{B}_f + \bar{\alpha}(1-\bar{\alpha})\bar{A}_0^T P\bar{B}_f \\ \bar{A}_d^T P & \bar{\alpha}\bar{A}_d^T P\bar{B}_f \\ P & \bar{\alpha}P\bar{B}_f \\ * & \bar{\alpha}\bar{B}_f^T P\bar{B}_f \end{bmatrix} < 0.$$

Moreover, if follows from (2) that

$$h^T(\eta(k))h(\eta(k)) \leq 2\theta\eta^T(k)\bar{G}^T\bar{G}\eta(k). \tag{17}$$

Furthermore, it follows from (16) and (17) that

$$\mathbb{E}\{\Delta V(k)\} \leq \mathbb{E}\{\zeta^T(k)\tilde{\Phi}_1\zeta(k) - \varepsilon_1[h^T(\eta(k))h(\eta(k))$$

$$- 2\theta\eta^T(k)\bar{G}^T\bar{G}\eta(k)]\}. \tag{18}$$

Considering the event-triggered condition (3), we have

$$\mathbb{E}\{\Delta V(k)\} \leq \mathbb{E}\{\zeta^T(k)\tilde{\Phi}_1\zeta(k) - \varepsilon_1[h^T(\eta(k))h(\eta(k))$$

$$- 2\theta\eta^T(k)\bar{G}^T\bar{G}\eta(k)] - \sigma^T(k)\sigma(k) + \delta^2 y^T(k)y(k)\}$$

$$= \mathbb{E}\{\zeta^T(k)\Phi_1\zeta(k)\}. \tag{19}$$

According to Theorem 1, we have $\Phi_1 < 0$. Thus, for all $\zeta(k) \neq 0$, $\mathbb{E}\{\Delta V(k)\} \leq \mathbb{E}\{\zeta^T(k)\tilde{\Phi}_1\zeta(k)\} < 0$. Furthermore, similar to [13], system (8) can be proved to be exponentially mean-square stable. The proof is complete. Then, the H_∞ index will be analyzed.

Theorem 2 : Let A_f, B_f, and C_f and γ be given. Then, system(8) is exponentially stable in the mean-square and satisfies the H_∞ performance constraint (9) for any nonzero $\bar{w}(k)$ under zero initial condition, if there exist matrices $P > 0$, $Q > 0$ and positive constant scalar ε_1 satisfying

$$\Phi_2 < 0, \tag{20}$$

where

$$\Phi_2 = \begin{bmatrix} \varphi_{11} + 2\varepsilon_1\theta\bar{G}^T\bar{G} + \delta^2\bar{C}^T C + \bar{L}^T\bar{L} & \bar{A}^T P\bar{A}_d & \bar{A}^T P \\ * & \varphi_{22} & \bar{A}_d^T P \\ * & * & P - \varepsilon_1 I \\ * & * & * \\ * & * & * \end{bmatrix}$$

$$\begin{bmatrix} \bar{\alpha}\bar{A}^T P\bar{B}_f \\ +\bar{\alpha}(1-\bar{\alpha})\bar{A}_0^T P\bar{B}_f & \bar{A}^T P\bar{B}_1 + \bar{\alpha}(1-\bar{\alpha})\bar{A}_0^T P\bar{B}_2 + \delta^2\bar{C}^T\bar{D} \\ \bar{\alpha}\bar{A}_d^T P\bar{B}_f & \bar{A}_d^T P\bar{B}_1 \\ \bar{\alpha}P\bar{B}_f & P\bar{B}_1 \\ \bar{\alpha}\bar{B}_f^T P\bar{B}_f - I & \bar{\alpha}\bar{B}_f^T P\bar{B}_1 + \bar{\alpha}(1-\bar{\alpha})\bar{B}_f^T P\bar{B}_2 \\ * & \bar{B}_1^T P\bar{B}_1 + \bar{\alpha}(1-\bar{\alpha})\bar{B}_2^T P\bar{B}_2 - r^2 I + \delta^2\bar{D}^T\bar{D} \end{bmatrix} < 0,$$

$$\bar{D} = \begin{bmatrix} 0 & D \end{bmatrix}.$$

Proof : It is clear that (20) implies (11). From Theorem 1, system (8) is exponentially stable. $\square$

Then, we will analysis the H_∞ performance.

$$\mathbb{E}\{\Delta V(k)\} \leq \bar{\zeta}^T(k)\tilde{\Phi}_2\bar{\zeta}(k), \tag{21}$$

where

$$\bar{\zeta}(k) = \begin{bmatrix} \zeta^T(k) & \bar{w}^T(k) \end{bmatrix}^T,$$

$$\tilde{\Phi}_2 = \begin{bmatrix} \Phi_1 & U^T \\ * & \bar{B}_1^T P\bar{B}_1 + \bar{\alpha}(1-\bar{\alpha})\bar{B}_2^T P\bar{B}_2 + \delta^2\bar{D}^T\bar{D} \end{bmatrix},$$

$$U = \begin{bmatrix} \bar{B}_1^T P\bar{A} + \bar{\alpha}(1-\bar{\alpha})\bar{B}_2^T P\bar{A}_0 + \delta^2\bar{D}^T\bar{C} & \bar{B}_1^T P\bar{A}_d \\ \bar{B}_1^T P & \bar{\alpha}\bar{B}_1^T P\bar{B}_f + \bar{\alpha}(1-\bar{\alpha})\bar{B}_2^T P\bar{B}_f \end{bmatrix}.$$

To handle with H_∞ performance, the following index is introduced:

$$J(n) = E\sum_{k=0}^{n}\{\|\tilde{z}(k)\|^2 - \gamma^2\|\bar{w}(k)\|^2\}, \tag{22}$$

where n is a nonnegative integer.

Under the zero initial condition, we have

$$J(n) = E \sum_{k=0}^{n} \{\| \tilde{z}(k)\|^2 - \gamma^2 \|\bar{w}(k)\|^2 + \Delta V(k)\} - \mathbb{E}\{\Delta V(n+1)\}$$

$$\leq E \sum_{k=0}^{n} \{\| \tilde{z}(k) \|^2 - \gamma^2 \| \bar{w}(k) \|^2 + \Delta V(k)\}$$

$$\leq E \sum_{k=0}^{n} \{\eta^T(k)\bar{L}^T\bar{L}\eta(k) - \gamma^2 \bar{w}^T(k)\bar{w}(k) + \bar{\zeta}^T(k)\tilde{\Phi}_2\bar{\zeta}(k)\}$$

$$= E \sum_{k=0}^{n} \{\bar{\zeta}^T(k)\Phi_2\bar{\zeta}(k)\}.$$

(23)

According to Theorem 2, we have $\Phi_2 < 0$, $J(n) < 0$. When $n \to \infty$, there is

$$\sum_{k=0}^{n} \mathbb{E}\{\| \tilde{z}(k) \|^2\} < \gamma^2 \sum_{k=0}^{\infty} \| \bar{w}(k) \|^2 .$$

(24)

The proof is complete.

4.2   Event-based H_∞ filter design
Here, the H_∞ filtering algorithm will be solved in Theorem 3.

Theorem 3 *Let the disturbance attention level $\gamma > 0$ be given. Then, for sensor network system (1) and filter (7), the H_∞ performance constraints (9) and exponential stability are guaranteed, if there exist positive matrices $P > 0$, $Q > 0$, and $\varepsilon_1 > 0$ and matrices X and C_f satisfying*

$$\Lambda = \begin{bmatrix} \Lambda_{11} & \Lambda_{12} & \Lambda_{13} \\ * & \Lambda_{22} & \Lambda_{23} \\ * & * & \Lambda_{33} \end{bmatrix} < 0,$$

(25)

where

$$\Lambda_{11} = \sum_{i=1}^{\alpha} \bar{A}_i^T P \bar{A}_i + (d_M - d_m + 1)Q + \varepsilon_1 2\theta \bar{G}^T \bar{G} + \delta^2 \bar{C}^T \bar{C} - P,$$

$$\Lambda_{12} = \begin{bmatrix} 0 & 0 & 0 & \delta^2 \bar{C}^T \bar{D} \end{bmatrix},$$

$$\Lambda_{13} = \begin{bmatrix} \hat{A}^T P + \bar{\alpha}\hat{C}^T X^T & \sqrt{\bar{\alpha}(1-\bar{\alpha})}R^T X^T & \hat{L}^T - H_2^T C_f^T \end{bmatrix},$$

$$\Lambda_{22} = diag\{\sum_{j=1}^{\beta} \bar{A}_{dj}^T P \bar{A}_{dj} - Q, -\varepsilon_1 I, -I, -\gamma^2 I + \delta^2 \bar{D}^T \bar{D}\},$$

$$\Lambda_{23} = \begin{bmatrix} \bar{A}_d^T P & 0 & 0 \\ P & 0 & 0 \\ \bar{\alpha}H_1^T X^T & \sqrt{\bar{\alpha}(1-\bar{\alpha})}H_1^T X^T & 0 \\ \hat{B}_1^T P + \bar{\alpha}\hat{D}^T X^T & \sqrt{\bar{\alpha}(1-\bar{\alpha})}\hat{D}^T X^T & 0 \end{bmatrix},$$

$$\Lambda_{33} = diag\{-P, -P, -I\},$$

$$\hat{A} = \begin{bmatrix} A & 0 \\ 0 & 0 \end{bmatrix}, H_0 = \begin{bmatrix} 0 \\ I \end{bmatrix}, K = \begin{bmatrix} B_f & A_f \end{bmatrix},$$

$$\hat{C} = \begin{bmatrix} C & 0 \\ 0 & \frac{1}{\bar{\alpha}}I \end{bmatrix}, R = \begin{bmatrix} C & 0 \\ 0 & 0 \end{bmatrix}, H_1 = \begin{bmatrix} I \\ 0 \end{bmatrix},$$

$$\hat{B}_1 = \begin{bmatrix} B & 0 \\ 0 & 0 \end{bmatrix}, \hat{D} = \begin{bmatrix} 0 & D \\ 0 & 0 \end{bmatrix},$$

$$\hat{L} = \begin{bmatrix} L & 0 \end{bmatrix}, H_2 = \begin{bmatrix} 0 & I \end{bmatrix}.$$

Furthermore, if (P, Q, X, C_f, ε_1) is a feasible solution of (25), then the filter matrices (A_f, B_f, C_f) could be obtained by means of matrices X and C_f, where

$$\begin{bmatrix} B_f & A_f \end{bmatrix} = K = (H_0^T P H_0)^{-1} H_0^T X.$$

(26)

Proof : Rewrite Φ_2 as follows:

$$\Phi_2 = \hat{\Phi}_2 + V_1^T P^{-1} V_1 + V_2^T P^{-1} V_2 + V_3^T V_3,$$

(27)

where

$$V_1 = \begin{bmatrix} P\bar{A} & P\bar{A}_d & P & \bar{\alpha}P\bar{B}_f & P\bar{B}_1 \end{bmatrix},$$
$$V_2 = \begin{bmatrix} \sqrt{\bar{\alpha}(1-\bar{\alpha})}P\bar{A}_0 & 0 & 0 & \sqrt{\bar{\alpha}(1-\bar{\alpha})}P\bar{B}_f & \sqrt{\bar{\alpha}(1-\bar{\alpha})}P\bar{B}_2 \end{bmatrix},$$
$$V_3 = \begin{bmatrix} \bar{L} & 0 & 0 & 0 & 0 \end{bmatrix},$$

$$\hat{\Phi}_2 = \begin{bmatrix} \Lambda_{11} & 0 & 0 & 0 & \delta^2 \bar{C}^T \bar{D} \\ * & \sum_{j=1}^{\beta} \bar{A}_{dj}^T P \bar{A}_{dj} - Q & 0 & 0 & 0 \\ * & * & -\varepsilon_1 I & 0 & 0 \\ * & * & * & -I & 0 \\ * & * & * & * & -\gamma^2 I + \delta^2 \bar{D}^T \bar{D} \end{bmatrix}.$$

□

According to Lemma 1, (27) is equivalent to

$$\begin{bmatrix} \hat{\Phi}_2 & V_1^T & V_2^T & V_3^T \\ V_1 & -P & 0 & 0 \\ V_2 & 0 & -P & 0 \\ V_3 & 0 & 0 & -I \end{bmatrix} < 0.$$

(28)

Moreover, rewrite the parameters in (8):

$$\bar{A} = \hat{A} + \bar{\alpha}H_0 K \hat{C}, \bar{A}_0 = H_0 KR, \bar{B}_f = H_0 K H_1,$$
$$\bar{B}_1 = \hat{B}_1 + \bar{\alpha}H_0 K \hat{D}, \bar{B}_2 = H_0 K \hat{D}, \bar{L} = \hat{L} - C_f H_2, PH_0 K = X.$$

(29)

Thus, (28) is equivalent to (25). Then, from Lemma 2, we obtain (9), and system (8) is exponentially stable. The proof is complete.

Remark 4 *The sufficient conditions guaranteeing the event-based filter satisfy Q1 and Q2 are proposed in Theorem 2. The design problem of desired filter is addressed in Theorem 3. It is easy to find that all the relevant information is contained in the LMI, such as system parameters, nonlinearity, and the threshold of event-triggered function.*

5 Numerical simulations

The system (1) is as follows:

$$A = \begin{bmatrix} 0.3 & -0.2 & 0 \\ 0 & 0.4 & -0.1 \\ -0.2 & 0.1 & 0.25 \end{bmatrix}, A_1 = \begin{bmatrix} 0.1 & 0.05 & 0 \\ 0 & 0.15 & 0.1 \\ 0 & -0.1 & -0.01 \end{bmatrix}, A_2 = \begin{bmatrix} 0.1 & -0.05 & 0 \\ 0 & 0.15 & 0.05 \\ 0.05 & -0.05 & 0.1 \end{bmatrix},$$

$$A_d = \begin{bmatrix} 0.05 & 0 & 0 \\ 0.1 & 0.1 & -0.1 \\ 0 & 0 & -0.1 \end{bmatrix}, A_{d1} = \begin{bmatrix} 0.1 & 0.05 & 0 \\ 0.02 & 0.05 & 0 \\ 0 & 0 & 0.1 \end{bmatrix}, A_{d2} = \begin{bmatrix} 0.1 & 0 & 0 \\ 0 & 0.05 & 0.05 \\ 0 & 0 & 0 \end{bmatrix},$$

$$C = \begin{bmatrix} 0.3 & -0.2 & 0.1 \\ 0 & 0.35 & 0.2 \end{bmatrix}, B = \begin{bmatrix} 0.2 \\ 0.15 \\ 0.4 \end{bmatrix}, D = \begin{bmatrix} 0.3 \\ 0.1 \end{bmatrix}, L = \begin{bmatrix} 0.5 & 0.2 & 0.3 \end{bmatrix}.$$

$f(k, x(k))$ and disturbance $w(k)$ and $v(k)$ are chosen as

$$f(k, x(k)) = \begin{bmatrix} \frac{(0.1x_1)}{1+2x_3^2} \\ \frac{0.1\sin(x_2)}{\sqrt{x_1^2+2}} \\ 0.2x_3 \end{bmatrix},$$

$$w(k) = \begin{bmatrix} \frac{5}{k+14} * cos(k) \end{bmatrix}, v(k) = \begin{bmatrix} \exp(-0.05k)\sin(k) \end{bmatrix}.$$

where $x_i (i = 1, 2, 3)$ denotes the ith element of the system state $x(k)$. Then, the constraint (2) can be met with

$$G(k) = \begin{bmatrix} 0.1 & 0 & 0 \\ 0 & 0.1 & 0 \\ 0 & 0 & 0.2 \end{bmatrix}, \theta = 1.$$

The initial value of state is $x(0) = [0.3 \quad 0.25 \quad -0.5]^T$. The initial value of state estimation is $\hat{x}(0) = [0 \quad 0 \quad 0]^T$. The probability of stochastic variable $\alpha(k)$ is taken as $\bar{\alpha} = 0.9$. Delay is $d_M = 3, d_m = 1$. Choose the event threshold $\delta = 0.3$. The disturbance attenuation level is $\gamma = 0.95$.

The filter parameters can be obtained as follows:

$$A_f = \begin{bmatrix} -0.0846 & 0.0548 & 0.0444 \\ 0.0025 & -0.1229 & 0.1027 \\ 0.0952 & -0.0107 & -0.0516 \end{bmatrix}, B_f = \begin{bmatrix} 0.4851 & 0.1995 \\ 0.2279 & 0.4215 \\ 0.2858 & 0.3967 \end{bmatrix},$$

$$C_f = \begin{bmatrix} 0.2714 & 0.1160 & 0.1708 \end{bmatrix}.$$

Figures 1, 2, 3, 4, 5, 6, and 7 show the simulation results. When setting the threshold $\delta = 0.3$, the results are described in Figs. 1, 2, 3, and 4. Figure 1 depicts the state

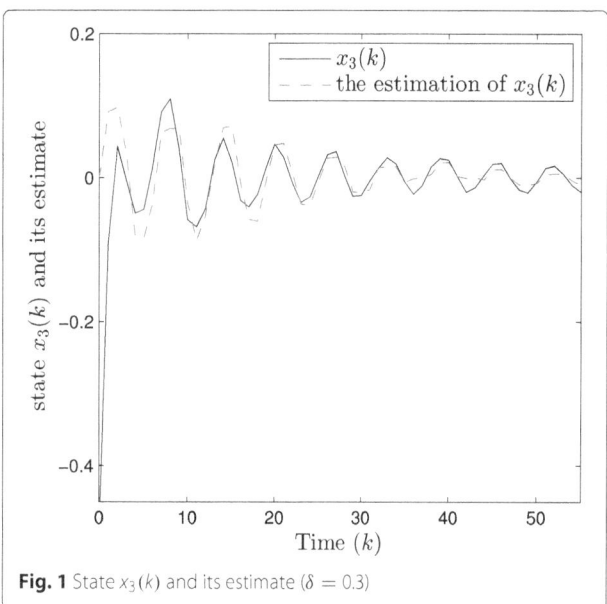

Fig. 1 State $x_3(k)$ and its estimate ($\delta = 0.3$)

variables $x_3(k)$ and its estimate $\hat{x}_3(k)$, and Fig. 2 plots the output $z(k)$ and its estimation $\hat{z}(k)$, whereas the estimation error $z(k) - \hat{z}(k)$ is shown in Fig. 3. Event-triggered times are plotted in Fig. 4, whereas one represents the times that event-triggered condition is satisfied and sensor signal is transmitted and zero represents times that event-triggered condition is not satisfied. It follows from Fig. 4 that the event-triggered communication mechanism can reduce the transmission frequency of the measurement output, which is energy efficient. According to Figs. 1, 2, and 3, it is easy to find that the proposed filter can estimate the state of the system well, and the energy-efficient filtering strategy has satisfying filtering

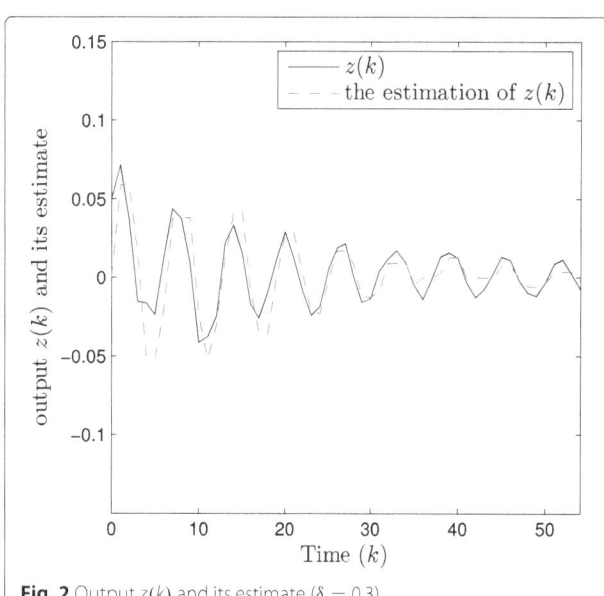

Fig. 2 Output $z(k)$ and estimate ($\delta = 0.3$)

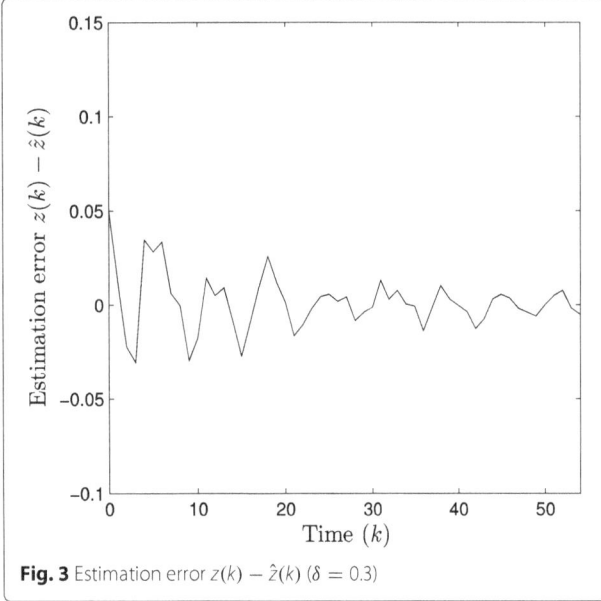

Fig. 3 Estimation error $z(k) - \hat{z}(k)$ ($\delta = 0.3$)

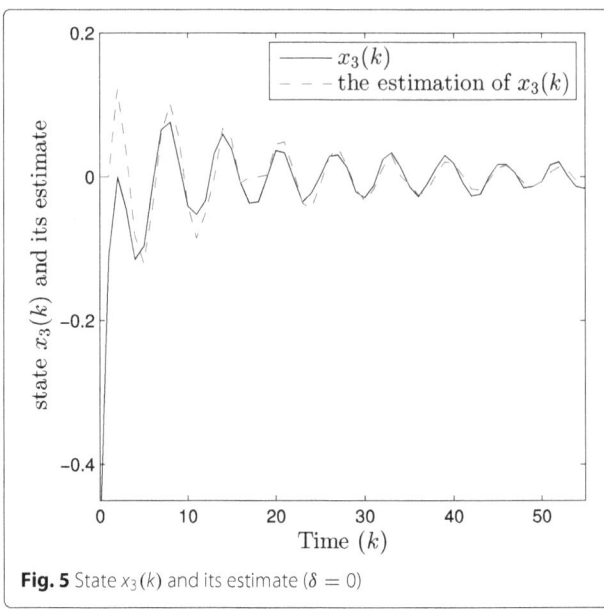

Fig. 5 State $x_3(k)$ and its estimate ($\delta = 0$)

performance. Next, we will compare the event-triggered mechanism with the time-triggered mechanism. When setting the threshold $\delta = 0$, e.g., the time-triggered mechanism, the corresponding results are depicted in Figs. 5, 6, and 7. Corresponding to Figs. 1, 2, and 3, Fig. 5 describes $x_3(k)$ and its estimate $\hat{x}_3(k)$, and Fig. 6 plots $z(k)$ and its estimation $\hat{z}(k)$, whereas the estimation error $z(k) - \hat{z}(k)$ is shown in Fig. 7. Compared with the simulation results between $\delta = 0$ and $\delta = 0.3$, we conclude that, with suitable threshold δ, the event-triggered mechanism could reduce the network burden while ensuring certain system performance. The results confirm the proposed filter design method which could well achieve the desired filtering requirement.

6　Conclusions

In this paper, based on the event-triggered mechanism, we have designed the energy efficiency H_∞ filter for a class of industrial sensor network system with time-varying delay, packet dropouts, and multiplicative noises. The event-triggered communication mechanism is adopted to improve energy efficiency. It could not only reduce the transmission frequency of the measurement output, but also guarantee the prescribed filtering performance. The time-varying delay is considered with event-triggered strategy, which has seldom been studied. Sufficient conditions are found through stochastic analysis technique.

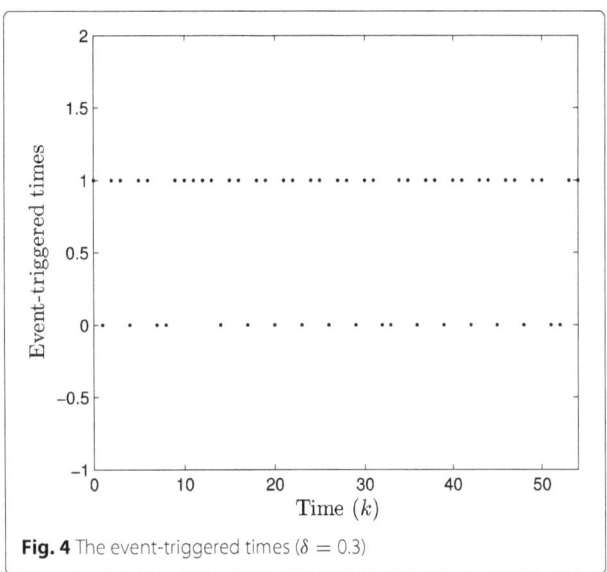

Fig. 4 The event-triggered times ($\delta = 0.3$)

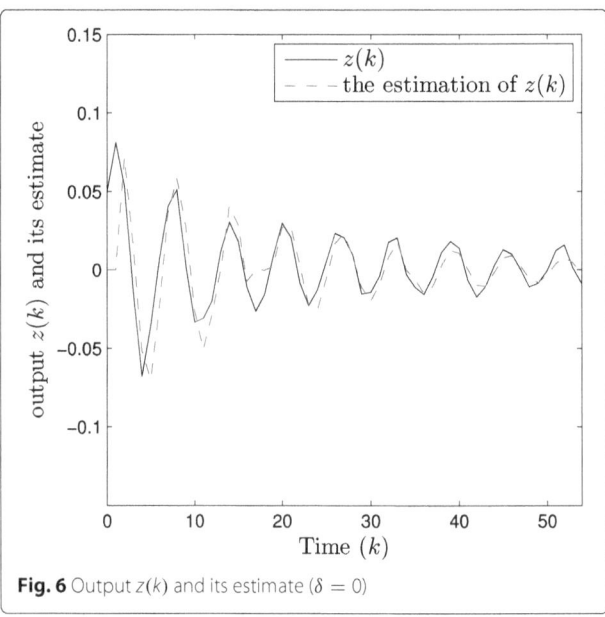

Fig. 6 Output $z(k)$ and its estimate ($\delta = 0$)

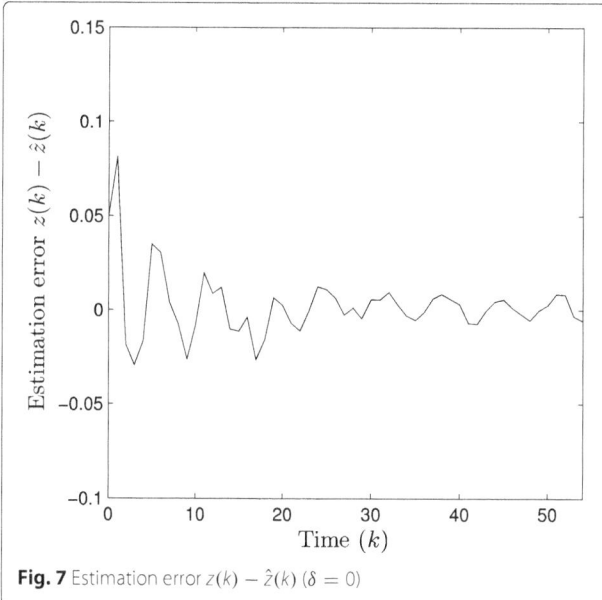

Fig. 7 Estimation error $z(k) - \hat{z}(k)$ ($\delta = 0$)

The filter parameters could be obtained by solving the certain LMI. Finally, the simulation confirms the proposed method.

Abbreviations
LMI: Linear matrix inequality

Authors' contributions
ML carried out the literature analysis and raised and refined the proposed issue in this paper. Meanwhile, she gave the mathematical description of the proposed issue. HL analyzed and designed the filter. JZ verified the analysis and design of the filter by simulation experiments. BZD and YMB checked, reviewed the manuscript, and gave valuable suggestions on the structure of the paper. All authors have read approved the final manuscript.

Authors' information
Hui Li received his BSc degree in Electrical Engineering and Automation in 2010 from Yangtzue University, JingZhou, China, his MSc degree in Control Theory and Control Engineering in 2013 from University of ShangHai for Science and Technology, ShangHai, China. From April 2016, he studied in Nanjing University of Science and Technology for PhD degree. His current research fields include networked filtering systems, multimedia big data processing, and sensor network systems.
Ming Lyu was born in Taizhou, China, in June 1980. She received her BSc degree in Automatic Control in 2002, her MSc degree in Automatic Control in 2004, and her PhD degree in Control Theory and Control Engineering in 2007, all from Nanjing University of Science and Technology, Nanjing, China. She is currently a research fellow in the School of Automation, Nanjing University of Science and Technology, Nanjing, China. Her current research interests include filtering, networked systems, multimedia big data processing, and sensors network systems.
Jie Zhang received his BSc degree in Automatic Control in 2002, his MSc degree in Automatic Control in 2004, and his PhD degree in Control Theory and Control Engineering in 2011, all from Nanjing University of Science and Technology, Nanjing, China. From April 2013 to March 2014, he was an Academic Visitor in the Department of Information Systems and Computing, Brunel University, UK. He is currently an associate research fellow in the School of Automation, Nanjing University of Science and Technology. His current research interests include stochastic systems, networked systems, wireless sensor network systems, and neural networks.

Yuming Bo received the Ph.D. degree in control theory and control engineering from Nanjing University of Science and Technology, Nanjing, China, in 2005. His research interests are focused on filtering and system optimization.

Baozhu Du received the B.S. in Information and Computing Science, and M.S. degree in Operational Research and Cybernetics from Northeastern University, Shenyang, Liaoning Province, China, in 2003 and 2006, respectively. She obtained the Ph.D. degree in Mechanical Engineering from The University of Hong Kong in 2010. Her current research interests include stability analysis and robust control/filter theory of time-delay systems, positive systems, Markovian jump systems, and networked control systems.

Funding
This work was supported by the Natural Science Foundation of Jiangsu Province of China (Grant No. BK20180467), the Research Start-up Funds of Nanjing University of Science and Technology, and the Alexander von Humboldt Foundation.

Competing interests
The authors declare that they have no competing interests.

Author details
[1]School of Automation, Nanjing University of Science and Technology, Nanjing 210094, China. [2]Institute for Automatic Control and Complex Systems, University of Duisburg-Essen, 47057 Duisburg, Germany.

References
1. S. Xu, T. Chen, J. Lam, Robust H_∞ filtering for uncertain Markovian jump systems with mode-dependent time delays. IEEE Trans. Autom. Control. **48**(5), 900–907 (2003)
2. H. Gao, T. Chen, H_∞ Estimation for uncertain systems with limited communication capacity. IEEE Trans. Autom. Control. **52**(11), 2070–2084 (2007)
3. S. Yin, L. Yu, W. Zhang, A switched system approach to networked H infinity filtering with packet losses. Circ. Syst. Signal Proc. **30**(6), 1341–1354 (2011)
4. Z. Wang, Y. Liu, X. Liu, H_∞ filtering for uncertain stochastic time-delay systems with sector-bounded nonlinearities. Automatica. **44**(5), 1268–1277 (2008)
5. R. Lu, H. Li, A. Xue, J. Zheng, Q. She, Quantized H-infinity filtering for different communication channels. Circ. Syst. Signal Proc. **31**(2), 501–519 (2012)
6. H. Gao, C. Wang, Delay-dependent robust H_∞ and $L_2 - L_\infty$ filtering for a class of uncertain nonlinear time-delay systems. IEEE Trans. Autom. Control. **48**(9), 1661–1666 (2003)
7. L. Ma, Z. Wang, Q. Han, H. K. Lam, Envelope-constrained H-infinity filtering for nonlinear systems with quantization effects: the finite horizon case. Automatica. **93**, 527–534 (2018)
8. L. Xie, L. Lu, D. Zhang, H. Zhang, Improved robust H_2 and H_∞ filtering for uncertain discrete-time systems. Automatica. **40**(5), 873–880 (2004)
9. J. Qiu, G. Feng, J. Yang, A new design of delay-dependent robust filtering for discrete-time T–S fuzzy systems with time-varying delay. IEEE Trans. Fuzzy Syst. **17**(5), 1044–1058 (2009)
10. B. Shen, Z. Wang, Y. Hung, Distributed H_∞-consensus filtering in sensor networks with multiple missing measurements: the finite-horizon case. Automatica. **46**(10), 1682–1688 (2010)
11. Z. Duan, J. Zhang, C. Zhang, E. Mosca, Robust H_2 and H_∞ filtering for uncertain linear systems. Automatica. **42**(11), 1919–1926 (2006)
12. H. Gao, Y. Zhao, J. Lam, K. Chen, H_∞ Fuzzy filtering of nonlinear systems with intermittent measurements. IEEE Trans. Fuzzy Syst. **17**(2), 291–300 (2009)
13. J. Zhang, Z. Wang, D. Ding, H-infinity state estimation for discrete-time delayed neural networks with randomly occurring quantizations and missing measurements. Neurocomputing. **148**, 388–396 (2015)
14. W. Zhang, J. Chang, F. Xiao, Y. Hu, N. N. Xiong, Design and analysis of a persistent, efficient, and self-contained WSN data collection system. IEEE Access. **7**, 1068–1083 (2019)

15. J. Tan, W. Liu, T. Wang, N. N. Xiong, H. Song, A. Liu, Z. Zeng, An adaptive collection scheme-based matrix completion for data gathering in energy-harvesting wireless sensor networks. IEEE Access. **7**, 6703–6723 (2019)

16. J. Zhao, J. Huang, N. N. Xiong, An effective exponential-based trust and reputation evaluation system in wireless sensor networks. IEEE Access. **7**, 33859–33869 (2019)

17. R. Wan, N. N. Xiong, Q. Hu, H. Wang, J. Shang, Similarity-aware data aggregation using fuzzy c-means approach for wireless sensor networks. EURASIP J. Wirel. Commun. Netw. **2019**, 59 (2019)

18. X. Liu, S. Zhao, A. Liu, N. N. Xiong, A. V. Vasilakos, Knowledge-aware proactive nodes selection approach for energy management in internet of things. Futur. Gener. Comput. Syst. **92**, 1142–1156 (2019)

19. Y. Liu, K. Ota, K. Zhang, M. Ma, N. N. Xiong, A. Liu, J. Long, QTSAC: an energy-efficient MAC protocol for delay minimization in wireless sensor networks. IEEE Access. **6**, 8273–8291 (2018)

20. J. Tang, A. Liu, J. Zhang, N. N. Xiong, Z. Zeng, T. Wang, A trust-based secure routing scheme using the traceback approach for energy-harvesting wireless sensor networks. Sensors. **18**(3), 751 (2018)

21. M. Wu, Y. Wu, C. Liu, Z. Cai, N. N. Xiong, A. Liu, M. Ma, An effective delay reduction approach through a portion of nodes with a larger duty cycle for industrial WSNs. Sensors. **18**(5), 1535 (2018)

22. N. N. Xiong, L. Zhang, W. Zhang, A. V. Vasilakos, M. Imran, Design and analysis of an efficient energy algorithm in wireless social sensor networks. Sensors. **17**(10), 2166 (2017)

23. H. Cheng, Z. Su, N. N. Xiong, Y. Xiao, Energy-efficient node scheduling algorithms for wireless sensor networks using Markov random field model. Inf. Sci. **329**, 461–477 (2016)

24. A. Shahzad, M. Lee, N. N. Xiong, G. Jeong, Y. K. Lee, J.-Y. Choi, A. W. Mahesar, I. Ahmad, A secure, intelligent, and smart-sensing approach for industrial system automation and transmission over unsecured wireless networks. Sensors. **16**(3), 322 (2016)

25. X. Wang, M. D. Lemmon, Event-triggering in distributed networked control systems. IEEE Trans. Autom. Control. **56**(3), 586–601 (2011)

26. C. Peng, T. Yang, Event-triggered communication and control co-design for networked control systems. Automatica. **49**(5), 1326–1332 (2013)

27. D. V. Dimarogonas, E. Frazzoli, K. H. Johansson, Distributed event-triggered control for multi-agent systems. IEEE Trans. Autom. Control. **57**(5), 1291–1297 (2012)

28. Y. Fan, G. Feng, Y. Wang, C. Song, Distributed event-triggered control of multi-agent systems with combinational measurements. Automatica. **49**(2), 671–675 (2013)

29. L. Ma, Z. Wang, H. K. Lam, Event-triggered mean-square consensus control for time-varying stochastic multi-agent system with sensor saturations. IEEE Trans. Autom. Control. **62**(7), 3524–3531 (2017)

30. D. Ding, Z. Wang, B. Shen, H. Dong, Event-triggered distributed H_∞ state estimation with packet dropouts through sensor networks. IET Control Theory Appl. **9**(13), 1948–1955 (2015)

31. Y. Tan, D. Du, Q. Qi, State estimation for Markovian jump systems with an event-triggered communication scheme. Circ. Syst. Signal Proc. **36**(1), 2–24 (2017)

32. D. Zhang, P. Shi, Q. Wang, L. Yu, Distributed non-fragile filtering for T-S fuzzy systems with event-based communications. Fuzzy Sets Syst. **306**(1), 137–152 (2017)

33. J. Zhang, C. Peng, Event-triggered H_∞ filtering for networked Takagi-Sugeno fuzzy systems with asynchronous constraints. IET Signal Proc. **9**(5), 403–411 (2015)

34. S. Hu, D. Yue, Event-based H_∞ filtering for networked system with communication delay. Signal Proc. **92**(9), 2029–2039 (2012)

35. H. Gao, C. Wang, Delay-dependent robust H_∞ and $L_2 - L_\infty$ filtering for a class of uncertain nonlinear time-delay systems. IEEE Trans. Autom. Control. **48**(9), 1661–1666 (2003)

36. W. Zhang, B. Chen, C. Tseng, Robust H_∞ filtering for nonlinear stochastic systems. IEEE Trans. Signal Proc. **53**(2), 589–598 (2005)

37. H. Gao, Y. Zhao, J. Lam, K. Chen, Fuzzy filtering of nonlinear systems with intermittent measurements. IEEE Trans. Fuzzy Syst. **17**(2), 291–300 (2009)

38. H. Dong, Z. Wang, H. Gao, Robust filtering for a class of nonlinear networked systems with multiple stochastic communication delays and packet dropouts. IEEE Trans. Signal Proc. **58**(4), 1957–1966 (2010)

39. B. Shen, Z. Wang, H. Shu, G. Wei, H_{linfty} filtering for nonlinear discrete-time stochastic systems with randomly varying sensor delays. Automatica. **45**(4), 1032–1037 (2009)

40. M. Wu, N. N. Xiong, L. Tan, An intelligent adaptive algorithm for environment parameter estimation in smart cities. IEEE Access. **6**, 23325–23337 (2018)

41. W. Ke, C. Wu, Y. Wu, N. N. Xiong, A new filter feature selection based on criteria fusion for gene microarray data. IEEE Access. **6**, 61065–61076 (2018)

42. E. Gershon, U. Shaked, I. Yaesh, H_∞ control and filtering of discrete-time stochastic systems with multiplicative noise. Automatica. **37**(3), 409–417 (2001)

43. S. Wen, Z. Zeng, T. Huang, H_{linfty} filtering for neutral systems with mixed delays and multiplicative noises. IEEE Trans. Circ. Syst. II: Express Briefs. **59**(11), 820–824 (2012)

44. Z. Wang, F. Yang, D. W. C. Ho, X. Liu, (2006) Robust H_∞ filtering for stochastic time-delay systems with missing measurements. IEEE Trans. Signal Proc. **54**(7), 2579–2587 (2006)

45. H. Gao, C. Wang, A delay-dependent approach to robust H_∞ filtering for uncertain discrete-time state-delayed systems. IEEE Trans. Signal Proc. **52**(6), 1631–1640 (2004)

46. E. Fridman, U. Shaked, A new H_∞ filter design for linear time delay systems. IEEE Trans. Signal Proc. **49**(11), 2839–2843 (2011)

Cluster tree topology construction method based on PSO algorithm to prolong the lifetime of ZigBee wireless sensor networks

Yang Yu[*] (ID), Bo Xue, Zhuyang Chen and Zhiwen Qian

Abstract

For wireless sensor networks (WSNs) based on ZigBee technology, the network topology plays an important role for improving the energy efficiency and the network lifetime. An appropriate construction method of network topology should be designed for saving the energy of the battery in each network node so as to prolong the lifetime of ZigBee WSNs. In this paper, a novel cluster tree topology construction method based on particle swarm optimization (PSO) algorithm is proposed. In order to transform the network topology construction problem into an energy consumption optimization problem, an evaluation function reflecting the network energy consumption is designed, and the network topology is mapped into a particle population individual suitable for the PSO algorithm. In order to prolong the network lifetime as long as possible, two network topology reconstruction methods with the PSO algorithm based on fixed and variable energy thresholds are further proposed, respectively. The simulation results show that the proposed methods can prolong the lifetime of ZigBee WSNs effectively.

Keywords: Cluster tree topology, PSO algorithm, ZigBee, Wireless sensor networks, Lifetime

1 Introduction

With the development of information industry, the application of information technology is more and more extensive. Especially with the arrival of the era of big data, the acquisition and processing of massive information has attracted more and more attention. Wireless sensor networks (WSNs) are a self-organizing network composed of a large number of sensor nodes with functions of information collection, processing, and transmission. Each sensor node can collect, process, and transmit data in the sensing area, which can cooperate to collect massive information in the target area [1, 2]. As one of the key technologies of WSNs, ZigBee technology based on the standard of IE802.15.4 has been widely used in smart home [3], intelligent agriculture [4], smart grid [5], health monitoring [6], industrial control [7], and other fields because of its low power consumption, low speed, low cost, and ad hoc network characteristics.

There are three types of nodes in ZigBee networks: coordinator, router, and end-device [8]. In addition to the

coordinator nodes powered by stable power supply, routers and end-devices are generally powered by dry batteries or storage batteries. The end-devices send data directly or through routers to the coordinator, and the low power characteristic allows it to turn off/turn on the transceiver periodically to save energy. The randomness of data forwarding makes routers have to turn on receivers at all times. Moreover, in application scenarios of WSNs such as industrial and farm, data is collected and transmitted frequently, which causes the routing nodes to consume much energy and premature failure to work because the battery is exhausted. As a result, the network is divided into isolated parts which are difficult to connect, and the network life time is greatly reduced [9, 10]. Therefore, how to use the limited energy to make WSNs work as long as possible, and reduce the cost of battery has become a hot issue in the current research of WSNs based on the ZigBee technology.

For WSNs with limited energy, a good network topology construction method can save a lot of energy, reduce the energy consumption of nodes, and effectively prolong the network lifetime [11]. In recent years, the research on WSN topology control algorithm mainly

* Correspondence: dxyy@jsut.edu.cn
School of Electric Information Engineering, Jiangsu University of Technology, Changzhou 213001, China

focuses on power control and hierarchical topology construction algorithms [12, 13]. In terms of power control-based topology construction method, the typical algorithms that have been proposed are node-based local mean algorithm and local mean neighbors algorithm [14], relative neighborhood graph algorithm based on neighbor graphs [15], etc. In terms of hierarchical topology construction method, typical algorithms include topological control algorithms based on uniform clustering, such as the low-energy adaptive clustering hierarchy algorithm and hybrid energy-efficient distributed clustering approach, etc. [16, 17].

From these research results, it is known that the network topology construction that reduces the number of routers as much as possible is considered to be an effective method for suppressing power consumption. In addition, in order to prolong the network lifetime, it is important to use the power of all nodes evenly. Currently, there are no good known methods for analyzing the network structure that minimizes the number of routers. Furthermore, it is not possible to study all possible network structures of large-scale networks during processing time. For problems that cannot be solved by graph theory or combinatorial mathematical analysis, the particle swarm optimization (PSO) algorithm can find suboptimal solutions for these problems in real time [18, 19]. Therefore, in this paper, we use the PSO

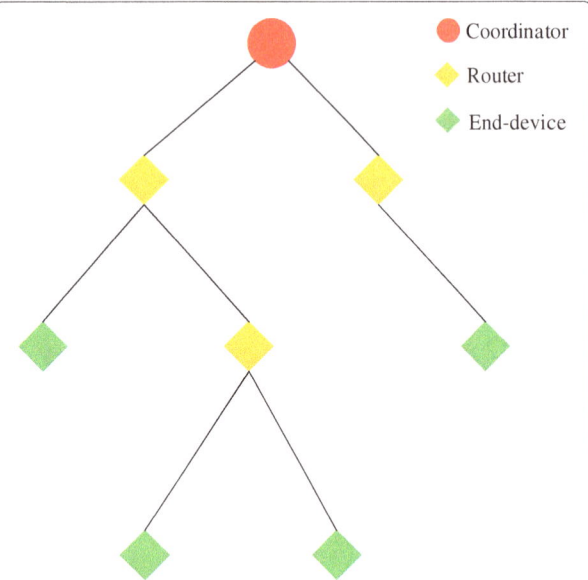

Fig. 1 Cluster tree topology of WSNs. This figure shows the simple and typical cluster tree network topology for wireless sensor networks

algorithm to study the construction method of the cluster tree topology for ZigBee WSNs with low power consumption and long network lifetime.

The rest of this paper is organized as follows. The system model is given in Section 2. In Section 3, the principle

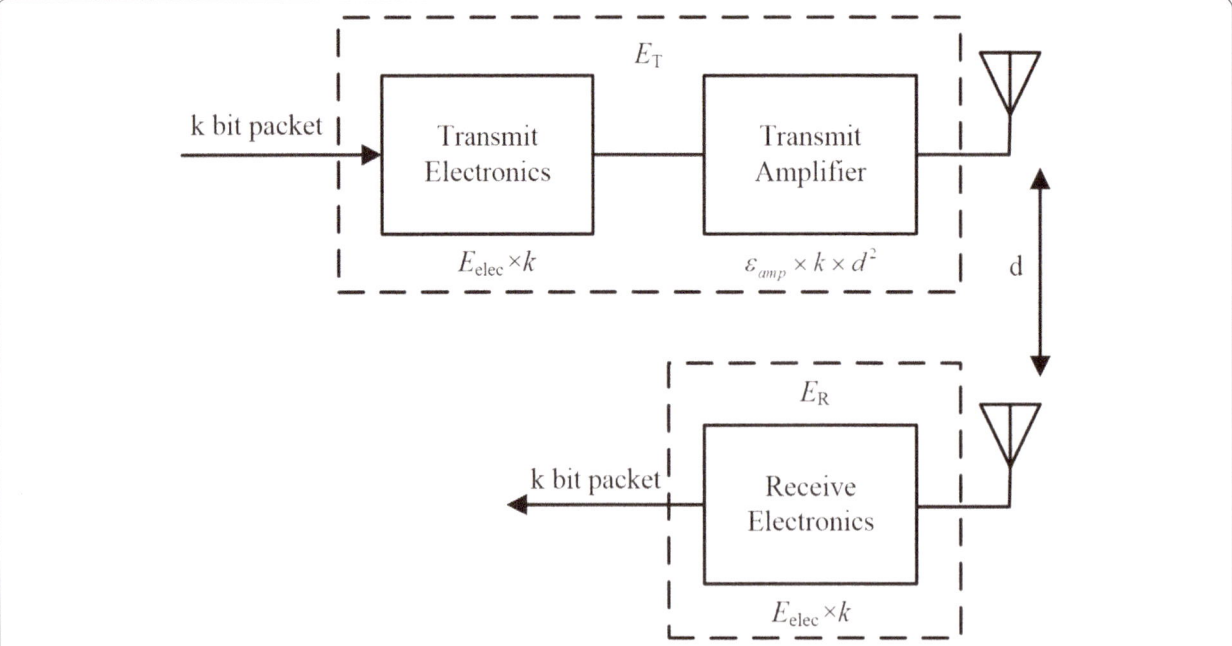

Fig. 2 Energy consumption model. This figure shows the first order radio model which is used as the energy consumption model that is considered in this paper. E_{elec} is the amount of energy required to transmit and receive one bit message, and ε_{amp} is the amount of energy required to amplify the transmitter signal. The energy consumption of transmission and reception are denoted as E_T and E_R, respectively

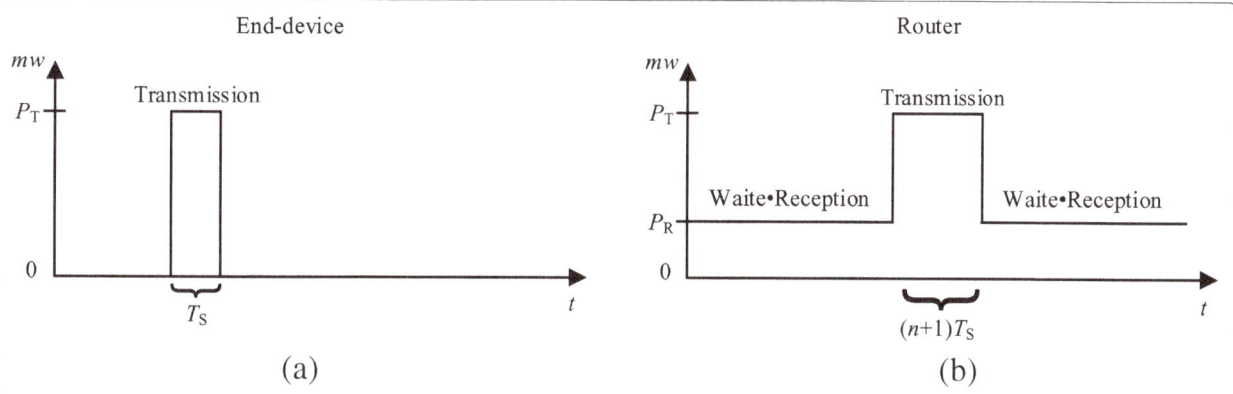

Fig. 3 Time series of network operation. This figure shows the time series of the operation of WSNs, where **a** shows the time series of the operation of the end-device, **b** shows the time series of the operation of the router, T_s is the transmission time, P_T is the transmission power, P_R is the reception power, and n is the number of child nodes of the router. Here, the end-device enters a sleep mode when no information is transmitted, while the router waits for a receivable state when no information is transmitted

of the PSO algorithm is introduced first, and the proposed cluster tree topology construction method based on the PSO algorithm is described in detail. The effectiveness of the proposed method is examined in Section 4 by computer simulation compared from multiple perspectives. Finally, Section 5 concludes the paper.

2 Methods

In this paper, the construction method of the network topology for ZigBee WSNs is based on the PSO algorithm which is a swarm intelligence algorithm. The cost function reflecting the energy consumption of the network is fist designed. Then the real number sequence is designed to represent the possible solution of the network topology. Thus, the network topology construction problem studied in this paper is transformed into an

optimization problem that can be solved by PSO algorithm. To verify the effectiveness of the proposed scheme, we compare the performance of the proposed scheme with the traditional method in several aspects.

3 System model

3.1 Cluster tree topology

ZigBee-based WSNs have three network topologies: star topology, mesh topology, and cluster tree topology. In a star network since only the range from the central node to one hop can be communicated, the delay time of data transmission is short and collisions hardly occur. On the other hand, there is a disadvantage that the coverage of the network is relatively small and the communication reliability is low because only one-hop communication is possible. The merit of mesh network is that multi-hop

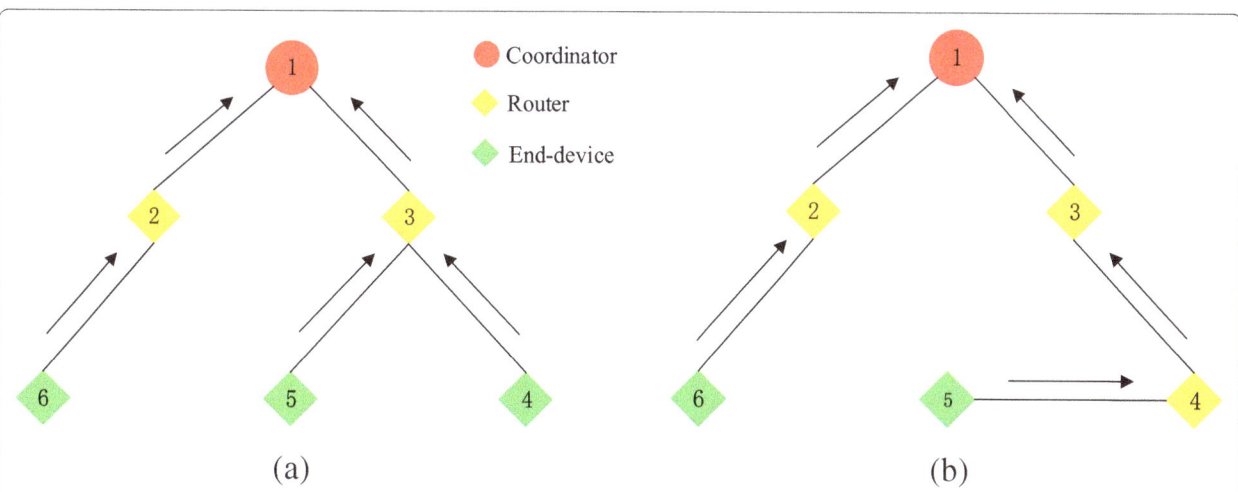

Fig. 4 Networks with different cluster tree topology. This figure shows an example of two WSNs with the same number of nodes but different cluster tree network topologies, where **a** corresponds to the topology with two routers and three end-devices; **b** corresponds to the topology with three routers and two end-devices

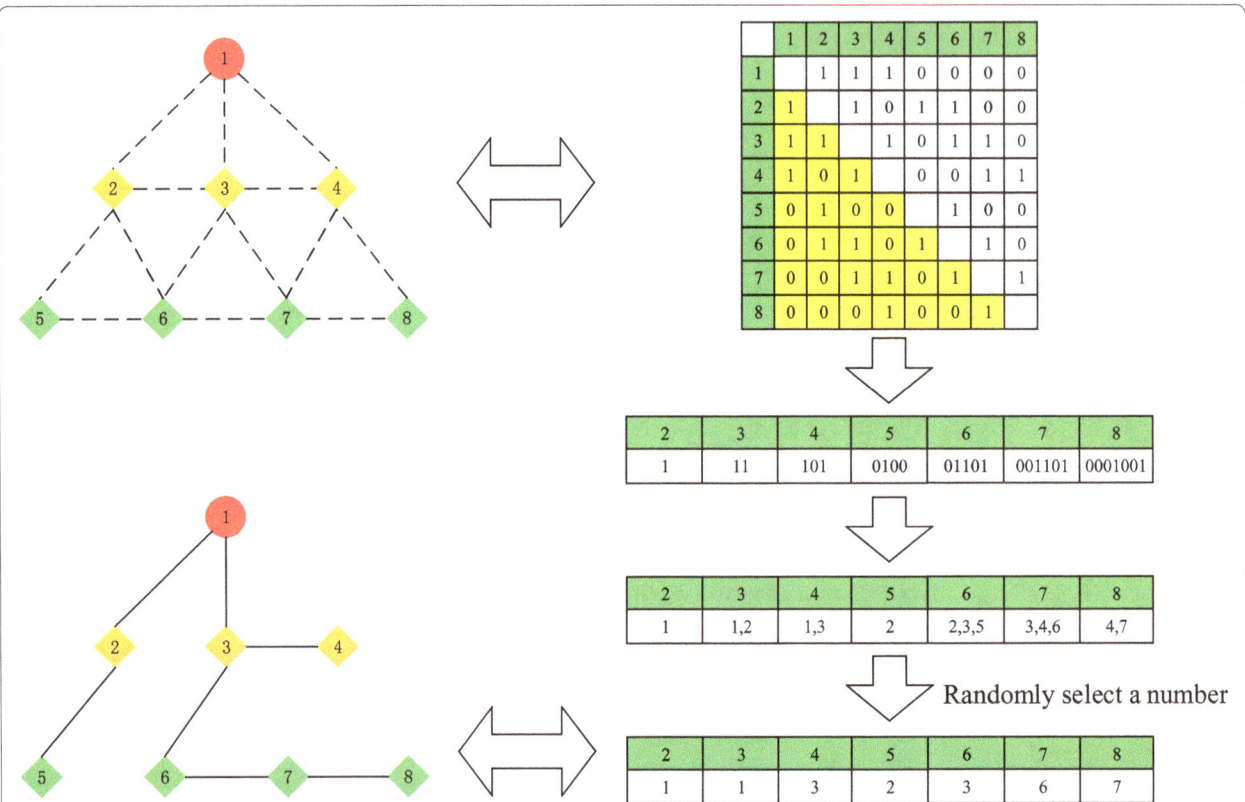

Fig. 5 Particle population initialization. We use the adjacency matrix to generate particles. As shown in Fig. 5, nodes that can communicate with each other are first connected by dashed lines to create an adjacency matrix. Here, 1 indicates a connection state, and 0 indicates a non-connection state. Next, the lower left half of the adjacency matrix is taken out and rewritten as one line as shown in Fig. 5. Finally, particles can be generated by randomly extracting a number from each part

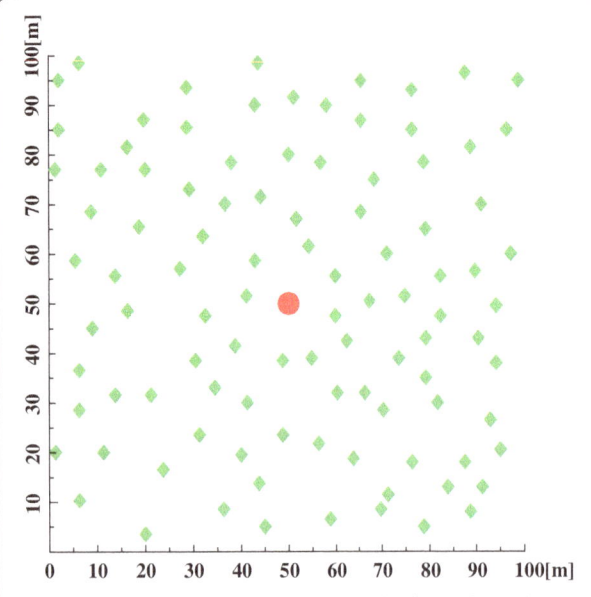

Fig. 6 100 nodes random ZigBee network. This figure shows that 100 nodes are placed in an observation area of 100 m × 100 m to collect information. Nodes are randomly distributed in the observation area, and the information of each node is managed by the coordinator

communication can be performed beyond the one-hop limit. In addition, communication reliability can be improved by making communication routes redundant. Cluster tree network has strong expansion ability, and has the advantages of star network and mesh network.

The cluster tree topology mainly used in WSNs is shown in Fig. 1. The cluster tree topology differs from the mesh topology in that the relationship between nodes is

Table 1 Parameters of simulation

Number of nodes	100
Observation area	100 m × 100 m
Node communication radius	30 m
Transmission power	35 mW
Received power	12.5 mW
Communication speed	250 kbps
Amount of information	1000 bits
Round time	30 s
battery capacity	100 J
c_1, c_2	2,2
population size	200
Number of generations	500

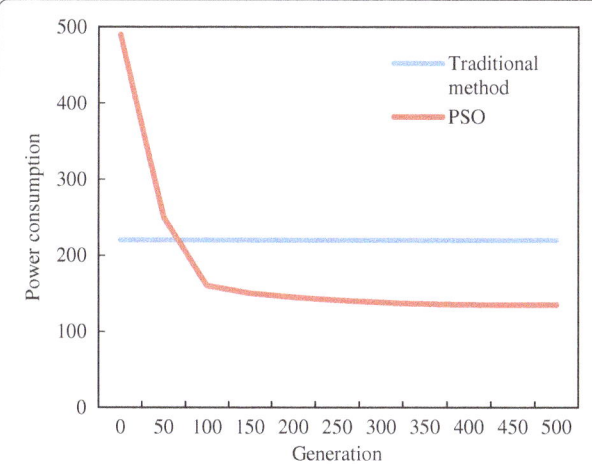

Fig. 7 Comparison of energy consumption. This figure shows a comparison of network power consumption. It can be seen that the power consumption of the network constructed with PSO is smaller than the energy consumption of the traditional method

not equal, and there is a clear parent-child relationship, and it is a pyramidal tree structure. In this structure, the top nodes of the network are the coordinator, the other parent nodes are routers, and the terminal nodes are end-devices. The advantages of cluster tree topology are as follows.

1. Since the network is hierarchical, the delay time of data transmission can be predicted.

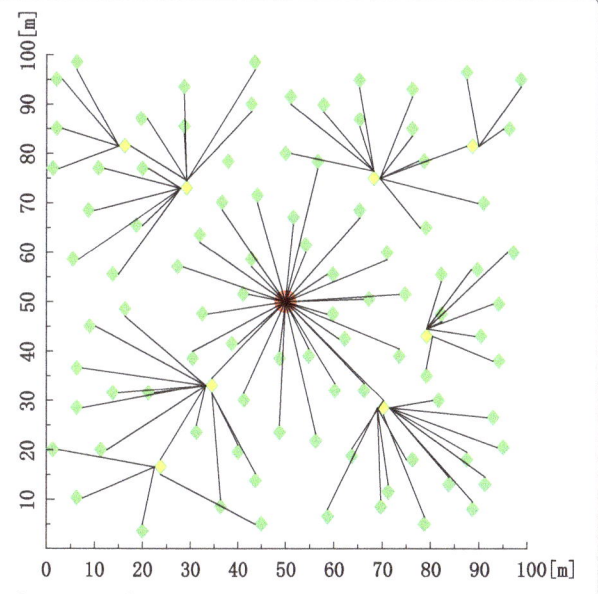

Fig. 8 Network topology construction results based on PSO. This figure shows an example of a network construction with the lowest power consumption obtained by the PSO algorithm. Here, the red node is the coordinator, the yellow node is the router, and the green node is the end-device. It can be observed that the number of routers is 8, and the number of end-devices is 91

2. The router always keeps tracking the status of its own child nodes, and the connection status of the network is aggregated along the tree root. Therefore, the coordinator can easily understand the connection status of the entire network.

On the other hand, there are some disadvantages of cluster tree topology.

1. Since there is only one communication route, sometimes information may not reach the destination node when a communication failure occurs.
2. Because information is aggregated to the coordinator, traffic jams occur closer to the coordinator, and the balance of network resource utilization is not good.

3.2 Energy consumption model

The energy consumption model considered in this paper is the first-order radio model as shown in Fig. 2. E_{elec} is the amount of energy required to transmit and receive one bit message, and ε_{amp} is the amount of energy required to amplify the transmitter signal. The energy consumption of transmission and reception denoted as E_T and E_R are calculated using (1) and (2), respectively [20].

$$E_T = E_{elec} \times k + \varepsilon_{amp} \times k \times d^2 \tag{1}$$

$$E_R = E_{elec} \times k \tag{2}$$

where the amount of communication is k bits and the communication radius of the node is d meters. When the communication time is t seconds and the communication rate is v [bps], the transmission power P_T and the reception power P_R are calculated using (3) and (4), respectively.

$$P_T = \frac{E_T}{t} = E_T \times \frac{v}{k} = \left(E_{elec} + \varepsilon_{amp} \times d^2 \right) \times v$$
$$= 35\,\text{mW} \tag{3}$$

$$P_R = \frac{E_R}{t} = E_R \times \frac{v}{k} = E_{elec} \times v = 12.5\,\text{mW} \tag{4}$$

It is assumed that E_{elec} is 50 (nJ/bit), ε_{amp} is 100 (pJ/bit/m^2), d is 30 m, v is 250 kbps, and there is no delay due to radio wave collision or communication error. The results of P_T and P_R can be calculated as 35 mW and 12.5 mW, respectively.

Figure 3 shows the time series of the operation of WSNs, where T_s is the transmission time, P_T is the transmission power, P_R is the reception power, and n is the number of child nodes of the router. Here, the end-device enters a sleep mode when no information is

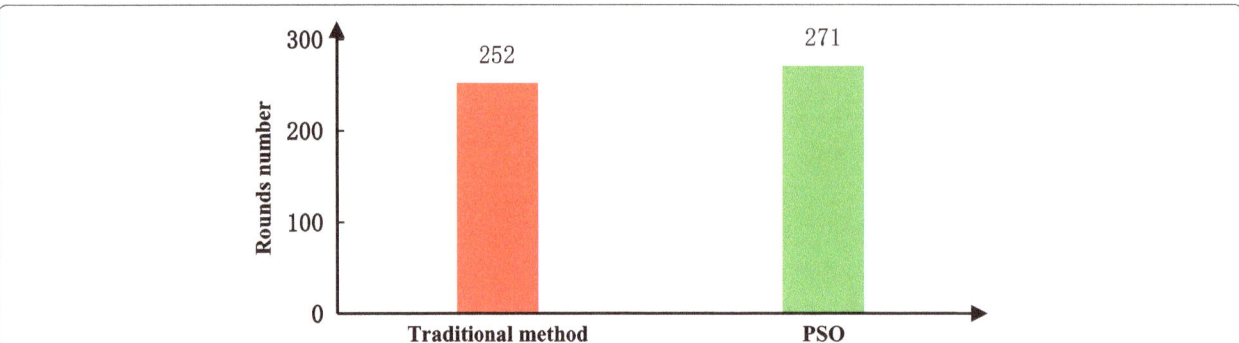

Fig. 9 Comparison of network lifetime. This figure shows the comparison of the performance in term of the network lifetime. It can be seen that there is little difference in the lifetime of the network, although the network constructed by the PSO algorithm consumes less energy in one round than traditional method

transmitted, while the router waits for a receivable state when no information is transmitted.

Figure 4 shows an example of two WSNs with the same number of nodes but different cluster tree topology networks. As shown in Fig. 4a, there are three end-devices that only transmit information. From Fig. 3a, the total energy consumption of end-devices can be calculated as shown in (5). The second and third node acts as routers to transmit information and also receives information, that is combining the data it owns with the data received from them child nodes and sends it to the coordinator. From Fig. 3b, the amount of energy consumption of routers can be calculated based on (6) in a similar way. Therefore, the power consumption of a network round is the sum of the power consumption of each node, and is calculated using (7).

$$E_{ed} = 3 \times T_t \times P_T \tag{5}$$

$$E_r = 2 \times T_t \times P_T + 3 \times T_t \times P_T + [2T - (2+3) \times T_t] \times P_R \tag{6}$$

$$P_{\text{WSN1}} = \frac{E_{ed} + E_r}{T} = \frac{8 \times T_t \times P_T + [2T - 5 \times T_t] \times P_R}{T} \tag{7}$$

As shown in Fig. 4b, it has one more routing node than Fig. 4a, and the power consumed can be calculated by the same method as described in (8).

$$P_{\text{WSN2}} = \frac{E_{ed} + E_r}{T} = \frac{9 \times T_t \times P_T + [3T - 6T_t] \times P_R}{T} \tag{8}$$

Substituting the values of P_T and P_R calculated in (3) and (4), it can be found that the value of P_{WSN2} is greater than that of P_{WSN1}. This indicates that different network topologies affect the power consumption of the network due to the difference in the number of routes, thereby affecting the lifetime of the network.

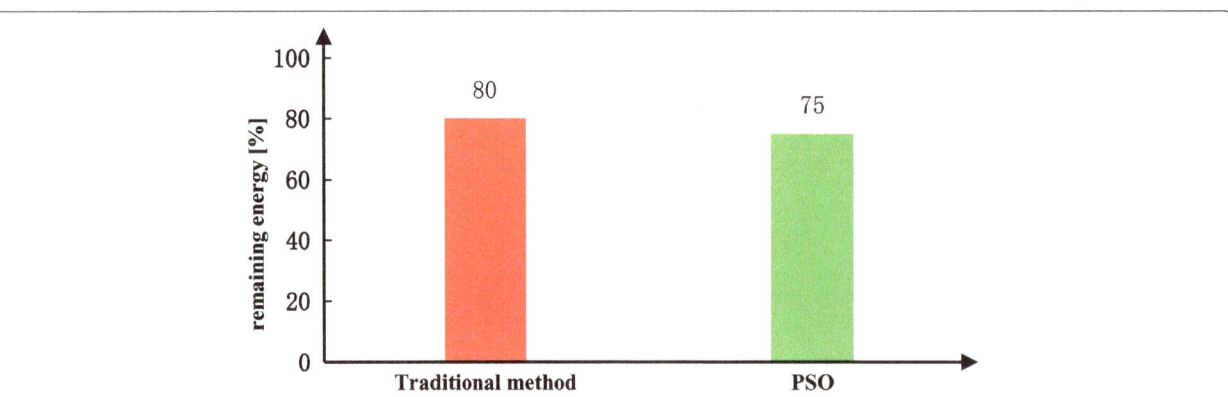

Fig. 10 Comparison of remaining energy. This figure shows the comparison of the performance in term of the remaining energy at the end of network lifetime

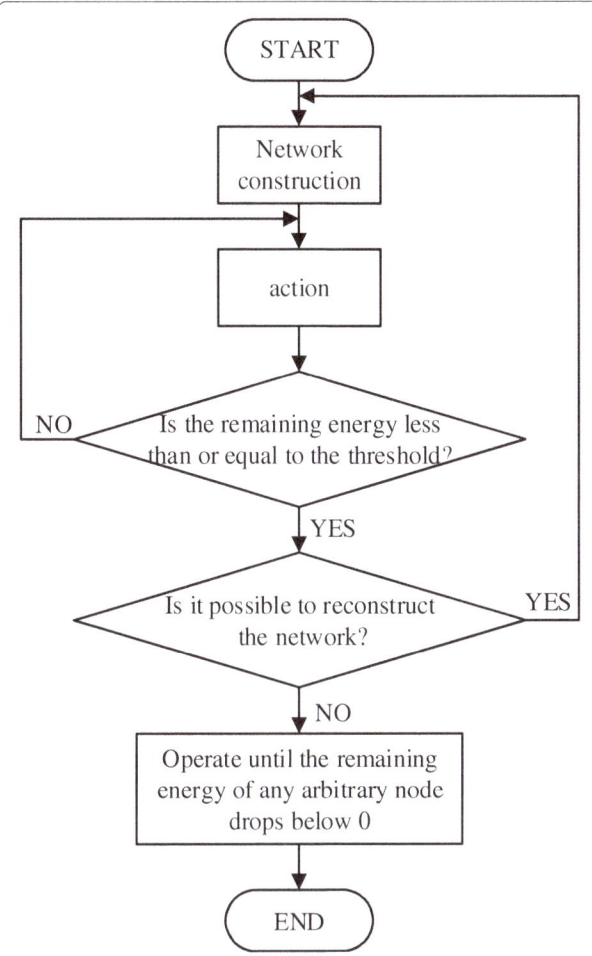

Fig. 11 Flow chart of reconstruction method with fixed threshold. This figure shows the flow chart of a network reconstruction method based on a fixed power threshold. As shown in the figure, configure the network and then operate it first. Next, network reconstruction is performed when the remaining rate of battery energy of any router falls below the threshold that is set to 10%. In this case, it is required that the node whose remaining rate of battery energy is lower than the threshold cannot be a router. If reconstruction is not possible, leave the network structure remains unchanged and run until the remaining energy of any node is below zero

4 Cluster tree topology construction method based on particle swarm optimization algorithm

4.1 Particle swarm optimization algorithm

Particle swarm optimization (PSO) algorithm was first proposed for solving the continuous optimization problem in 1995 by Kennedy and Eberhart [19]. Its basic concept stems from the study of predation behavior of birds. On this basis, they proposed a discrete binary version of PSO in 1997 to solve the combinatorial optimization problem in engineering practice.

Similar to other evolutionary algorithms, the PSO algorithm also adopts the concepts of "group" and "evolution." The difference is that the PSO algorithm does not use evolutionary operators for individuals like other evolutionary algorithms, but treats each individual as a non-weight and non-volume particle flying at a certain speed in the search space. The speed of the flight is dynamically adjusted by the individual's flight experience and the group's flight experience.

Suppose that in a D-dimensional target search space, the population size is m particles, and the position of the ith particle in the D-dimensional search space is represented as a D-dimensional vector

$$\mathbf{x}_i = (x_{i1}, x_{i2}, \cdots, x_{iD})^T, i = 1, 2, \cdots, m \qquad (9)$$

The position of each particle is a potential solution for the problem being optimized. By substituting a vector $\mathbf{x}_i$ into an objective function, the fitness value of the objective function can be calculated, and then the quality of the particle $\mathbf{x}_i$ can be measured according to the fitness value. The flight speed of the ith particle is also a D-dimensional vector

$$\mathbf{v}_i = (v_{i1}, v_{i2}, \cdots, v_{iD})^T, i = 1, 2, \cdots, m \qquad (10)$$

The optimal position vector that the ith particle has searched so far is

$$\mathbf{pbest}_i = (pbest_{i1}, pbest_{i2}, \cdots, pbest_{iD})^T \qquad (11)$$

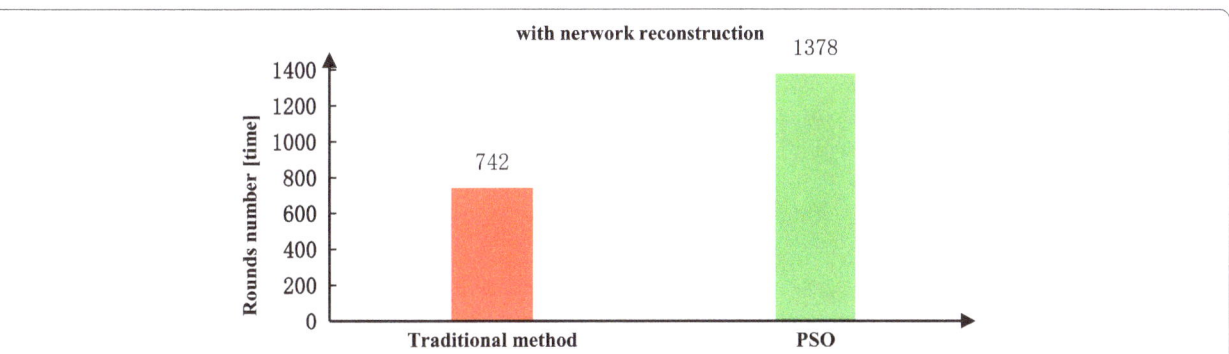

Fig. 12 Comparison of network lifetime. This figure shows the comparison of the performance in term of the network lifetime when the reconstruction method with fixed threshold is used

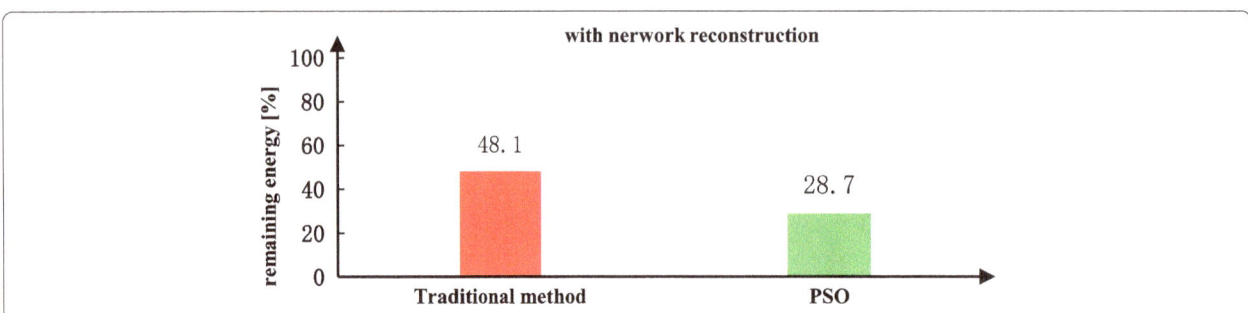

Fig. 13 Comparison of remaining energy. This figure shows the comparison of the performance in term of the remaining energy at the end of network lifetime when the reconstruction method with fixed threshold is used

where **pbest**$_i$ is called an individual extreme point. The optimal position vector that the entire particle swarm has searched so far is

$$\mathbf{gbest} = (gbest_1, gbest_2, \cdots, gbest_D)^T \qquad (12)$$

where **gbest** is called a global extreme point. In each iteration, every particle updates its velocity and position according to the above two optimal values. The renewal equation is

$$
\begin{aligned}
v_{id}^{k+1} = v_{id}^k &+ c_1\, rand_1^k \left(pbest_{id}^k - x_{id}^k\right) \\
&+ c_2\, rand_2^k \left(gbest_d^k - x_{id}^k\right)
\end{aligned}
\qquad (13)
$$

$$
x_{id}^{k+1} = \begin{cases} 1, sig\left(v_{id}^{k+1}\right) > \rho_{id}^{k+1} \\ 0, \quad \text{otherwise} \end{cases}
\qquad (14)
$$

where v is the dth velocity of the ith particle in the kth iteration; c_1, c_2 are learning factors used to adjust the maximum step size of the flight toward the global optimal particle and the individual optimal particle; if they are too small, the particles may be far away from the target area, if they are too large, the particles may fly over the target area. The appropriate value of c_1, c_2 can accelerate the convergence and not easily fall into local optimum, usually let $c_1 = c_2 = 2$; $rand_{1,\ 2}$ is a random number between [0, 1]. x_{id}^k is the current position of the dth dimension of the ith particle in the kth iteration; $pbest_{id}$ is the position of the ith particle at the individual extreme point of the dth dimension; $gbest_d$ is the dth dimensional position of the global extreme point of the entire population.

In (14), $sig\left(v_{id}^{k+1}\right)$ is the threshold function that converts the velocity to a value between [0, 1], which is defined as

$$sig\left(v_{id}^{k+1}\right) = \frac{1}{1 + e^{-a\left(v_{id}^{k+1} - c\right)}} \qquad (15)$$

where usually take $a = 1$, $c = 0$. The individual components ρ_i of the vector ρ_{id} are random numbers between [0, 1]. It can be seen that if the value of v_{id} is larger, the position of the particle x_{id} is more likely to select as 1;

otherwise, x_{id} is more likely to select as 0. Therefore, v_{id} is equivalent to a probability threshold, indicating the tendency of the particle i to take 1 or take 0 at the d-dimensional position.

During the running of the PSO algorithm, the individual extreme of a single particle and the global extreme of the entire particle swarm are continuously updated, and the global extreme output at the end of the algorithm is the optimal solution of the problem.

4.2 Cluster tree topology construction for ZigBee WSNs using PSO

Construction method of cluster tree topology based on the PSO algorithm actually uses the global search and combination optimization capabilities of the PSO algorithm to determine the optimal combination of routers and end-devices in the solution space for ZigBee WSNs.

Based on the introduction and analysis of Section 2.2, we define the evaluation function as in (16) by extending (7) to general cases. Fitness P_{WSN} represents the power consumption of one round of WSNs. One round is the period during which all nodes in WSNs complete once data transmission.

$$P_{WSN} = \frac{H \times T_t \times P_T + [T \times R - (k + R) \times T_t] \times P_R}{T}$$

$$(16)$$

Here, T_t is the transmission time, T is one round time of the network, R is the number of routers, H is the total number of hops of the network, P_T is the transmission power, P_R is the reception power, and k represents the sum of the number of child nodes of each router.

The process steps of the proposed network topology construction method using PSO is described as follows.

1. Particle population initialization

The generation of the initial population is an operation that randomly generates a preset number of particles. Here, we use the adjacency matrix to generate particles.

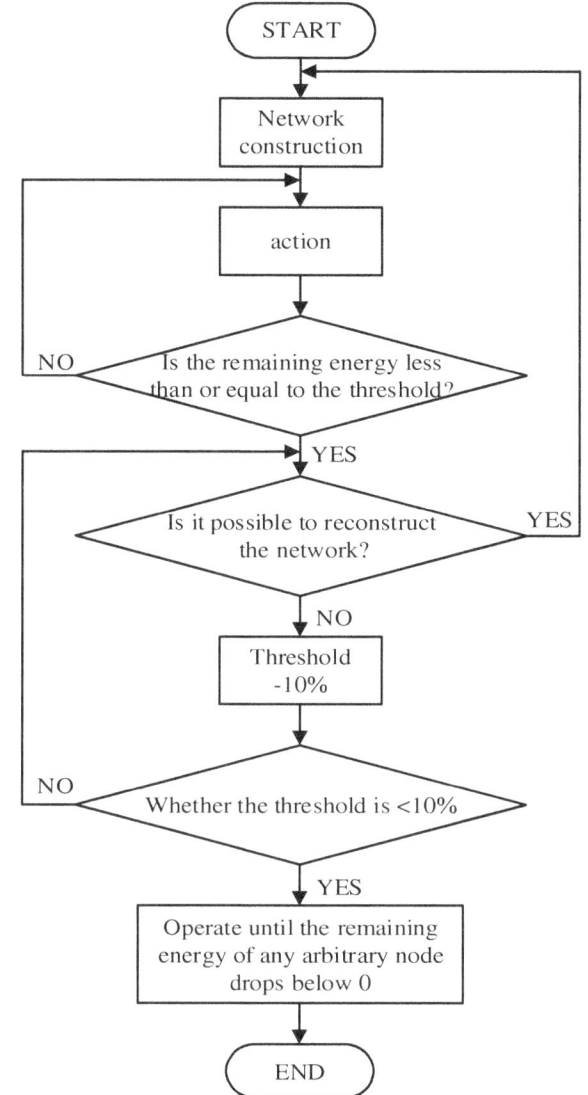

Fig. 14 Flow chart of reconstruction method with variable threshold. This figure shows the flow chart of a network reconstruction method based on a variable power threshold. As shown in the figure, when the remaining rate of battery energy of any router drops below the threshold that is set to 80%, network reconstruction is performed. It is required that nodes whose remaining power is below the threshold cannot be set as a router. If network reconstruction becomes impossible in the current threshold, the threshold is reduced by 10%. In this case, the node with the remaining power higher than the threshold can be set as the router, and the node below the threshold can only be set as the end-device. In this way, the threshold decreases with 10% step size, and network reconstruction is carried out in a circular way. When the threshold becomes less than 10%, the network structure is maintained and the operation of the network is performed until the remaining energy of any node is below zero

As shown in Fig. 5, nodes that can communicate with each other are first connected by dashed lines to create an adjacency matrix. Here, 1 indicates a connection state, and 0 indicates a non-connection state. Next, the

lower left half of the adjacency matrix is taken out and rewritten as one line as shown in Fig. 5. Finally, particles can be generated by randomly extracting a number from each part.

2. Evaluation of particle fitness

Particle fitness evaluation is an operation of calculating the fitness of each particle in the population using the evaluation function of (16). The value of fitness can reflect the quality of the particle. In order to prolong the lifetime of the network, it is desirable that the smaller the value of the fitness (the smaller the power consumption) the better.

In the first iteration of the particle population, the initial population of N particles is randomly generated according to the method in the step (1). The individual extreme value *pbest* is set to the current particle position. The positional coordinates of each particle are substituted into the evaluation function (16) to calculate the fitness value, and the position of the particle with the largest value is set as the global extreme value *gbest*.

If it is not the first iteration of the particle population, current position vector for each particle is substituted into the evaluation function to calculate the fitness value of the particle. If it is greater than the current individual extreme value of the particle, *pbest* will be set to the current position of the particle. If the largest of the individual extreme value of all particles is greater than the current global extreme value, *gbest* will be set to the position of the best particle.

3. Velocity and position of particle updating

For each bit on each particle code string, the velocity and position are updated according to formula (13) and (14).

4. Judgment of stop condition

If the current number of iterations reaches the predetermined maximum number of generations, the iteration is stopped, and the result of the cluster tree topology is outputted; otherwise, the process proceeds to step 2, and the algorithm is continued.

5 Simulation results and discussion

In this study, environmental monitoring in a wide area is assumed. Nodes are randomly distributed in the observation area, and the information of each node is managed by the coordinator. Specifically, as shown in Figs. 6, 100 nodes are placed in an observation area of 100 m × 100 m to collect information.

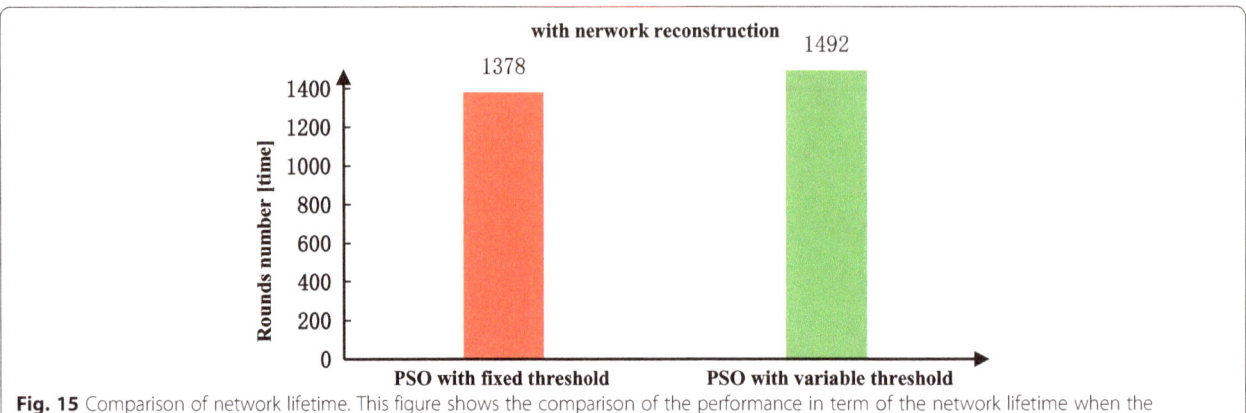

Fig. 15 Comparison of network lifetime. This figure shows the comparison of the performance in term of the network lifetime when the reconstruction method with variable threshold is used

The communication radius of the node is 30 m. The position of the coordinator is located at the center of the viewing area (50, 50). The coordinator is powered by an external power source, while the node is powered by the battery.

Here, the roles of routers and end-devices can be exchanged with each other, and the movement of the node is not considered. In addition, it is assumed that data collisions do not occur during wireless communication. The experimental parameters are shown in Table 1. In the following section, the cluster tree topology constructed by the traditional method and the PSO algorithm are experimented respectively, and the performance is compared and analyzed.

5.1 Construction result of the network topology based on the PSO algorithm

Experiments are carried out using the traditional method and PSO. The experimental results are the average of ten simulation results. Figure 7 shows a comparison of network power consumption. It can be seen that the power consumption of the network constructed with PSO algorithm is smaller than that of the traditional method.

An example of a network construction with the lowest power consumption obtained by the PSO algorithm is shown in Fig. 8. Here, the red node is the coordinator, the yellow node is the router, and the green node is the end-device. It can be observed that the number of routers is 8, and the number of end-devices is 91.

5.2 Simulation and analysis of network lifetime

Based on the network topology with lowest power consumption obtained in Section 4.1, we study its network lifetime in this section. We compare the network lifetime when the network is not reconfigured and the network is reconstructed, respectively. To calculate the lifetime of the network, a capacity of 100 J battery power is supplied to nodes other than the coordinator.

After the network is constructed, the network starts running. We assume that the remaining energy of any node is less than 0, the lifetime of the network stops. The comparison of the performance in term of the network lifetime and the remaining energy at the end of network lifetime is shown in Figs. 9 and 10, respectively.

It can be seen that there is little difference in the lifetime of the network, although the network constructed

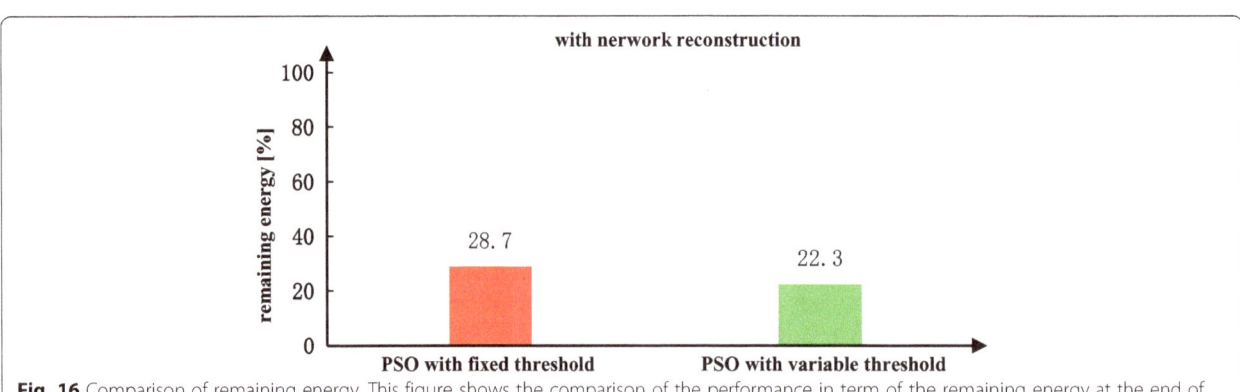

Fig. 16 Comparison of remaining energy. This figure shows the comparison of the performance in term of the remaining energy at the end of network lifetime when the reconstruction method with variable threshold is used

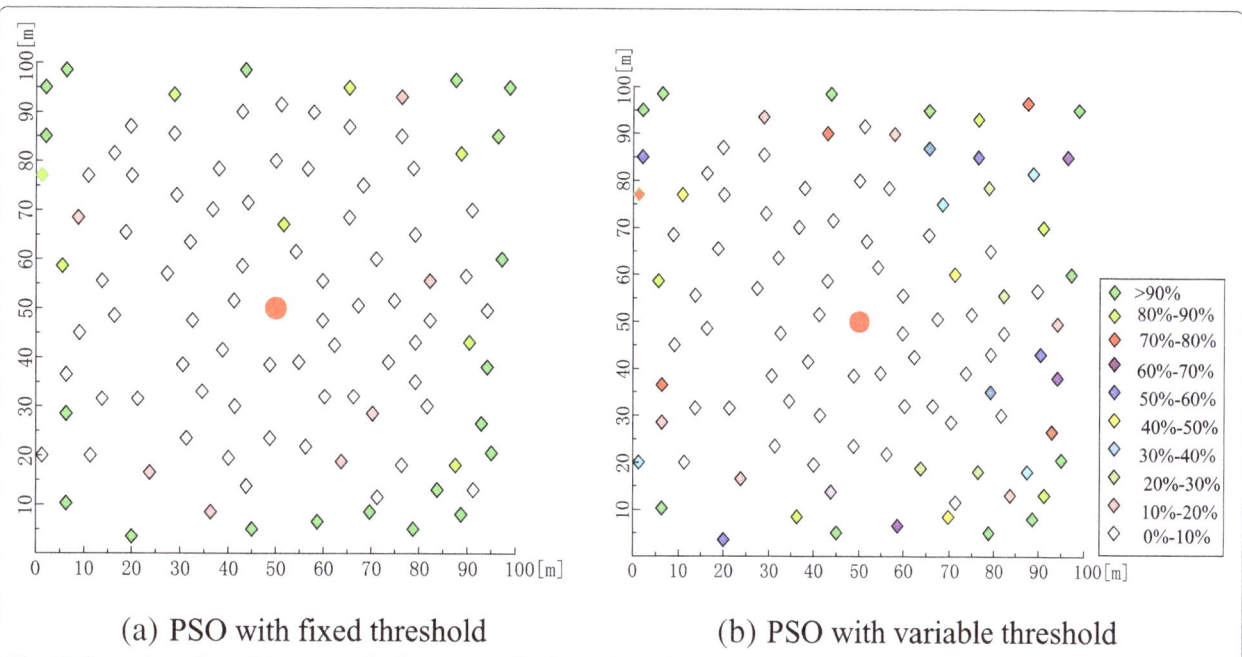

(a) PSO with fixed threshold   (b) PSO with variable threshold

Fig. 17 Comparison of remaining energy distribution map. This figure shows the distribution map of the energy consumption of the network constructed by the proposed reconstruction method, where **a** corresponds to the topology construction method using PSO algorithm and reconstruction method with fixed threshold; **b** corresponds to the topology construction method using PSO algorithm and reconstruction method with variable threshold

by the PSO algorithm consumes less energy in one round than traditional method. In addition, the energy consumption is lower but the remaining energy is still kept at about 70% using the proposed method. This means that the network should be reconstructed to deplete the remaining power and prolong the lifetime of the network.

5.2.1 Network reconstruction with fixed power threshold

The flow chart of the network reconstruction method based on a fixed power threshold is shown in Fig. 11. First, configure the network and then operate it. Next, network reconstruction is performed when the remaining rate of battery energy of any router falls below the threshold that

is set to 10%. In this case, it is required that the node whose remaining rate of battery energy is lower than the threshold cannot be a router. If reconstruction is not possible, leave the network structure remains unchanged and run until the remaining energy of any node is below zero. Here, we do not consider the power consumed during network reconstruction for simplicity.

The performance comparison in term of the network lifetime and the remaining energy is shown in Figs. 12 and 13, respectively. Comparing Figs. 10 and 13, in the case of the proposed PSO scheme, it is confirmed that the remaining energy of the network with reconstruction is 46.3% less than that of the network without reconstruction. And as can be seen from Figs. 12 and 13, it is

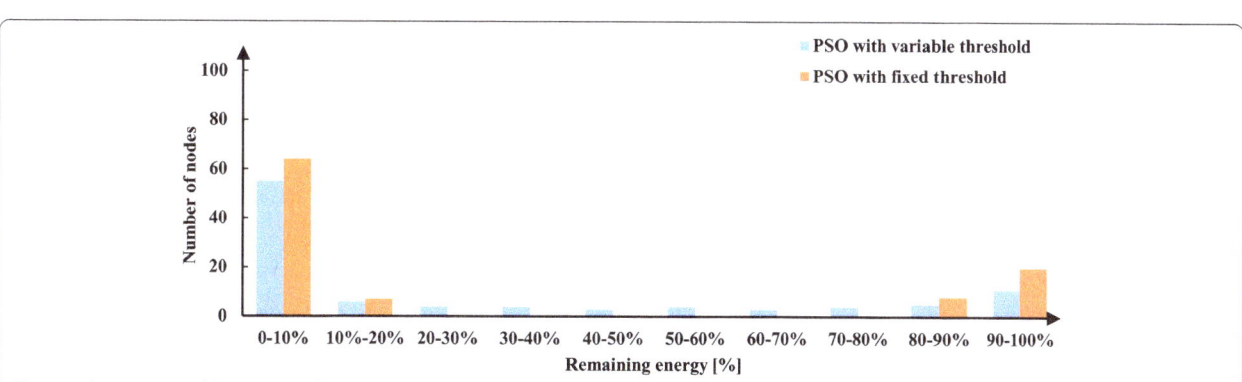

Fig. 18 Comparison of histogram of remaining energy distribution. This figure shows the histogram of the energy consumption of the network constructed by the proposed reconstruction method

shown that it is successful to reduce the network remaining energy and prolong the network lifetime by reconstruction the network compared to the case without reconstruction. In addition, the lifetime of the network constructed with PSO algorithm is about 1.9 times that of the traditional method, which proves the effectiveness of our proposed PSO scheme.

5.2.2 Network reconstruction with variable power threshold

The flow chart of the network reconstruction method with the variable power threshold is shown in Fig. 14. When the remaining rate of battery energy of any router drops below the threshold that is set to 80%, network reconstruction is performed. It is required that nodes whose remaining power is below the threshold cannot be set as a router. If network reconstruction becomes impossible in the current threshold, the threshold is reduced by 10%. In this case, the node with the remaining power higher than the threshold can be set as the router, and the node below the threshold can only be set as the end-device. In this way, the threshold decreases with 10% step size, and network reconstruction is carried out in a circular way. When the threshold becomes less than 10%, the network structure is maintained and the operation of the network is performed until the remaining energy of any node is below zero.

Figures 15 and 16 show the comparison of the network lifetime and the remaining energy using the proposed network reconstruction method with the variable power threshold. It can be seen that the proposed reconstruction method with the variable power threshold can improve the performance of reconstruction method with the fixed power threshold.

Figures 17 and 18 show the distribution of the energy consumption of the network constructed by the proposed reconstruction method. We note that the energy of the nodes near the coordinator is almost exhausted. This is because all the data are transmitted to the coordinator via the router directly connected to the coordinator, so the power consumption of the router closer to the coordinator is eliminated first.

In addition, we can observe that compared to the fixed threshold method, the number of nodes with remaining energy of 90% is decreased significantly using the variable threshold method. It is understood that the remaining power of the battery can be used uniformly for constructed network using the PSO algorithm. This is because by using the variable threshold method, the number of network reconstructions is increased, and the data transmission load can be prevented from being concentrated on the router and the power of all nodes can be uniformly used, thereby prolonging the network life time.

6 Conclusions

This paper presents new methods to construct the cluster tree topology for ZigBee WSNs. Firstly, based on the number of routers, the number of end-devices and the total number of hops, an evaluation function is constructed to reflect the energy consumption of the network, thus converting the problem of network topology construction into an optimization problem. Then, in order to make the optimization problem suitable for solving with the PSO algorithm, the network topology is transformed into particle individuals of the PSO algorithm. On these bases, the construction method of the cluster tree network topology based on the PSO algorithm is proposed. The simulation results show that the proposed method can construct a network topology with lower energy consumption. From simulation results, we also find that it is necessary to reconstruct the network to further prolong the network lifetime. Therefore, two network topology reconstruction methods with the PSO algorithm based on fixed and variable energy threshold are further proposed, respectively. The simulation results prove that the proposed network topology reconstruction methods can prolong the network lifetime effectively.

Abbreviations
PSO: Particle swarm optimization; WSNs: Wireless sensor networks

Acknowledgements
Authors would like to express the sincere thanks to the National Natural Science Foundation of Jiangsu province and China for their funding support to carry out this project.

Authors' contributions
YY and BX conceived the main idea and proposed the algorithm. ZYC performed the experiments and analyzed the simulation results. YY and ZWQ wrote the paper. All authors have read and approved the final manuscript.

Funding
This work is supported in part by the Natural Science Foundation of Jiangsu Province (Grant No.BK20160294) and National Natural Science Foundation of China (Grant No. 61601208).

References
1. J. Yick, B. Mukherjee, D. Ghosal, Wireless sensor network survey. Comput. Netw.. **52**(12), 2292–2330 (2008)
2. Z. Fei, B. Li, S. Yang, C. Xing, H. Chen, L. Hanzo, A survey of multi-objective optimization in wireless sensor networks: metrics, algorithms, and open problems. IEEE Commun. Surv. Tutorial. **19**(1), 550–586 (2017)
3. C. Kaiwen, A. Kumar, N. Xavier, S.K. Panda, in *2016 International Conference on Sustainable Energy Technologies (ICSET)*. An intelligent home appliance control-based on WSN for smart buildings (2016), pp. 282–287
4. T. Ojha, S. Misra, N.S. Raghuwanshi, Wireless sensor networks for agriculture: the state-of-the-art in practice and future challenges. Comput. Electron. Agric. **118**(3), 66–84 (2015)
5. P. Yi, A. Iwayemi, Z. Chi, Developing ZigBee deployment guideline under WiFi interference for smart grid applications. IEEE Trans. Smart Grid. **2**(1), 110–120 (2011)

6. B. Pandya, F.K. Chuang, C.H. Tseng, T.D. Chiueh, An energy-efficient communication system using joint beamforming in multi-hop health monitoring sensor networks. EURASIP J. Wirel. Commun. Netw. **2017**, 172 (2017)

7. S. Tummalapalli, M.V.D. Prasad, ZIGBEE operated FPGA based nodes in wireless industrial automation monitoring and control. Int. J. Eng. Trends Technol. **4**(5), 1569–1572 (2013)

8. P. Baronti, P. Pillai, V.W.C. Chook, S. Chessa, A. Gotta, Y.F. Hu, Wireless sensor networks: a survey on the state of the art and the 802.15.4 and ZigBee standards. Comput. Commun. **30**(7), 1655–1695 (2007)

9. G. Omojokun, A survey of zigbee wireless sensor network technology: topology, applications and challenges. Int. J. Comp. App. **130**(9), 47–55 (2015)

10. M. Ouadou, O. Zytoune, D. Aboutajdine, Y.E. Hillali, A. Menhaj-Rivenq, Improved cluster-tree topology adapted for indoor environement in ZigBee sensor network. Procedia Comput. Sci. **94**, 272–279 (2016)

11. P. Nayak, A. Devulapalli, A fuzzy logic-based clustering algorithm for WSN to extend the network lifetime. IEEE Sensors J. **16**(1), 137–144 (2015)

12. A.A. Aziz, Y.A. Sekercioglu, P.G. Fitzpatrick, M.V. Ivanovich, A survey on distributed topology control techniques for extending the lifetime of battery powered wireless sensor networks. IEEE Commun. Surv. Tutorial. **15**(1), 121–144 (2013)

13. J. Esch, A survey on topology control in wireless sensor networks: taxonomy, comparative study, and open issues. Proc. IEEE **101**(12), 2538–2557 (2013)

14. M. Ashouri, H. Yousefi, J. Basiri, A.M.A. Hemmatyar, A. Movaghar, PDC: prediction-based data-aware clustering in wireless sensor networks. J. Para. Distri. Comp. **2**(12), 24–36 (2015)

15. J. Szurley, A. Bertrand, M. Moonen, Distributed adaptive node specific signal estimation in heterogeneous and mixed topology wireless sensor networks. Signal Process. **7**(4), 44–61 (2015)

16. W. Heinzelman, A. Chandrakashan, H. Balakrishnan, An application-specific protocol architecture for wireless microsensor networks. IEEE Trans. Wirel. Commun. **1**(4), 660–670 (2002)

17. O. Younis, S. Fahmy, HEED: a hybrid, energy-efficient, distributed clustering approach for ad hoc sensor networks. IEEE Trans. Mob. Comput. **3**(4), 366–379 (2004)

18. Y. Zhang, S. Wang, G. Ji, A comprehensive survey on particle swarm optimization algorithm and its applications. Math. Probl. Eng. **2015**, 1),1–1)38 (2015)

19. J. Kennedy, R.C. Eberhart, in *Systems, man, and cybernetics conference. A discrete binary version of the particles swarm algorithm* (1997), pp. 4104–4108

20. W.R. Heinzelman, A.P. Chandrakasan, H. Balakrishnan, *33rd Annual Hawaii International Conference on System Sciences. Energy-efficient communication protocol for wireless sensor networks, (2006)* (2006), pp. 1–10

6

Realistic propagation effects on wireless sensor networks for landslide management

Nattakarn Shutimarrungson and Pongpisit Wuttidittachotti[*] ⓘ

Abstract

This paper presents the development of propagation models for wireless sensor networks for landslide management systems. Measurements of path loss in potential areas of landslide occurrence in Thailand were set up. The effect of the vegetation and mountain terrain in the particular area was therefore taken into account regarding the measured path loss. The measurement was carried out with short-range transmission/reception at 2400 MHz corresponding to IEEE 802.15.4 wireless sensor networks. The measurement setup was divided into two main cases, namely, the transmitting and receiving antennas installed on the ground and 1-m high above the ground. The measurement results are shown in this paper and used to develop propagation models suitable for operation of short-range wireless sensor networks of landslide management systems. The propagation model developed for the first case was achieved by fitting the averaged experimental data by the log-normal model plus the standard deviation. For the second case, the model was derived from the ray tracing theory. The mountain-side reflection path was added into the model which contained the reflection coefficient defined for the soil property. Furthermore, the resulting propagation models were employed in order to realistically evaluate the performance of wireless sensor networks via simulations which were conducted by using Castalia. In the simulations, the sensor nodes were placed as deterministic and random distributions within square simulated networks. The comparison between the results obtained from the deterministic and random distributions are discussed.

Keywords: Propagation models, Wireless sensor networks, Landslide management, Path loss measurement

1 Introduction

A landslide, which is a globally widespread and short-lived phenomenon, causes not only a number of human losses of life and injury but also extensive economic damage to private and public properties. The main factors of landslide occurrences are steep slope angles along with accumulated rainfall, moisture, and pore pressure saturation in the soil [1]. Thailand, located at the center of peninsular Southeast Asia and covered by a number of mountainous plateau areas, is one of the countries that most face rainfall-induced landslides every year [2]. In order to avoid or reduce the loss due to landslide disasters, there is a need for a landslide management system that can monitor and/or predict landslide occurrence.

A landslide management system is an essential key to reducing losses due to landslides by generating early warning for people living in potential landslide areas. In order to achieve an underlying system, sensors such as rain gauges, moisture sensors, piezometers, tiltmeters, geophones, and strain gauges can be installed in the potential landslide areas in order to collect the essential information needed to perform data analysis for landslide monitoring and prediction. Some examples of the use of sensors to monitor and/or predict landslides can be seen in [1, 3–6]. Besides sensor technologies, a communication network is also required for sending the information collected by sensors.

Wireless sensor networks have received considerable interest in the research area of landslide monitoring and prediction, as seen in examples [1, 5–7]. Another example was proposed to use a wireless sensor network in order to collect ambient data for general applications including landslide monitoring and prediction [8]. The performance of IEEE 802.15.4 commonly known as Zigbee, leading technology of short-range wireless sensor networks, was measured to verify that it can be used for

* Correspondence: pongpisit.w@it.kmutnb.ac.th
Faculty of Information Technology, King Mongkut's University of Technology North Bangkok, Bangkok, Thailand

a wide variety of applications [9]. In [10], the wireless sensor network was developed and then installed on the landslide area in Italy in order to monitor and manage the risks of landslides. Several parameters collected by sensors equipped with the coordinator of the wireless sensor network were employed to assess the possible risks and to provide useful information for an early warning system. Furthermore, an open-source wireless sensor system, called SMARTCONE, was designed and implemented for detection of the occurrence of the slope movement and debris flow of the hillside [11]. The performance of the proposed system was elevated via experimentation.

From the literature review, another interest is that the deployment of a wireless sensor network inside a potential landslide area requires knowledge of the node distance relevant to the path loss in order to provide full connectivity. An appropriate propagation model employed to predict such a path loss and the received-signal coverage is therefore essential for network planning. Extensive research has been conducted on propagation models, including theoretical and empirical models of wireless sensor networks. Some empirical models have been proposed in order to determine the path loss for a wide range of operating frequencies [12–18]. Although these proposed models are simple, there is no parameter that controls the relationship between the models and the forest environments. In [19], a half-space model for dealing with wave propagation at the frequency of 1–100 MHz in forest areas was proposed. In this approach, the associated phenomenon dominated by a lateral wave mode of propagation was also discussed. Subsequently, this approach was extended to the dissipative dielectric slab model in order to take the ground effect into account for the wave propagation at the frequency of 2–200 MHz in the forest environment [20]. Recently, near-ground wave propagation was examined in a tropical plantation as seen in the example [21], where the experiment was conducted at very high frequency (VHF) and ultra-high frequency (UHF) bands. In this approach, the ITU-R model was slightly modified by taking the lateral wave effect into account. Moreover, the ITU-R model was further improved with considering the effect of the rain attenuation that was measured in Malaysia [22].

As mentioned, although those approaches are simple and valid for wave propagation in forest areas, they do not consider the realistic effects due to the environment in the context of landslide areas, especially in Thailand. In this paper, the measurement of path loss in one of the potential landslide areas in Thailand was examined. The measurement results were then employed to develop appropriate propagation models for particular landslide-monitoring/prediction applications. Simulations with realistic propagation effects were conducted in order to evaluate the performance of applying the wireless sensor network to the landslide management systems.

The main contributions of this paper are summarized as follows.

1. The measurement of the path loss was set up in accordance with the practical situation of applying the wireless sensor network to the landslide management of Thailand, whose climate and terrain are unique. In the measurement, the transmitting and receiving antennas of the short-range wireless sensor network at an operating frequency of 2400 MHz were placed on the ground and at a 1-m height above the ground. This was done in order to investigate the effect of the antenna height on the propagation model.
2. The propagation models were then developed for the two cases, namely, the transmitting and receiving antennas installed on the ground and 1-m high above the ground. The first model was achieved by fitting the averaged experimental data by the developed model with the log-normal model plus the standard deviation. The second model was derived from the ray tracing theory. The mountain-side reflection path was added to the model which contained the reflection coefficient defined for the soil property.
3. The evaluation of the performance of the wireless sensor networks was presented via simulations, where the realistic wave propagation was taken into account.

Following this, the propagation prediction models are discussed along with the derivation of their equations in Section 2. These models are employed to compare with the model being proposed in this paper. The measurement and estimation of the path loss of the wireless sensor network are presented in Section 3. The simulation setup for evaluating the performance of the wireless sensor network for landslide management systems is discussed in Section 4, where their results and discussion are presented. Finally, the conclusions are drawn in Section 5. All results are discrbied by the data set of simulation and experimental results in Section 6 Additonal file 1.

2 Basic principles of propagation prediction models

In this section, we give an overview of the basic principles of propagation prediction models which were employed to compare with the proposed ones. There are several propagation models that have been employed to estimate path loss. One of the most simple and popular models is the free space loss as given by [23]:

$$\mathrm{PL}_{\mathrm{free}}(\mathrm{dB}) = -27.56 + 20\,\log_{10}(f) + 20\,\log_{10}(d) \tag{1}$$

where f and d, respectively, are the frequency in mega-hertz and the distance between the isotropic transmitting and receiving antennas in meters. This theoretical propagation model is practicable for operation in the far-field region when there are no obstacles in the first ellipsoid of the Fresnel zone.

In [12], Weissberger developed a new empirical model that can estimate excess attenuation due to vegetation as written in the following:

$$\mathrm{PL}_{\mathrm{weissberger}}(\mathrm{dB}) = \begin{cases} 0.45f^{0.284}d & d < 14\mathrm{m} \\ 1.33f^{0.284}d^{0.588} & 14\mathrm{m} < d \le 400\mathrm{m} \end{cases} \tag{2}$$

where f and d are in gigahertz and meters, respectively. The COST 235 models are proposed in [16] based on measurements conducted with a millimeter wave band between 9.6–57.6 GHz through a grove of trees. The model divides the propagation scenario into two different conditions as follows:

$$\mathrm{PL}_{\mathrm{COST}}(\mathrm{dB}) = \begin{cases} 15.6f^{-0.009}d^{0.26} & \text{in leaf} \\ 26.6f^{-0.2}d^{0.5} & \text{out of leaf} \end{cases} \tag{3}$$

where f and d are in megahertz and meters, respectively. In [15], the FITU-R model was developed based on the ITU-R recommendation [13]. The optimization method for the numerical parameters using the least squared error that fit several sets of measurement data at the frequency of 11.2 and 20 GHz was presented. The model was written as:

$$\mathrm{PL}_{\mathrm{FITU\text{-}R}}(\mathrm{dB}) = \begin{cases} 0.39f^{0.39}d^{0.25} & \text{in leaf} \\ 0.37f^{0.18}d^{0.59} & \text{out of leaf} \end{cases} \tag{4}$$

where f and d are in megahertz and meters, respectively. The FITU-R model was also modified for the VHF and UHF bands from the measurements in a palm plantation at the frequency of 240 MHz and 700 MHz. The modification takes the excess foliage loss with the lateral wave effect into account. This model is called a lateral ITU-R model [24], which can be used for long-range propagation in foliage areas and is defined by:

$$\mathrm{PL}_{\mathrm{LITU\text{-}R}}(\mathrm{dB}) = 0.48f^{0.43}d^{0.13} \tag{5}$$

where f and d are in megahertz and meters, respectively. This model is valid for in-leaf case.

In addition to the theoretical and empirical propagation models discussed above, a log-normal model, one of the most popular models based on the probabilistic distribution of the additional attenuation, has been widely used to predict path loss [25]. This model is defined by:

$$\mathrm{PL}_{\text{log-normal model}}(\mathrm{dB}) = \mathrm{PL}(d_0) + 10n\,\log_{10}\frac{d}{d_0} + X_\sigma \tag{6}$$

where n is the path loss exponent indicating the rate at which the signal attenuates with the distance. Generally, n is equal to 2 for free space. $\mathrm{PL}(d_0)$ is the path loss at a known reference distance d_0 in the far-field region. X_σ denotes a zero-mean Gaussian-distributed random variable (in dB) with standard deviation σ (in dB). Experimentally, it has been found that the path loss in cluttered multipath environments is log-normally distributed, involving shadowing effects.

In order to guarantee the suitability of the models, the root means square error (RMSE) [15] between the data obtained from the predicted models and measurement should be determined. The RMSE is a useful tool that can be used to measure the difference between the path loss predicted by a model and measured by the radio frequency (RF) equipment. The RMSE is given by:

$$\mathrm{RMSE} = \sqrt{\frac{\sum_{i=1}^{k}\left(X_{\mathrm{obs},i} - X_{\mathrm{model},i}\right)^2}{k}} \tag{7}$$

where X_{obs} and X_{model} are measured and predicted data, while k denotes the number of samples.

3 Measurement and propagation estimation

In this section, we discuss the setup of the measurement of the path loss. The measurement results are shown and then discussed as well. The propagation models suitable for the operation of wireless sensor networks of landslide management systems in a particular area of Thailand are introduced based on the existing models and our measurement results.

3.1 Measurement setup

In order to develop an appropriate propagation model for the short-range wireless sensor network of the landslide management systems, measurements were conducted in a potential landslide area on the small mountain of Nakhon Ratchasima province in Thailand during the rainy season, when the average monthly rainfall was 80 mm. The area chosen for the measurement was similar to a bald mountain where the terrain mainly consists of soil and sand. Figure 1 shows a basic block diagram of our measurement setup. The RF propagation

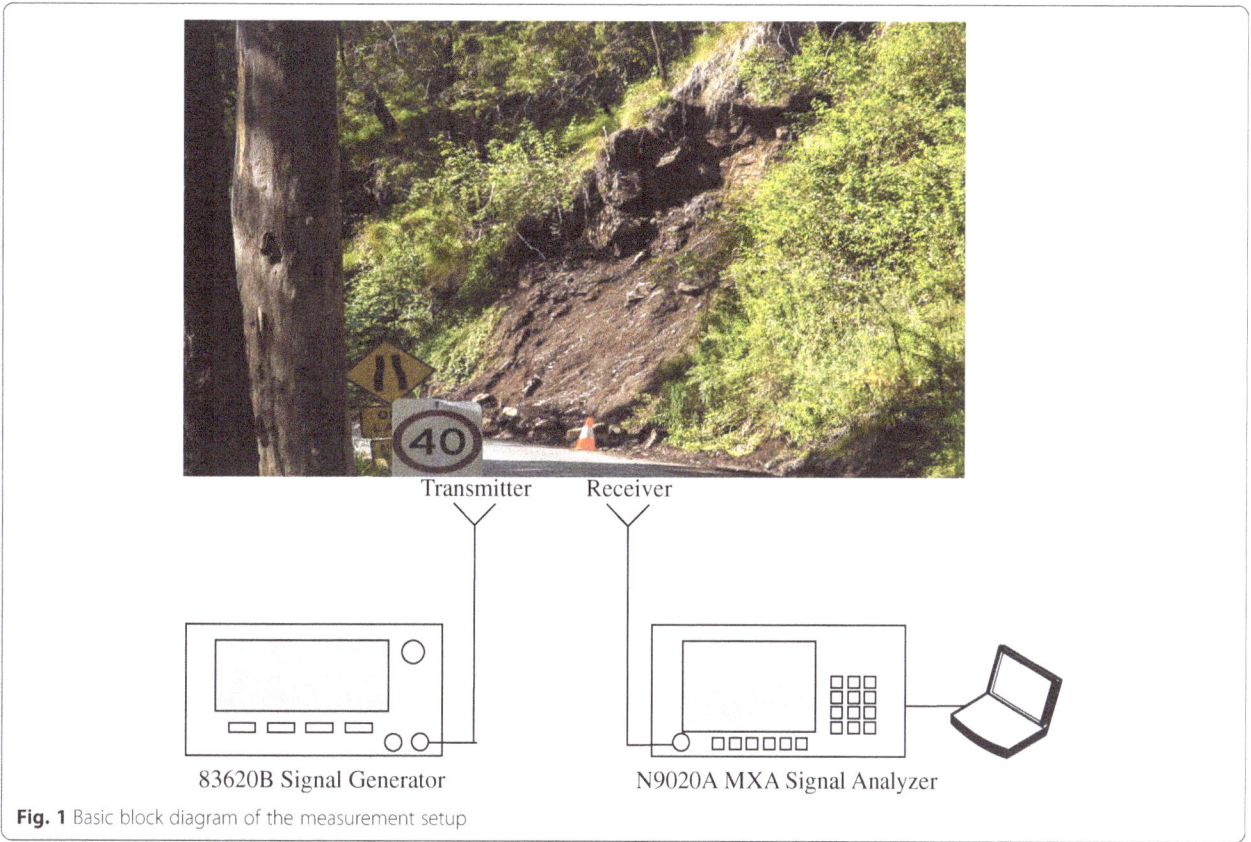

Transmitter   Receiver

83620B Signal Generator   N9020A MXA Signal Analyzer

Fig. 1 Basic block diagram of the measurement setup

measurements were performed at 2400 MHz by using RF equipment, including a signal generator (83620B Hewlett Packard), a spectrum analyzer (N9020A MXA Agilent technologies), and transmitting and receiving antennas. The continuous wave (CW) was generated by using a signal generator at 2400 MHz with the power of 17 dBm. The vertically polarized omnidirectional antennas with a typical gain of 5 dBi were employed for the measurement. The loss in the RF cable was 5.6 dB. The measurement data were captured by using a spectrum analyzer and stored into a control computer via GPIB interface for post-processing. In this paper, the measurement was divided into two different main cases. First, the transmitting and receiving antennas were placed on the ground since it is easy to place many small sensors, including transmitters and receivers, on ground in order to measure the data, such as seismic vibrations, average monthly rainfall, and relative humidity, which are used for landslide detection in the practical situation of wireless sensors networks. Second, the height of the receiving and transmitting antennas was 1 m above the ground. This was done in order to determine the effect of the antenna height on the propagation model. In some scenarios, a 1-m antenna tower can be probably installed in the landslide area. Figure 2 shows the RF-propagation measurements done in the potential

landslide area. In order to study the short-range wave propagation in such an area, the distance d between the transmitting and receiving antennas was varied from 0.5 to 50 m with a step of 0.5 m, resulting in 100 measured positions. In an individual position, the measurement was repeated 30 times in order to achieve accurate results. The system calibration of the measurement data was performed by the removal of the antenna gain and the cable loss of the transmitter and receiver. Note that our measurement was done horizontally on the mountain. The vertical direction was no longer under consideration because of the limitation of measurement setup. However, this problem will be resolved and discussed in a future publication.

3.2 Measurement results and developed propagation models

In order to evaluate the possibility of the use of the well-known propagation models, including COST235, FITU, LITU-R, Weissberger, log-normal, and free space models, to achieve the appropriate path-loss prediction in the landslide area, the measured path loss versus the distance for the first case of the antennas on the ground were plotted, together with its average and COST235, FITU, LITU-R, Weissberger, log-normal, and free space models, as shown in Fig. 3. In the figure, the ability to

Fig. 2 Measurement setup. **a** First case. **b** Second case

predict the path loss using COST235, FITU, LITU-R, Weissberger, and free space models becomes poor when the distance increased. These models underestimated the path loss significantly by up to 68 dB at 25 m under the measured data. On the other hand, the log-normal model, whose path loss exponent was initially deduced as $n = 2.4$, was more suitable to be used for curve fitting. The path loss exponent was then varied as $n = 1.5$, 1.6, 1.7, 1.8, and 1.9. Figure 4a shows the path loss of the log-normal model with varying the path loss exponent. The root mean square (RMS) error given in (7), between the measured data and path loss predicted by the log-normal model, was calculated in order to investigate the performance of the fitted curve. The RMS errors obtained from the log-normal models with $n = 1.5$, 1.6, 1.7, 1.8, and 1.9 were 3.16, 2.63, 3.02, 4.08, and 5.43, respectively. It can be seen that the path loss of the log-normal model with $n = 1.6$ was closest to that of the averaged

measurement data. In this paper, we developed the propagation model from our measurement results of wireless sensor networks, specifically for landslide management systems. Reconsidering the measurement results, the measured path loss at 1 m was 52.53 dB and was then chosen as the reference distance path loss $\text{PL}(d_0)$. Thus, the propagation model developed for the wireless sensor network for the landslide management systems was achieved on the basis of the log-normal model and our measurement data as given by:

$$\text{PL}_{\text{on ground}}(\text{dB}) = \text{PL}(d_0) + 10n \log_{10}d + \text{SD} \qquad (8)$$

where SD denotes the measured standard deviation of the received power around the average. Here, the standard deviation SD of this measurement case was 8.42. The path loss $\text{PL}(d_0)$ at a known reference distance $d_0 =$

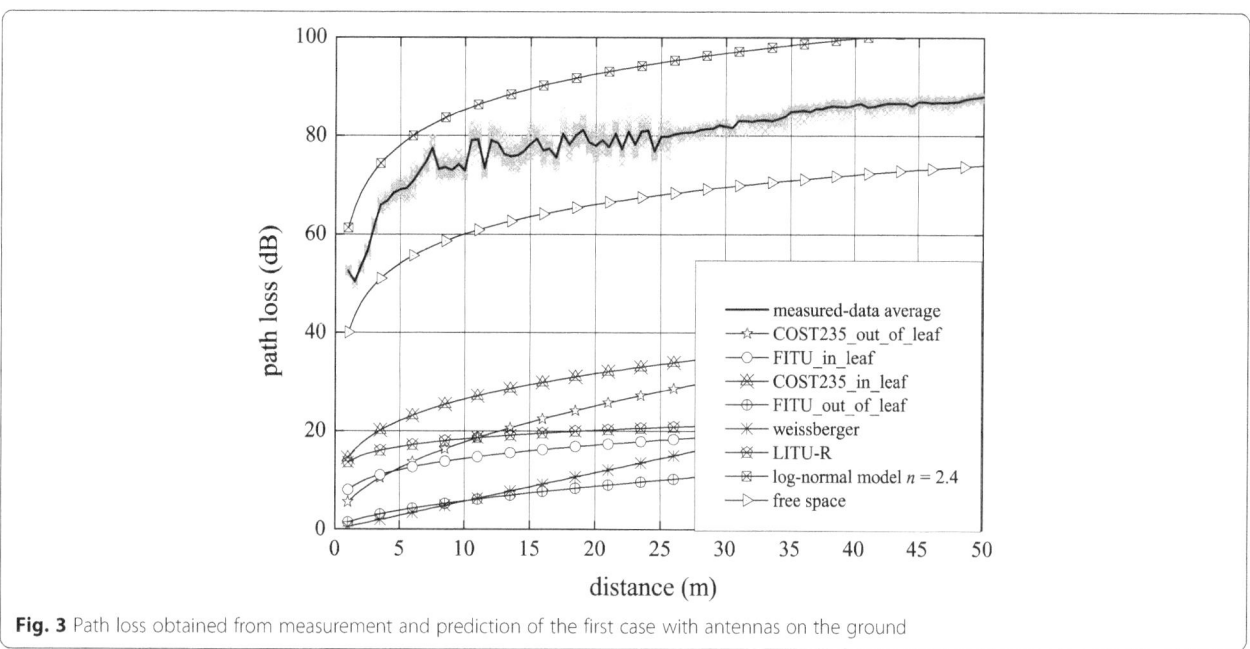

Fig. 3 Path loss obtained from measurement and prediction of the first case with antennas on the ground

1 m was 52.53 dB. Figure 4b shows the path loss of the developed propagation model with varying its path loss exponent as $n = 1.4$, 1.5, 1.6, 1.7, 1.8, 1.9, and 2. The RMS errors between the developed propagation model and averaged measurement data were 3.20, 3.07, 3.48, 4.30, 5.33, 6.47, and 7.68 when $n = 1.4$, 1.5, 1.6, 1.7, 1.8, 1.9, and 2, respectively. Note that the path loss of the developed propagation model with SD = 8.42 and $n = 1.5$ was almost identical to that of the predicted conventional log-normal model. This implies that we can use our developed propagation model instead of the conventional log-normal model for the wireless sensor network for the landslide management systems.

Figure 5 shows the predicted and measured path losses obtained from the second case of a 1-m antenna height above the ground. In the figure, the log-normal and free space models were employed to predict the path loss. It was seen that the log-normal models with different path loss exponents, i.e., $n = 1.8$, 2.0, and 2.4, overestimated the measured path loss significantly. The RMS errors of the use of the log-normal models with $n = 1.8$, 2.0, and 2.4 were 8.50, 11.42, and 17.63, respectively. On the other hand, the RMS error of the use of the free space model was 3.21. Although the use of this model to predict the measured path loss was more suitable than using the log-normal model, the RMS error was somewhat high as compared with the former case. Note that there were some significant fluctuations which appeared at the distance 21 to 26 m of the measured data. A propagation model specific for the wireless sensor network of the landslide management systems should be considered when employing 1-m-high antennas above the ground. In this paper, we introduced the

development of the propagation model by using a multiple-ray tracing model to fit the measured data.

In case of the 1-m-high antennas, we determined a basic geometry model for the measurement setup again, as depicted in Fig. 6. The transmitting and receiving antennas were placed on the mountain at heights of h_T and h_R, respectively. The d denotes the direct distance between transmitting and receiving antennas. In this paper, the propagation model developed by using the multi-ray tracing considers three major coexisting transmission paths, namely, line-of-sight (LOS), ground reflection, and mountain-side reflection paths. Note that the mountain-side reflection path was added to the developed model instead of the conventional two-ray ground-reflection model. The total received power was calculated from the combination of the individual received power from LOS, ground reflection, and mountain-side reflection paths. Based on the Friis transmission formula, we derived the equation of the received power from three-ray tracing including line-of-sight (LOS), ground reflection, and mountain-side reflection paths, as written by:

$$P_r = P_t \left(\frac{\lambda}{4\pi}\right)^2 \left(\frac{\sqrt{G_{\mathrm{LOS}}}}{d_{\mathrm{LOS}}} - \Gamma_{\mathrm{gr}} \frac{\sqrt{G_{\mathrm{gr}}}e^{-j2\pi\left(d_{\mathrm{gr}}-d_{\mathrm{LOS}}\right)/\lambda}}{d_{\mathrm{gr}}} \right.$$
$$\left. - \Gamma_m \frac{\sqrt{G_m}e^{-j2\pi(d_m-d_{\mathrm{LOS}})/\lambda}}{d_m}\right)^2$$

(9)

where P_r and P_t denote the received and transmitted powers, respectively. G_{LOS}, G_{gr}, and G_m are the

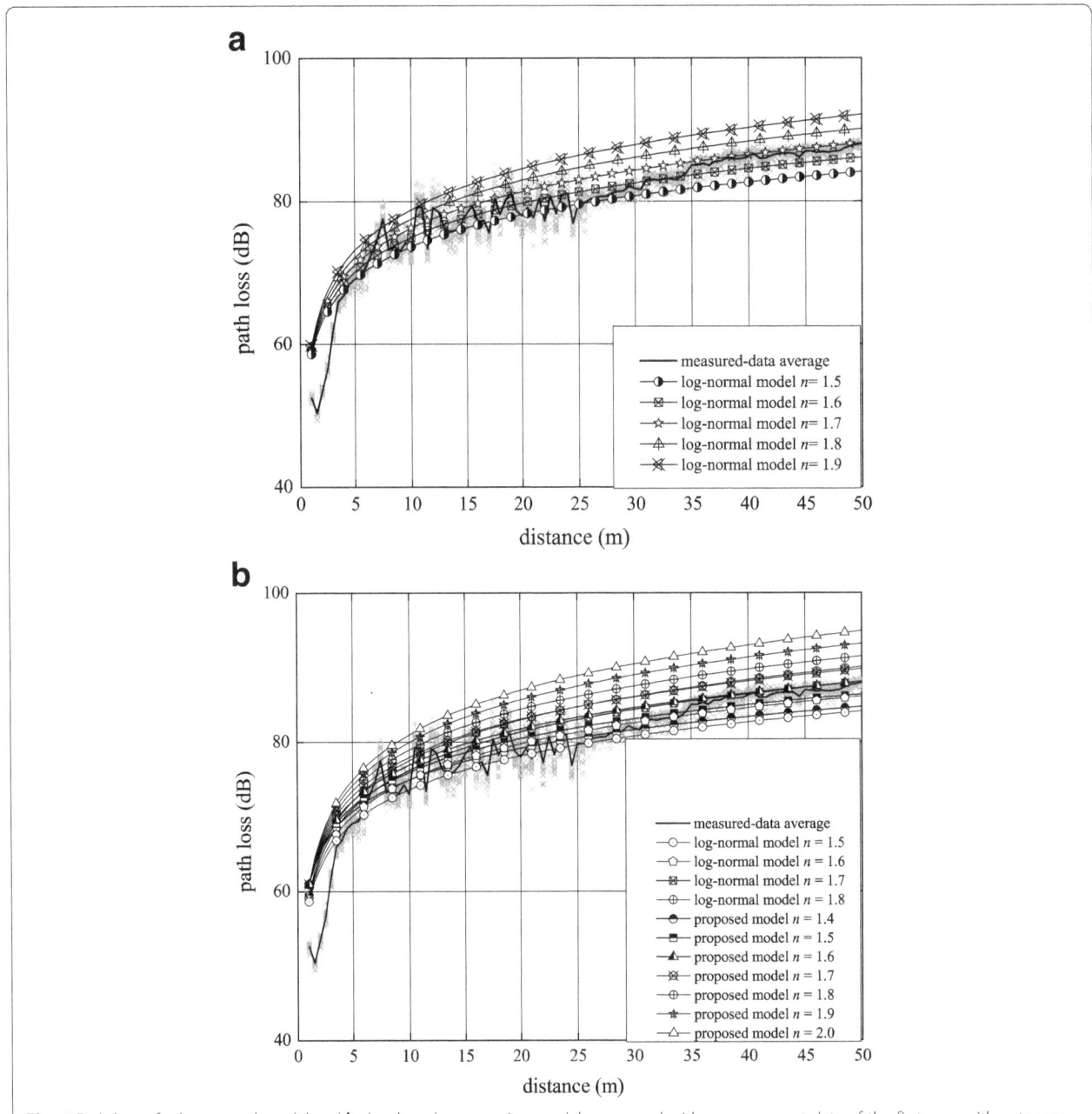

Fig. 4 Path loss of **a** log-normal model and **b** developed propagation model, compared with measurement data of the first case with antennas on the ground

combined antenna gains along the LOS, ground reflection, and mountain-side reflection paths. Since the pattern of all antennas employed in our measurement was omnidirectional, the G_{LOS}, G_{gr}, and G_m were therefore equal. The Γ_{gr} and Γ_m were reflection coefficients of the ground and mountain side, respectively. Since the measurement setup was conducted on a mountain, the Γ_{gr} and Γ_m were equal as well, which were re-denoted as Γ. Practically, the reflection coefficient Γ can be achieved by measurement

of the soil property. The use of the reflection coefficient for path loss prediction is very useful when there is a need to change the considered landslide area having different soil properties. The distances of the wave travel from the transmitting antenna to the receiving antennas along the LOS, ground reflection, and mountain-side reflection paths were denoted as d_{LOS}, d_{gr}, and d_m, respectively. Applying trigonometry to this problem, the distances d_{LOS}, d_{gr}, and d_m can be given by:

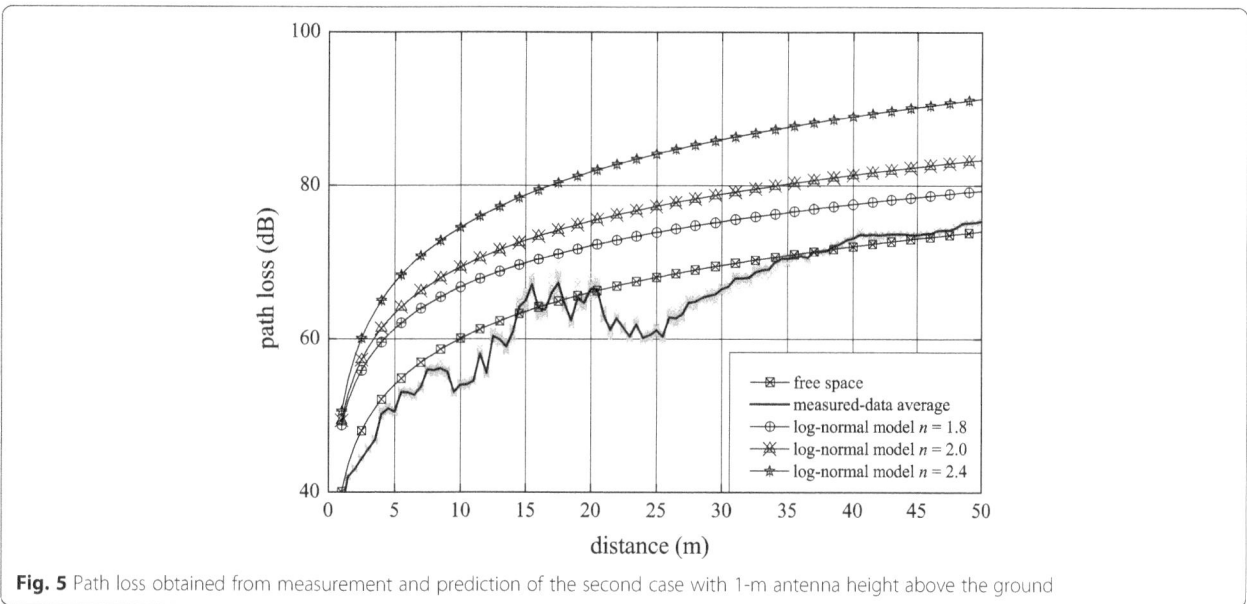

Fig. 5 Path loss obtained from measurement and prediction of the second case with 1-m antenna height above the ground

$$d_{\mathrm{LOS}} = \sqrt{(h_T - h_R)^2 + d^2} \qquad (10)$$

$$d_{\mathrm{gr}} = \sqrt{(h_T + h_R)^2 + d^2} \qquad (11)$$

and

$$d_m = \sqrt{(h_T \cot\theta + h_R \cot\theta)^2 + (d/2)^2} \qquad (12)$$

These equations indicate that the distances d_{LOS}, d_{gr}, and d_m depended upon the antenna height, mountain slope θ, and direct distance d between the transmitting and receiving antennas.

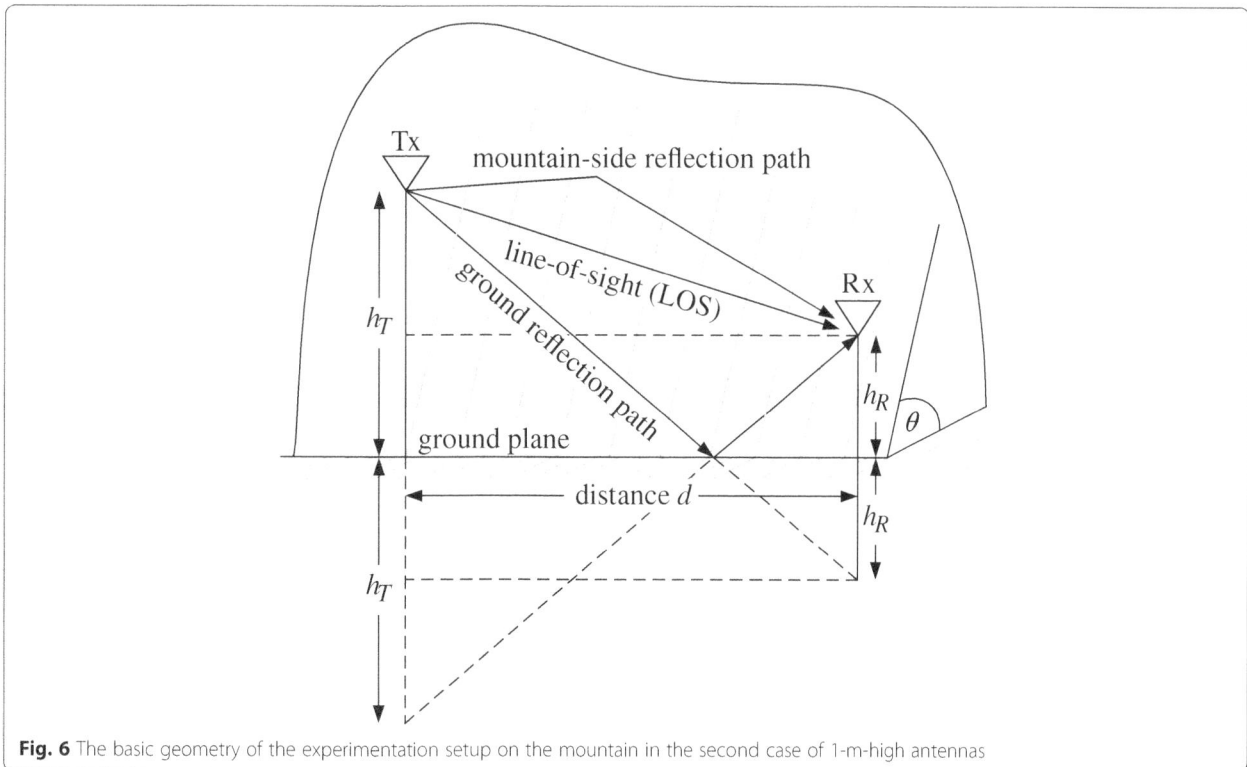

Fig. 6 The basic geometry of the experimentation setup on the mountain in the second case of 1-m-high antennas

To simplify (9) and to fit the experimental data with the path loss $P_L(d_0)$ at a known reference distance d_0, the total path loss can be rewritten as

$$PL_{\text{above ground}}(\text{dB}) = PL(d_0)$$
$$+20 \log\left[\frac{1}{d_{\text{LOS}}} - \Gamma \frac{e^{-j2\pi\left(d_{\text{gr}}-d_{\text{LOS}}\right)/\lambda}}{d_{\text{gr}}} - \Gamma \frac{e^{-j2\pi(d_m-d_{\text{LOS}})/\lambda}}{d_m}\right]$$
$$(13)$$

Here, the path loss $P_L(d_0)$ at a known reference distance $d_0 = 1$ m is 37.54 dB. Since the short distances d_{LOS}, d_{gr}, and d_m were not approximately equal because the determined distance was short.

Figure 7 shows the path loss of the developed propagation model based on the multi-ray tracing, compared with that of the free space model. Here, the slope of the mountain was set as $\theta = 30°$. The reflection coefficient Γ, which mainly depends upon the soil property, was varied in order to investigate the appropriation of the developed propagation model. $\Gamma = -1$ indicates that the soil reflects all of the transmitted power to the receiving antennas. The RMS errors obtained from the developed propagation model were 3.19, 4.48, 2.94, and 3.22 when $\Gamma = 0$, -0.4, -0.8, and -1, respectively. The developed propagation model based the multi-ray tracing and our measurement data achieved the smallest RMS error when $\Gamma = -0.8$. Moreover, it should be noted that the path loss obtained from the developed propagation model when $\Gamma = 0$ which means the transmitted power was completely absorbed by the soil, was almost identical to that of the free space model. This indicates that there was no reflection from the ground and

mountain-side paths. There exists only the LOS path in the path loss of the predicted results. This also reveals that our developed propagation model can be applied to other mountains that possess different soil properties by choosing the appropriate reflection coefficient Γ.

4 Practical simulations for wireless sensor networks

4.1 Simulation scenario

Simulations were conducted using Castalia, an extension of OMNET++, in order to evaluate the performance of the wireless sensor networks for landslide management systems. In our simulations, the measurement results presented in the previous section were employed instead of the default radio model. The realistic propagation effects on the wireless sensor networks were therefore taken into account. The IEEE 802.15.4 standard was chosen as Physical (PHY) and Medium Access Control (MAC) layers with an operating frequency of 2400 MHz, corresponding to that of our measurement setup. The routing protocol deployed for the wireless sensor networks of the landslide management systems was an Ad hoc On-Demand Distance Vector (AODV). The simulations were run at a fixed duration of 500 s with a packet generated at a constant packet rate of 0.5 packets per second. The wireless channel bit rate was 250 kbps. All sensor nodes placed in the simulations were configured to set their transmission power at 57.42 mW and their sensitivity value at -100 dBm based on the Chipcon CC2420 transceiver chip. Omnidirectional antennas with a typical gain of 5 dBi were employed for signal transmission and reception.

The simulations were set up corresponding to the practical situation in the potential landslide area of

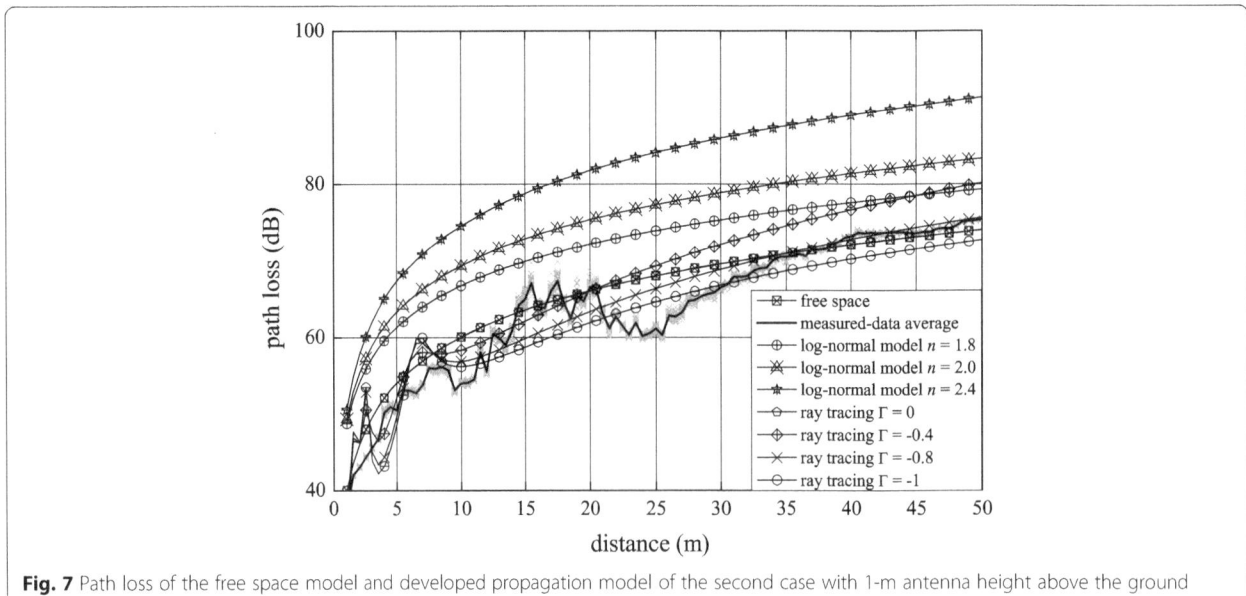

Fig. 7 Path loss of the free space model and developed propagation model of the second case with 1-m antenna height above the ground

Thailand. Simulations were distinguished into two main catalogs in accordance with the measurement cases. First, the path loss obtained from the measurement with the antennas placed on the ground was employed as a radio model in the simulations in order to investigate the realistic propagation effects due to the antenna position and other environmental parameters. Second, we employed the measurement results of the case of 1-m-high antennas above the ground in order to demonstrate the effect of the antenna height on the performance of the wireless sensor networks.

Figure 8 shows the network simulation models. In each catalog of simulations, the node positions of the simulated networks were deterministically and randomly distributed

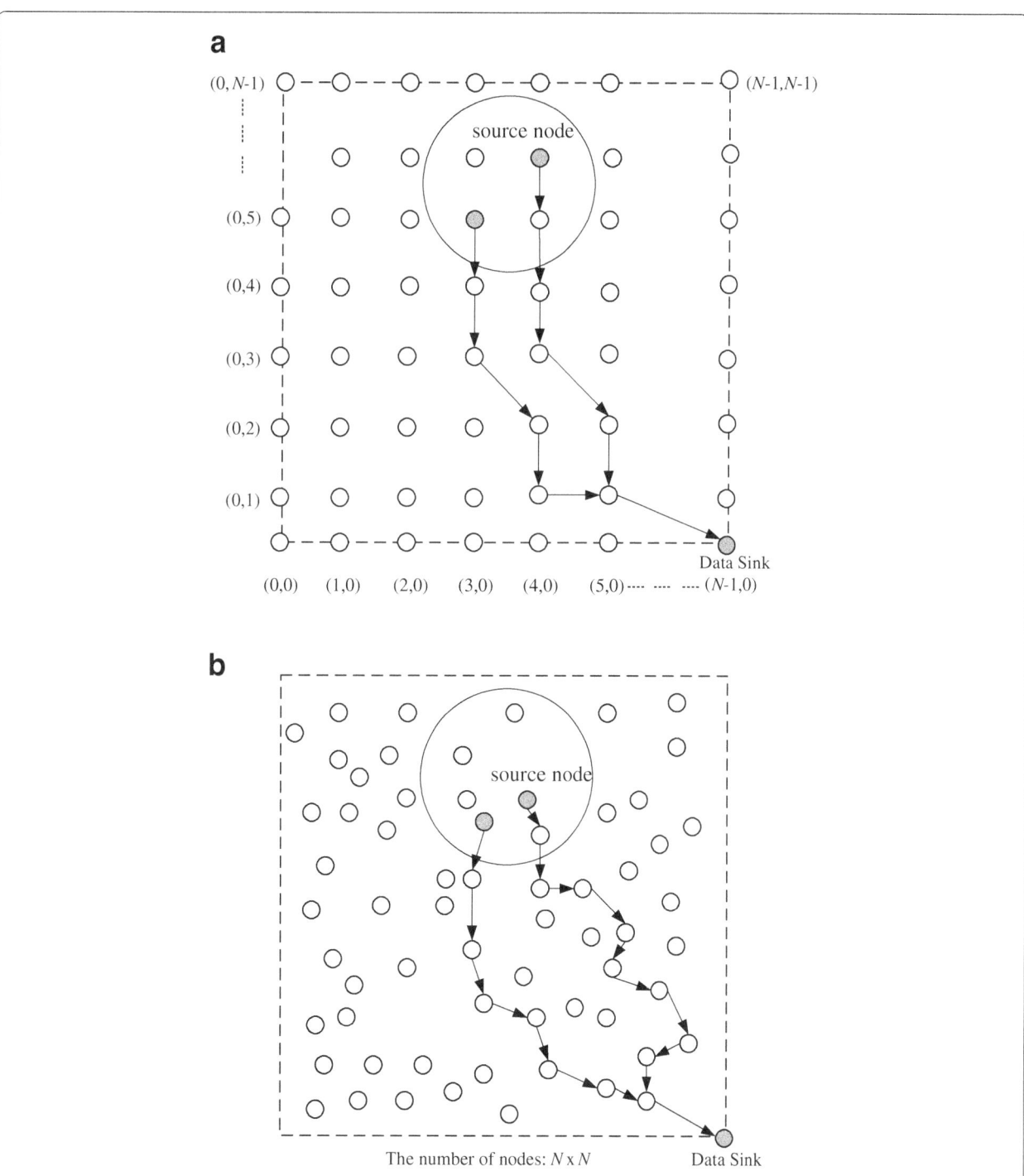

Fig. 8 Network simulation models with **a** deterministic and **b** random distributions

within a 300 m × 300 m square as shown in Fig. 8a and b, respectively. In practical situations, sensors installed to predict a landslide occurrence are probably deployed either randomly or deterministically depending on the selection of the user. Thus, the deterministic and random distribution of the node placement should be determined. This simulation consisted of $N \times N$ sensor nodes. The distance between two adjacent sensor nodes in our simulation model depended upon the number of placed sensor nodes. All of the sensor nodes were stationary, and the sink was located at the bottom right of the network area

as shown in the figure. It should be noted that the sink can send information gathered by the sensor nodes to the gateway node, which may be installed outside the potential landslide area. The gathered data received from the gateway node through communication networks such as Internet or mobile networks can be used to calculate/predict the landslide occurrence at the monitoring center.

4.2 Results

In order to evaluate the performance of the wireless sensor networks for the landslide management systems, the

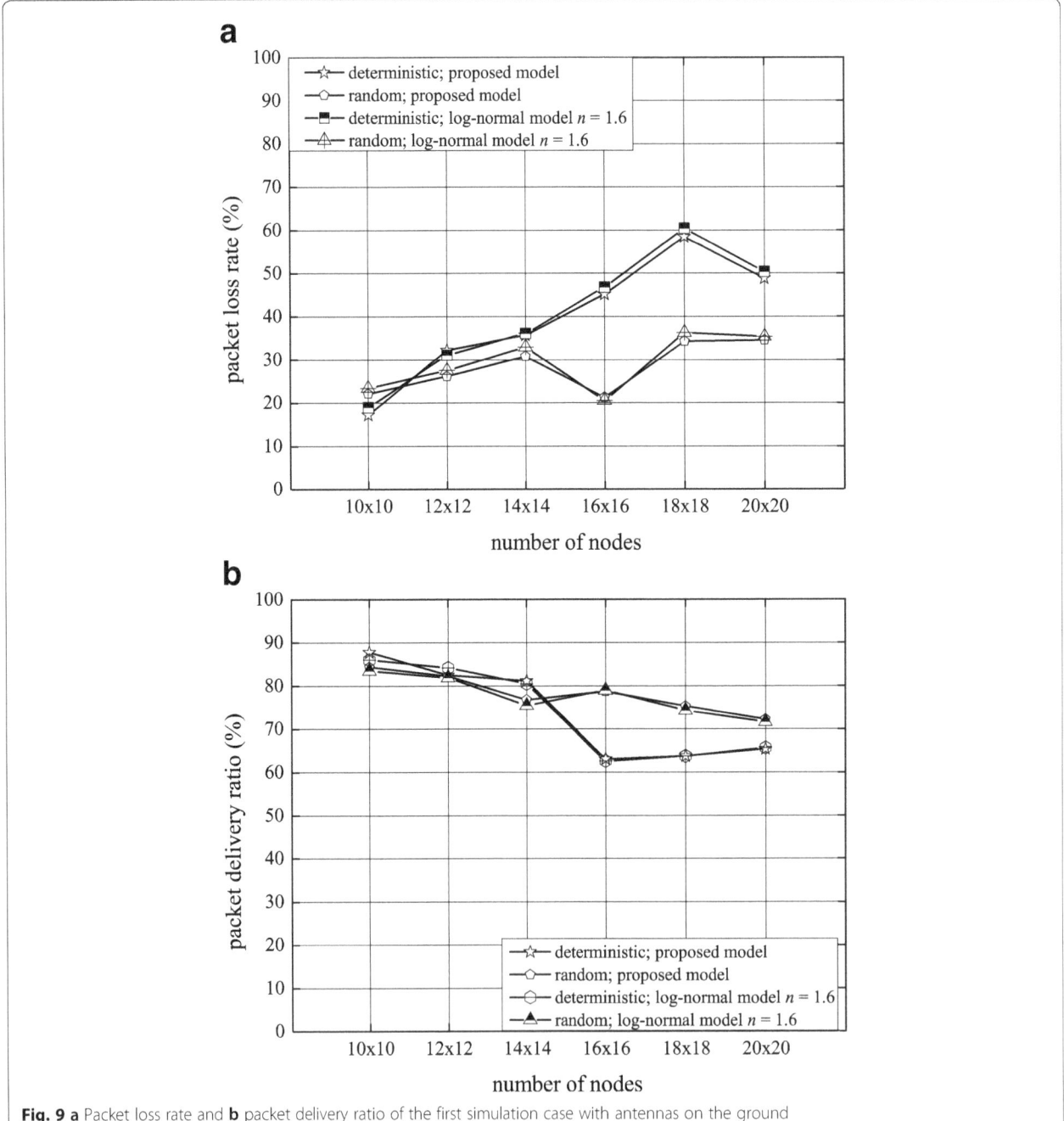

Fig. 9 a Packet loss rate and **b** packet delivery ratio of the first simulation case with antennas on the ground

matrices, that are the packet loss rate and packet delivery ratio, are determined in this section. The number of sensor nodes was varied from 10 × 10 to 20 × 20. Figure 9 shows the packet loss rate and packet delivery ratio versus the number of sensor nodes when the antennas were placed on the ground. It was seen that the packet loss rate increased proportionally to the number of nodes. The collision probably occurs when the density of nodes is high. The packet loss rate slightly decreased at $N = 20 \times 20$ since the probability of success of finding a communication path was high. The packet delivery ratio

decreased when the number of nodes increased. Although the packet loss rate suddenly decreased when $N = 16 \times 16$, compared with that when $N = 15 \times 15$, the packet delivery ratio still was high. In the case of the deterministic node distribution, the distance between the two neighbor nodes of $N = 16 \times 16$ and $N = 14 \times 14$ was 23 and 20 m, respectively. The packet loss rate of the case of the random node distribution was less than 40% for all node numbers under consideration while that of the deterministic node distribution was greater than 40% for $N = 16 \times 16$ to $N = 20 \times 20$. On the other hand, the

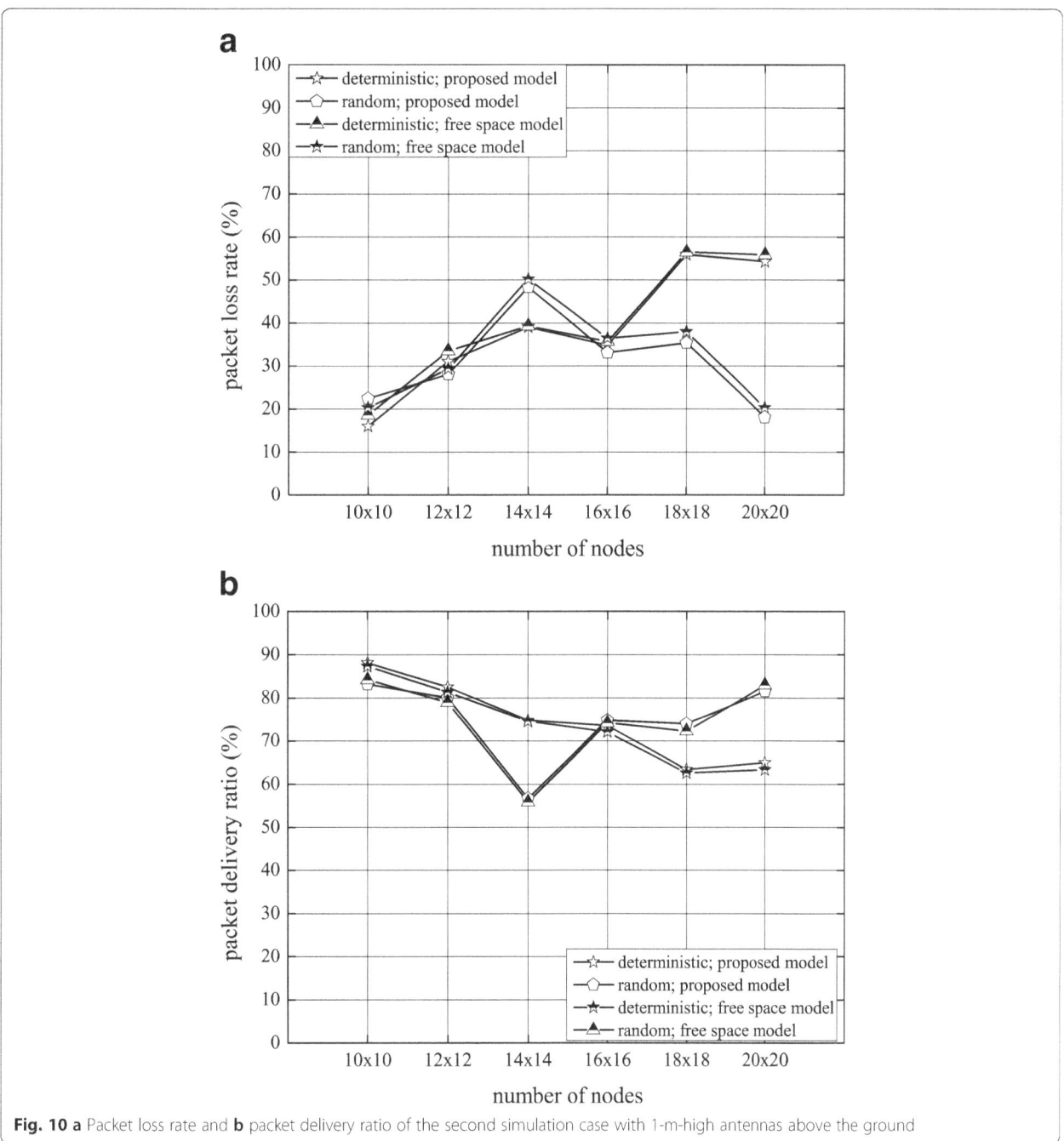

Fig. 10 a Packet loss rate and **b** packet delivery ratio of the second simulation case with 1-m-high antennas above the ground

packet delivery ratio of the random node distribution was greater than 70% of all node numbers under consideration while that of the deterministic node distribution was less than 70% for $N = 16 \times 16$ to $N = 20 \times 20$. Note that simulation results obtained from the use of the proposed model and the log-normal model were almost identical.

Figure 10 shows the packet loss rate and packet delivery ratio versus the number of sensor nodes when the antennas were placed at a 1-m height above the ground. The packet loss rate and packet delivery ratio of this case were similar to those when the antenna was placed on the ground. The random node distribution had a slightly better packet loss rate and packet delivery ratio than the deterministic node distribution. The simulation results obtained from the use of the proposed model and free space model were almost identical as well.

5 Conclusion

In this paper, the development of propagation models has been proposed for wireless sensor networks for landslide management systems. The propagation models were developed based on the measurement data and existing propagation models. For the development, the measurement was set up using two main cases—transmitting and receiving antennas installed on the ground and 1-m high above the ground. The propagation models for the first and second scenarios were developed based on the log-normal model and multi-ray tracing models along with our experimental data, respectively. The path loss versus distance was shown to validate that the developed propagation models were suitable for the operation of landslide management systems using a wireless sensor network. Furthermore, the resulting propagation models were employed in order to realistically evaluate the performance of wireless sensor networks via simulations, which were conducted using Castalia. In the simulations, the sensor nodes were placed as deterministic and random distributions. The simulation results have been shown to confirm that the short-range wireless sensor network at an operating frequency of 2400 MHz can be employed for landslide management systems.

Abbreviations

AODV: Ad hoc on Demand Distance Vector; FITU: Fitted ITU; IEEE: The Institute of Electrical and Electronics Engineers; ITU-R: The International Telecommunication Union-Radiocommunications Sector; LITU-R: Lateral ITU; MAC: Media access control; PHY: Physical layer; RF: Radio frequency; RMSE: Root mean square error; UHF: Ultra high frequency; VHF: Very high frequency

Funding

This research project was financial supported by National Research Council of Thailand through the 2018 Graduate Research Scholarships.

Authors' contributions

NS is the Ph.D. candidate who performed all the work in this paper. She is the main writer of this paper. PW is the main research supervisor of NS who helped her in fine-tuning the proposed scheme. All authors proposed the main idea, read and approved the final manuscript.

Authors' information

Nattakarn Shutimarrungson received the M.Sc. degree in Information Technology Management from Mahidol University, Bangkok, Thailand in 2004. She is currently a Ph.D. candidate in information technology, Department of Information Technology, at Faculty of Information Technology, King Mongkut's University of Technology North Bangkok (KMUTNB), Thailand. Her research interests include wireless sensor networks and propagation models.

Pongpisit Wuttidittachotti is currently an associate professor and head of Department of Data Communication and Networking, Faculty of Information Technology at the Faculty of Information Technology, King Mongkut's University of Technology North Bangkok (KMUTNB), Thailand. He received his Master of Science in Information Technology from KMUTNB, in 2003. He obtained a scholarship to study in France and then received a Master of Research and Ph.D. in Networks, Telecommunications, Systems and Architectures from INPT-ENSEEIHT, in 2005 and 2009 respectively. Also, he was awarded a Postdoctoral scholarship from University of Paris XI in 2009. His research interests include eHealth/mHealth, MANET, information security, 3G/4G/5G, networks performance evaluation, VoIP quality measurement and QoE/QoS.

Competing interests

The authors declare that they have no competing interests.

References

1. M.V. Ramesh, V.P. Rangan, Data reduction and energy sustenance in multisensor networks for landslide monitoring. IEEE Sensors J. **14**(5), 1555–1563 (2014)
2. O. HJ, S. Lee, W. Chotikasathien, C.H. Kim, J.H. Kwon, Predictive landslide susceptibility mapping using spatial information in the Pechabun area of Thailand. Environ Geol **57**(3), 641 (2009)
3. S. Biansoongnern, B. Plungkang, S. Susuk, Development of low cost vibration sensor network for early warning system of landslides. Energy Procedia **89**, 417–420 (2016)
4. M.V. Ramesh, in *IEEE, SENSORCOMM'09. Third International Conference on Sensor Technologies and Applications, 405-409*. Real-time wireless sensor network for landslide detection (2009)
5. G.R. Teja, V.K.R. Harish, D.N.M. Khan, R.B. Krishna, R. Singh, S. Chaudhary, in *2014 IEEE International Advance Computing Conference (IACC)*. Land slide detection and monitoring system using wireless sensor networks (wsn) (2014), pp. 149–154
6. A. Terzis, A. Anandarajah, K. Moore, I. Wang, in *Proceedings of the 5th ACM International Conference on Information Processing in Sensor Networks*. Slip surface localization in wireless sensor networks for landslide prediction (2006), pp. 109–116
7. Y. Wang, Z. Liu, D. Wang, Y. Li, J. Yan, Anomaly detection and visual perception for landslide monitoring based on a heterogeneous sensor network. IEEE Sensors J. **17**(13), 4248–4257 (2017)
8. F. Wang, J. Liu, L. Sun, Ambient data collection with wireless sensor networks. EURASIP J. Wirel. Commun. Netw. (2010). https://doi.org/10.1155/2010/698951
9. P.R. Casey, K.E. Tepe, N. Kar, Design and implementation of a testbed for IEEE 802.15.4 (Zigbee) performance measurements. EURASIP J Wirel Commun Netw **23** (2010). https://doi.org/10.1155/2010/103406
10. A. Giorgetti, M. Lucchi, E. Tavelli, M. Barla, G. Gigli, N. Casagli, M. Chiani, D. Dardari, A robust wireless sensor network for landslide risk analysis: system design, deployment, and field testing. IEEE Sensors J. **16**(16), 6374–6386 (2016)
11. H.C. Lee, K.H. Ke, Y.M. Fang, B.J. Lee, T.C. Chan, Open-source wireless sensor system for long-term monitoring of slope movement. IEEE Trans Instrumentation and Measurement **66**(4), 767–776 (2017)
12. M.A. Weissberger, *An initial critical summary of models for predicting the attenuation of radio waves by trees* (Electromagnetic compatibility analysis center, Annapolis, MD, 1981) ECAC-TR-81-101

13. CCIR, Influences of terrain irregularities and vegetation on troposphere propagation. Geneva, Switzerland, pp. 235–236. CCIR Rep. (1986)
14. A. Seville, K.H. Craig, Semi-empirical model for millimetre-wave vegetation attenuation rates. Electron. Lett. **31**(17), 1507–1508 (1995)
15. M.O. Al-Nuaimi, R.B.L. Stephens, Measurements and prediction model optimization for signal attenuation in vegetation media at centimetre wave frequencies. IEE Proceedings-Microwaves, Antennas and Propagation. **145**(3), 201–206 (1998)
16. COST 235, Radio Propagation Effects on Next Generation Fixed-Service Terrestrial Telecommunication Systems. Luxembourg. Final Rep. (1996)
17. H.Y. Chen, Y.Y. Kuo, Calculation of radio loss in forest environments by an empirical formula. Microw. Opt. Technol. Lett. **31**(6), 474–480 (2001)
18. J. Liang, Q. Liang, S.W. Samn, A propagation environment modeling in foliage. EURASIP J. Wirel. Commun. Netw. (2010). https://doi.org/10.1155/2010/873070
19. T. Tamir, On radio-wave propagation in forest environments. IEEE Trans. Antennas Propag. **15**(6), 806–817 (1967)
20. T. Tamir, Radio wave propagation along mixed paths in forest environments. IEEE Trans. Antennas Propag. **25**(4), 471–477 (1977)
21. D. TR Rao, N.T. Balachander, M.V.S.N. Prasad, Ultra-high frequency near-ground short-range propagation measurements in forest and plantation environments for wireless sensor networks. IET Wireless Sensor Systems **3**(1), 80–84 (2013)
22. R.M.D. Islam, Y.A. Abdulrahman, T.A. Rahman, An improved ITU-R rain attenuation prediction model over terrestrial microwave links in tropical region. EURASIP J Wirel Commun Netw **189** (2012)
23. J.D. Parsons, *The mobile radio propagation channel* (Wiley, 2000)
24. Y.S. Meng, Y.H. Lee, B.C. Ng, Empirical near ground path loss modeling in a forest at VHF and UHF bands. IEEE Trans. Antennas Propag. **57**(5), 1461–1468 (2009)
25. T.S. Rappaport, *Wireless Communications: Principles and Practice (Vol. 2)* (New Jersey, prentice hall PTR, 1996)

7

Market segmentation of wireless sensor system in network commodity selection

Fuxiang Liu

Abstract

Online product recommendation and market segmentation are directly related to the consumer experience and the healthy development of the e-commerce market. In order to improve the selection of network products and the effect of market segmentation, this study introduced a wireless sensing system based on the recommendation of traditional network products and combined the operational conditions of the network commodity market to construct a personalized recommendation e-commerce system. Meanwhile, a series of representative WSN congestion control strategies are designed by dynamically adjusting the network topology, routing algorithm, channel allocation protocol, and channel resource utilization, and the system structural components and functions are implemented. The research shows that the proposed algorithm and system have certain practical effects and can provide theoretical reference for subsequent related research.

Keywords: Wireless sensing system, Network commodity, Application, Market segmentation, Commodity selection

1 Introduction

Similar to shopping on e-commerce sites, many organizations are now launching online product recommendation services. Faced with a large number of online products, how to choose the most suitable product recommendation service for users is the key research direction of e-commerce development. In the e-commerce website, the user's interest and preferences are predicted by collecting and analyzing the user's purchase record, browsing history, and other related information, thereby recommending the products that he may be interested in [1]. Similarly, the user-specific intelligent push system proposed in this paper will focus on how to intelligently push suitable online commodity services for users. In this system, the user profile data collected in the Internet of Things and some other personal information of the user will be combined with the idea of data mining to analyze the health services and commodities required by the target users, and finally push the recommendation results to the user browsing interface.

In the e-commerce industry, providing products according to customer needs and enabling customers to easily find the products they want to buy is the key to gaining competitive advantage. The practice of foreign e-commerce enterprises shows that category management is a very effective management method for e-commerce enterprises to gain competitive advantage and increase market share. Although category management has only been born for 20 years, the effectiveness of category management has attracted the interest of many domestic and foreign scholars. Through continuous practice and research, many representative viewpoints have been proposed. Fader and Lodish [2] found that consumer characteristics such as the frequency of purchase have a certain explanatory power for the pricing and promotion environment of supermarket goods. Hoch [3] and others studied the relationship between the market price elasticity of various commodities and the demographic characteristics of customer groups. Raiu [4] studied the difference in sales of different types of goods and established its relationship with category characteristics and marketing groups and variables. The results of Poel [5] and others show that the complementarity between commodities is directly proportional to the interaction of promotions. These studies strongly support the rational identification of commodity classification theories based on consumer demand. Compared with European and American countries, China's research is later. In 1997, the Hong Kong Supply Chain Management Advisory Committee was established, and the

Correspondence: liufms@ctgu.edu.cn
Science College and Three Gorges Mathematical Research Center, China Three Gorges University, Yichang 443002, China

category management experiment was carried out. Then, the Taiwan area was actively promoted. The category management in the mainland is promoted by the China ECR (Efficient Consumer Response) committee established in 2001. Zhang Hongxia and Zhang Songjie [6] first analyzed the significance, basic steps, and key factors of product category management in category management in 1999, and the application prospects of category management are promising. He Yun [7] discussed the optimization management of category, and he believes that the purpose of the optimization management of the real category is to make the shelf most valuable, and the product portfolio placed on the shelf is the customer's favorite product. Xia Weichao [8] deeply analyzed the relationship between satisfying customer needs and supplier management and believed that category management should be based on category accounting, which is designed to meet consumer needs and manage suppliers' new management methods. Wu Zikai [9] emphasized the importance of category tactics.

On the whole, the focus of the above research is different. It provides a reference for the research of the category management model from various angles, which is very helpful for the next step. In the enterprise management category management process, because there is no corresponding category management software system to help enterprises to manage the process, system, and actualization, the enterprise lacks effective information technology support, which leads to the time-consuming and unsatisfactory decision-making process of the category management decision-making process. The category management model system is a business information management system. From the perspective of developed countries, the business information management system is divided into six stages: automatic settlement management of cash registers, sales analysis management of POS machines, inventory management, storage and transportation management, integrated management, data warehousing, and data mining [10]. In other words, the computer information management system can be roughly divided into three levels: one is the POS system, the second is the point-of-sale sales management system, and the third is the decision support system [11]. At present, domestic supermarket enterprises have realized the coverage of POS machines and can monitor the sales situation, inventory situation, and storage and transportation situation of goods in real time, and can monitor and dynamically manage the quantity, location, and price of goods. That is to say, the computer management system of most supermarket retail enterprises still stays in the collection of basic data. However, with the improvement of marketing management level, the current point-to-point sales service of the system can no longer meet the needs of supermarket

managers. Therefore, managers want the system to automatically analyze sales data, mine customer preferences, and provide decision support. However, in practice, enterprises lack confidence in information systems, and they believe that information systems cannot solve the practical problems of category management. As a result, companies are struggling to deploy a category management system. On the other hand, it can be seen in the enterprises that have already been implemented that based on the information management software currently used, most enterprises can collect basic internal data required for category management. However, the system has not been used in the analysis of data or does not have the ability to analyze data and report data required by category management [12].

In summary, current network products require a category management software that meets the actual characteristics. The software provides systematic and scientific services that improve the relationship between suppliers and retailers, improve customer satisfaction, reduce costs, and increase profits.

2 Classical congestion algorithm of wireless sensor networks

WSN is a wireless network with no data center, its nodes form a network through white organization and adaptive, and there is no control center. In order to improve the quality of network services and prolong the service life of the network, based on the above-mentioned causes of congestion caused by WSN, researchers around the world have designed a series of representative WSN congestion control strategies by dynamically adjusting network topology, routing algorithms, and introducing channel allocation protocols to improve channel resource utilization [13].

2.1 CODA

Congestion detection avoidance (CODA) also known as congestion prediction and avoidance algorithm, is a classic WSN congestion control scheme. The program is mainly divided into three major operational mechanisms (Fig. 1) [14].

The CODA algorithm timely realizes the prediction of network congestion, and achieves network congestion mitigation, avoidance, and cancellation by reducing the data transmission rate of the source node for a long time. However, during the running process, the upstream congestion node performs the local congestion control policy by using the AIMD policy (discarding the cached data packet). Moreover, it adjusts the transmission rate of the multi-source node that sends the data packet to the receiving end through back pressure adjustment, and the rate is different, which is not conducive to maintaining the WSN quality of service (QOS

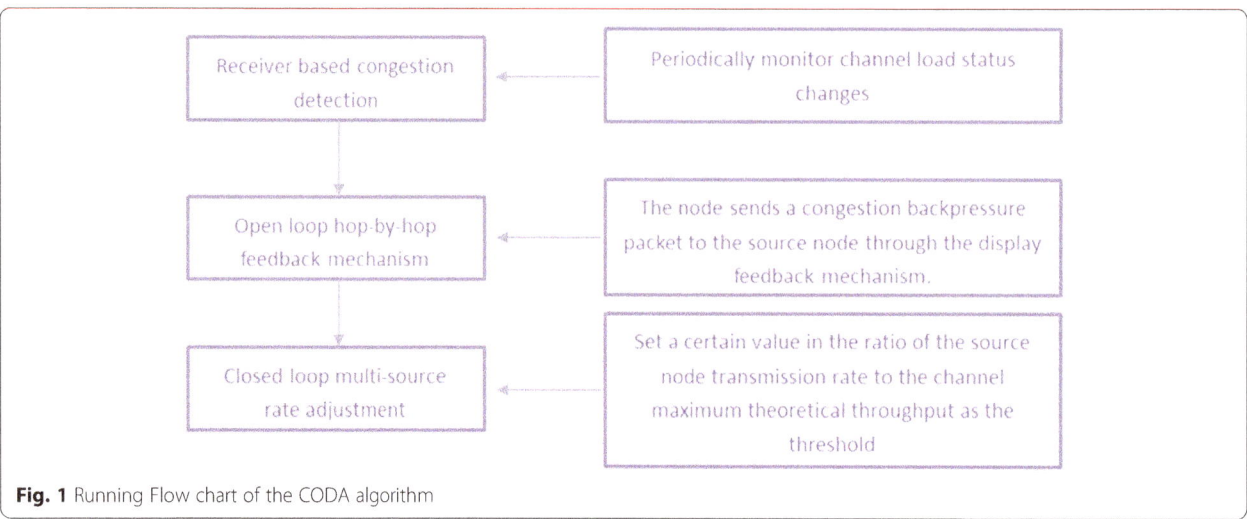

Fig. 1 Running Flow chart of the CODA algorithm

performance) and fairness, and the channel sensing also consumes a lot of energy [15].

2.2 ESRT

Event-to-sink reliable transport (ESRT) is a reliable event transmission protocols. The congestion control strategy of ESRT realizes the reliability of network data transmission and the effectiveness of congestion control by changing and transforming the network state. The sensor nodes and the aggregation nodes in the network directly transmit data. The ESRT acts on the SINK node to obtain the network congestion detection by adjusting the network transmission status to obtain the source node transmission rate, and the number of packets received by the aggregation node in each period determines the reliability of the network transmission. The ESRT congestion control policy operation relies on the following five network states: Reliability states $R = $ (LR, HR), network congestion $C = (NC, C)$

$$S = R^T \otimes C$$

network states $S_i \in \{(\text{LR}, \text{NC}), (\text{LR}, \text{C}), (\text{HR}, \text{NC}), (\text{HR}, \text{C}), \text{OOR}\}$, LR—low reliability, HR—high reliability, NC—no congestion, C—congestion, and OOR—best state. The ESRT strategy can guarantee the reliability of the network and the quality of network services. However, the SINK node communicates directly with the source node and does not support multi-hop clustering routing and topology mechanisms and is not suitable for large wireless sensor networks [16].

2.3 Congestion control strategy of adaptive resource

When detecting that the network is in a congested state, waking up the dormant node (increase network resources), changing the network topology, and forming a multi-emission route can alleviate the many-to-one data

transmission mode, and achieve network congestion cancellation through multi-channel communication. After the congestion is released, the node without data transmission continues to sleep (reducing network resources), and the multi-route and multi-channel communication modes are released. The algorithm improves energy utilization, extends network lifetime, and ensures network stability and OOS performance.

2.4 DCCP-data packet congestion control protocol

The DCCP congestion control protocol establishes, maintains, and disassembles the reliable handshake negotiation mechanism at both ends and users can adopt different congestion control mechanisms according to different applications. Different congestion control policies are described by the congestion control flag (CCID), and CCID2 implements congestion mitigation and release by adopting a cache queue date packet of a drop node similar to the AIMD protocol. CCID3 implements congestion control by adjusting the transmission rate of source node data, which is similar to the TFRC protocol. However, the DCCP protocol cannot guarantee the high reliability of the network. The high bit error rate leads to an increase in network data packet loss rate and reduces the throughput of the DCCP congestion control protocol.

2.5 An algorithm for avoiding node congestion in line sensor networks

The strategy adopted by the algorithm is to allocate the congestion avoidance of the sending window to the source node and to set the priority for the data packet scheduling. The upstream node allocates a transmission window for the downstream node according to a certain policy, and the node sets the priority for the data packet in the cache queue, and the data packet with the highest priority is always transmitted in the effective transmission window.

Although the algorithm effectively achieves congestion avoidance and improves energy utilization and improves the fairness of network transmission and reduces network transmission delay, it still lacks efficient congestion detection mechanism and time synchronization algorithm for end-to-end data transmission.

2.6 COMUTl is a cluster-based data traffic congestion control algorithm

During the operation of the COMUT, the node detects its own data transmission traffic and reports it to the cluster head node within a certain time interval. The cluster head node determines whether the local traffic is greater than the threshold according to the received data. If it is greater than the threshold, the AIMD mechanism is used to adjust the transmission rate of the upstream node to implement congestion control and release. This algorithm affects the stability of network throughput and improves data transmission delay [17].

Based on the specific application scenarios, these algorithms establish an efficient WSN congestion control strategy system for congestion detection, congestion avoidance, and congestion cancellation.

(1) Nodes close to the SINK node area are used as relay nodes for network-wide data transmission and become "hot spots" for power consumption and congestion. (2) Due to the limited bandwidth of nodes and the low level of end-to-end communication links, network congestion may occur in some cluster head nodes or common nodes.

3 Congestion control strategy based on CCBDC-RSUL algorithm

This paper is a wireless sensor network model based on clustering topology. In the case of a sudden increase in data volume, it is easy to cause two phenomena of in-cluster and cluster head congestion and cluster head congestion in SINK area. For the problem of congestion in the intra-cluster, a sub-cluster head VCH (Vice Cluster Head) is selected in the non-cluster head node CM to share the data stream for the cluster head node CH, and the congestion in the intra-cluster can be implemented. For the problem of inter-cluster congestion (SINK area congestion problem), the network area is non-uniformly layered, and the dormant nodes near the congested nodes are awakened and become cluster heads. Then, the two major strategies of the LEMPT optimized communication path implement inter-cluster congestion cancellation.

The network determines whether it is in a congested state by using periodic congestion detection. If congestion is not detected in the period, it enters congestion detection in the next network period. If a congestion trend is detected, the network determines whether the congestion is intra-cluster or inter-cluster by the size of the P. When $P \in (0, 0.5)$ is detected, the network is in an intra-cluster congestion state, and the clustering-based data traffic congestion control algorithm (COMUT) algorithm is used in the intra-cluster to obtain the sub-cluster head (VCH). Moreover, the nodes in the intra-cluster choose which cluster head to join according to the minimum value of the communication link energy consumption and update the node route to clustered. After that, it is judged whether there is congestion of inter-cluster. If there is, it will enter the inter-cluster congestion control link, otherwise it will detect whether the congestion is released. If it is judged that the network is in a state of inter-cluster congestion, the network is first non-uniformly layered, and a LEMPT tree-like optimization path is established between the clusters and the cluster head. Next, it is checked whether congestion has been alleviated or released. If the congestion is not alleviated, the node (CM) that is adjacent to the congestion cluster head (CH) and has the highest energy is awaked and become the cluster head, and the LEMPT is updated. After that, it is detected whether the congestion is released. If it is not released, it enters the congestion control loop again, and if congestion has been removed, the congestion control resource is deleted [18].

According to the congestion detection of the network by combining the queue length of the node buffer and the congestion degree, it can be seen that when $B_{p1} < B_p \leq B_{p2} \cap (C(n) > 1)$ or $B_p > B_{p2}$, the network CH node will be congested, and the congestion control link needs to be entered. The congestion type is determined according to the ratio P from the intra-cluster data traffic to the inter-cluster data traffic. When the data traffic from the cluster is much larger than the data traffic from the cluster, it is determined that the congestion belongs to intra-cluster congestion; when $0 < P < 0.5$, the data traffic from the inter-cluster is much larger than the data traffic from the inter-cluster, and the congestion is determined to belong to intra-cluster congestion. When $0.5 \leq P \leq 1.5$, the network is judged to be intra-cluster and inter-cluster congestion. In both cases, the network must first enter the congestion control phase in the intra-cluster.

We divide the nodes of the whole network into important, minor and general three levels according to their priorities. The areas where the node timers randomly select the set values are sequentially increased. The division of each area is shown in equation (1) [19]:

$$T_i \in \begin{cases} (0.8, 1]t, i \in A \\ (1, 1.4]t, i \in B \\ (1.4, 2]t, i \in C \end{cases} \tag{1}$$

In the formula, t represents the average value set by the previous time, A is a set of important nodes, B is a set of secondary nodes, and C is a set of general nodes.

According to the principle of formula (1), the set value is randomly selected for each node timer, and if the node data times out in the node cache queue, the node becomes a VCH.

In the clustering process of nodes, the coordinates of CH are (x_{ch}, y_{ch}), the coordinates of VCH are (x_{vc}, y_{vc}), and the coordinates of CM are (x_i, y_i). The principle of adding the corresponding cluster heads and clustering is shown in Eqs. (2), (3), and (4).

$$d_{i\text{-ch}} = \sqrt{(x_i - x_{ch})^2 + (y_i - y_{ch})^2} \quad (2)$$
$$d_{i\text{-vch}} = \sqrt{(x_i - x_{vc})^2 + (y_i - y_{vc})^2} \quad (3)$$

$$U_i = \begin{cases} U_{ch}, d_{i\text{-ch}} < d_{i\text{-vch}} \\ U_{vch}, d_{i\text{-vch}} < d_{i\text{-ch}} \end{cases} \quad (4)$$

In formula (4), U_{ch} denotes a cluster set with CH as a cluster head, U denotes a cluster set with VCH as a cluster head, and U_i denotes a node element within the cluster. $d_{i\text{-ch}}$ and $d_{i\text{-vch}}$ respectively represent the distance between the nodes CM to CH and VCH. The two cluster regions are segmented as shown in Fig. 2 [20].

First, the distances $d_{i\text{-ch}}$ and $d_{i\text{-vch}}$ of the nodes i to CH and VCH in the cluster are respectively determined, and the sizes of the two are determined. Which of the distance values is small, then node i is added to the cluster head. If the distance between the two is equal, the remaining energy of the two cluster heads is judged, and the node i is added to the cluster head with the larger remaining energy.

When $(B_{p1} < B_p \leq B_{p2}) \cap (C(n) > 1) \cap (P > 1.5)$ or $B_p > B_{p2} \cap (P > 1.5)$, network congestion belongs to inter-cluster congestion, and the network directly enters the inter-cluster congestion control phase.

When $(B_{p1} < B_p \leq B_{p2}) \cap (C(n) > 1) \cap (0.5 \leq P \leq 1.5)$ or $B_p > B_{p2} \cap (0.5 \leq P \leq 1.5)$, network congestion belongs to cluster-to-cluster congestion.

After the network releases, the congestion within the cluster through intra-cluster congestion control, the network can enter the cluster congestion control phase (congestion mitigation and congestion cancellation).

Our work introduces the ULEE algorithm to achieve non-uniform stratification of the network, aiming to solve the "hot spot" problem near the SINK node area. Li Chengfa and others proposed the EEUC agreement, whose main ideas are: Since multi-hop communication between WSN clusters is used to realize data transmission between the cluster head and the SINK node and the nodes in the network are equivalent, the closer to the nodes in the SINK node area, the more data relay pressure is assumed. Furthermore, it is completed by introducing two major parameters of C and R_{cmax}.

For the non-uniform layered design of the network, the closer to the SINK node, the smaller the layer, the smaller the cluster size, and the fewer the number of nodes in the intra-cluster. By alleviating the load of data acquisition, processing, and transmission of nodes in the area, more energy and transmission resources are used for data relay transmission of the whole network, and congestion reduction is implemented by using UL (uneven layer). Formula (5) is

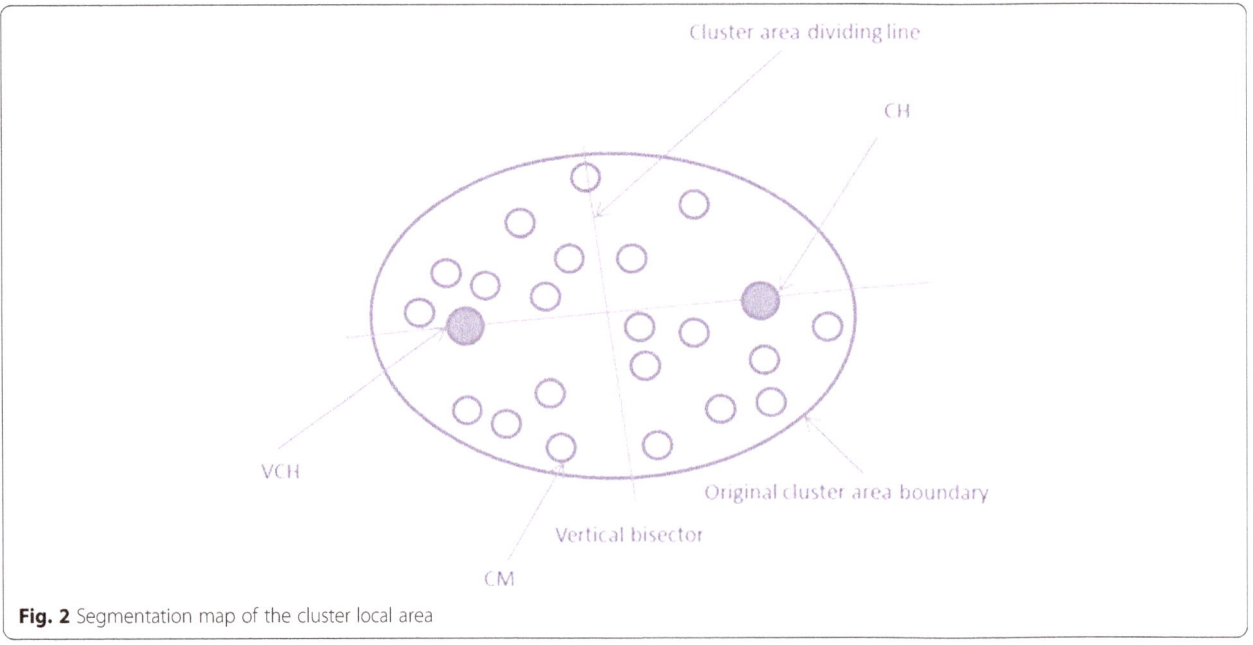

Fig. 2 Segmentation map of the cluster local area

the formula for calculating the competition radius (boundary) of network nodes.

$$R_r = \left\{1 - c \frac{d_{\max} - d(s_i, \text{BS})}{d_{\max} - d_{\min}}\right\} R_{c\,\max} \tag{5}$$

In formula (5), g is the node competition radius, $c = d_{\min}/d_{\max}$, $c \in (0, 1)$ and c is the degree of non-uniformity. That is, the smaller the value of c, the greater the difference between the two. $d_{\max}$ and $d_{\min}$ are the maximum and minimum values of the nodes to the BS in the WSN, respectively. The distance from node s_i to the base station is $d(s_i, \text{BS})$, and $R_{c\max}$ is the maximum competition radius of the node set by the network. The above algorithm lacks the theoretical derivation of each hierarchical boundary (node competition radius). Therefore, a theoretical derivation method for obtaining the node competition radius by the energy consumption of each layer node is proposed. The energy consumption of each layer node is shown in Eq. (6).

$$\begin{cases} E_n = (\pi \times r_n^2 \times \rho) \times \left[E_{\text{elec}} \times k + \varepsilon_{fs} \times k \times (r_n + r_{n-1})^2\right] \\ E_{n-1} = (\pi \times r_{n-1}^2 \times \rho) + \left[\pi \times r_n^2 \times \rho \times \left(\frac{L}{2 \times r_n}\right) \times \left(\frac{2 \times r_{n-1}}{L}\right)^2\right] \\ \quad \times \left[E_{\text{elec}} \times k + \varepsilon_{fs} \times k \times (r_{n-1} + r_{n-2})^2\right] \\ \quad \cdots \\ E_1 = \left\{\pi \times r_1^2 \times \rho + \left[\sum_{i=2}^n \pi \times r_i^2 \times \rho \times \left(\frac{L}{2 \times r_i}\right) \times \left(\frac{2 \times r_i}{L}\right)\right]\right\} \\ \quad \times \left[E_{\text{elec}} \times k + \varepsilon_{fs} \times k \times \overline{R}^2\right] \end{cases} \tag{6}$$

In Eq. (6), E_i is the total energy consumption of the ith layer, k is the data packet size sent by each node, and $\overline{R}$ is the average distance from the first layer to the sink node (base station). When $E_n \approx E_{n-1} \approx \cdots \approx E_2 \approx E_1$, the competition radius of each layer node can be approximated, thus dividing the network. The closer the network layer is to the base station, the smaller the area of the layer, that is, the fewer the number of nodes, the less the local energy consumption, and communication resources are occupied, so that more energy and communication resources are used for relay transmission. After the network layering ends, the cluster head creates a LEMPT optimized path that passes through multiple hops k to all cluster heads in the network.

Topology-Aware Resource Adaptation to Alleviate Congestion in Sensor Networks (TARA) is shown in Fig. 3

In the low-energy routing design, by controlling the minimum distance $r_{\min}$ between the cluster head nodes, the cluster head nodes can be evenly distributed in the network with the greatest possibility, and the variance of the cluster size is minimized. Therefore, the hot spot phenomenon in which cluster head nodes are too concentrated in Fig. 3a is unlikely to occur in the WSN network designed in this paper. Figure 3d indicates that the cross-hotspots with overlapping traffic are unlikely to occur in more obvious application occasions of various areas. Figure 3c shows the cross-hotspot situation of traffic surge, which is mainly caused by the fact that the number of child nodes (neighbor nodes) of some nodes is large due to multi-hop within the cluster. Since the node bandwidth is narrow and the number of available channels with small interference is small, the amount of cached data of the node is large. It is highly likely to cause loss and congestion caused by packet overflow and packet wait timeout, which is explained and resolved. Figure 3b indicates a hot spot situation in the base station (sink node) area. The main reason for this phenomenon is that the network cluster head node transmits data to the base station based on multi-hop multi-path transmission. The nodes near the base station have to bear the energy consumption and communication resources (channel allocation, occupation) pressure of data acquisition, processing, and transmission of nodes within the cluster. At the same time, it also bears the energy consumption of the relay transmission of almost all network data between other cluster heads and the base station and the load of the communication resources.

Figures 4 and 5 respectively show a schematic diagram of a communication model of a four-layer network and a two-hop inter-cluster and a communication diagram of a four-layer network and four-hop inter-cluster. The cluster head labeled by the triangle in the figure is the wake-up node. Some cluster head nodes need to accept and output more nodes' data. These nodes are consumed too much, and the link quality is degraded due to communication interference, which is likely to cause hot spot-congestion. To solve this problem, the base station becomes a cluster head by waking up the hot spot neighbor sleeping node, and the base station creates and updates the LEMPT path, and the newly added cluster head will be awakened and share part of the data traffic of the congested cluster head, which can largely alleviate or even cancel the network congestion.

4 Network commodity selection system based on wireless sensor system

On the server, the system deploys and adopts B/S three-tier architecture design, including data layer, functional layer, and presentation layer. Through layered design, each layer only needs to complete the functions of this layer and complete the functions provided for other layers, and the layers are independent of each other.

Figure 6 is a general structural diagram of the personalized commodity intelligent push system described by this paper.

4.1 Presentation layer

It provides an interface for users to interact with the system. Users can register, log in, browse services,

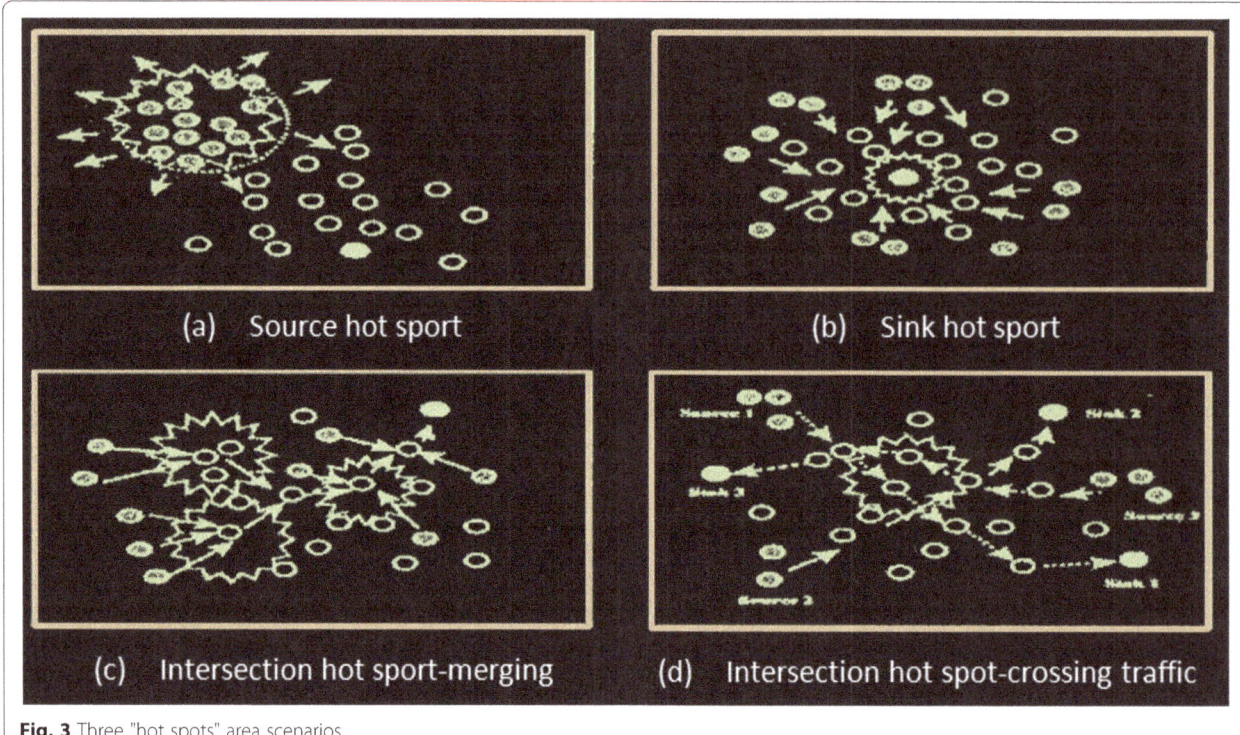

Fig. 3 Three "hot spots" area scenarios

product information, view personal feature information, view push services, and product information on the browser page.

4.2 Functional layer

According to the user's personal information and collected feature data, the data mining related algorithm is used to process and recommend appropriate services and products to the user. After the user logs in to the system, the calculated recommendation result is pushed to the interface browsed by the user.

4.3 Data layer

The system data layer mainly stores data information in the system. The data of the system includes (1) data related to the user: user name, login password, personal information filled in, user personal health characteristic data collected by the wireless sensor network, and health

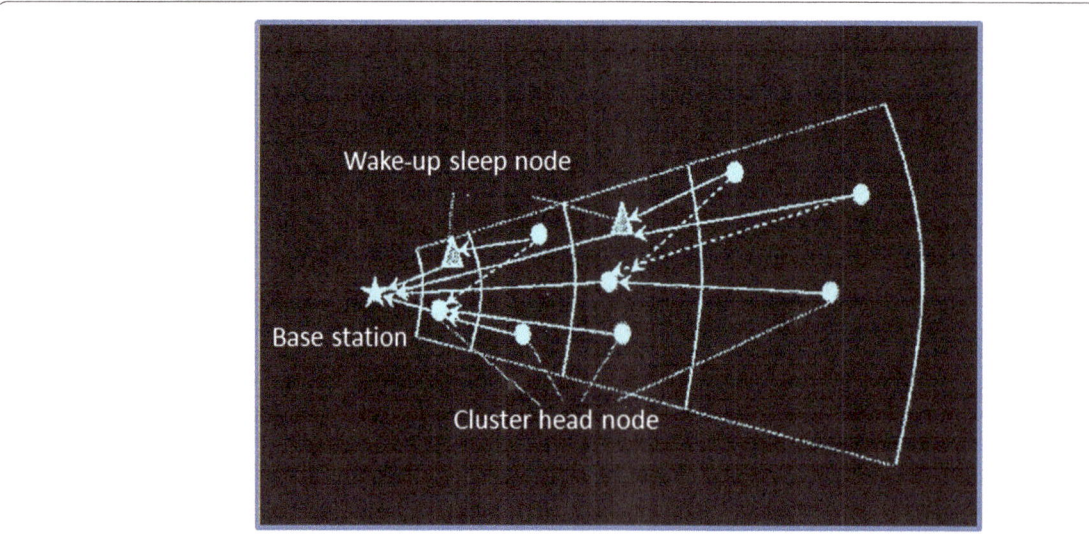

Fig. 4 Schematic diagram of intra-clusters multi-hop of four-layer and two-hop

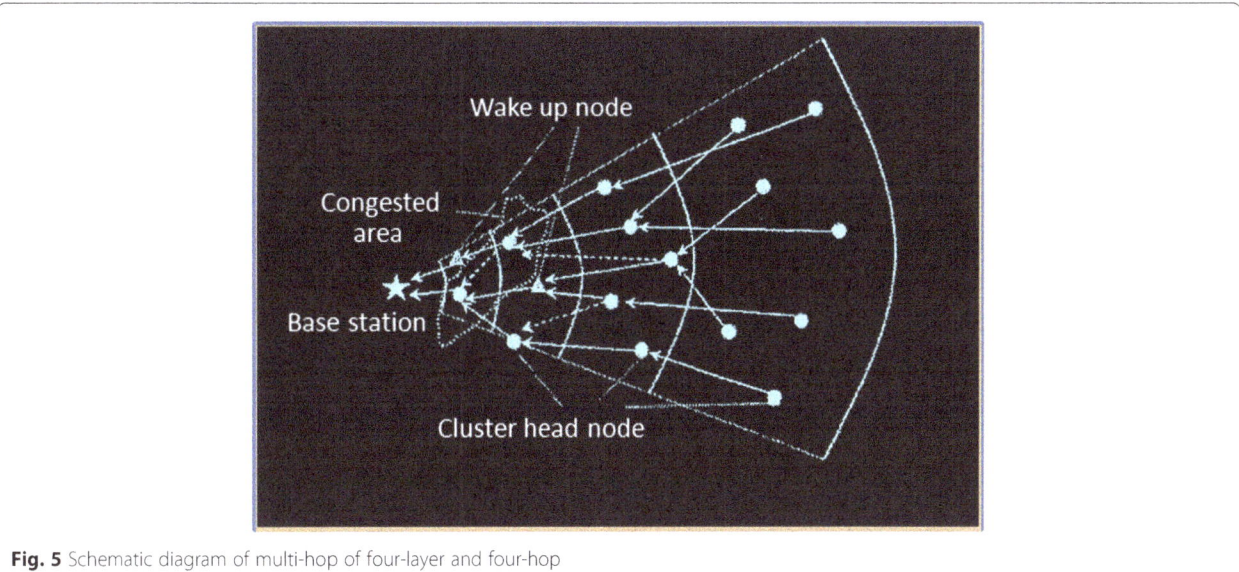

Fig. 5 Schematic diagram of multi-hop of four-layer and four-hop

service and product information previously selected by the user. (2) Information related to services and commodities. This type of information is mainly entered and maintained by the administrator in the background.

Figure 7 is a network architecture diagram of a consumer intelligent push system based on a wireless sensing system. The system mainly uses the idea of data mining to analyze and process the target user feature data collected by the sensor terminal and obtain the set of neighboring users that are close to the target user status, and recommends the goods to the target users according to the products selected by other users in the past. Finally, the calculated and obtained recommendation results are actively pushed from the server to the user browsing interface.

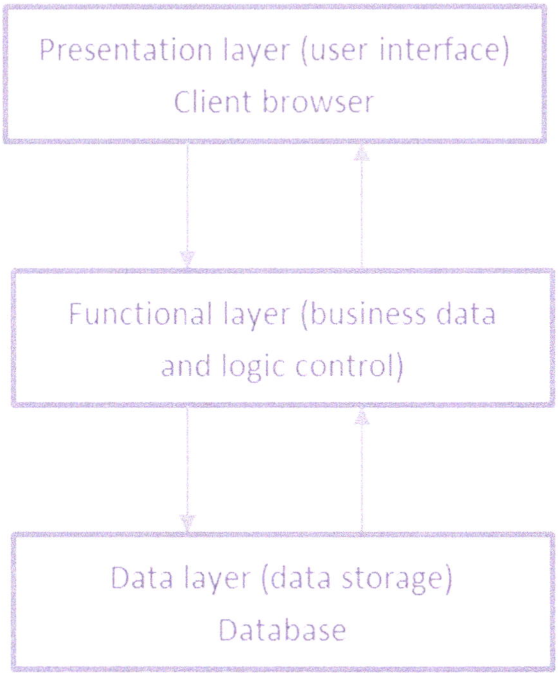

5 Analysis and discussion

The quality of wireless sensor network communication services is mainly affected by network congestion. Researchers often use a variety of metrics such as cache queue length, cache rate, throughput, packet loss rate, network transmission delay, and network efficiency. However, these metrics are based on different application environments of wireless sensor networks. Therefore, there is always no single metric that can systematically and comprehensively evaluate the network congestion completely. In the general sense, network congestion means that the node accepts data packets faster than the sending rate, so that the data queue of the node buffer grows continuously. Network congestion will cause data throughput to drop, loss rate to increase, transmission delay to increase, network energy consumption to increase, and even more network failures and crashes.

Wireless sensor network congestion can seriously affect the reliability of network service quality (QOS) and data transmission, reduce energy utilization, and accelerate the decline of the network. Therefore, network congestion control strategy has become one of the key technologies in wireless sensor network design. While designing the network congestion algorithm, we must also consider the network transmission quality, dynamic adaptability, robustness, energy utilization, timely prediction, network

Presentation layer (user interface)
Client browser

Functional layer (business data and logic control)

Data layer (data storage)
Database

Fig. 6 Overall structure of the intelligent commodity push system

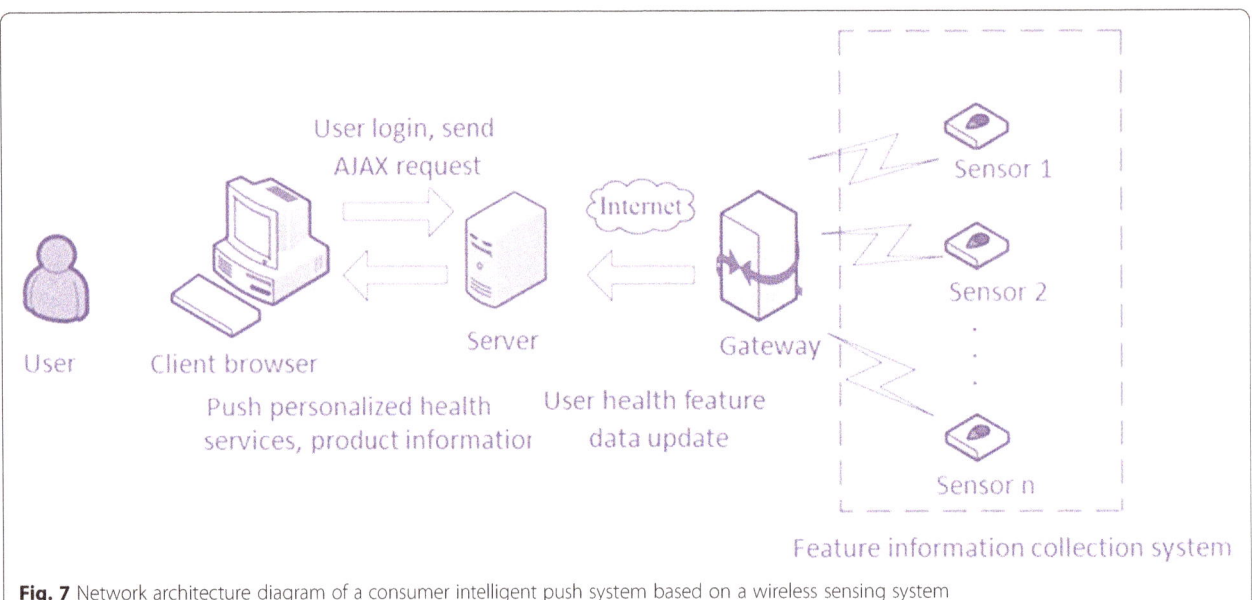

Fig. 7 Network architecture diagram of a consumer intelligent push system based on a wireless sensing system

congestion avoidance, and cancellation in different application scenarios. Each type of congestion control strategy uses congestion detection to achieve network congestion prediction and timely discovery, and further achieves network congestion mitigation through congestion avoidance. Finally, network congestion or local congestion elimination is achieved through data transmission mode adjustment during congestion cancellation.

5.1 Congestion detection

In different application scenarios of the WSN, the congestion prediction of the network congestion is often implemented by monitoring parameters such as the length and rate of the cache queue, the time at which the node CPU processes the data packet, the ratio of the data packet input rate, and the ratio of the packet processing or forwarding rate. In addition, congestion detection can be realized by monitoring parameters such as increased network packet loss rate, network delay time increase, channel communication load increase, or interference enhancement.

5.2 Congestion avoidance

After the WSN is discovered, the congestion control data packet is sent to the upstream node or directly to the node where the entire network is congested. The network has the following two types of congestion feedback: one is that the sensor network generates congestion control packets and is sent directly to the congestion source node, and the other is that the sensor network uses the control packet or the congestion control bit in the data packet and feeds back it to the congestion source node. After that, the congestion source node judges the congestion state by receiving different data packets or control packets, and then executes the corresponding congestion control strategy.

5.3 Congestion lifted

The upstream node that receives the congestion packet uses a series of congestion control methods to implement congestion mitigation and release. In order to eliminate the congestion caused by the wireless link layer, WSN mostly adopts strategies such as optimizing the topology and introducing a wireless channel allocation protocol such as *CSMA/CA*. For node-level network congestion, the congestion control strategy basically adopts methods such as reducing the data transmission rate or dropping the congestion data packet. In addition, some congestion control strategies are designed to achieve congestion mitigation and release by designing corresponding routing algorithms.

6 Conclusion

Wireless sensors are used to collect product information viewed by consumers. After late data processing and data mining, more potential information can be obtained, such as color preference and commodity turnover rate. This information can provide data support, quantitative analysis, and effective estimates for the merchants to do the next marketing plan and strategy, store layout planning, and other initiatives. Moreover, this paper focuses on the wireless sensor network data acquisition system in the network commodity sales system and combines the operating conditions of the network commodity store to develop wireless sensor network hardware design, routing design, network congestion control strategy, database design, and construction. In addition, the system mainly uses the idea of data mining to analyze and process the target user feature data collected by the sensor terminal and obtain a set of neighboring users close to the target user. After that, the target user is recommended according to the previously selected product status of other users in the neighborhood user group. Finally, the calculated recommendation results are actively pushed from the server to the user browsing interface.

Acknowledgements
The authors gratefully acknowledge the helpful comments and suggestions of the reviewers, which have improved the presentation.

Authors' contributions
All authors take part in the discussion of the work described in this paper.

Funding
This work was supported by Humanity and Social Science foundation of MOE of China (No.20171304).

Competing interests
These no potential competing interests in our paper.

References
1. A. Hentati, E. Driouch, J.F. Frigon, et al., Fair and low complexity node selection in energy harvesting wireless sensor networks [J]. IEEE Systems Journal **2017**, 1–11
2. A. Al-Baz, A. El-Sayed, A new algorithm for cluster head selection in LEACH protocol for wireless sensor networks [J]. International Journal of Communication Systems **31**(1), 01–13 (2018)
3. F. Haleem, J. Huma, J. Bilal, et al., Analytical network process based optimum cluster head selection in wireless sensor network [J]. PLOS ONE **12**(7), e0180848 (2017)
4. K. Nitesh, M. Azharuddin, P. Jana, Minimum spanning tree-based delay-aware mobile sink traversal in wireless sensor networks: delay-aware mobile sink traversal in WSN [J]. International Journal of Communication Systems **30**(13), e3270 (2017)
5. A. Keshavarz-Mohammadiyan, H. Khaloozadeh, Interacting multiple model and sensor selection algorithms for manoeuvring target tracking in wireless sensor networks with multiplicative noise [J]. International Journal of Systems Science **48**(5), 10 (2017)
6. M. Taherian, M. Maeen, M. Haghparast, Promoting the quality level of signaling in railway transportation system taking advantage from wireless sensor networks technology [J]. Computers **6**(3), 26 (2017)
7. Y.C. Chou, M. Nakajima, A clonal selection algorithm for energy-efficient mobile agent itinerary planning in wireless sensor networks [J]. Mobile Networks and Applications (2017)
8. V. Talla, M. Hessar, B. Kellogg, et al., LoRa Backscatter [J]. Proceedings of the ACM on interactive, mobile, wearable and ubiquitous technologies **1**(3), 1–24 (2017)
9. S.D. Trapasiya, H.B. Soni, Energy efficient policy selection in wireless sensor network using cross layer approach [J]. IET Wireless Sensor Systems **7**(6), 191–197 (2017)
10. H. Zhu, F. Xiao, L. Sun, et al., R-TTWD: Robust device-free through-the-wall detection of moving human with WiFi [J]. IEEE Journal on Selected Areas in Communications, 1–1 (2017)
11. GPU Parallel Implementation of Spatially Adaptive Hyperspectral Image Classification [J]. IEEE Journal of Selected Topics in Applied Earth Observations & Remote Sensing, 2017, PP(99):1-13.
12. M. Hosseini, H. Mcnairn, Using multi-polarization C- and L-band synthetic aperture radar to estimate biomass and soil moisture of wheat fields [J]. International Journal of Applied Earth Observation and Geoinformation **58**, 50–64 (2017)
13. J. Yang, H. Zou, H. Jiang, et al., CareFi: sedentary behavior monitoring system via commodity WiFi infrastructures [J]. IEEE Transactions on Vehicular Technology **67**(8), 7620–7629 (2018)
14. X. Zheng, J. Wang, L. Shangguan, et al., Design and implementation of a CSI-based ubiquitous smoking detection system [J]. IEEE/ACM Transactions on Networking **25**(6), 3781–3793 (2017)
15. H. Deng, J. Wu, L. Zhu, et al., Texture edge-guided depth recovery for structured light-based depth sensor [J]. Multimedia Tools and Applications **76**(3), 4211–4226 (2017)
16. Y. Lu, J. Zhang, B. Li, et al., Harnessing commodity wearable devices to capture learner engagement [J]. IEEE Access, 99):1–99):1 (2019, PP)
17. G. Yu, Z. Yifan, L. Jie, et al., Sleepy: wireless channel data driven sleep monitoring via commodity WiFi devices [J]. IEEE Transactions on Big Data, 1–1 (2018)
18. R. Zhang, N. Zhang, C. Du, et al., From electromyogram to password [J]. ACM Transactions on Intelligent Systems and Technology **9**, 1):1–1)20 (2017)
19. M. Karar, S. Paul, A. Mallick, et al., Interaction behavior between active hydrogen bond donor-acceptors as a binding decoration for anion recognition: experimental observation and theoretical validation [J]. ChemistrySelect **2**(9), 2815–2821 (2017)
20. H. Rueda, C. Fu, D.L. Lau, et al., Single aperture spectral+ToF compressive camera: toward hyperspectral+depth imagery [J]. IEEE Journal of Selected Topics in Signal Processing **11**(7), 992–1003 (2017)

A spanning tree construction algorithm for industrial wireless sensor networks based on quantum artificial bee colony

Yuanzhen Li[*] , Yang Zhao and Yingyu Zhang

Abstract

In industrial Internet, many intelligent applications are implemented based on data collection and distribution. Data collection and data distribution in the wireless sensor networks are very important, where the node topology can be described by the spanning tree for obtaining an efficient transmission. Classical algorithms in graph theory such as the Kruskal algorithm or Prim algorithm can only find the minimum spanning tree (MST) in industrial wireless sensor networks. Swarm intelligence algorithm can obtain multiple solutions in one calculation. Multiple solutions are very helpful for improving the reliability of industrial wireless sensor networks.

In this paper, we combine quantum computing with artificial bee colony and design a spanning tree construction algorithm for industrial wireless sensor networks. Quantum computations are introduced into the onlooker bees search. Food source replacement strategy is improved. Finally, the algorithm is simulated and evaluated. The results show that the new proposed algorithm can obtain more alternative solutions and has a better performance in search efficiency.

Keywords: Industrial wireless sensor network, Minimum spanning tree, Artificial bee colony, Quantum computing

1 Introduction

WSN (wireless sensor networks) [1, 2] is an autonomous measurement and control network system that consist of a large number of ubiquitous, small sensor nodes with communication and computing capabilities that are densely deployed in unattended monitoring areas [3]. WSN is a new information acquisition and processing technology [4]. Due to its advantages of low communication cost, flexible networking, and ease of use, it has a wide application prospect in the industry, military, environment, medical, and other fields [5]. IWSN (industrial wireless sensor networks) [6–8] is used to control and monitor various industrial tasks and is an emerging application of WSN.

Due to its high flexibility, low node cost, no wiring, and relatively easy maintenance, IWSN has been fa-vored by more and more companies [9]. A large number of wireless sensors are deployed in industrial parks [10]. These sensor nodes form a multi-hop network in the form of ad hoc networks, which are very sensitive to the equipment, production lines, and environmental information of the industrial site and transmitted to the control center in real time [11]. Through the calculation and analysis of data, the control center can monitor the operating conditions of the equipment, find problems in a timely manner, issue control commands, and reduce safety problems in the production process.

IWSN faces more challenges than ordinary wireless sensor networks [12], which have the following features: (l) The sensor node deployment of the IWSN is related to the industrial environment. It needs to be manually installed on the plant equipment that needs to be monitored [13], emphasizing reliable monitoring of designated points [14]. The nodes of the WSN are generally deployed in a random and

* Correspondence: liyuanzhen@163.com
School of Computer Science, Liaocheng University, Liaocheng, Shandong 252059, People's Republic of China

dense manner, focusing on the overall coverage of the monitoring area. (2) In IWSN, once the sensor node is installed, it is generally no longer moving, unless the node failure needs to be replaced or the plant equipment is moved [15]; in contrast, the WSN node mobility is strong. (3) In addition to sensor nodes in IWSN, there are routers, handheld devices, and other nodes [16]. Different types of nodes perform different functions, forming a heterogeneous network. The WSN generally refers to homogeneous networks, and the status of nodes is equal. (4) The factory environment is complex, and the interference is severe [17]. Therefore, the IWSN wireless network protocol design aims at high reliability and real-time performance. WSNs generally work in unattended areas. Node energy is limited, so the protocol design focuses on energy saving.

Field equipment production process data and network status data are collected periodically [18]. IWSN analyzes and processes this data to monitor factory equipment and networks. In the production process, it is often necessary to obtain some statistical information, such as the total number of network nodes and the temperature of the production site [19]. At present, the data collection of industrial wireless networks is basically centralized. That is, the original data is uniformly collected and processed by the central node. This is generally called the data collection protocol [20, 21]. The control information such as commands for monitoring the industrial field devices is performed through another protocol. This is generally referred to as data distribution protocol [22]. The data distribution process is one-to-many communication, and the data collection process is many-to-one communication. The above two processes are preferably implemented using trees to save network resources, especially the minimum spanning tree. Therefore, in industrial wireless sensor networks, how to construct an effective minimum spanning tree (MST) is a critical issue [23]. Classical algorithms in graph theory, such as Kruskal algorithm and Prim algorithm, utilize greedy strategies to generate only one MST. In IWSN, a series of spanning trees are required to improve reliability and deal with network changes [24]. Even if the found spanning tree is not optimal, but suboptimal, the found suboptimal spanning tree has important practical significance for responding to the dynamic changes of the network.

In the evolutionary process of swarm intelligence algorithm [25–27], a series of solutions are generated, which is very suitable for the solution of the spanning tree problem in Industrial Wireless Sensor Networks environment. At the same time, the swarm intelligence optimization algorithm can search without prior knowledge to find a solution to the optimization problem [28, 29]. Artificial bee colony algorithm is a kind of swarm intelligence algorithm, which was put forward by Karabogay in order to solve the problem of multivariable function optimization. Artificial bee colony algorithm is an optimization method proposed to imitate bee behavior. It is a specific application of swarm intelligence thought. Its main characteristic is that it does not need to know the special information about the problem, but only needs to compare the pros and cons of the problem. Through the local optimization behavior of individual artificial bees, the global optimal value is finally emerging in the group, and it has a faster convergence speed. Quantum computation is a new computational model that follows quantum mechanics regulation to regulate quantum information units. From the point of view of computational efficiency, due to the existence of the superposition of quantum mechanics, some known quantum algorithms are faster than conventional general-purpose computers when dealing with certain problems. In this paper, quantum computation and artificial bee colony algorithm are combined and a quantum artificial bee colony algorithm is proposed to solve multicast tree construction problem in industrial wireless sensor networks.

This paper is organized as follows: In Section 2, we first describe IWSN architecture and the minimum spanning tree problem in IWSN. We then present the proposed algorithm in Section 3. Sections 4 and 5 report the simulation results and discussion. Finally, Section 6 concludes the paper.

2 IWSN architecture and minimum spanning tree problem

Figure 1 shows the common IWSN architecture [30]. In IWSNs, sensor nodes are used as field devices to form industrial wireless networks. Nodes have the dual functions of data collection and routing. Sensor nodes distributed in the factory transmit the collected field data to servers through multi-hop routing. The access router is responsible for connecting the wireless network and the wired network and forwarding the field data from the wireless network to the wired network. There can be one or more access routers in the IWSNs. The gateway is responsible for the protocol conversion between the IWSNs and the existing factory network, so that both networks can send packets to each other. The management server is responsible for managing all network devices and in charge of storing topology information,

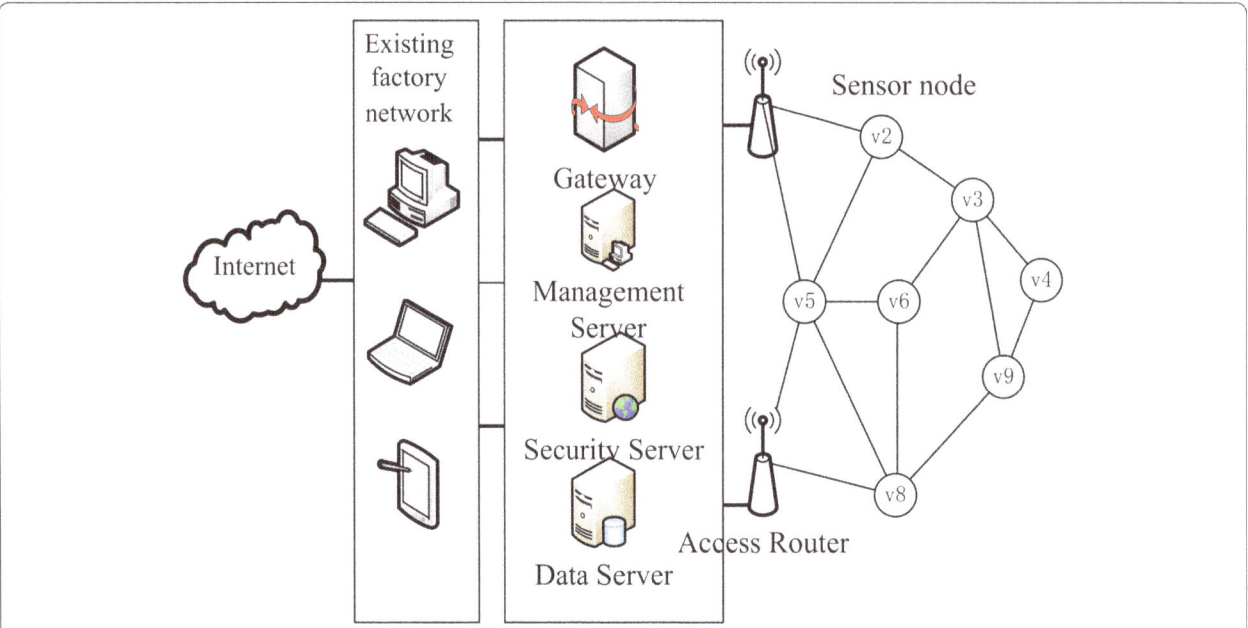

Fig. 1 The architecture of IWSN and minimum spanning tree problem. The figure is used to describe the architecture of industrial wireless sensor networks and the minimum spanning tree problem in industrial wireless sensor networks

node information, and neighbor relationships, as well as the link status of the entire network. In addition, the management server is also responsible for processing network information, managing network communication processes, and interacting with industrial applications. The security server [31] is responsible for the security management of the network and supports data integrity verification, data encryption, identity authentication, and replay protection. The data server stores field device configuration data, process flow data parameters, and data generated during the production process [32]. Gateways, management servers, security servers, and file servers are logically differentiated and can actually be deployed on the same network device as IWSNs controllers [33]. Finally, IWSN may need to connect to the Internet, depending on the specific conditions and needs of the factory.

A very important area of IWSN is data distribution and data collection. The effective implementation of data distribution and data collection is to use a minimal spanning tree. The minimum spanning tree problem is a basic problem in the areas of graph theory, optimization, and network optimization. Let graph $G = (N,E,W)$ be a connected undirected weighted graph, where N is the set of nodes, E is the set of edges, and W is the weight defined on the edge. $W = \sum_{e \in E} w_e$ is the weight function.

In graph theory, a tree is defined as an acyclic connected graph. If a subgraph of a connected graph G is a tree and contains all vertices of G, the subgraph T is called a spanning tree of G. If G has n vertices, its spanning tree

has n vertices and $n - 1$ edges. $W_T = \sum_{e \in E_T} w_e$ is the weight function of the tree T. The spanning tree of a graph G is not unique. A graph can have many different spanning trees. Among these trees, the spanning tree with the least sum of edge weight is the minimum spanning tree (MST).

Because the minimum spanning tree problem is very important in network optimization, researchers have conducted a detailed study of this problem. At present, there are already some classical algorithms for solving the minimum spanning tree in graph theory, such as Kruskal algorithm or Prim algorithm. These algorithms can only get one solution at a time. The characteristic of industrial wireless sensor networks determines that multiple solutions are also necessary and meaningful. These different solutions are mutually complementary and backup. Swarm intelligence algorithm to obtain multiple solutions at a time just can effectively solve this requirement of industrial wireless sensor networks. This paper will use the artificial bee colony algorithm and use the idea of quantum computing to solve this problem.

3 Quantum artificial bee colony
3.1 Standard artificial bee colony
The artificial bee colony (ABC) algorithm [34], originally proposed by Karaboga in 2005, is based on the bee family's foraging behavior. The bee is a social insect. Although the behavior of individual

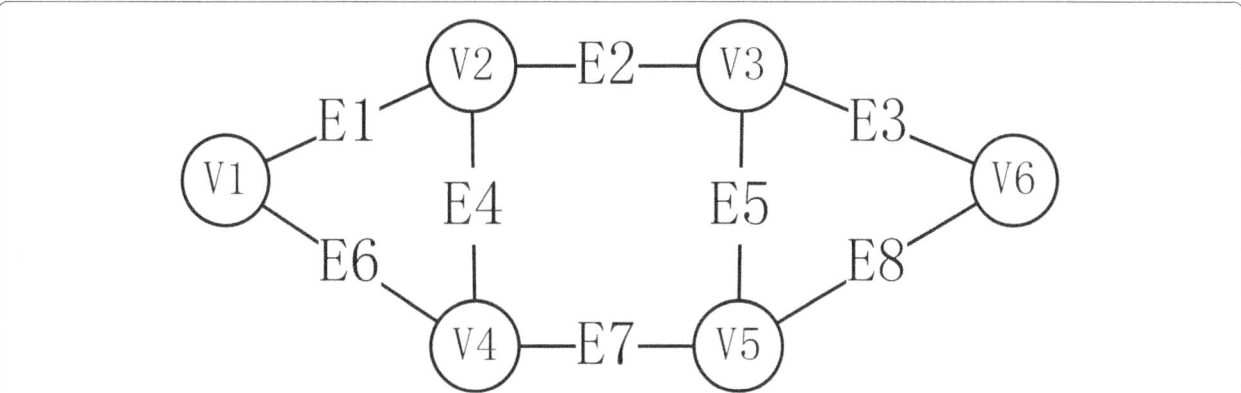

Fig. 2 Graph G. This figure is used to illustrate the coding technique used in this paper. This figure and Figs. 3 and 4 together illustrate the coding technique used in this article

insects is extremely simple, the group of individuals shows extremely complex behavior. Bees can collect nectar from food sources with great efficiency in any environment; at the same time, they can adapt to changes in the environment. As a group organism of nature, the bee colony has a more rigorous foraging system within its organization. The artificial bee colony algorithm is based on the division of labor and cooperation of different types of work groups in the bee colony, thereby more effectively searching for the global optimal solution.

The minimum search model for swarms that generate swarm intelligence includes the basic three components [35]: food sources, employed bees, and unemployed bees. There are also two basic behavioral models: recruiting bees for food sources and giving up food sources. (1) Food sources: the value of food sources is determined by many factors, such as the distance from the hive, the richness of the nectar, and the ease of

obtaining nectar. (2) Employed bees: for each food source, there is only one employed bee, that is, the number of employed bees is equal to the number of food sources. The employed bee stores information about food sources and shares this information with other bees with a certain probability.(3) Unemployed bees: their main task is to find and mine food sources. There are two types of unemployed bees: the scout bees and the onlooker bees. Scout bees search for new food sources nearby. The onlooker bees wait inside the hive and find food sources by sharing information with the employed bee.

The ABC algorithm randomly generates initial populations containing *PS* solutions (food sources). The employed bee conducts a neighborhood search on the corresponding food source, compares the new food source with the original food source, and selects a solution with a high degree of fitness as a candidate solution. When searching work finished, the employed bees share the food source information with the onlooker bees. The onlooker

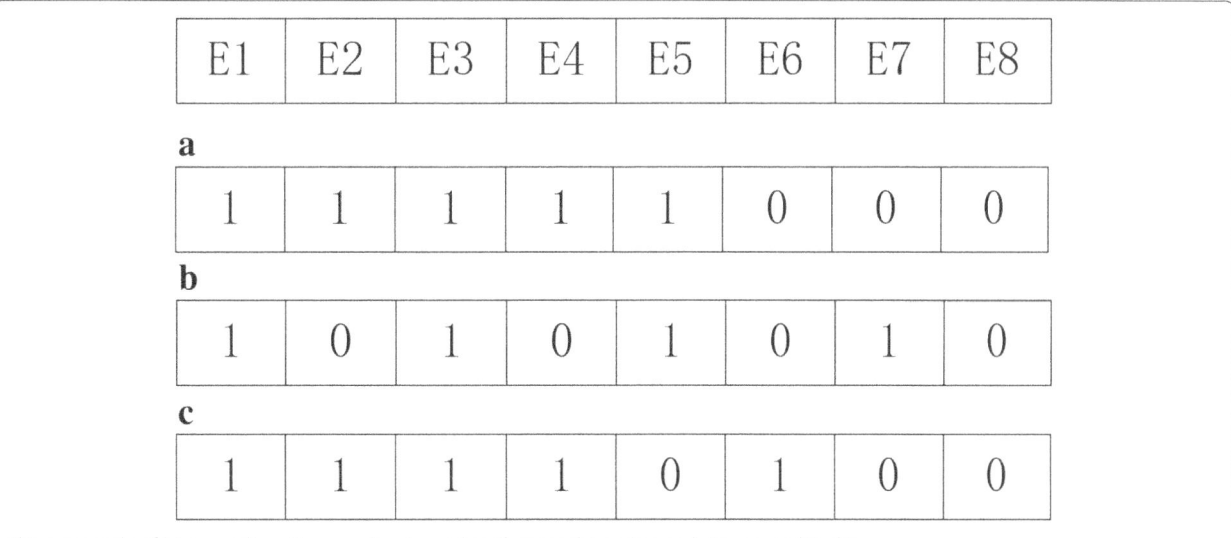

E1	E2	E3	E4	E5	E6	E7	E8
a							
1	1	1	1	1	0	0	0
b							
1	0	1	0	1	0	1	0
c							
1	1	1	1	0	1	0	0

Fig. 3 Example of binary coding. Binary coding is used to illustrate the coding techniques used in this paper

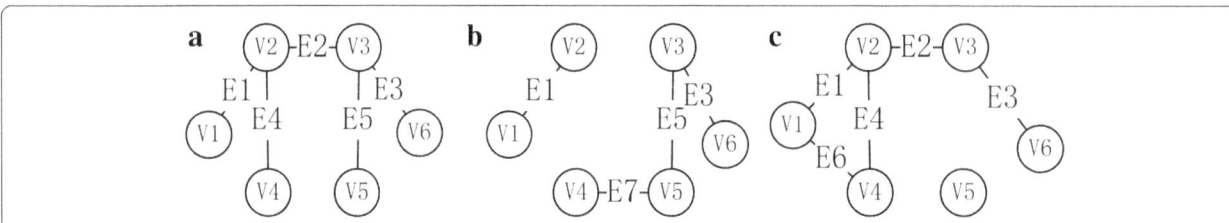

Fig. 4 Tree structure represented by a binary string as in Fig. 3. This figure is used to illustrate the coding technique used in this paper. Tree structure represented by a binary string as in Fig. 3

bees choose the food source according to probability p_i. The higher the food source's fitness value, the greater the probability of being selected.

$$p_i = \frac{f_i}{\sum\limits_{i=1}^{PS} f_i} \tag{1}$$

where f_i is the fitness value of the i-th solution x_i. Then, the onlooker bees also conduct a neighborhood search and choose a better solution. If a solution is not improved for consecutive NC cycles, it is discarded and a random solution is randomly generated by the scout bee. The main steps of the artificial bee colony algorithm are as follows.

(1) Initialize the colony population;

(2) Employed bees search for new honey source near their associated food sources;

(3) The onlooker bee to select the food source using formula (1) and search for a new honey source near the selected food source;

(4) Scout bees to search for new honey sources

(5) Memorize the best food source found so far

(6) If the maximum number of iterations is not reached, repeat the above steps (2–5). The final best honey position is the global optimal solution to be searched.

Artificial bee colony algorithm has shown good performance [36, 37] in the solution of complex optimization problems due to its advantages of simple principles, convenient implementation, good applicability, and favorable cooperation between population division and labor [38]. The original artificial bee colony algorithm is mainly aimed at solving the problem of continuous space function optimization. In order to solve many combinatorial optimization problems in practical engineering, many discrete artificial bee colony algorithms are proposed. The flow of the discretized artificial bee colony algorithm is the same as the artificial bee colony algorithm.

3.2 Quantum artificial bee colony algorithm

Quantum computing is a new cross discipline combining information science and quantum mechanics. Quantum computations represented by quantum algorithms have a high degree of parallelism, exponential storage capacity, and exponential acceleration of classical heuristic algorithms. It has great superiority and contains great vitality.

Quantum computing has become a frontier field for scholars from all over the world. By using quantum computing in traditional intelligent optimization, quantum computing and intelligent computing are combined. This will change the traditional optimization

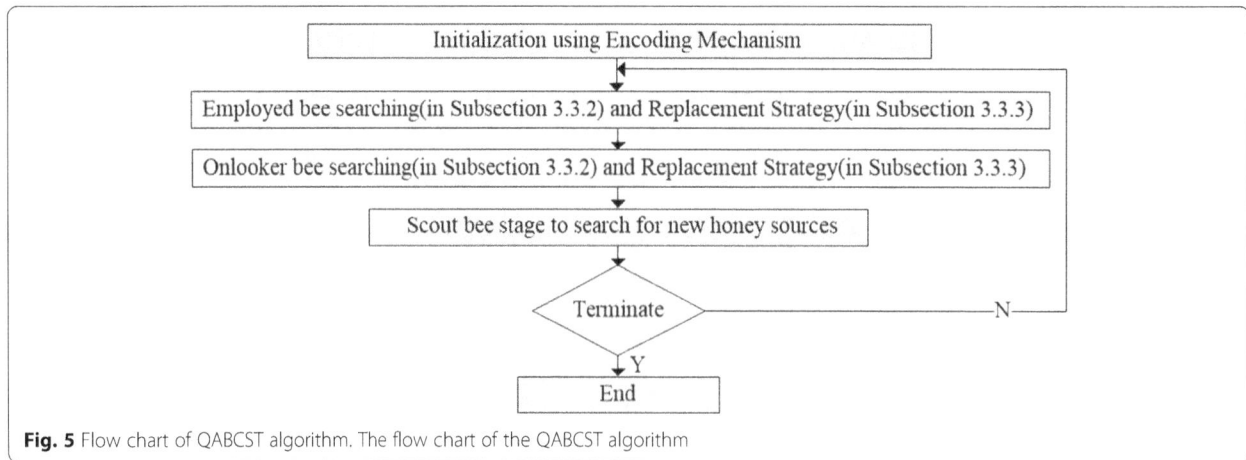

Fig. 5 Flow chart of QABCST algorithm. The flow chart of the QABCST algorithm

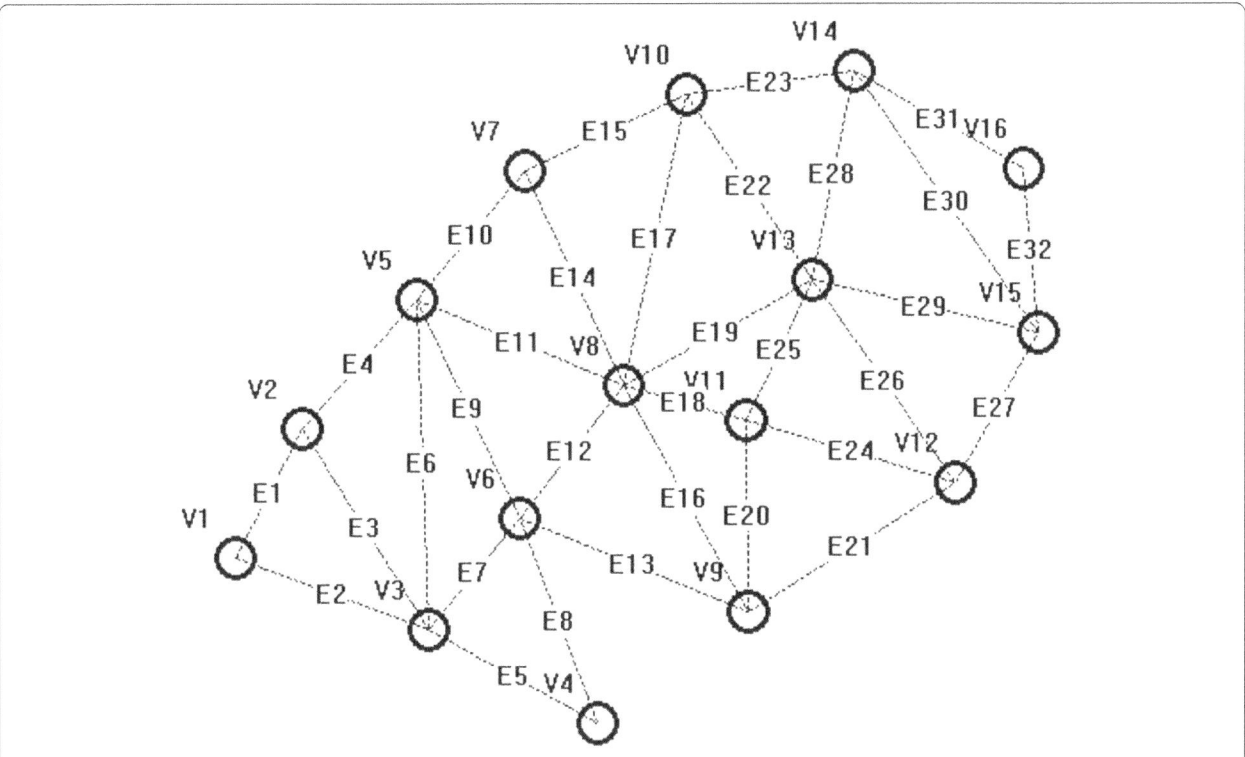

Fig. 6 An IWSN example with 16 nodes and 32 edges. An industrial wireless sensor network diagram for performance simulation. The diagram contains 32 nodes with 16 nodes

methods of intelligent computing and improve the performance of search optimization and convergence speed. Using quantum computing to improve the particle swarm algorithm, Sun et al. proposed quantum particle swarm optimization (QPSO) algorithm [39]. It is assumed that the particle's behavior has quantum characteristics. Under the role of quantum mechanics, particles no longer have limitations on the trajectory and velocity, and the ability of particles to find the optimal solution is greatly improved. Regarding the specific implementation of the algorithm, QPSO uses (Eqs. 2–7) to update the entire population.

$$mb = \frac{\sum_{i=1}^{N} pb_i}{N} \tag{2}$$

$$a = rand(0,1) \tag{3}$$

$$p = a * pb_i + (1-a) * gb \tag{4}$$

$$b = 1 - \frac{t}{2 * t_{\max}} \tag{5}$$

$$u = rand(0,1) \tag{6}$$

$$p_i = \begin{cases} p - b* \mid mb-p_i \mid * \ln\left(\frac{1}{u}\right), & u \geq 0.5 \\ p + b* \mid mb-p_i \mid * \ln\left(\frac{1}{u}\right), & u < 0.5 \end{cases} \tag{7}$$

Here, N is the population size; pb_i is the individual optimal position of the i-th particle; gb is the optimal location of the entire population; mb is the mean of best position, which is the average of the individual optimal positions of all particles; $rand(0,1)$ is a function whose return value is a random decimal between [0, 1]; t is the current evolutionary generation; $t_{\max}$ the maximum evolutionary generation of the algorithm; b is called the contraction expansion coefficient, which gradually decreases with the iteration of the algorithm; p_i is the position of the i-th particle.

Inspired by the QPSO algorithm, we use a similar approach in the QABC (quantum artificial bee colony) algorithm. We only use the idea of quantum computing in the onlooker stage. A quantum representation of solutions is used to enhance the diversity of the basic ABC. In addition, the exploitive capability of the ABC is boosted through the use of the quantum interference concept.

Table 1 The coordinates of the nodes

Node	X	Y	Node	X	Y
V1	40.94	292.97	V9	309.2	1.1. 320.0
V2	75.53	226.48	V10	277.73	1.2. 55.13
V3	141.22	329.18	V11	308.81	1.3. 222.7
V4	230.90	377.83	V12	419.35	1.4. 254.59
V5	135.32	160.54	V13	343.43	1.5. 150.27
V6	189.61	272.43	V14	365.85	1.6. 43.24
V7	192.76	94.59	V15	464.2	1.7. 177.83
V8	244.29	204.32	V16	145.33	1.8. 93.51

3.3 Quantum artificial bee colony algorithm for constructing spanning trees

3.3.1 Encoding mechanism

The standard ABC algorithm for solving continuous optimization problems is not directly suitable for solving the minimum spanning tree problem. Because the minimum spanning tree problem is an optimization problem, binary representation is used. The number of edges in the graph G is denoted as $|E|$. $|E|$ bit binary code is used to represent a solution. Each binary bit corresponds to an edge in the graph G and takes value 0 or 1, where 1 indicates that the corresponding edge is contained in the spanning tree T, and 0 means the opposite. The number of nodes in the graph G is denoted as $|N|$. According to the characteristics of the spanning tree, the spanning tree contains only $|N|$ -1 edges. In a feasible solution, only $|N|$ -1 binary bit is 1. The other binary bits are all 0. For example, given a graph, G = (N, E)

Table 2 The weight of the edges

Edge	Vertex	Vertex	Weight	Edge	Vertex	Vertex	Weight
E1	V1	V2	74.96	E17	1.9. V8	1.10. V10	1.11. 152.89
E2	V1	V3	106.65	E18	1.12. V8	1.13. V11	1.14. 67.09
E3	V2	V3	121.91	E19	1.15. V8	1.16. V13	1.17. 112.92
E4	V2	V5	89.01	E20	1.18. V9	1.19. V11	1.20. 97.3
E5	V3	V4	102.04	E21	1.21. V9	1.22. V12	1.23. 128.11
E6	V3	V5	168.74	E22	1.24. V10	1.25. V13	1.26. 115.62
E7	V3	V6	74.58	E23	1.27. V10	1.28. V14	1.29. 88.92
E8	V4	V6	113.1	E24	1.30. V11	1.31. V12	1.32. 115.05
E9	V5	V6	124.37	E25	1.33. V11	1.34. V13	1.35. 80.28
E10	V5	V7	87.46	E26	1.36. V12	1.37. V13	1.38. 129.02
E11	V5	V8	117.44	E27	1.39. V12	1.40. V15	1.41. 88.9
E12	V6	V8	87.34	E28	1.42. V13	1.43. V14	1.44. 109.35
E13	V6	V9	128.7	E29	1.45. V13	1.46. V15	1.47. 123.87
E14	V7	V8	121.23	E30	1.48. V14	1.49. V15	1.50. 166.69
E15	V7	V10	93.69	E31	1.51. V14	1.52. V16	1.53. 103.51
E16	V8	V9	132.65	E32	1.54. V15	1.55. V16	1.56. 84.69

with $|N|$ = 6 and $|E|$ = 8 as shown in Fig. 2, where the edges and nodes are numbered in order so that each bit of a solution could be decoded to an edge. Figure 3 shows three coding schemes. The corresponding graphs for the three coding schemes are shown in Fig. 4. The number of 1 in code (b) is 4, and (b) is an infeasible string. The number of 1 in code (a) and code (c) is 5. In Fig. 4, after decoding (c), there is a loop (v1-v2-v4-v1) and isolated node $v5$. Binary string (a) is a feasible code; binary string (c) is not an infeasible code.

3.3.2 Search mechanism

In the initialization phase, for each food source, $|N|$ – 1 position is randomly selected and set to 1 and other positions are set to 0. The calculation of fitness is calculated according to formula (8). If the binary string is a feasible code (Figs. 3a and 4a), the fitness value is $fit = \sum_{e \in E_T} w_e$. If the binary string is infeasible (Figs. 3c and 4c), the fitness is ∞.

$$fit = \begin{cases} \sum_{e \in E_T} w_e, & \text{feasible} \\ \infty, & \text{infeasible} \end{cases} \quad (8)$$

Employed bees use the following techniques when searching. A position is randomly chosen from the elements with 1 and denoted as i_1. Another position is randomly chosen from the elements with 0 and denoted as i_2. The values of positions i_1 and i_2 are interchanged. After such changes, the number of elements with 1 does not change. This technique can guarantee that the number of 1 in the binary string is constant. The fitness of the newly generated solution is calculated according to Eq. (8). The replacement strategy, described in detail in Section 3.3.3, is executed.

In the process of onlooker bee searching, the quantum computing technique introduced in Section 3.2 is used and improved. For the calculation of the average best position, we use the elite strategy. Food sources are sorted in order of fitness value from small to large. For the sorted food source, the mean value of the front half food source is calculated (formula (9)). Because the particle swarm algorithm and the artificial bee swarm algorithm are essentially different, the meaning of the variable has also changed. In order to apply to the artificial bee colony algorithm, the previous formulas (2)–(7) are modified. The new calculation methods are formulas (9)–(15).

$$nmb_j = \frac{\sum_{i=1}^{N/2} fc_{ij}}{N/2} \quad (9)$$

$$a = rand(0, 1) \quad (10)$$

$$p = a * fc_{ij} + (1-a) * gb_j \qquad (11)$$

$$b = 1 - \frac{t}{2 * t_{max}} \qquad (12)$$

$$u = rand(0,1) \qquad (13)$$

$$q_j = \begin{cases} p - b * \mid nmb - fc_{ij} \mid * \ln\left(\dfrac{1}{u}\right), & u \geq 0.5 \\[3mm] p + b * \mid nmb - fc_{ij} \mid * \ln\left(\dfrac{1}{u}\right), & u < 0.5 \end{cases} \qquad (14)$$

$$p_j = \begin{cases} 1, & q_j \geq 0.5 \\ 0, & q_j < 0.5 \end{cases} \qquad (15)$$

Here, N is the population size; fc_{ij} is the j-th component of the i-th smallest food source according to fitness; gb is the first small food source according to fitness; gb_j is the j-th component of gb; nmb is the mean after improvement; $Rand(0,1)$ is a function whose return value is a random decimal between [0, 1]; t is the current evolutionary generation; t_{max} the maximum evolutionary generation of the algorithm; b is called the contraction expansion coefficient, which gradually decreases with the iteration of the algorithm.

If p_j and the j-th component of the current food source are equal, nothing is done. Otherwise, a location, denoted as k, is randomly selected from the elements with $(1 - p_j)$. Then, the elements of position k and position j are interchanged. That is, the element at position j is assigned p_j, and the element at position k becomes $1 - p_j$. The above operation is the same as the operation of the employed bee, and it also ensures that the number of elements with 1 does not change. The search process for onlooker bees is shown in Algorithm 1.

Algorithm 1 OnlookerBees

1:   Select a food source randomly according to formula (1), which is labeled fc_i.

2:   Select a location at random for the selected food source, which is labeled j

3:   Use Equation (9)-(15) to calculate the value of p_j

4:   If p_j is not equal to fc_{ij} then

5:   a location, denoted as k, is randomly selected from the elements with $(1 - p_j)$.

6:   the elements of position k and position j are interchanged.

7:   End if

If no better food source is found in the search for employed bees and onlooker bees, the food source is not updated. If it is not updated after the set number of times,

the food source is discarded. The scout bee will randomly generate a new food instead. The procedure is similar to the initialization step.

3.3.3 Replacement strategy

In the previous part of this article, we have mentioned that multiple solutions can be obtained in one calculation. Therefore, the first principle of our replacement strategy is to ensure the diversity of food sources. Under the guidance of such principles, if the newly found food source is better than the original food source, but is the same as any other food source, it will not be updated and will be discarded. In addition, as described in Section 3.3.1, the initial food source and new food source search may be infeasible. Therefore, another principle of our replacement strategy is to replace as much of the infeasible food source as possible.

After employed bee and onlooker bees search for new food sources, the specific algorithm for food source replacement is shown in Algorithm 2.

Algorithm 2 Replacement

1:   The solution corresponding to the new food source is infeasible, return. Otherwise, turn to 2;

2:   If any other food source is not feasible, an infeasible food source is randomly selected and replaced, return. Otherwise, turn to 3;

3:   If the new food source is better than the original food source, the new food source replaces the original food source.

To this position in this paper, based on quantum computing and artificial bee colony, the algorithm for solving the spanning tree of industrial wireless sensor networks has been introduced. We will use QABCST to represent this algorithm later. The main steps of QABCST are similar to the ABC algorithm described in Section 3.1. The flow chart of the QABCST algorithm is shown in Fig. 5.

The QABCST algorithm can generate multiple solutions in one calculation, and these solutions are backups of each other. The tree decoded from optimal solution can be used for data distribution or data collection. When the link failure causes the optimal tree to be unavailable, one of the backup solutions is used. Firstly, the

Table 3 Minimum spanning tree found by the Kruskal algorithm

ST	Weight	Edge series
Optimal solution	1326.409668	1,2,4,5,7,10,12,15,18,20,23,25,27,31,32

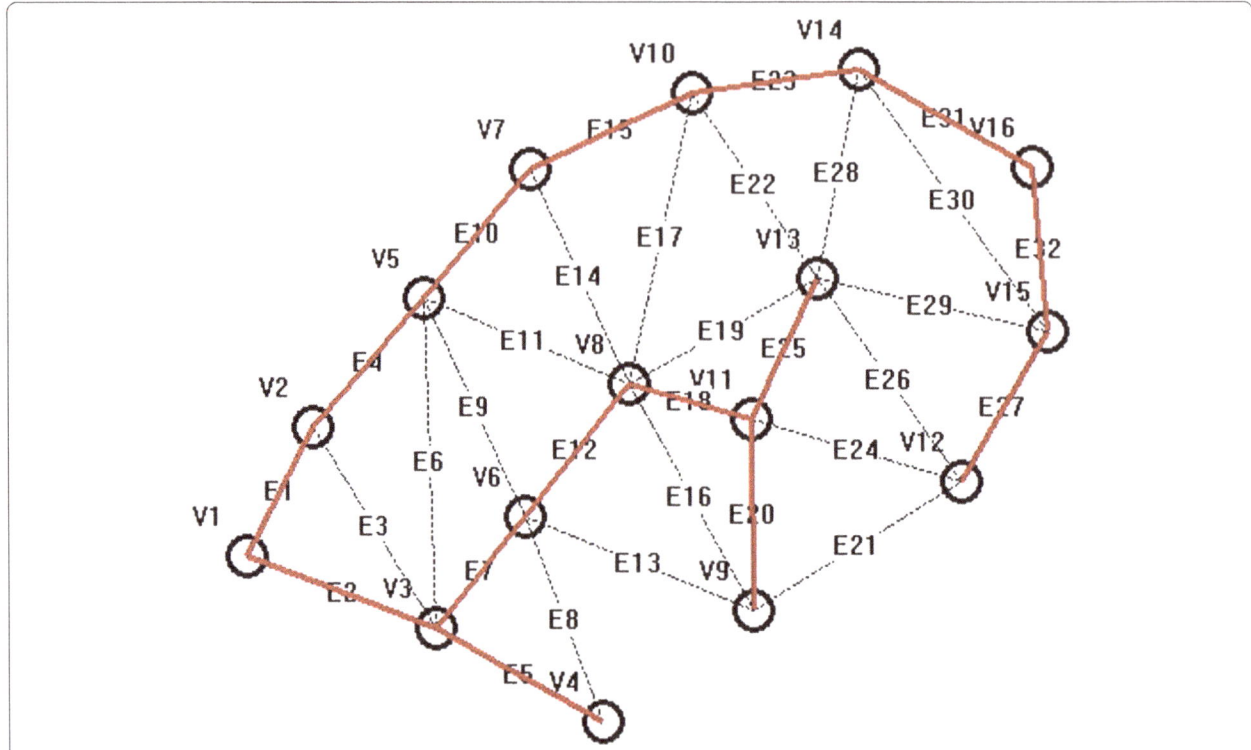

Fig. 7 Minimum spanning tree found by the Kruskal algorithm. The minimum spanning tree obtained by the Kruskal algorithm during performance simulation

candidate solutions containing the failed link can be marked as invalid. Then, choose the optimal one from the remaining solutions.

4 Experimental

In this section, simulation experiments are conducted on the newly proposed algorithm to verify the validity of our proposed algorithm.

The industrial wireless sensor network shown in Fig. 6 is used as an example to simulate. The

Table 4 Spanning tree found by BABC

ST	Weight	Edge series
1	1431.365601	1,7,8,10,11,15,18,20,24,25,27,28,29,31,32
2	1505.983398	1,3,5,7,10,12,13,15,18,23,24,28,30,31,32
3	1452.080566	1,4,5,7,8,10,13,18,23,24,25,26,27,28,31
4	1595.697266	2,4,7,8,9,10,12,16,17,19,20,21,22,23,32
5	1447.339722	1,2,5,7,9,10,18,19,20,23,25,26,27,28,31
6	1423.100586	1,2,5,7,8,10,12,15,18,20,21,23,27,28,31
7	1501.466919	1,7,8,9,10,15,17,18,20,23,24,26,27,28,32
8	1424.996460	1,5,7,10,15,18,19,20,21,23,24,25,26,27,32
9	1465.025146	2,4,8,9,10,12,13,14,15,18,23,25,27,31,32
10	1449.954712	1,2,4,5,7,10,12,16,19,22,25,27,28,31,32
Optimal solution	1326.409668	1,2,4,5,7,10,12,15,18,20,23,25,27,31,32

Table 5 Spanning tree found by QABCST

ST	Weight	Edge series
1	1647.062378	1,3,6,8,10,13,17,18,19,20,23,24,28,29,32
2	1668.069214	1,2,3,4,6,8,13,14,15,21,24,27,28,29,32
3	1411.012573	1,5,7,10,12,15,16,18,20,22,23,24,25,28,32
4	1463.207642	2,3,4,5,8,10,12,18,19,20,23,24,25,28,32
5	1424.814697	1,4,5,7,8,10,12,17,18,19,20,23,27,31,32
6	1435.052612	1,2,5,7,9,12,15,18,19,20,21,23,27,31,32
7	1442.719849	1,2,3,4,5,7,12,15,18,20,22,23,24,29,32
8	1452.438354	1,2,5,7,10,14,16,18,22,23,25,27,29,31,32
9	1525.517456	1,4,5,10,11,12,15,16,18,22,23,26,27,30,32
10	1346.926025	1,4,5,7,10,12,15,18,20,22,23,24,25,27,32
11	1396.752075	1,3,4,5,7,8,10,12,18,19,20,23,24,25,32
12	1432.078125	2,4,5,7,9,12,15,18,19,21,23,25,27,31,32
13	1476.490845	1,2,3,4,5,7,10,12,15,20,23,24,26,29,32
14	1383.834351	1,4,5,7,9,12,15,18,20,22,23,24,25,27,32
15	1360.906860	1,3,4,5,7,10,12,18,19,20,23,25,27,31,32
16	1443.100220	2,3,4,5,7,10,12,15,18,20,23,24,29,31,32
17	1495.755859	1,4,5,11,12,14,17,18,21,23,25,27,28,31,32
18	1576.713135	1,2,4,6,8,12,13,14,15,18,21,24,27,28,32
19	1455.923950	1,4,5,8,10,11,12,15,16,18,19,22,23,27,32
20	1372.569336	1,4,5,7,8,10,12,18,19,20,23,25,27,29,32
Optimal solution	1326.409668	1,2,4,5,7,10,12,15,18,20,23,25,27,31,32

industrial wireless sensor network, represented as G, has 16 nodes and 32 edges. These nodes are deployed in a 500 × 400 rectangular area. Table 1 shows the coordinates of the nodes.The weight function on each edge is the Euclidean distance between two nodes. Table 2 lists the correspondence between edges and nodes, as well as the weight of the edges.

Our new proposed algorithm will be compared with Kruskal algorithm and BABC [40] algorithm. BABC uses the basic artificial bee colony algorithm to solve the minimum spanning tree. It can also obtain multiple spanning tree construction schemes in one calculation. The population size is 20, and the algorithm loops 3000 times. All the algorithms have been coded using C++ in VS 2010. We run all the configurations on an Intel (R) Core (TM) i7-2600 CPU @ 3.40 GHz with 8.00 GB RAM in the Windows 10 Operation System.

5 Results and discussion

Each of the three algorithms runs one time, and the experimental results are compared. First, the Kruskal algorithm is used to handle this example. The Kruskal algorithm can only obtain one solution. The obtained ST (spanning tree) is shown in Table 3. The resulting spanning tree has a weight of 1326.409668. The

Table 6 Average of the 10 minimum spanning tree weights

Kruskal	BABC	QABCST
1326.409668	1328.747522	1327.755135

resulting minimum spanning tree is shown in Fig. 7. The Kruskal algorithm's calculation time is about 0.05 s.

Tables 4 and 5 are the results obtained after the BABC and QABCST algorithms are run once. As can be seen from Tables 4 and 5, the QABCST algorithm has got more solutions. This is due to the diversity of our replacement strategies. The reason for this result is the persistence of diversity in QABCST's replacement strategy. Figure 8 is an illustration of the other one (10th in Table 5) of the solutions obtained by the QABCST algorithm.

The following is a comparison of the average performance of the QABCST and BABC algorithms running multiple times. The algorithm is performed 10 times independently to obtain its average performance. Table 6 shows the average of the 10 minimum spanning tree weights. The result of Table 6 about Kruskal is the result of running once. As can be seen from the table, the QABCST algorithm has better performance than BABC.

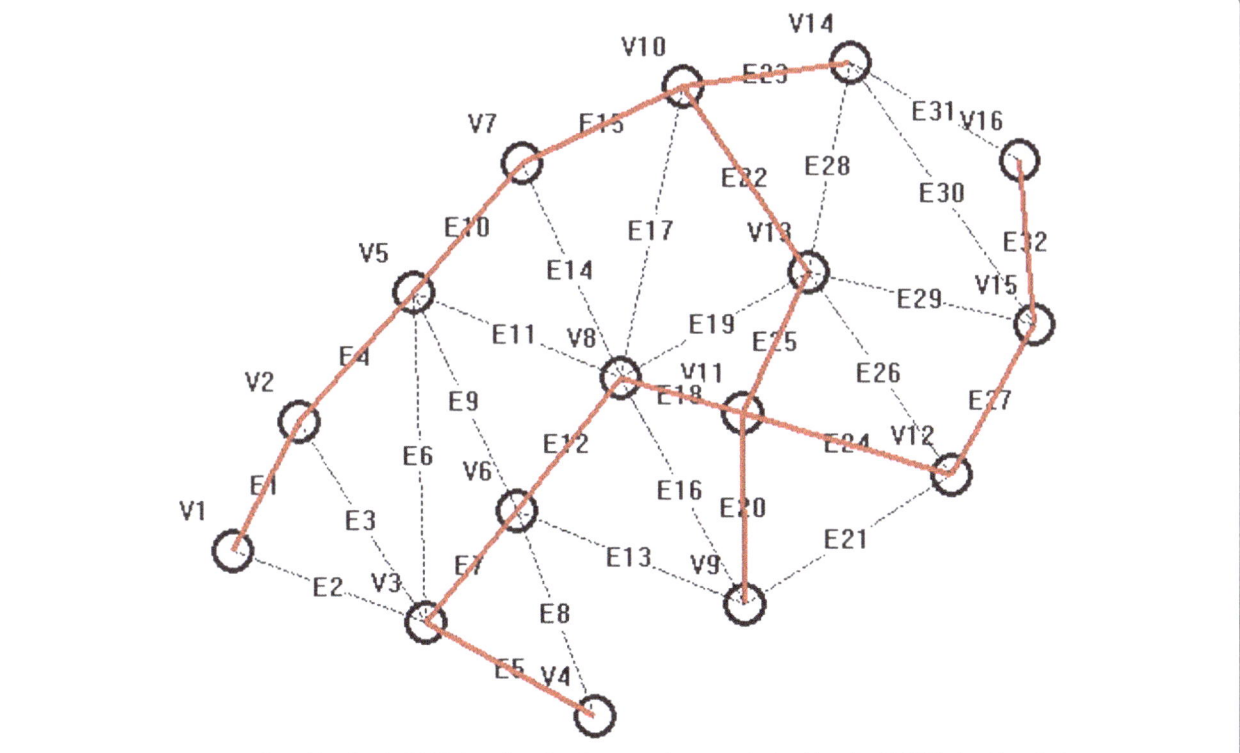

Fig. 8 Another spanning tree found by the QABCST algorithm. Another spanning tree obtained by the QABCST algorithm during performance simulation

6 Conclusion

With the rapid development of wireless sensor networks, wireless devices are increasingly deployed in industrial environments. Compared with common wireless sensor networks, industrial wireless sensor networks have higher requirements for the determinism and reliability of data communications. Therefore, it is particularly important to design reasonable mechanisms for the data aggregation/distribution of IWSNs to ensure the certainty and reliability of the data transmission process. Data collection and distribution in industrial wireless sensor networks can be described using the spanning tree problem in graph theory. Existing classical algorithms such as Kruskal and Prim algorithm can only get one solution at a time. In order to improve reliability, industrial application scenarios need to provide multiple solutions for mutual backup. This paper improves the artificial bee colony algorithm based on the idea of quantum computing and proposes a spanning tree construction algorithm for industrial wireless sensor networks based on quantum artificial bee colony. Finally, our algorithm was verified by experiments. The experimental results show that the algorithm can achieve better performance and can obtain more solutions at the same time. Future work includes increasing a priori knowledge of the network structure to improve search efficiency.

Abbreviations
MST: Minimum spanning tree; WSN: Wireless sensor networks; IWSN: Industrial wireless sensor networks; ABC: Artificial bee colony; QPSO: Quantum particle swarm optimization; QABC: Quantum artificial bee colony; ST: Spanning tree

Acknowledgements
Not applicable.

Authors' contributions
The first author conducted the experiments and wrote the first draft of the paper. Other co-authors helped to revise the paper and polished the paper. All authors read and approved the final manuscript.

Authors information
Not applicable.

Funding
This research was supported by National Natural Science Foundation of China (No. 61773192).

Competing interests
The authors declare that they have no competing interests.

References
1. H. Cheng, Z. Su, N. Xiong, Y. Xiao, Energy-efficient node scheduling algorithms for wireless sensor networks using Markov random field model. Inf. Sci 329(C), 461–477 (2016)
2. X. Jiang, Z. Fang, N.N. Xiong, et al., Data fusion-based multi-object tracking for unconstrained visual sensor networks. IEEE. Access. 6, 13716–13728 (2018)
3. J. Liu, J. Wan, Q. Wang, P. Deng, K. Zhou, Y. Qiao, A survey on position-based routing for vehicular ad hoc networks. Telecommun. Syst. 62(1), 15–30 (2016)
4. M. Wu, L. Tan, N. Xiong, Data prediction, compression, and recovery in clustered wireless sensor networks for environmental monitoring applications. Inf. Sci. 329(SI), 800–818 (2016)
5. Y. Liu, K. Ota, K. Zhang, et al., QTSAC: an energy-efficient MAC protocol for delay minimization in wireless sensor networks. IEEE Access. 6, 8273–8291 (2018)
6. V. C G, G. P H, Industrial wireless sensor networks: challenges, design principles, and technical approaches. IEEE Trans. Ind. Electron. 56(10), 4258–4265 (2009)
7. D.E. Boubiche, A.S. Pathan, J. Lloret, H. Zhou, S. Hong, S.O. Amin, M.A. Feki, Advanced industrial wireless sensor networks and intelligent IoT. IEEE. Commun. Mag 56(2), 14–15 (2018)
8. D.V. Queiroz, M.S. Alencar, R.D. Gomes, I.E. Fonseca, C. Benavente-Peces, Survey and systematic mapping of industrial wireless sensor networks. J. Netw. Comput. Appl. 97, 96–125 (2017)
9. M. Gidlund, S. Han, E. Sisinni, A. Saifullah, U. Jennehag, From industrial wireless sensor networks to industrial Internet of things. IEEE. Trans. Ind. Inf. 14(5), 2194–2198 (2018)
10. T. Liang, B. Zeng, J. Liu, L. Ye, C. Zou, An unsupervised user behavior prediction algorithm based on machine learning and neural network for smart home. IEEE. Access. 6, 49237–49247 (2018)
11. J. Liu, J. Wan, B. Zeng, Q. Wang, H. Song, M. Qiu, A scalable and quick-response software defined vehicular network assisted by mobile edge computing. IEEE. Commun. Mag. 55(7), 94–100 (2017)
12. Cheffena, industrial wireless sensor networks: channel modeling and performance evaluation. EURASIP. J. Wirel. Commun. Netw. 297 (2012)
13. C. Wang, J. Li, B. Wang, Face synthesis based on parts-based sparse component analysis face representation. Optik. Int. J. Light. Electron. Opt 140, 843–852 (2017)
14. M. Kumar, R. Tripathi, S. Tiwari, QoS guarantee towards reliability and timeliness in industrial wireless sensor networks. Multimed. Tools. Appl. 77(4), 4491–4508 (2018)
15. S. Wu, W. Chou, J. Niu, M. Guizani, Delay-aware energy-efficient routing towards a path-fixed mobile sink in industrial wireless sensor networks. SENSORS. 18(3), 899 (2018)
16. J. Tan, A. Liu, M. Zhao, H. Shen, M. Ma, Cross-layer design for reducing delay and maximizing lifetime in industrial wireless sensor networks. EURASIP J. Wirel. Commun. Netw. 50 (2018)
17. M. Huang, A. Liu, N.N. Xiong, et al., A low-latency communication scheme for mobile wireless sensor control systems. IEEE. Trans. Syst. Man. Cybern. Syst. 49(2), 317–332 (2019)
18. W. Zhang, J. Chang, F. Xiao, et al., Design and analysis of a persistent, efficient, and self-contained WSN data collection system. IEEE. Access. 7, 1068–1083 (2019)
19. J. Tan, W. Liu, T. Wang, et al., An adaptive collection scheme-based matrix completion for data gathering in energy-harvesting wireless sensor networks. IEEE. Access. 7, 6703–6723 (2019)
20. H. Zheng, W. Guo, N. Xiong, A Kernel-based compressive sensing approach for mobile data gathering in wireless sensor network systems. IEEE. Trans. Syst. Man. Cybern. Syst. 48(12), 2315–2327 (2018)
21. X. He, S. Liu, G. Yang, et al., Achieving efficient data collection in heterogeneous sensing WSNs. IEEE. Access. 6, 63187–63199 (2018)
22. K. Huang, Q. Zhang, C. Zhou, N. Xiong, Y. Qin, An efficient intrusion detection approach for visual sensor networks based on traffic pattern learning. IEEE Trans. Syst. Man. Cybern. Syst. 47(10), 2704–2713 (2017)
23. H. Cheng, Y. Chen, N. Xiong, et al., Layer-based data aggregation and performance analysis in wireless sensor networks. J. Appl. Math. 502381 (2013)
24. S. Montero, J. Gozalvez, M. Sepulcre, Neighbor discovery for industrial wireless sensor networks with mobile nodes. Comput. Commun. 111, 41–55 (2017)
25. Z. Zheng, J. Li, Optimal chiller loading by improved invasive weed optimization algorithm for reducing energy consumption. Energ. Buildings. 161, 80–88 (2018)
26. J. Li, Q. Pan, S. Xie, An effective shuffled frog-leaping algorithm for multi-objective flexible job shop scheduling problems. Appl. Math. Comput. 218(18), 9353–9371 (2012)

27. H. Sang, Q. Pan, J. Li, et al., Effective invasive weed optimization algorithms for distributed assembly permutation flowshop problem with total flowtime criterion. Swarm. Evol. Comput. **44**(6), 64–73 (2019)
28. Z. Zheng, J. Li, P. Duan, Optimal chiller loading by improved artificial fish swarm algorithm for energy saving. Math. Comput. Simul. **155**(SI), 227–243 (2019)
29. H. Sang, Q. Pan, P. Duan, et al., An effective discrete invasive weed optimization algorithm for lot-streaming flowshop scheduling problems. J. Intell. Manuf. **29**(6), 1337–1349 (2018)
30. J. Zhao, Y. Qin, D. Yang, J. Duan, Reliable graph routing in industrial wireless sensor networks. Int. J. Distrib. Sens. Netw. **9**(12), 758217 (2013)
31. J. Akerberg, M. Gidlund, T. Lennvall, J. Neander, M. Bjorkman, Efficient integration of secure and safety critical industrial wireless sensor networks. EURASIP. J. Wirel. Commun. Netw. **100** (2011)
32. J. Li, P. Duan, H. Sang, et al., An efficient optimization algorithm for resource-constrained steelmaking scheduling problems. IEEE. Access. **6**, 33883–33894 (2018)
33. C. Pei, Y. Xiao, W. Liang, X. Han, Trade-off of security and performance of lightweight block ciphers in Industrial Wireless Sensor Networks. EURASIP. J. Wirel. Commun. Netw. **117** (2018)
34. J. Li, Q. Pan, P. Duan, An improved artificial bee colony algorithm for solving hybrid flexible flowshop with dynamic operation skipping. IEEE. Trans. Cybern. **46**(6), 1311–1324 (2016)
35. K.Z. Gao, P.N. Suganthan, Q.K. Pan, et al., Artificial bee colony algorithm for scheduling and rescheduling fuzzy flexible job shop problem with new job insertion. Knowl. Based. Syst. **109**, 1–16 (2016)
36. Y.Y. Han, Q.K. Pan, J.Q. Li, et al., An improved artificial bee colony algorithm for the blocking flowshop scheduling problem. Int. J. Adv. Manuf. Technol. **60**, 1149–1159 (2012)
37. Y. Han, J.J. Liang, Q. Pan, et al., Effective hybrid discrete artificial bee colony algorithms for the total flowtime minimization in the blocking flowshop problem. Int. J. Adv. Manuf. Technol. **67**, 397–414 (2013)
38. J. Li, Q. Pan, Solving the large-scale hybrid flow shop scheduling problem with limited buffers by a hybrid artificial bee colony algorithm. Inf. Sci. **316**, 487–502 (2015)
39. Jun S, Wenbo X, Bin F, in Proceedings of 2005 IEEE International Conference on Systems, Man and Cybernetics. Adaptive parameter control for quantum-behaved particle swarm optimization on individual level (IEEE 2005), pp. 3049-3054.
40. X. Zhang, X. Zhang, A binary artificial bee colony algorithm for constructing spanning trees in vehicular ad hoc networks. Ad. Hoc. Networks. **58**(4), 198–204 (2017)

9

Node importance evaluation method based on multi-attribute decision-making model in wireless sensor networks

Rongrong Yin[1,2]*, Xueliang Yin[1,2], Mengdi Cui[1] and Yinghan Xu[1]

Abstract

Identifying important nodes is very crucial to design efficient communication networks or contain the spreading of information such as diseases and rumors. The problem is formulated as follows: given a network, which nodes are the more important? Most current studies did not incorporate the structure change as well as application features of a network. Aiming at the node importance evaluation in wireless sensor networks, a new method which ranks nodes according to their structural importance and performance impact is proposed. Namely, this method considers two aspects of the network, network structural characteristics and application requirements. This method integrates four indicators which reflect the node importance, namely, node degree, number of spanning trees, delay, and network energy consumption. Firstly, the changes in the four indicators are analyzed using the node deletion method. Then, the TOPSIS multi-attribute decision-making method is applied to merge these four evaluation indicators. On this basis, a more comprehensive evaluation method (MADME) for node importance is obtained. Theory study reveals MADME method saves computational time. And the simulation results show the superiority of the MADME method over various algorithms such as the N-Burt method, betweenness method, DEL-Node method, and IE-Matrix method. The accuracy of the evaluation can be improved, and the key nodes determined by the MADME method have a more obvious effect on the network performance. Our method can provide guidance on influential node identification in the network.

Keywords: Wireless sensor networks, Node importance, Multi-attribute decision-making, Node deletion, Structural importance, Application performance

1 Introduction

Wireless sensor networks are composed of a large number of sensors equipped with radio communication capabilities [1]. Owing to their simple deployment and flexible and fast distribution, they have been widely applied in intelligent home, agricultural production, and other fields. Wireless sensor networks have the capability of self-organizing, where a large number of nodes are used to make up multi-hop ad-hoc networks for information transmission by means of initial communications and negotiation [2]. The failure of some nodes in the network usually causes changes to the network structure and performance. Especially, the failure of the key nodes [3] in the network often leads to the collapse of the whole network. Therefore, in the dynamic and complex network environment, current wireless sensor network research must urgently address methods of determining the key node quickly and accurately and provide targeted protection, thus ensuring the reliability and stability of the network [4].

The problem of node importance evaluation [5] originates from the complex network, and the problem of node importance evaluation is mainly studied in terms of the structural characteristics and application requirements of the network. For example, previous researches [6, 7] used the structural characteristic indicators such as the degree and the K-shell, respectively, to quantify the importance of a node. Furthermore, the application requirement indicators such as the network transmission efficiency and the

* Correspondence: yrr@ysu.edu.cn
[1]School of Information Science and Engineering, Yanshan University, Qinhuangdao, China
[2]The Key Laboratory of Special Fiber and Fiber Sensor of Hebei Province, Yanshan University, Qinhuangdao, China

load flow, respectively, were also used to assess the importance of nodes in a complex network [8, 9].

However, for the actual wireless sensor network, the performance indicator of the network application requirements is the direct target of the network optimization. In terms of indicators of the network structure importance, there are some limitations when analyzing the actual application requirements for network timeliness or network lifetime. But at the same time, network performance is essentially information transfer processes along nodes or links. Thus, network topology plays an important role. The local and global structure characteristics of a node directly reflect the efficiency of information transmission and the energy loss in the network. If we only consider the performance indicators of the network application requirements, and the influence of structural importance on network performance is ignored, it would lead to a weak robustness of the network topology. From the above discussion, we can easily find that node importance evaluation in a network is a result of the structural importance and performance impact jointly. Therefore, the node importance in wireless sensor networks cannot be evaluated with only a single indicator and requires a combination of the structural characteristics and application requirements as the two aspects for comprehensive evaluation.

Our main contributions are summarized as follows.

1. In terms of the structural characteristics of the network, consider the change in the sum of the node degree and the change in the number of spanning trees before and after removing nodes as evaluation indicators.
2. With regard to the application requirements of the network, consider the amount of delay changes and the amount of energy consumption changes when a node is removed as evaluation indicators.
3. The TOPSIS multi-attribute decision-making model is constructed, which combines the above four indicators. A novel method for node importance evaluation in wireless sensor network is proposed based on the model.
4. Through the simulation analysis, the method is more comprehensive and accurate than other methods based on a single indicator, and some key nodes with small differences can be discovered. In addition, it is also verified that this method can improve the evaluation validity and is significant for deliberate attacks.

The rest of the paper is organized as follows. Section 2 describes the key design methods of MADME and the experiments. Section 3 describes the related works. Section 4 introduces the attributes determination of the decision model and respectively takes two evaluation indicators from the structural characteristics and the application requirements as the attributes. Section 5 mainly introduces the construction process of the TOPSIS multi-attribute decision-making model and the steps of the node importance evaluation method based on this decision model. Section 6 presents the experiment results and discussion. Finally, Section 7 concludes the paper.

2 Methods/experimental

The aim of this paper is to identify the key nodes in wireless sensor networks. To solve this problem, an evaluation model for node importance based on multi-attribute decision-making is proposed. This model incorporates the structural features as well as the application requirements of a node in wireless sensor networks. In addition, based on information entropy method, this model analyzes the weight influence of the evaluation indicators. Finally, a comprehensive evaluation method MADME, which jointly considers node degree, the number of spanning trees, delay, and energy consumption, is designed to reflect the node importance. To analyze the performance of MADME, extensive simulations are carried out. The simulations consider two aspects, accuracy and validity. Simulation results confirm that MADME can distinguish the important nodes with slight difference, and according to the key nodes obtained by the MADME method to deliberately attack the network, the network is quickly disintegrated.

3 Related works

The structural characteristics and application requirements are the two main aspects of the network. Herewith, we present previous research on the node importance evaluation from these two aspects.

Based on the structural characteristics of the network, Wang [10] considers the node degree and proposes a method for evaluating the importance based on the local characteristic of a node. This method states that the greater the degree of both nodes and neighbor nodes, the more important the node, which is simple and effective. In reference [11], considering the relationship between the betweenness [12] and the node degree, a new indicator for node importance is defined, the greater the value of the indicator, the more important the node. Although the method is more accurate than that of the single indicator (e.g., betweenness and node degree), the time complexity is high. In reference [13], the node importance is quantified by the global influence of a node. On the basis of node deletion method, the number of spanning trees is proposed. The most important node is defined that the number of spanning trees is the smallest after being removed, but the time complexity is not reduced. In reference [14], the graph

Fourier transform centrality (GFT-C) is introduced to quantify how important a particular node is to other nodes in a network. GFT-C utilizes not only the local properties, but also the global properties of a network topology. However, the time complexity is still higher. To lower the time complexity, reference [15] introduces the least square support vector machine (LS-SVM) method to establish the evaluation model. LS-SVM method selects four complicated importance indicators and finds the relationship of simple attributes from local properties and node importance obtained from global attributes. The computational complexity is decreased significantly.

Based on the application requirements of the network, reference [16] evaluates node importance from the perspective of delay. It shows that removing the most important node always results in a maximum increase in the shortest distance from the source node to the sink node and has the largest transmission delay of network. Reference [17] proposes an approach based on an energy field model which evaluates the node importance by analyzing the status of data transmission among associated nodes. In reference [18], a weighted minimum path tree is used as the metric. It determines whether the node is important based on the weighted path tree function. The method extends the life of the network to a certain extent, but each node has to rebuild the shortest path tree of the whole network after removing nodes, resulting in a waste of energy. To solve this problem, reference [19] comprehensively considers the remaining lifetime of nodes and the increase in energy consumption caused by removing a node. And a method is proposed based on an energy indicator. It can find the key nodes with faster energy consumption and important position in the network, which is of great significance to enhance the invulnerability and prolong the life of the network. On the basis of a "no return" node deletion method, reference [20] uses the network efficiency, largest component size, and network flow as the indicators of network performance. Evaluating node importance is done by comparing the change of network overall performance before and after deleting nodes.

In the research of node importance evaluation, the concepts of "structural characteristics" and "application requirements" have been well studied, but these research results mentioned for node importance evaluation only consider the impacts of structural characteristics or application requirements, which ignore the compositive influence of both aspects. In recent years, considering there is a relation between the location of a node in the network and its influence in network performance, some scholars have also considered comprehensively both aspects of the evaluation method. In reference [21], someone believes that the important node in road traffic

network is related to the traffic flow through the node and the location of the node. And a method based on node contraction is obtained by combining two indicators. Reference [22] gives a new centrality called density centrality to identify and rank the node importance. The density centrality is computed by considering the degree and the distance between two nodes. Reference [23] uses the node efficiency and the node degree to evaluate the node importance. Reference [24] studies the bi-objective critical node detection problem and finds a set of solutions which minimize the pairwise connectivity of the induced graph and the cost of removing these critical nodes at the same time. These methods improve the comprehensiveness of critical node judgment, but ignore the weight problem of the various indicators, which makes the obtained important nodes far from reality. The accuracy of their evaluation requires improvement. Therefore, to research the node importance evaluation, the evaluation model should switch from the unilateral indicators to combined indicators, and the weight of the various indicators on the node importance should be considered.

Taking into account the above description, in this paper, we propose a multi-attribute decision-making model to evaluate the node importance from both these two perspectives. There are two goals need to be reached, one is to improve the evaluation comprehensiveness, and the other is to increase the evaluation accuracy. This study takes the node degree and the number of spanning trees as the structural characteristics, and takes the delay and energy consumption as the application requirements, and the information entropy method is adopted to obtain the weight of each indicator. Then, we integrate the contribution degree of the four indicators to the node importance and construct the multi-attribute decision-making model. Finally, a more comprehensive method for node importance evaluation is proposed. This method not only overcomes the limitations of using a unilateral evaluation, but also takes into account the weight influence of the evaluation indicators, which makes the node importance evaluation more comprehensive and accurate.

4 Attribute determination of the decision model

Aiming at the node importance evaluation, the evaluation indicators are divided into two aspects: structural characteristics and application requirements. Usually, the structural characteristics of a network have a node degree, betweenness, spanning tree, and so on. Network application requirements include throughput, delay, energy consumption, and others. This study combines the structural characteristics and application requirements of a node. In order to quantify the node importance, the changes in indicators of both the structural characteristics and application requirements before and after removing nodes are analyzed. And four indicators are used as the attributes of a decision model.

4.1 Structural characteristic indicators of the network

The structural characteristics of the network affect the robustness of the network topology. By analyzing the relationship between the node itself and the location information in the network, the local and global indicators are obtained. This study takes the node degree and the number of spanning tree as the evaluation indicators of the structural characteristics. The changes in the node degree and the number of spanning trees are calculated when a node is removed, and are chosen as the attributes in the decision model.

4.1.1 Node degree

The node degree refers to the local attribute of the structural characteristics, which indicates the number of neighbor nodes. By analyzing the relationship between the nodes, it can reflect the direct influence of a node on other nodes in the network.

Set up a network $G = (V, E)$, where $V = \{v_1, v_2, ..., v_n\}$ corresponds to the set of nodes and $E = \{e_1, e_2, ..., e_m\}$ corresponds to the collection of edges. It has a total of n nodes and m edges and is a non-looped non-connected map.

Its total node fully associative matrix is defined as $A_c = [a_{ij}]_{n \times m}$, where n corresponds to the number of nodes and m corresponds to the number of edges in the graph. The elements a_{ij} can be expressed as [25]

$$a_{ij} = \begin{cases} 1 & \text{node } i \text{ is associated with edge } j \\ 0 & \text{node } i \text{ is not associated with edge } j \end{cases} \quad (1)$$

The node degree of node i is calculated as

$$k(i) = \sum_{j \in m} a_{ij} \quad (2)$$

where the change in node degree for the entire network before and after removing node i depends on the sum of the changes both the node i and its neighbor nodes.

$$K_i = 2k(i) \quad (3)$$

4.1.2 Number of spanning trees

The number of the spanning tree is considered to be the global attribute of the structural characteristics. This means that some edges of the connected graph are removed, the nodes in the graph can be connected, and the whole graph does not appear in the ring structure. By analyzing the number of spanning trees after removing nodes and related edges, the node importance based on network topology is reflected.

According to the matrix theory of the Binet-Cauchy theorem [13], the formula for the number of spanning trees can be obtained. Set G as an undirected graph and τ is the number of spanning trees for graph G. For the associative matrix A_c, each row corresponds to a node and each column corresponds to an edge. Arbitrarily removing the i row of A_c (node i is used as the reference node) will obtain the matrix A. The number of spanning trees is

$$\tau(G) = \det(AA^T) \quad (4)$$

Begin to delete the node in the matrix A_c, for the ith node, removing the ith row and the columns where the element is not zero in ith row. A new matrix B_c is formed. Matrix B is obtained by removing any row from the matrix B_c as a reference node. The number of spanning trees can be expressed as

$$\tau(G-v_i) = \det(BB^T) \quad (5)$$

Thus, the change in the number of spanning trees of the network is

$$\tau_i = \tau(G) - \tau(G-v_i) \quad (6)$$

4.2 Application requirements indicators of the network

The application requirements of the network are mainly based on the actual network performance. The analysis shows that the performance changes of the nodes in the network information transmission process affect the network performance. Therefore, this study selects the delay and energy consumption of the two performance indicators to quantify the node importance. We calculate the amount of network delay and network energy consumption based on a node deletion method. And we use them as attributes in the decision model.

4.2.1 Delay

The delay can effectively reflect the transmission timeliness of the actual network. It depends on the transmission rate and transmission distance of the nodes in the network. After the deletion of a node, the delay is increased mainly owing to the increase in the shortest path distance of each node in the network. It leads to a larger information transmission distance. Therefore, the change in the network delay before and after removing nodes can be reflected by the change in the shortest path distance. After deleting a node, the greater the amount of delay changes, the greater the node importance.

In the undirected graph $G = (V, E)$, the weight of each edge is known, and the shortest path distance among nodes is found according to the Floyd algorithm [12]. Before removing the node i, calculate the sum of the shortest path distance among n node pairs, and record each as d_i, where the ith should not contain the sum of the shortest path distance from node i to the remaining nodes. After removing the node i, recalculate the sum of the shortest path distance d'_i, so the amount of change in the shortest path distance when node i is removed (that is, the amount of delay changes) is

$$D_i = d_i - d'_i \qquad (7)$$

4.2.2 Energy consumption

For energy-constrained wireless sensor networks, energy consumption affects the lifetime of the network. The energy consumption of the network includes the energy consumption of the receiving data and transmitting data. Generally, the node is in an important position in the network, where the greater the receiving and transmitting data, the larger the energy consumed. Thus, the amount of the energy consumption changes, to a certain extent, can reflect the node importance.

To quantify the energy consumption of the network, the nodes are stratified according to the number of minimum hops form the node to the sink node (located in the center of the monitoring area). This is shown in Fig. 1, where numbers 1–24 represent the nodes, the dotted circles represent the network layers, from the inner to the outer for the first to the third layer. The

node away from the sink is the sub-node, and the node near the sink is the parent node.

When node i is removed, the path from the sub-node of node i to the sink node (where the path is the minimum number of hops from sub-node to the sink node) becomes longer, resulting in the increased energy consumption. The energy consumed (time t) before removing node i is denoted by $E_t(i)$, and the energy consumed (time $t + 1$) after removing node i is denoted by $E_{t+1}(i)$. We will have the increased energy consumption E_{ADD}

$$E_{\mathrm{ADD}i} = E_{t+1}(i) - E_t(i) \qquad (8)$$

The energy consumption adopts the first-order radio model [26], where $E_t(i)$ is the energy consumption of all sub-nodes for node i; these sub-nodes transmit data by the node with the same layer at t time. The energy consumption at time t is

$$E_t(i) = \sum_{s=1}^{j} P_t v_E(s) \left(2E_{\mathrm{elec}} + \varepsilon_{\mathrm{amp}} d^2 \right) \qquad (9)$$

where E_{elec} is the RF transmission coefficient, $\varepsilon_{\mathrm{amp}}$ is the amplification factor of the transmitting device, d is the data transmission distance between nodes, j is the number of sub-nodes that need transmit data through the same layer node, P_t is the normalized probability that the sub-node s of node i transmits the data through the same layer node at time t, and $v_E(s)$ is the energy

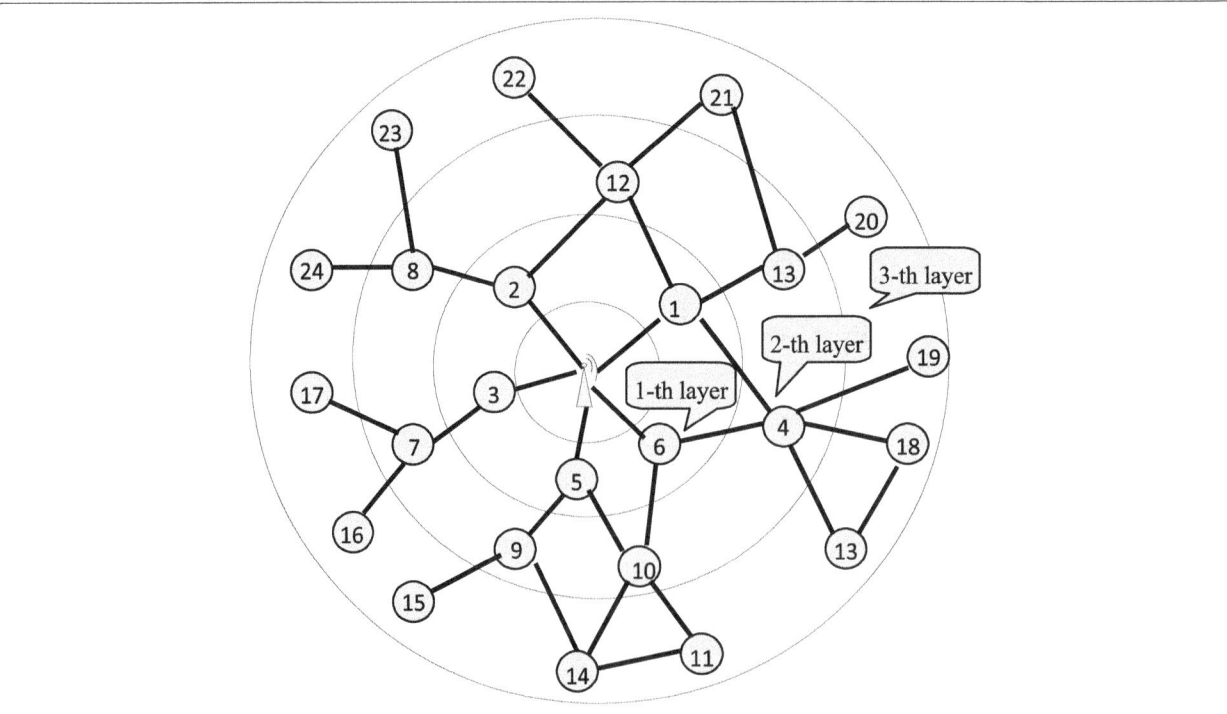

Fig. 1 Schematic diagram of network structure. An example of a network model, where 1–24 represent the nodes, the center is the sink node, and the dotted circles represent the network layers from the inner to the outer for the first to the third layer

consumption rate of node s. The energy consumption at $t + 1$ (after removing node i) can be given by $E_{t+1}(i)$

$$E_{t+1}(i) = \sum_{s=1}^{j} P_{t+1} v_E(s) \left(2E_{\text{elec}} + \varepsilon_{\text{amp}} d^2\right) \tag{10}$$

Substituting Eq. (9) and Eq. (10) into Eq. (8), we can derive the following result

$$E_{\text{ADD}i} = \sum_{s=1}^{j} P_{t+1} v_E(s) \left(2E_{\text{elec}} + \varepsilon_{\text{amp}} d^2\right)$$
$$- \sum_{s=1}^{j} P_t v_E(s) \left(2E_{\text{elec}} + \varepsilon_{\text{amp}} d^2\right) \tag{11}$$

Finally,

$$E_{\text{ADD}i} = \sum_{s=1}^{j} (P_{t+1} - P_t) v_E(s) \left(2E_{\text{elec}} + \varepsilon_{\text{amp}} d^2\right) \tag{12}$$

Equation (12) shows that when $(P_{t+1} - P_t)$ is not zero, the sub-node s has the same layer node for data transmission. That is, the removal of node i increases the energy consumption of the network owing to the increased energy consumption of node s.

As shown in Fig. 1, we calculate the increased energy consumption of the network when node 10 is removed. The added value depends on the increased energy consumption of the sub-node 11 during the data transmission. Since the number of the shortest path from node 14 (which is also node 10's sub-node) to the sink node does not change, the energy consumption of node 14 does not increase because of the removal of node 10.

5 Evaluation method for node importance based on the multi-attribute decision model

Multi-attribute decision-making is generally the use of existing decision-making information in a certain manner to sort a set of limited options and merit. Based on the structural characteristics and application requirements of the network, this study analyzes synthetically the changes in node degree, spanning tree, delay, and energy consumption, and evaluates the node importance for the network. Thus, the node importance evaluation is a multi-attribute decision-making problem.

This study uses a common multi-attribute decision-making method—TOPSIS method [27]. TOPSIS is the sorting method of approximating the ideal solution, which sorts the schemes by the ideal solution and the negative solution of the multiple attributes. The idea of node importance evaluation method based on TOPSIS is that every node is regarded as a scheme and the evaluation indicators are regarded as the attributes of each

scheme, and evaluating the importance of each scheme is the criteria [28] for decision-making.

The following section lists the multi-attribute decision model and the steps of the node importance evaluation method based on this decision model.

5.1 The construction of the multi-attribute decision-making model

Assuming that there are m nodes in the network, the corresponding set of decision schemes can be expressed as $A = \{a_1, a_2, ..., a_m\}$. In this paper, when the node i is removed, there are four indicators to evaluate its importance, including the amount of change in the node degrees K_i, the amount of change in the number of spanning trees τ_i, the amount of change in the delay D_i, and the amount of change in the energy consumption $E_{\text{ADD}i}$. Its matrix is $X = (x_{ij})_{m \times 4}$, where x_{ij} is the jth indicator of the ith node, and $i = 1,2,...,m$, $j = 1,2,3,4$.

$$X = \begin{bmatrix} x_{11} & x_{12} & x_{13} & x_{14} \\ x_{21} & x_{22} & x_{23} & x_{24} \\ \vdots & \vdots & \vdots & \vdots \\ x_{m1} & x_{m2} & x_{m3} & x_{m4} \end{bmatrix} = (K_i, \tau_i, D_i, E_{\text{ADD}i})_{m \times 4} \tag{13}$$

There are intricate relationships between the indicators that can be divided into interest attributes (the larger the value is, the more important the node is) and cost attributes (the smaller the value is, the more important the node is). According to the analysis of the attributes, we can see that the four indicators (e.g., K_i, τ_i, D_i, $E_{\text{ADD}i}$) are all interest attributes. As the dimension of each indicator is different, for the sake of comparison, the original decision matrix $X = (x_{ij})_{m \times 4}$ is processed to obtain the dimensionless decision matrix $Y = (y_{ij})_{m \times 4}$.

The interest attributes are standardized as below

$$y_{ij} = \frac{x_{ij} - \min\limits_{i} x_{ij}}{\max\limits_{i} x_{ij} - \min\limits_{i} x_{ij}} \tag{14}$$

The weight vector of the four indicators K_i, τ_i, D_i, $E_{\text{ADD}i}$ is recorded as $w = (w_1, w_2, w_3, w_4)^T$ and conforms to $\sum_{j=1}^{4} w_j = 1$, which constitutes a weighted normalized matrix $Z = (z_{ik})_{m \times 4}$

$$z_{ik} = w_k y_{ik}, i = 1, 2, ..., m; k = 1, 2, 3, 4 \tag{15}$$

The information entropy method [29] belongs to the objective weighting method. It is used to calculate the weight vector $w = (w_1, w_2, w_3, w_4)^T$. The closer the value of each attribute in different schemes, the greater the entropy, thus the weight of the indicator is more objective.

First, the matrix Y is normalized to obtain a normalized matrix $H = \{h_{ij}\}_{m \times 4}$

$$h_{ij} = y_{ij} / \sum_{i=1}^{m} y_{ij} \qquad (16)$$

Then, the information entropy of the indicator is given by

$$E_j = -\frac{1}{\ln m} \sum_{i=1}^{m} h_{ij} \ln h_{ij}; \; j = 1, 2, 3, 4 \qquad (17)$$

And the weight vector of the four indicators is

$$w_j = \frac{1 - E_j}{\sum_{k=1}^{4} (1 - E_k)} \qquad (18)$$

Furthermore, the positive and negative ideal solutions of the decision scheme (n nodes) are determined according to the matrix Z, we will have

$$A_k^+ = \max_i(z_{ik}) = \{z_1^{max}, z_2^{max}, ..., z_n^{max}\} \qquad (19)$$

$$A_k^- = \min_i(z_{ik}) = \{z_1^{min}, z_2^{min}, ..., z_n^{min}\} \qquad (20)$$

And then the translation matrix $T = (t_{ik})_{m \times 4}$ is used to translate the matrix Z, we obtain t_{ik}

$$t_{ik} = z_{ik} - A_k^+ \qquad (21)$$

By translating, the positive ideal solution becomes $\{0, 0, ..., 0\}$, and the negative ideal solution becomes $A_k'^- = t_{lk}$, that $|t_{lk}| \geq |t_{ik}|$, $1 \leq l \leq m$.

Now, we will calculate the vertical distance VD_i. VD_i reflects the degree that the scheme approaches the ideal solution. The smaller the value of VD_i, the better the scheme. We will have

$$VD_i = \left| A_k'^- \cdot T_{ki} \right| = \sum_{k=1}^{4} \left(A_k'^- \times t_{ik} \right) \qquad (22)$$

Finally, obtain the reciprocal of the vertical distance VD_i and do the normalization process, VDD_i will be given by

$$VDD_i = \frac{1/VD_i}{\max\{1/VD_i\}} \qquad (23)$$

VDD_i is used for evaluating the node importance. Sort the node importance according to the order from large to small, and ultimately the evaluation of node importance based on multi-attributes will be realized.

5.2 The node importance evaluation method MADME
On the basis of the TOPSIS multi-attribute decision model, the steps of the node importance evaluation method (named MADME method) are given below (Fig. 2):

From the above method, it can be seen that the time complexity of the method depends on the calculation of the amount of change in the node degree $T = (t_{ik})_{n \times 4}$, the amount of change in the number of spanning trees VD_i, the amount of change in the delay VD_i, and the amount of change in the energy consumption E_{ADDi}. Calculating the amount of change in the node degree needs to consider the degree of each node in the network, and time complexity is $O(n)$. Calculating the amount of change in the number of spanning trees requires n cycles of the fully associative matrix A_C, and its time complexity is $O(n^3)$. Calculating the amount of delay needs to calculate the shortest path distance between nodes by using the Floyd algorithm, then the time complexity is $O(n^3)$. Calculating the amount of energy consumption based on the CNDBE algorithm, the time complexity is $O(n^2)$. Therefore, the time complexity of MADME method is $O(n^3)$.

6 Results and discussion
We use Matlab to implement our simulations. In the simulations, the sensor nodes are randomly deployed in the simulation area. For the purpose of demonstrating the efficiency of the method proposed, the BA ($m_0 = 3$, $m = 2$) scale-free network commonly used in wireless sensor networks is generated as a test bed. The network is inherently robust and efficient [30], and it has good fault tolerance and survivability against random node failure.

6.1 Accuracy verification of the method
First, we consider the accuracy of the method; the following is analyzed from the two aspects of both the method itself and comparison with other methods.

6.1.1 Analysis of the method itself
In this section, we analyze the accuracy of the method itself. This study uses the following simulation environment. The nodes are distributed in a square area of 1000 m × 1000 m, the initial energy of each node is 2 J, and the number of nodes is 100. The specific parameters of the experiment are shown in Table 1.

We use the simulation environment in Table 1 to obtain a BA scale-free network (as shown in Fig. 3). Then, we use the MADME method to calculate the node importance. The result is shown in Fig. 4.

It is easy to see from Fig. 4 that the node numbers 42, 97, and 100 have a greater node importance. The network topology (see Fig. 3) shows that the connectivity among these three nodes and other nodes is relatively large. Obviously, these three nodes are the key nodes for the network.

Then, we analyze the changes to the four indicators of the node, and the influence of the different indicator

INPUT: Completely associative matrix A_C with n nodes

OUTPUT: The importance VDD_i of node i

BEGIN

1.　FOR (i from 1 to N) do
2.　　　Calculate the node degree $k(i)$, the shortest path distance matrix between all nodes $D_{BE} = \lfloor d_{ij} \rfloor$, the number of spanning trees for the entire network $\tau(G)$, and the energy consumption of the sub-nodes $E_t(i)$

3.　END FOR
4.　FOR (i from 1 to N) do
5.　　　Delete node i in turn
6.　　　　Calculate the amount of change in node degree K_i, the amount of change in the number of spanning trees τ_i, the shortest path distance matrix $D_{AF}^{(i)} = \left[d_{st}^{(i)} \right]$ between all nodes and get the amount of change in the delay D_i, and calculate the energy consumption $E_{t+1}(i)$ of the sub-nodes of node i and get the amount of change in energy consumption E_{ADDi}

7.　END FOR
8.　FOR (j from 1 to 4) do
9.　　　Calculate the information entropy E_j and weight vector w_j of above four indicators

10.　END FOR
11.　Construct a decision matrix $X = \left(x_{ij} \right)_{n \times 4}$, and normalize it to get matrix $Y = \left(y_{ij} \right)_{n \times 4}$, and then get a weighted normalization matrix $Z = \left(z_{ij} \right)_{n \times 4}$. Calculate the positive ideal solution A_k^+ and the negative ideal solution A_k^- of the matrix, and get the translation matrix $T = \left(t_{ik} \right)_{n \times 4}$

12.　FOR (i from 1 to N) do
13.　　　Calculate the vertical distance VD_i of node i
14　　　Get the reciprocal of VD_i, do the normalization process and get VDD_i as the importance of each node

15 END FOR

Fig. 2 The proceeding of the MADME method. The pseudo code of the MADME method that is an evaluation method for node importance based on a multi-attribute decision-making model

value on the node importance. The nodes are sorted according to the importance from low to high, and the corresponding four indicators of each node follow the node importance from high to low. The corresponding results can be obtained (as shown in Fig. 5).

Figure 5 shows that the four indicators decrease with the decrease in node importance. This is consistent with the obtained conclusions that the four indicators belong to the interest indicators. On the other hand, Fig. 5 also shows that the four indicators of individual node do not conform to the overall trend of change. It means that the importance of various indicators for individual node is different. Each indicator weight value reflects the influence of the indicator on the node importance. Thus, the result also indicates that the node importance is related to the weight of the four indicators.

To clarify the advantages of the MADME method, the first 20% key nodes are refined in the overall network. Table 2 gives the importance of the first 20% key nodes and the corresponding indicator values.

From Table 2, the structural characteristics and application requirements of nodes have different values for each indicator, which affects the node importance. For example, the increase in network delay and energy consumption when node V42 is removed is littler than that of V65. The change in node degree and number of spanning trees is considerably greater after deleting node V42 than V65, and node V42 is more important than node V65. It means that node V42 is more important in the network structure than node V65. As another example, after removing node V88 and V98, respectively, the increase in the sum of node degree and the increase in delay are not large. Although the change in the number of spanning trees after removing node V98 is greater than that of node V88, the change in the energy consumption is littler, and node V88 is more important than node V98. It means that node V88 is more important than node V98 in the network application. As we explained above, various indicators identify different nodes as the important nodes. Our MADME, on the contrary,

Table 1 Experimental parameters

Parameter	Value	Parameter	Value	Parameter	Value
Node distribution area A	1000 m × 1000 m	Node initial energy	2 J	Node maximum transmission radius R	300 m
Number of nodes	100	Transmission rate	20 kbps	The amplification factor of transmitting device	100 pJ/bit/m^2
Sink node coordinates	(500,500)	Radio frequency transmission coefficient	50 nJ/bit	The node itself generates data L	4000 bits

takes care of both structural characteristics and application requirements of all nodes and identifies nodes V42, V65, V88, and V98 as the important nodes. These results demonstrate the reasonableness of the MADME method.

6.1.2 Comparison with other algorithms

To further evaluate the accuracy of the MADME method, the comparing methods include the N-Burt method [31], betweenness method [12], DEL-Node method [13], and the IE-Matrix method [23]. The N-Burt method improves the structural holes indicator from the perspective of local importance. It can find the important nodes with both the degree attribute and the bridging attribute in the network. The betweenness method analyzes the node position from the perspective of global importance and measures the node importance. The DEL-Node method evaluates the node importance based on the number of spanning trees after deleting a particular node. It can quantify the important nodes within the global network from the perspective of network structure. The IE-Matrix method integrates the

node degree and efficiency and comprehensively evaluates the importance from the perspective of both the structure and the performance of a node. The comparing results are listed in Table 3.

By analyzing the network topology (see Fig. 3) and the node importance evaluation results (see Table 3), it can be observed that the connection degree of node V100 is the largest, whereas the connection degree of nodes V97 and V92 is 27. Obviously, their importance is different. Thus, it can be seen that if we take the ranking of node degree as the result, it is likely to ignore the difference of each node. So the node importance cannot be accurately evaluated only by the node degree.

The N-Burt method is used to quantify the key node, and the structural hole indicator is improved by analyzing the node degree and its neighborhood structure. As shown in the network topology (see Fig. 3), although node V89 has more structural holes than node V92, the connection degree of node V92 is greater than that of node V89. In addition, the delay and energy consumption of node V89 are smaller than those of node V92. It is obvious that the node importance is inaccurate by

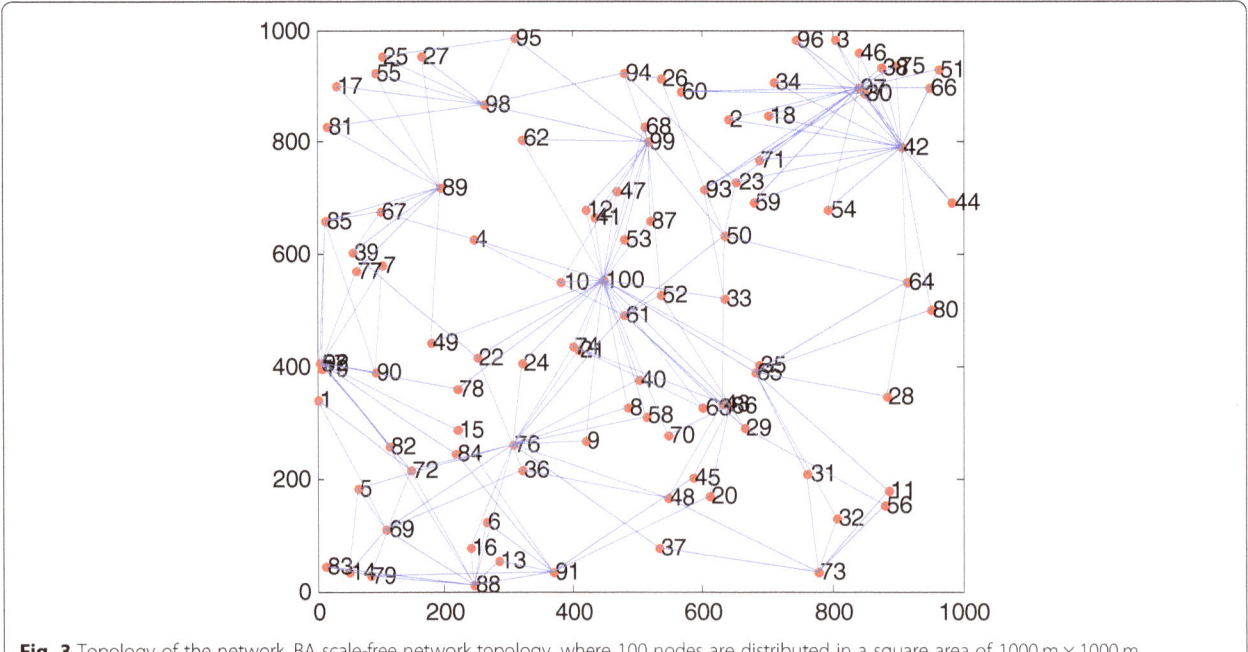

Fig. 3 Topology of the network. BA scale-free network topology, where 100 nodes are distributed in a square area of 1000 m × 1000 m

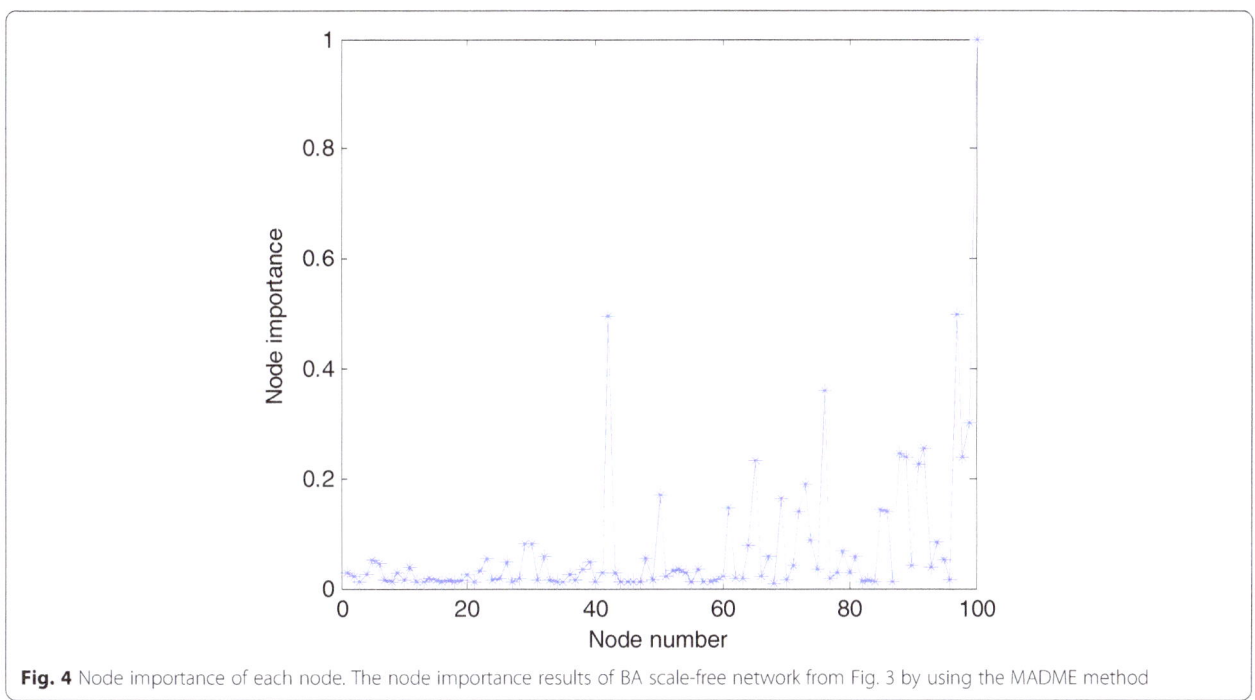

Fig. 4 Node importance of each node. The node importance results of BA scale-free network from Fig. 3 by using the MADME method

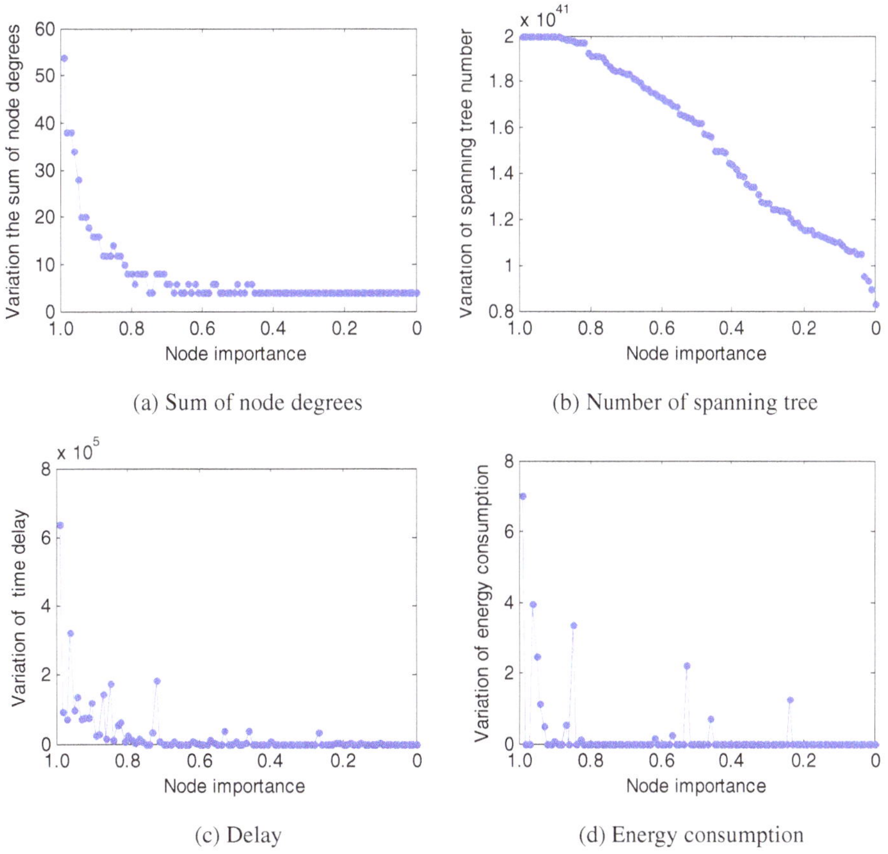

(a) Sum of node degrees

(b) Number of spanning tree

(c) Delay

(d) Energy consumption

Fig. 5 Changes in the four indicators. The changes in the four indicators of the key nodes, and the influence of four indicators on the node importance. **a** Sum of node degrees. **b** Number of spanning tree. **c** Delay. **d** Energy consumption

Table 2 Calculation results of each indicator for the first 20% key nodes in the network

Node number	VDD_i	K_i	τ_i	D_i	E_{ADDi}
V100	1	54	1.9978e+41	6,367,736	7.0150
V97	0.4993	38	1.9978e+41	92,853	0
V42	*0.4953*	*38*	*1.9978e+41*	*72,759*	*0*
V76	0.3607	34	1.9977e+41	323,705	3.9294
V99	0.2998	28	1.9977e+41	99,352	2.4558
V92	0.2555	20	1.9975e+41	136,308	1.1236
V88	*0.2435*	*20*	*1.9948e+41*	*71,560*	*0.4788*
V89	0.2398	18	1.9955e+41	76,375	0
V98	*0.2374*	*16*	*1.9973e+41*	*74,538*	*0*
V65	*0.2322*	*16*	*1.9959e+41*	*118,073*	*0.0543*
V91	0.2268	16	1.9955e+41	22,348	0
V73	0.1897	12	1.9900e+41	26,618	0
V50	0.1693	12	1.9825e+41	144,279	0.5175
V69	0.1640	12	1.9806e+41	13,175	0
V61	0.1473	14	1.9743e+41	175,905	3.3506
V85	0.1439	12	1.9708e+41	8831	0
V72	0.1406	12	1.9686e+41	56,043	0.1041
V86	0.1383	10	1.9693e+41	62,593	0
V74	0.0883	8	1.9246e+41	5171	0
V94	0.0825	8	1.9153e+41	23,628	0

considering the network structure only. As shown in Table 3, our MADME method recognizes the importance of nodes V92 and V89 (in descending order) by taking care of both the structural importance and performance impact of all nodes, which overcomes the defect of the N-Burt method.

The betweenness method reflects the node importance well in the process of information transmission. However, some nodes with the similar importance position, such as nodes V78 and V98, still need to be further quantified by combining with the application indicators. In the DEL-Node method, after deleting the nodes V100, V97, V42, V76, and V99, the network is divided into different areas. The number of spanning trees is zero, and the corresponding node importance is one. We also need to further distinguish the importance of these nodes. As shown in Table 3, our MADME method takes care of both local and global influences of all nodes and distinguishes the differences of nodes V78 and V98 and nodes V100, V97, V42, V76, and V99, and overcomes the shortcomings of the betweenness and DEL-Node methods.

The IE-Matrix method considers the structure indicators and performance indicators simultaneously. However, because the weight of these indicators is unknown, IE-Matrix fails to estimate influential nodes. For example, from the network performance consideration, node V99 is more important than node V97. But the weight of the structural characteristics is greater than that of the application requirements. From the two perspectives, node V97 is more important than node V99. As shown in Table 3, our MADME method takes into account the weight difference of the indicators, detects nodes V97 and V99 (in descending order) as the most influential nodes, and overcomes the weakness of the IE-Matrix method.

These comparison results show the superiority of our MADME method over other methods. By incorporating network structural characteristics and the application requirements, and taking into account the weight value of various indicators, the accuracy of the node importance evaluation can be further improved. And the importance of the special nodes with slight difference can be further distinguished remarkably than the existing methods.

6.2 Validity verification of the method

Next, we consider the validity of the MADME method; this study attempts to purposefully remove the important nodes in the network, which simulates the intentional attack. We use the maximum number of connected branch nodes as the performance measure. The effect of intentional attack based on node importance on the robustness of the network is analyzed. First, in the 100-node network, the top ten nodes with the highest importance evaluated by each method are removed, as shown in Fig. 6.

From Fig. 6, we observe that compared with other methods, the number of the maximum connected branch nodes based on the MADME method decreases most greatly. When the first six key nodes are removed according to the ranking of the MADME method, the number of maximum connected branch nodes is less than 50, while according to the rankings of other four methods, the performance is more than 50. It is obvious that the overall performance of the MADME method declined much quicker than that of the four other methods, which tells us that the key nodes discovered by the MADME method are more crucial to the network. This is because the MADME method takes care of both structure role and application features of all nodes. That is to say, if the key node obtained by the MADME method is under attack, the network will collapse rapidly, which verifies the validity of the MADME method.

Furthermore, we also investigated the network with different sizes. After removing the first 10% of the key nodes identified by the above methods, the proportions of the maximum connected branch node accounts for the all nodes in the network are shown in Fig. 7.

By analyzing the curves in Fig. 7, for different network sizes, when removing the first 10% of the key nodes

Table 3 Evaluation results of node importance

Node ranking	MADME		N-Burt		IE-Matrix		Betweenness		DEL-Node	
	Node number	Importance value	Node number	Importance value	Node number	Importance value	Node number	Importance value	Node number	Importance value
1	V100	1	V100	0.3564	V100	0.5534	V100	0.4236	V100	1
2	V97	0.4993	V76	0.2703	V61	0.5231	V23	0.2749	V97	1
3	V42	0.4953	V89	0.2672	V99	0.4757	V61	0.2669	V42	1
4	V76	0.3607	V92	0.2411	V97	0.4585	V50	0.2665	V76	1
5	V99	0.2998	V73	0.2179	V30	0.4478	V76	0.2523	V99	1
6	V92	0.2555	V50	0.2013	V76	0.4352	V97	0.2309	V92	0.9999
7	V88	0.2435	V88	0.1947	V74	0.3871	V92	0.1402	V98	0.9998
8	V89	0.2398	V61	0.1892	V42	0.3586	V99	0.1299	V65	0.9991
9	V98	0.2374	V42	0.1828	V10	0.3253	V65	0.1089	V91	0.9989
10	V65	0.2322	V65	0.1810	V53	0.3147	V72	0.1042	V89	0.9989
11	V91	0.2268	V85	0.1791	V68	0.3090	V74	0.0970	V88	0.9985
12	V73	0.1897	V99	0.1731	V21	0.3083	V42	0.0931	V73	0.9961
13	V50	0.1693	V97	0.1691	V41	0.3059	V88	0.0915	V50	0.9923
14	V69	0.1640	V98	0.1670	V65	0.2790	V4	0.0871	V69	0.9914
15	V61	0.1473	V23	0.1667	V87	0.2560	V78	0.0859	V61	0.9883
16	V85	0.1439	V72	0.1651	V47	0.2334	V98	0.0859	V85	0.9865
17	V72	0.1406	V91	0.1647	V12	0.2285	V67	0.0796	V86	0.9858
18	V86	0.1383	V64	0.1630	V24	0.2127	V89	0.0792	V72	0.9854
19	V74	0.0883	V67	0.1624	V22	0.2032	V94	0.0620	V74	0.9634
20	V94	0.0825	V94	0.1564	V35	0.1926	V86	0.0545	V94	0.9587

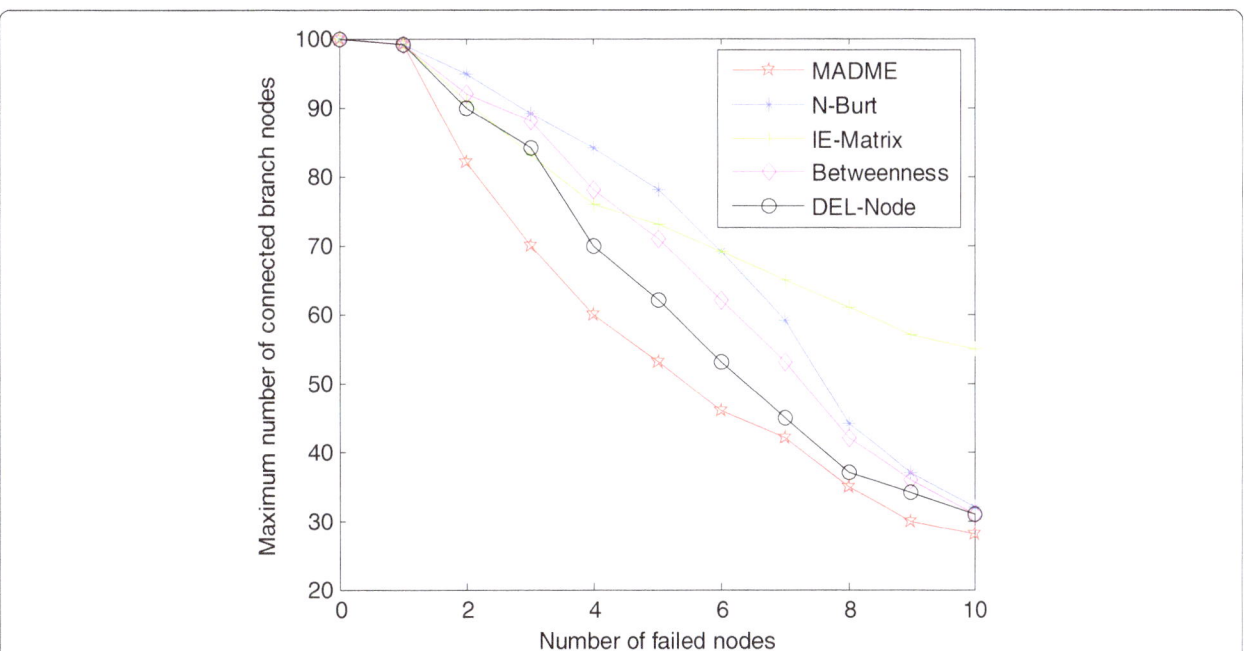

Fig. 6 The maximum connected branch after removing the key nodes. The effect of attack on the network robustness under the five methods, such as the MADME, N-Burt, IE-Matrix, betweenness, and DEL-Node methods, in which the top ten nodes with the highest importance evaluated by each method are removed in turn

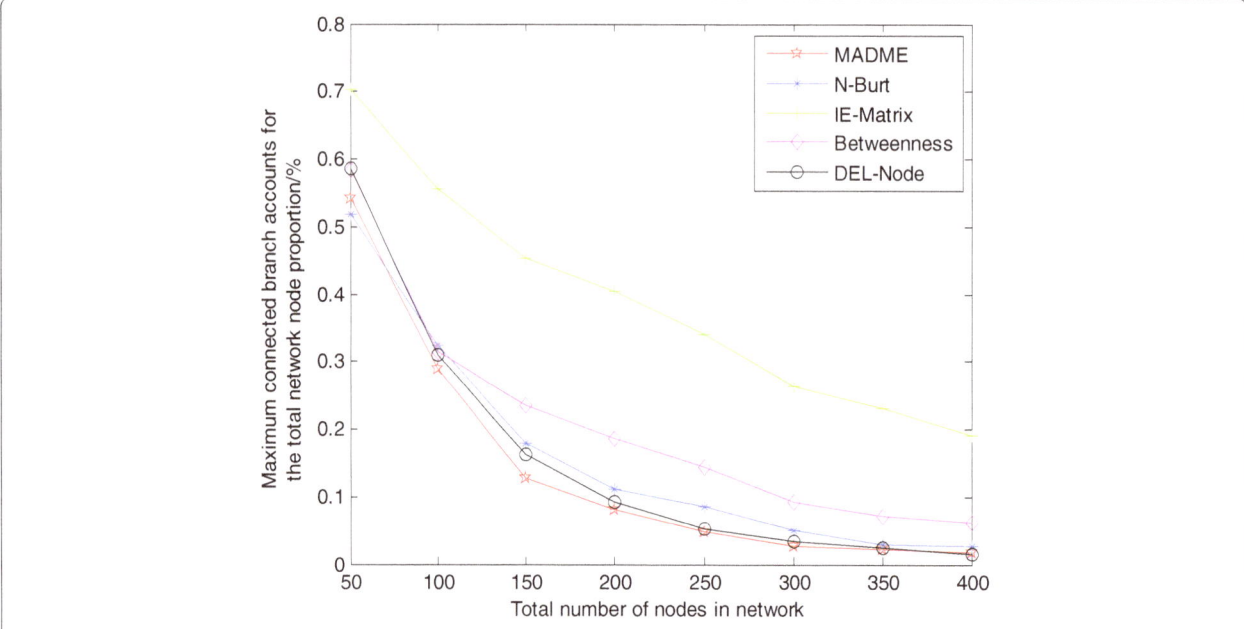

Fig. 7 Maximum connected branch proportion of the network after removing the key nodes under different network scales. For different network sizes, the maximum connected branch proportion of the network after removing the first 10% of the key nodes identified by the five methods is analyzed, i.e., MADME, N-Burt, IE-Matrix, betweenness, and DEL-Node methods

identified by the MADME method, the proportion of the maximum connected branch of the network is always less than that of the betweenness and IE-Matrix methods. This is mainly because the advantage of the MADME method is that it considers not only the node attributes from both structure importance and performance impact, but also the attributes' weight; it is more objective when applied to the evaluation of node importance. This result also indicates that the key nodes obtained by our MADME method have a greater impact on the robustness of the network. In addition, although the impact of the MADME method is closer than that of the N-Burt and DEL-Node methods, the MADME method is better in small-scale networks (such as a less than 150-node network). This is because MADME can recognize the potential key nodes more quickly; the ranking of MADME method is quite different from that of other methods in small-scale networks. But with increasing nodes, although the orders are different, the key nodes are almost the same.

From the above analysis, the impact on the robustness of the network by deliberately attacking the key nodes according to the MADME method is greatest, and even the whole network is paralyzed. This means that according to the MADME method to provide more targeted protection for the key nodes, the deliberate attack on the network can be effectively resisted. The experiment results further verify the validity of the MADME method.

The node importance evaluation method MADME based on structural characteristics and application

requirements shows its accuracy and validity; thus, the evaluating efficiency can be improved.

7 Conclusions
In this article, we introduce a multi-attribute decision-making function to integrate the four indicators including the structural characteristics and application requirements of a network. In the meantime, we also take into account the weight of these indicators. Then, the MADME method for evaluating the node importance is proposed. The main feature of the MADME method is that the application requirement indicator is consistent with reality and the structural characteristic indicator reflects the robustness of the structure, which improves the evaluating efficiency. Theoretical analysis indicates that the MADME method saves computational time. The simulation results show that the MADME method is reasonable from the two aspects, structure and performance, which overcomes the shortcomings of the N-Burt, betweenness, and DEL-Node methods only in terms of the network structure. In addition, it takes into account the weight of the four indicators and overcomes the weakness of the IE-Matrix method which combined the two aspects structure and performance. Moreover, compared with various methods such as N-Burt, betweenness, DEL-Node, and IE-Matrix, according to the key nodes obtained by the MADME method to deliberately attack the network, the network is quickly disintegrated. The result shows that the MADME method is more effective for the node importance evaluation. The key nodes discovered by the MADME method

have a more obvious impact on the network performance. Therefore, using the MADME method to quantify the key nodes, designing a network protection strategy, the networks' invulnerability will be improved. The proposed model and method are expected to be applied to the analysis and design of wireless sensor networks and to discover some potential key nodes quickly and accurately.

Acknowledgements
The authors are thankful to the chief editor and anonymous reviewers for their valuable comments in reviewing this paper. Special thanks also to professor Zabih Ghassemlooy of Northumbria University, UK. He gave some important suggestions for the improving of this paper.

Authors' contributions
RY proposed the idea of the identified key nodes. Moreover, she participated in the initial writing and revision of this paper. XY contributed to the relationship analysis of these indicators, and he wrote the initial version of this paper. MC and YX have provided advice and feedback. All authors read and approved the final manuscript.

Funding
This work is supported by National Natural Science Foundation of China under grant No. 61802333, State Scholarship Fund of China Scholarship Council under grant No. 201808130258, and Science and Technology Research Project of Colleges and Universities in Hebei Province under grant No. QN2018029.

Competing interests
The authors declare that they have no competing interests.

References
1. Y. Zhang, W. Li, Modeling and energy consumption evaluation of a stochastic wireless sensor network. EURASIP J. Wirel. Commun. Netw. **282**, 2012 (2012)
2. J. Yick, B. Mukherjee, D. Ghosal, Wireless sensor network survey. Comput. Netw **52**(12), 2292–2330 (2015)
3. W. Asif, H.K. Qureshi, M. M Rajarajan, M. Lestas, Combined Banzhaf & Diversity Index (CBDI) for critical node detection. J. Netw. Comput. Appl. **64**(C), 76–88 (2016)
4. K. Liu, S. Liu, Novel sensor node importance evaluation method based on the agglomeration contraction principle. J. Xidian Univ. **42**, 90–96 (2015)
5. X.L. Ren, L.Y. Lv, Review of ranking nodes in complex networks. Chin. Sci. Bull. **59**, 1175–1197 (2014)
6. D.B. Chen, L.Y. Lu, M.S. Shang, Y.C. Zhang, T. Zhou, Identifying influential nodes in complex networks. Physica A. **391**, 1777–1787 (2012)
7. A. Zeng, C.J. Zhang, Ranking spreaders by decomposing complex networks. Phys. Lett. A **377**, 1031–1035 (2013)
8. W.L. Fan, Z.G. Liu, Ranking method for node importance based on efficiency matrix. J. Southwest Jiaotong Univ. **49**, 337–342 (2014)
9. H.W. Yang, Y. Zhang, H.K. Wang, Y. Liu, New measure of node importance based on load flow in node-weighted complex networks. Appl. Res Comput. **30**, 134–137 (2013)
10. J.W. Wang, L.L. Rong, T.Z. Guo, A new measure method of network node importance based on local characteristics. J. Dalian Univ. Technol. **50**, 823–826 (2010)
11. C.H. Comin, L.D. Costa, Identifying the starting point of a spreading process in complex networks. Phys. Rev. E. **84**, 056105 (2011)
12. A. Dorfman, N. Kumar, J. Hahm, Highly sensitive biomolecular fluorescence detection using nanoscale ZnO platforms. Langmuir the Acs J Surfaces Colloids. **22**, 4890–4895 (2006)
13. Y. Chen, A.Q. Hu, X. Hu, Evaluation method for node importance in communication networks. J. China Inst. Commun. **25**, 129–134 (2004)
14. R. Singh, A. Chakraborty, B.S. Manoj, GFT centrality: A new node importance measure for complex networks. Physica A. **487**, 185–195 (2017)
15. X.X. Wen, C.L. Tu, M.G. Wu, X.R. Jiang, Fast ranking nodes importance in complex networks based on LS-SVM method. Physica A. **506**, 11–23 (2018)
16. E. Nardelli, G. Proietti, P. Widmayer, Finding the most vital node of a shortest path. Theor. Comput. Sci. **296**, 167–177 (2001)
17. Q.D. Sun, Y.M. Qiao, J.M. Wang, S. Shen, Node importance evaluation method in wireless sensor network based on energy field model. EURASIP J. Wirel. Commun. Netw. **2016**(1), 199 (2016)
18. W. Bechkit, M. Koudil, Y. Challal, et al., *A new weighted shortest path tree for converge cast traffic routing in WSN. 2012 IEEE Symposium on Computers and Communications (ISCC), Cappadocia, Turkey*, vol 2012 (2012), pp. 000187–000192
19. B. Liu, W.J. Wang, Y.Q. Li, R.R. Yin, T. Han, Crucial node decision algorithm based on energy in WSNs. J. Electron. Inf. Technol. **36**, 1728–1734 (2014)
20. X.X. Wen, C.L. Tu, M.G. Wu, Node importance evaluation in aviation network based on "no return" node deletion method. Physica A. **503**, 546–559 (2018)
21. Z.L. Hong, B.Y. Liu, Y.P. Zhang, Application of complex network in transportation network's node importance evaluation. J. Xi'an Technol. Univ. **34**(5), 404–410 (2014)
22. A. Ibnoulouafi, M.E. Haziti, Density centrality: Identifying influential nodes based on area density formula. Chaos Solitons Fractals **114**, 69–80 (2018)
23. X. Zhou, F.M. Zhang, K.W. Li, X.B. Hui, H.S. Wu, Finding vital node by node importance evaluation matrix in complex networks. Acta Phys. Sin. **61**, 050201 (2012)
24. J. Li, P.M. Pardalos, B. Xin, J. Chen, The bi-objective critical node detection problem with minimum pairwise connectivity and cost: Theory and algorithms. Soft. Comput. (2019). https://doi.org/10.1007/s00500-019-03824-8
25. P. Zhang, Z.Y. Dong, Z. Shen, Multi-parameter optimization algorithm for communication network node importance evaluation. Comput. Eng. **39**, 95–96 (2013)
26. W.R. Heinzelman, A. Chandrakasan, H. Balakrishnan, *Energy-Efficient Communication Protocol for Wireless Microsensor Networks*, vol 2000 (Proceedings of the 33rd Annual Hawaii International Conference on System Sciences, Hawaii, 2000), pp. 3005–3014
27. W. Yuan, Y. Wang, J. Wen, Establishment and application of the improved TOPSIS model based on game theory. J Water Resources Architectural Engineering **14**, 188–191 (2016)
28. H. Yu, Z. Liu, YJ. Li, Key nodes in complex networks identified by multi-attribute decision making method. Acta Phys. Sin. **62**, 020204 (2013)
29. S.W. Wang, J.X. Liu, B.Q. Cao, M.D. Tang, X. Wang, Recommended method of mashup services based on information entropy multi-attribute decision-making. Comput. Sci. **42**, 263–266 (2015)
30. A.L. Barabasi, R. Albert, Emergence of scaling in random networks. Science **286**, 509–512 (1999)
31. X.P. Su, R.R. Song, Leveraging neighborhood "structural holes" to identifying key spreaders in social networks. Acta Phys. Sin. **64**, 020101 (2015)

10

Ant colony optimization algorithm based on mobile sink data collection in industrial wireless sensor networks

Hong Zhang[1][*] [ID], Zhanming Li[1], Wanneng Shu[2] and Jarong Chou[3]

Abstract

Industrial wireless sensor network (IWSN) has changed the information transmission way for existing industrial control system. In mobile sink-based industrial wireless sensor networks, the energy consumption optimization for data collection has always been a hot research issue. To meet the delay requirements and minimize energy consumption, a data collection strategy based on ant colony optimization with mobile sink is proposed for industrial wireless sensor networks. Firstly, in order to reduce the number of nodes directly accessed by sink and shorten the traversed path, the selection of rendezvous nodes based on entropy weight method is introduced according to the density of nodes, relative residual energy, and the degree of uniformity of distribution. Then, secondly, an ant colony optimization algorithm is proposed to obtain the optimal access path for mobile sink, which can achieve a trade-off between the energy consumption of the network and transmission delay. The simulation results show that, compared with the existing algorithms, the proposed algorithm can minimize the delay and prolong the lifetime of the network.

Keywords: Industrial wireless sensor networks, Delivery latency, Energy efficiency, Mobile sink

1 Introduction

Although the wireless sensor networks have been proposed, studied, and developed for more than a decade of years, there are still a lot of challenging issues especially in various industrial scenarios. Industrial wireless sensor networks (IWSN) is composed of autonomous sensor nodes distributed in monitoring area, which can acquire physical and environmental data such as temperature, sound, vibration, pressure, and movement. It has demonstrated enormous applicable value and commercial potential in aspect of military, industrial, and civil fields, such as, battlefield monitoring, industrial control process, machine operation, home automation, and traffic monitoring [1]. Due to limited battery power and laborious replacement of sensor nodes, energy efficiency is one of the most important issues in wireless sensor networks. With the scenario of static sink node, data forwarding in multi-hop manner will cause traffic concentration and make sensor

nodes near the sink undertake more energy consumption. Hot-spot problem will worsen the network lifetime and even affect the operation of normal sensor nodes [2].

To deal with hot-spot problem, data collection using mobile sink in wireless sensor networks have been put forward in recent years. By using mobile sink node to travel in a certain path, it can reduce the energy consumption of sensor nodes and makes the energy consumption more balanced in the whole network [3]. Commonly, each node will be traversed by mobile sink node in turn, and the distance between them can be cut down to one hop. Then, the energy consumption for data forwarding will achieve the optimal resolution as well as high message delivery latency. For large-scale deployment, data collection efficiency will be very poor. A feasible solution is to tolerate a certain delay to compensate for the energy consumption and delay of data collection [4]. The whole network will be divided into several routing trees, and some sensor nodes act as rendezvous points. The other nodes forward the data to nearby rendezvous points, which can be traversed by mobile sink individually for data aggregation. Obviously, the mobile sink-based mode and path selection

* Correspondence: 532337136@qq.com; zhh@gszy.edu.cn
[1]College of Electrical and Information Engineering, Lanzhou University Technology, Lanzhou 730050, People's Republic of China
Full list of author information is available at the end of the article

directly affect the efficiency of data collection and the overall performance of the network. Therefore, how to improve the energy utilization and collect as much data as possible with certain time delay constraint has become an important evaluation factor of the system. In this paper, we focus the network lifetime problem for time-sensitive data gathering and study the optimal visiting points and a data collection path for a mobile sink.

The rest of this paper is organized as follows. Section 2 shows the methods in industrial wireless sensor networks. Section 3 discusses the related work of data acquisition. Section 4 designs an optimal energy consumption model for industrial wireless sensor. The ant colony optimization algorithm is proposed in Section 5. Section 6 shows the simulation experimental results and concludes the paper with summary and future research directions.

2 Methods

Since the mobile sink-based mode and path selection directly affect the efficiency of data collection and the overall performance of the network, we performed this study to improve the energy utilization and collect as much data as possible with certain time delay constraint. To reduce the number of nodes directly accessed by sink and shorten the traversed path, the selection of rendezvous nodes based on entropy weight method is introduced according to the density of nodes, relative residual energy, and the degree of uniformity of distribution. Next, ant colony optimization is applied to obtain the optimal access path for mobile sink so as to achieve a trade-off between the energy consumption of the network and transmission delay.

3 Related work

Data collection is one of the most basic applications in industrial wireless sensor networks. When the node senses interesting events and collects data, it sends them to sink through single or multiple hops. Due to the energy constraints of sensor nodes, it is necessary to maximize the lifetime of the network on the basis of acquire efficient data delivery [5, 6]. Accordingly, the trajectory of mobile sink may have a direct impact on the data delivery tasks in mobile sink-based data collection schemes. Sink mobility patterns in industrial wireless sensor networks can be classified as random mobility pattern, fixed-path mobility pattern, and controlled mobility pattern.

Recently, a number of data collection schemes have been proposed for mobile sink-based WSNs. Emre et al. [7] compared static and mobile data collection methods for sink nodes and establishes an optimization model of network lifetime. By being transformed into a linear model, the optimal solution of the trajectories of the mobile sink can be obtained. Guo et al. [8] proposed a converge-cast algorithm for efficient data collection in

MWSNs. The monitoring area is divided into several disks, and the rendezvous points (RPs) that are referred as specialized nodes working as sub-sinks will be chosen by employing quantum genetic algorithm. In [9], by conducting systematic analysis on the behaviors of the mobile sink in terms of both throughput capacity and lifetime of the network, Liu et al. presented a mobility-assisted data collection model and developed a comprehensive theoretical method to achieve beneficial throughput capacity and lifetime. Kumar et al. [10] proposed a clustering algorithm to divide all nodes into different clusters according to their location and employed classical traveling salesman algorithm to find the shortest path through all cluster centers. The time complexity of those algorithms increases sharply with the increase of the number of sensor nodes. Therefore, they are more suitable for scenarios with less number of sensor nodes and data transmission hops.

Other scholars study the selection method of sink node's mobile path in a distributed manner. Lee et al. [11] integrated the initial address, data collection routing, and dwell time of sink nodes to establish a linear programming model to obtain the optimal mobile sink sojourning pattern. Besides, by utilizing the variance of residual energy of neighboring sensor nodes, a simple practical heuristic algorithm for sink mobility is presented to alleviate the risk of disequilibrium of energy consumption among sensor nodes. Wang et al. [12] proposed a moving strategy called energy-aware sink relocation (EASR) for mobile sinks, which employs the information related to the node's residual energy and adjust the transmission range of sensor nodes and the relocating scheme for the sink adaptively. Wichmann et al. [13] employed the aircraft as a mobile sink node to collect the storage data of sensor nodes. According to its mobility characteristics, the mobile path based on TSP (traveling salesman problem) algorithm is reconstructed. In addition, the path is dynamically adjusted according to the estimated time of data collection to minimize the transmission delay. However, the algorithm does not consider the optimization of energy consumption of sensor nodes and network lifetime. Tashtarian et al. [14] established a network lifetime optimization model for data collection in a single-hop mode by using the location information of nodes and constructs a decision tree to solve the optimization model to obtain the mobile path of sink node. Ghosh et al. [15] decomposed the monitoring region into several triangulars, generating circles passing through three vertices of the triangle. Then, each circle center can be regarded as the RP. Sink node make use of greedy algorithm to determine the next residence location. However, due to the single-hop mode, the communication range of sink is limited. For the sake of data integrity of the whole monitoring region, it results in a longer mobile path and a higher data transmission delay. Shi et al. [16] proposed a routing protocol with dynamic layered to handle the overhead of mobile sink, which integrates Voronoi scoping and dynamic anchor selection to reduce the number of message interactions for updating routing tables.

The problem with sink mobility is concerned with the increase of data latency. Due to the mobility of sink nodes, some sensor nodes may have to wait for a long time consequently, which will cause the buffer overflow and increase the latency in data delivery. To achieve efficient data delivery to mobile sink, Kim et al. [17] designed an intelligent agent-based routing protocol for densely deployed and large wireless sensor networks. Pavithra et al. [18] proposed a weighted set programming algorithm based on rendezvous point (RP). By optimizing the data forwarding route from source node to fixed sink, the weight of each sensor node was calculated, and then the node was dynamically selected as the set point to realize that sink could traverse all the set points for efficient data collection. Wang et al. [19] proposed an optimal path construction method of variable-length coding parthenogenetic algorithm based on quadratic grid partitioning. Firstly, coarse-grained grids were used to partition the network area, and variable-length coding parthenogenetic algorithm was used to obtain the optimal path grids. Then, according to the initial optimal path constructed, fine-grained grids are used again for each passing grids to optimize the data collection path. Aiming at the problem of large delay caused by the speed constraints of mobile sink, Yu et al. [20] proposed a path optimization strategy. Based on the mathematical analysis of the relationship between delay and network energy consumption, a dynamically adjustable weight of Node's priority to data delivery is proposed. The optimal node weight is obtained by simulated annealing genetic algorithm, and the convergent node and the optimal mobile path finally can be resolved by multiple iterations.

However, none of the aforementioned approaches have considered minimization of both the stated objectives simultaneously. As this is an NP-hard problem [21], we are motivated to develop a data collection strategy based on ant colony optimization for wireless sensor networks with mobile sink.

4 Proposed methods

4.1 System model

Suppose that sensor nodes are randomly deployed in a rectangular monitoring area to form a multi-hop self-organizing network. The sensor network is fully connected, and all nodes have unique identities and are no longer mobile after deployment. Also, they are powered by limited energy and can not be supplemented. Each node should be configured with the same communication radius r and has the same initial energy E_0. Sink node is capable of mobility and has external power supply. To reduce the high transmission overhead caused by long-distance transmission, the sensor node can forward data packets to the destination by multi-hop mode.

The sensor nodes can be classified into two kinds: rendezvous node (RN) and ordinary node, and RNs can communicate directly with sink [22]. Comparatively, the ordinary node needs to send data to specific RN through a multi-hop mode. When sink node has not moved to the communication range, RN can put the aggregated data into the cache beforehand. Once the sink move along a certain trajectory and stay away with the RN within one hop, it can send to the sink directly. The first-order radio model is applied for energy depletion analysis, and the energy consumption in each round can be expressed as follows:

$$E_{\text{total}} = \sum_{i=1}^{n} \left(e_{tr} k_t^i + e_{\text{rec}} k_r^i \right) \tag{1}$$

In which, e_{tr} and e_{rec} represents the energy consumption for sending or receiving unit data. k_t^i and k_r^i represent the amount of data being received and sent by node i, respectively.

Assuming that the amount of data generated by each node in each round is q, there exists a relationship between the data received and the data forwarded by relay node i, i.e., $k_t^i = k_r^i + q$. Then, it can be deduced that the energy consumption of data transmission in the whole network is related to the number of hops during the data transmission, which can be given as

$$\sum_{i=1}^{n} k_r^i = \sum_{i=1}^{n} h_i q \tag{2}$$

where h_i represents the minimum hop number that the node i sends data to the fusion node.

Further, the total energy consumption can be expressed as the sum of the minimum hops among all sensor nodes, as shown in formula (3).

$$\begin{aligned} E_{\text{total}} &= \sum_{i=1}^{n} \left(e_{tr} k_t^i + e_{\text{rec}} k_r^i \right) \\ &= \sum_{i=1}^{n} \left[e_{tr} \left(k_r^i + q \right) + e_{\text{rec}} k_r^i \right] \\ &= q \left[n e_{tr} + \sum_{i=1}^{n} \left(e_{tr} + e_{\text{rec}} \right) k_r^i \right] \end{aligned} \tag{3}$$

It can be seen that optimization of the total energy consumption is equivalent to minimizing the sum of the hops from the ordinary nodes to fusion nodes. In the process of mobile sink's data collection, the number of hops depends on to the selection of rendezvous points of mobile sink. Therefore, the choice of sink mobile trajectory directly affects the energy consumption of the whole network [23].

4.2 Data collection strategy with delay constraint

According to the previous analysis, to achieve the delay requirements and minimize the overall energy consumption of the network, it is necessary to optimize the sink point trajectory in wireless sensor networks [24]. Furthermore, by considering delay constraints, the solution of optimal sink mobile path reduces the overall energy consumption. On the one hand, it is necessary to solve the optimal set of RNs to satisfy with the minimization the total energy consumption. On the other hand, the optimization of travel path selection by mobile sink needs to be fully considered to accord with the constraints for time delay [25].

For a given set of sensor nodes $S = \{s_1, s_2, \cdots, s_n\}$ and the set of RNs $V = \{v_1, v_2, \cdots, v_K\}$, an undirected graph $G = \{\Psi, E\}$ can be constructed with a collection of edges E, and $\Psi = S \cup V$. The problem of data collection with mobile sink can be transformed into Traveling Salesman Problem. By constructing the path set $PATH(G)$ for data collection based on the undirected graphs G, it will ensure that the hops from any ordinary node to RN can not exceed the limit value $h_{\max}$ as well as obtaining the minimization of the total energy consumption.

Let $P_i = \{p_{i,1}, p_{i,2}, \cdots, p_{i,k_i}\}$ denote the set of all paths from the sensor node i to its adjacent RN node set, in which $p_{i,j}$ is the j-th path from k_i data forwarding paths. Therefore, the location selection of RNs can be formulated as an optimization problem:

$$\min_{P \in P_i(G)} E_{total}(P)$$

$$s.t. \quad \forall P_i \in PATH, \exists p_{i,j} \in PATH, dist(i, j) + dist(j, v(i)) \leq h_{\max} r$$

$$\max_{i=1,2,\cdots,n} \{h(i, v(i))\} \leq h_{\max} \, 1 \leq i \leq n, 1 \leq j \leq |P_i|$$

$$(4)$$

where $dist(i, j)$ is the Euclidean distance between node i and j. $v(i)$ is the next-hop RN being chosen by node i, and the number of the corresponding hops can be defined by $h(i, v(i))$. Besides, r indicates the communication radius of the sensor node.

For any node j on the path P_i, the sum of the hops to the access path does not exceed the threshold of $h_{\max}$. Since the Euclidean distance between two adjacent nodes is less than the communication radius r, the sum of the distance from node j to $v(i)$ and the distance between node j and i can be no more than $h_{\max} r$.

4.3 Determination of rendezvous points

In order to ensure the overall energy depletion of the network being balanced, the selection of resident points

may take into account comprehensively about the factors, including the node's density, relative residual energy, and uniformity of distribution. As a regional center directly interacts with sink, RN is responsible for data fusion of other nodes within a certain range and has to undertake more energy consumption [26, 27]. The ratio of residual energy to initial energy of a node is defined as relative energy. In order to achieve better energy consumption balance, the larger the relative energy is, the node can be more competitive. In addition, the node density is defined as the number of nodes in the communication range. The higher the node density, the lower the average energy depletion of nodes will demonstrate during the course of inter-cluster communication. By taking into account of the balanced distribution of RNs in the whole network, it is beneficial to improve the overall transmission efficiency of the network. Therefore, the nodes with larger centroid degree, i.e., the degree of proximity between a node and its neighbors, should be selected as far as possible. Based on the above analysis, the relevant indicators are defined as follows:

$$\begin{cases} f_1 = \dfrac{E_{res}(i)}{E_0} \\ f_2 = \dfrac{\mathrm{cov}(i)}{\displaystyle\sum_{j \in neighbor(i)} \mathrm{cov}(j)} \\ f_3 = \dfrac{D(i)}{\max\{D(i)\}} \\ D_i = 1 - \dfrac{\sqrt{\left(x_i - \dfrac{1}{n}\displaystyle\sum_{j=1}^{n} x_j\right)^2 + \left(y_i - \dfrac{1}{n}\displaystyle\sum_{j=1}^{n} y_j\right)^2}}{r} \end{cases} \quad (5)$$

where $E_{res}(i)$ represents the residual energy of node i and E_0 is the initial energy. $\mathrm{cov}(i)$ denotes the number of nodes within the communication radius of node i, and $neighbor(i)$ represents the neighbor nodes set. In addition, $D(i)$ is defined as the centroid degree of node i.

The weight of each indicator is determined by entropy weight method. First of all, normalize the above indicators as:

$$\tilde{f}_m = \frac{f_m(i) - \min\{f_m(i)\}}{\max\{f_m(i)\} - \min\{f_m(i)\}} \quad (6)$$

where $\max\{f_m(i)\}$ and $\min\{f_m(i)\}$ denotes the original minimum and maximum value of the index function, respectively. $f_m(i)$ is the m-th index function of node i.

Next, the group vector entropy and weight values can be calculated as:

$$\begin{cases} U_m - \dfrac{1}{\ln K} \sum \tilde{f}_m(i) \, \ln \tilde{f}_m(i) \\ w_m = \dfrac{1-U_m}{\sum (1-U_m)} \end{cases} \tag{7}$$

According to the index function value and the weight value, the comprehensive evaluation of each node can be determined by formula (8). The sensor node with higher value will be chosen as the candidate resident point.

$$Cand\_RN(i) = \sum w_m \tilde{f}_m(i) \tag{8}$$

4.4 Path selection based on ant colony optimization

Ant colony optimization (ACO) is a kind of bionic technique inspired by the behavior of real ants from nature [28]. By finding optimal paths through graphs, it demonstrates good enough performance for solving combinatorial optimization problems and near-optimal solutions to the travelling salesman problem. Basically, ant colony optimization algorithm is a kind of positive feedback mechanism, which has the characteristics of randomness, adaptability, and distribution. It is suitable for parallel computation and accurate solution [29]. However, there are two disadvantages in traditional ACO algorithms. One is the operational inefficiency owing to the diversity of ants, the other is that it is easy to fall into local convergence and the obtained path usually can not meet with the optimal solution [30].

When sink obtains the set of RNs $\Psi = \{v_1, v_2, \cdots, v_K\}$, the planning of the travel path can be approximated to TSP problem, which can be resolved approximately by adopting ant colony algorithm. During the course of the current node ants choosing the next-hop node from the neighboring node set, the probability of selection is crucial. Denote $\tau_{ij}(t)$ as the pheromone concentration on the path from node i to j. The heuristic function $\eta_{ij}(t)$ can be defined as: $\eta_{ij}(t) = 1/|d(i, sink) - d(j, sink)|$, which indicates the expected degree of ants from i to j. $d(i, sink)$ and $d(j, sink)$ represent the distance from the mobile sink to either the current RN or the next one, respectively. N_{k_a} denotes the set of RNs to be visited by ant k_a. In addition, α, β are parameters to regulate the relative weight of the pheromone trail and heuristic value, which can be satisfied with $\alpha \geq 0$, $\beta \geq 0$, $\alpha + \beta = 1$. Hence, the probability of moving from RN i to j can be expressed as follows:

$$p_{ij}(t) = \begin{cases} \dfrac{[\tau_{ij}(t)]^\alpha \cdot [\eta_{ij}(t)]^\beta}{\sum\limits_{u \in N_{Ki}} [\tau_{iu}(t)]^\alpha \cdot [\eta_{iu}(t)]^\beta}, & k \in N_i \\ 0, & other \end{cases} \tag{9}$$

For the distance constraint, the gain of the access points set with minimum energy consumption is obviously optimal. Therefore, it is reasonable to choose the access node based on the energy cost and distance from mobile sink. By defining the effect of RN selection on global energy consumption in utility priority metric path, we introduce the concept of path superiority as a criterion for ant path evaluation after a round. The smaller the superiority is, the better the path quality will be. Then, the path superiority can be defined as:

$$PR(i) = \lambda_1 \frac{E(V_{k_a} \cup \{v_i\}) - E(V_{k_a})}{TSP(V_{k_a} \cup \{v_i\}) - L(V_{k_a})} + \lambda_2 \frac{E(V_{k_a}) - E(V_{k_a} - \{v'\})}{L(V_{k_a}) - TSP(V_{k_a} - \{v'\})},$$

$$TSP(V_{k_a} \cup \{v_i\}) \leq D_m V_m$$

$$\tag{10}$$

In which, V_{k_a} represents the RN set being visited, $L(V_{k_a})$ indicates the corresponding shortest path length, and $E(V_{k_a})$ represents the total energy consumption for transferring data to RNs.

According to the ratio between energy gain and distance cost, path superiority can evaluate the influence of global optimal solution for candidate node to be selected as next hop. Then, the determination of the relative can impact on the pheromone and the heuristic for the decision of the ant.

Furthermore, after the ants end their tours, the pheromone trail amount on every edge (i, j) is updated by

$$\tau_{ij}(t+1) = (1-\rho)\tau_{ij} + \rho \Delta \tau_{ij}(t) \tag{11}$$

where ρ is the local pheromone decay parameter and $\Delta \tau_{ij}(t)$ is the added pheromone trail amount at the point t.

The forward ants move from access point i to j for pheromone updating, and the updating rules are as follows:

$$\Delta \tau_{ij}(t) = \frac{Q_L E(j))}{E(i) + d(i, j)} \tag{12}$$

where Q_L represent the local pheromone concentration. $E(i)$ and $E(j)$ represent the energy consumption for data being migrated from the member nodes to RN i and j, respectively. $d(i, j)$ is the Euclidian distance.

Subsequently, the backward ants go back to the source along the reverse path and complete global pheromone updating, and the corresponding updating rules can be given as

$$\Delta \tau_{ij}^{k_a}(t) = \frac{Q_G}{PATH(G, k_a) * PR(i)} \tag{13}$$

where Q_G represents the Global pheromone concentration and $PATH(G, k_a)$ is the path length of the ant k_a.

Formula (13) shows that updating pheromones through path superiority ensures the energy requirement of the

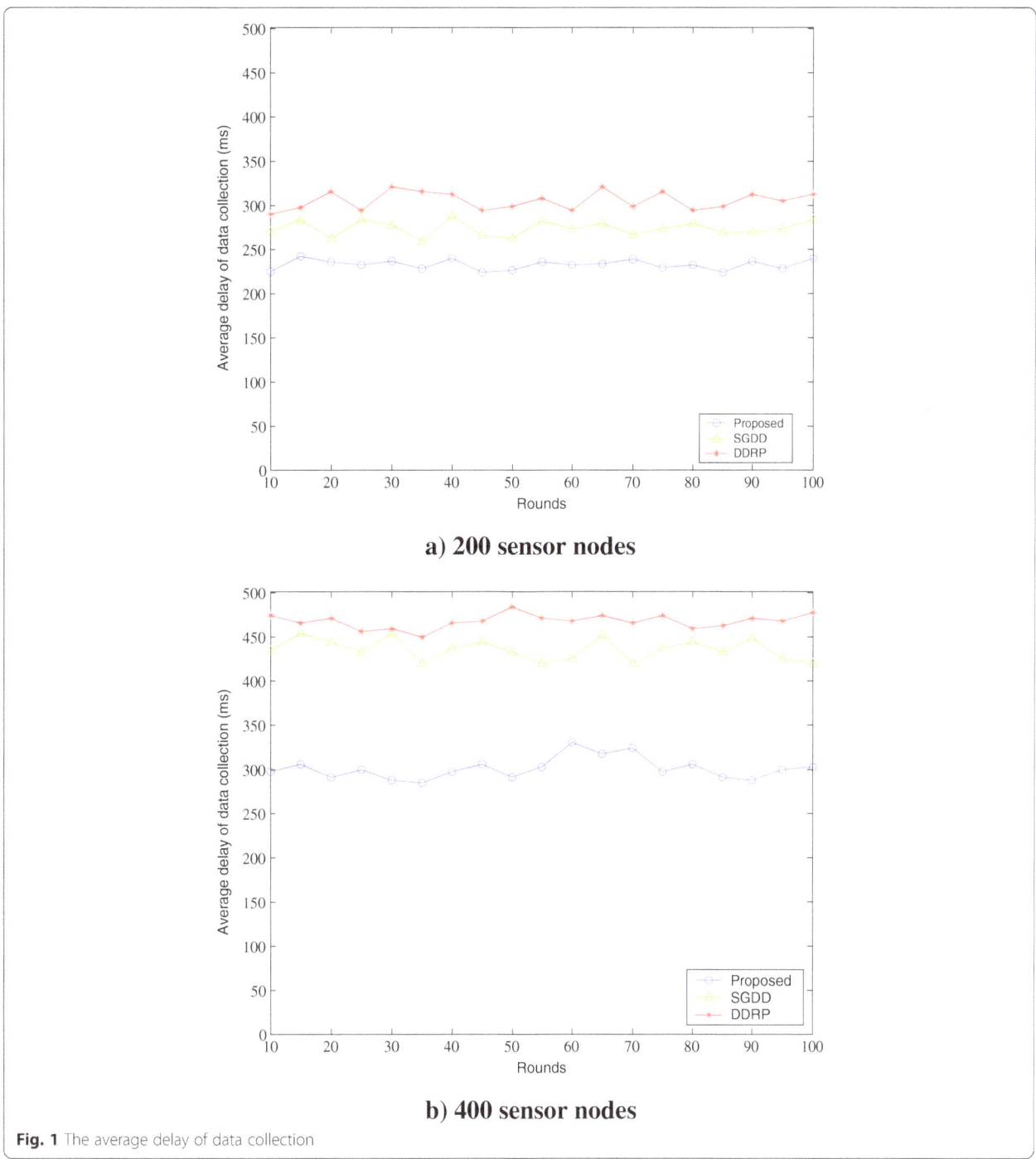

Fig. 1 The average delay of data collection

path and takes into account the energy balance, which is conducive to improving the efficiency of data collection.

5 Experimental results and analysis

In this section, several numbers of simulations are conducted and evaluated to verify the performance of our proposed method. The main parameters of the environment are as follows: the sensor nodes are randomly deployed in a rectangular monitoring area of 200×200 m, the initial energy of all sensor nodes is 2 J. The communication radius is equal to 50 m, and the delay requirement is set as 10 min. The energy consumption of transmitting and receiving circuits in the energy consumption model is set to 50 nJ/bit, and the moving speed of mobile sink is fixed to 2 m/s. Besides, λ_1 and λ_2 is equal to 0.5. The proposed method is compared with related works, data-driven routing protocol

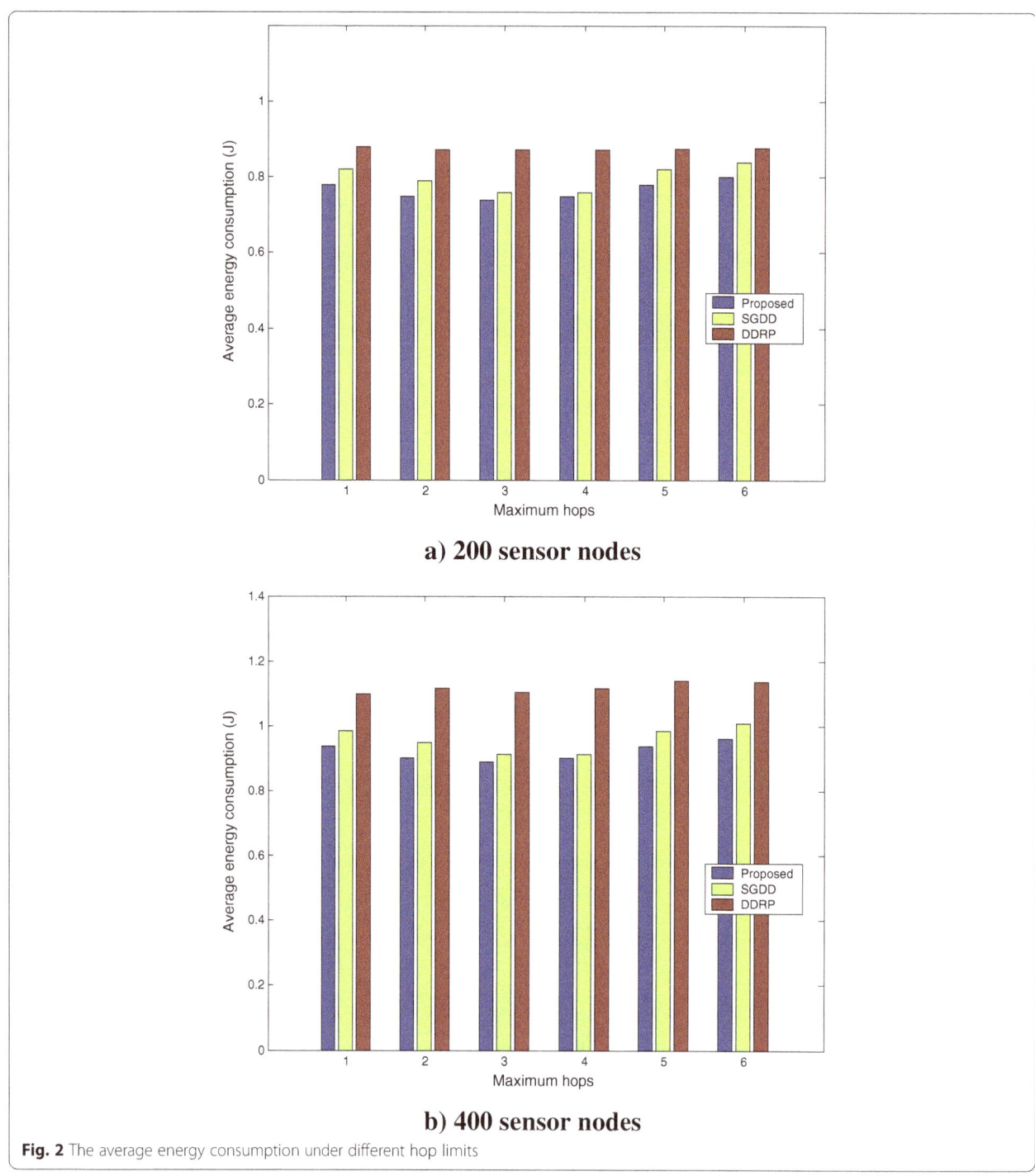

a) 200 sensor nodes

b) 400 sensor nodes

Fig. 2 The average energy consumption under different hop limits

(DDRP) [31] and self-managed grid-based data dissemination protocol (SGDD) [32], for data aggregation with mobile sink.

Figure 1 shows the comparisons of the average delay of data collection in mobile sink mode. In order to reflect the global performance and avoid the impact of energy-exhausted nodes in advance, the rounds without a dead node is selected. From the experimental results, it can be seen that DDRP takes the longest time to complete a cycle

of data acquisition, followed by SGDD and proposed method. The reason is that DDRP needs more nodes to access directly and the relay nodes in the network are dispersed, which bring about the transmission of data over long distance. In the proposed method, the priority of RN nodes can be defined to optimize the travel path of a mobile sink. The data collection efficiency of sink is improved and the data collecting latency can be reduced more sharply than other algorithms. Especially, when the

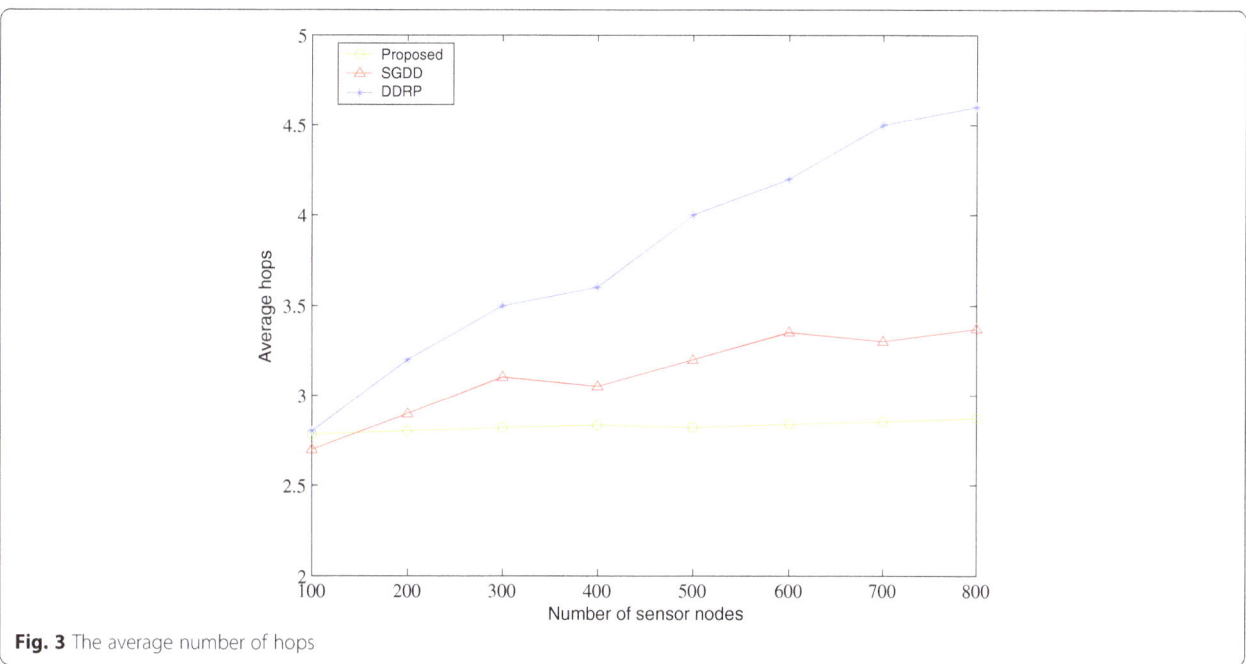

Fig. 3 The average number of hops

nodes are distributed densely, the difference of average delay is more obvious.

Figure 2 shows the comparisons of the average energy consumption of the three algorithms under different hop limits. Basically, when the maximum hop number is set as 3 or 4, SGDD and PROPOSED can achieve the optimal average energy consumption. It demonstrates that the overall energy consumption of the network can not reach the optimal state under the condition of single hop mode and excessive hops. This is because by employing irregular clustering, and the long forwarding path from some sub-nodes to the sink node in SGDD, will lead to a significant increase in aspect of the overall energy consumption when the number of hops increases. Similarly, if setting the maximum hops without restraint, it will lead to excessive energy consumption of RNs for data aggregation and be not conducive to the minimization of average energy consumption. Comparatively, the average energy consumption is not sensitive to maximum hop constraints in DDRP. However, the experimental results clearly show

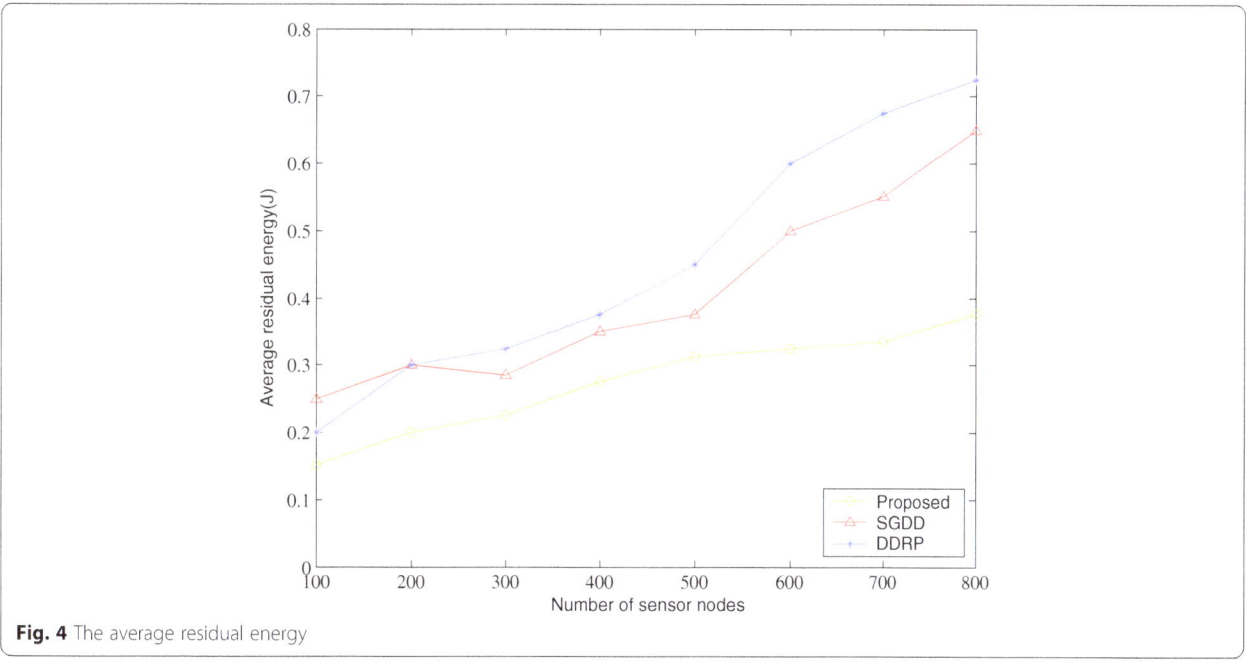

Fig. 4 The average residual energy

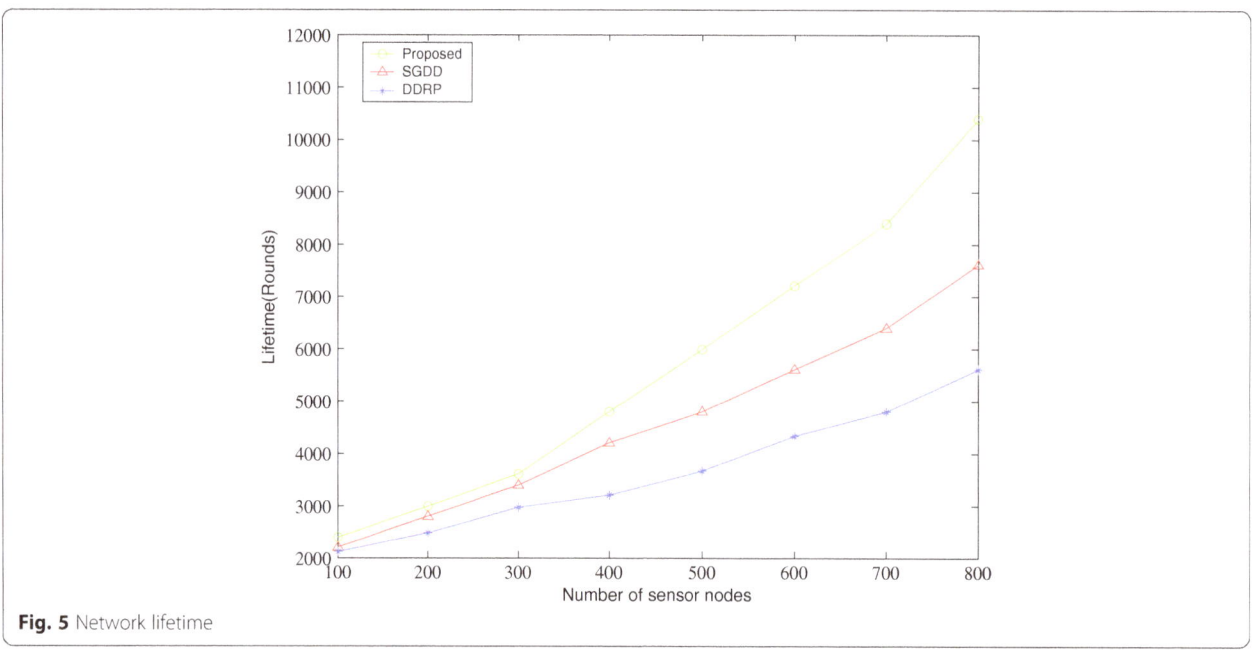

Fig. 5 Network lifetime

that the average energy consumption of DDRP is much higher than that of the other two algorithms.

Next, to verify the effectiveness of the proposed algorithm, its performance is compared with DDRP [31] and SGDD [32] in terms of the average hops, average residual energy, and network life time for different numbers of sensor nodes. Figure 3 illustrates the average number of hops from normal nodes to mobile sink for different numbers of sensor nodes. As it can be seen from the result, the average hops from source node to mobile sink demonstrates less variation by our proposed method, which is beneficial from a set measure used in multi-objective optimization to estimate the RNs. SGDD takes the nearest node from mobile sink as the root node to dynamically construct the number of routes, and lacks of global consideration, which affects the stability of average hops. By utilizing a random clustering method rather than probabilistic model to weaken the impact of "hot spot" problem, it makes DDRP's performance lag behind other algorithms.

Figure 4 illustrates the average residual energy of nodes when network ends. When the network is no longer able to support normal task, the average residual energy of surviving nodes reflects the utilization of node resources and whether the energy consumption of nodes is balanced in the process of operation. By considering the residual energy of candidate nodes, our proposed method handle the nodes near the resident point with priority and make the mobile sink change its moving trajectory to balance the energy consumption of the nodes. In contrast, the cluster heads usually are far away

from the mobile sink in DDRP, which affects the energy balance of the network negatively.

In Fig. 5, we compare the network lifetime for different numbers of sensor nodes. The network lifetime in this test is defined as the time when the first node runs out of its energy. From the simulation results, the lifetime of our proposed method can achieve higher than that of both SGDD and DDRP in all cases. The reason behind the increased network lifetime manifests the highly available data gathering strategy and moving trajectory strategy with ACO optimization, which can result in minimum energy consumption and better load balancing among the sensor nodes. In DDRP, the mobile sink dwelled more time, which caused the nodes near the mobile sink to waste higher energy consumption and die prematurely.

6 Conclusions and future work

The oncoming requirement on the industrial wireless sensor networks covers from the lower physics layers to the upper application layer. In this paper, we propose an ant colony optimization algorithm based on mobile sink data collection in industrial wireless sensor networks. To reduce the number of nodes directly accessed by sink and shorten the traversed path, the selection of RNs based on entropy weight method is introduced according to the density of nodes, relative residual energy, and the degree of uniformity of distribution. Furthermore, an ant colony optimization algorithm is introduced to obtain the optimal access path for mobile sink, which can achieve a trade-off between the energy consumption of the network and transmission delay. The simulation results show that, compared with the existing algorithms, the proposed algorithm can minimize the delay and prolong the lifetime of the network.

In our future research, we will try to optimize the scheme in a scenario with obstacles, and design a complete protocol for communication for data collection with mobile sink in industrial wireless sensor networks.

Abbreviations
ACO: Ant colony optimization; DDRP: An efficient data-driven routing protocol; IWSN: Industrial wireless sensor networks; RN: Rendezvous node; RP: Rendezvous point; SGDD: Self-managed grid-based data dissemination protocol; TSP: Traveling salesman problem

Acknowledgements
The authors acknowledged the anonymous reviewers and editors for their efforts in valuable comments and suggestions.

Authors' contributions
HZ proposed the innovation ideas and theoretical analysis, and ZL carried out the experiments and data analysis. WS conceived of the study, participated in its design and coordination, and helped to draft the manuscript. All authors read and approved the final manuscript.

Funding
This work was supported by the Project of Gansu Province for Industrial and Information Development (Grant No.23051358), Gansu Province for Guiding Scientific and Technological Innovation and Development (Grant No. 2018ZX-05), and National Natural Science Foundation of China (Grant No. 61603420).

Competing interests
The authors declare that they have no competing interests.

Author details
[1]College of Electrical and Information Engineering, Lanzhou University Technology, Lanzhou 730050, People's Republic of China. [2]College of Computer Science, South-Central University for Nationalities, Wuhan 430074, People's Republic of China. [3]Deparment of Electrical and Computer Engineering, Michigan State University, East Lansing, MI 48824, USA.

References
1. C. Tunca, S. Isik, M.Y. Donmez, C. Ersoy, IEEE Commun. Surv. Tutor. 16(2), 877–897 (2014)
2. S. Yang, U. Adeel, Y. Tahir, J.A. McCann, IEEE Trans. Mob. Comput. 16(5), 1420–1433 (2017)
3. N. Ghosh, R. Sett, I. Banerjee, Comput. Electr. Eng. 64, 288–304 (2017)
4. Z. Huang, G. Shan, J. Cheng, J. Sun, Neural Comput. Appl. (2018). https://doi.org/10.1007/s00521-018-3728-2
5. F. Tashtarian, M. Hossein, Y. Moghaddam, K. Sohraby, S. Effati, IEEE Trans. Veh. Technol. 64(7), 3177–3189 (2015).
6. M.E. Keskin, I.K. Altinel, N. Aras, Comput. J. 54(12): 1987-1999 (2011).
7. M.E. Keskin, I.K. Altinel, N. Aras, C. Ersoy, Comput. J. 54(12), 1987–1999 (2011)
8. J. Guo, LJ. Sun, WJ. Xu, J. Commun. 33(9), 176–184 (2012)
9. W. Liu, K. Lu, J. Wang, G. Xing, L. Huang, IEEE Trans. Veh. Technol. 61(6), 2777–2788 (2012)
10. A.K. Kumar, K.M. Sivalingam, A. Kumar, Wirel. Netw 19(3), 285–299 (2013)
11. K. Lee, Y.-H. Kim, H.-J. Kim, S. Han, Wirel. Netw 20(2), 303–318 (2014)
12. C.-F. Wang, J.-D. Shih, B.-H. Pan, T.-Y. Wu, IEEE Sensors J. 14(6), 1932–1943 (2014)
13. A. Wichmann, T. Korkmaz, Comput. Commun. 72(72), 93–106 (2015)
14. F. Tashtarian, M.H.Y. Moghaddam, K. Sohraby, S. Effati, Comput. Netw. 77, 128–143 (2015)
15. N. Ghosh, I. Banerjee, Comput. Electrical Engin. 48(10), 417–435 (2015)
16. L. Shi, Z. Yao, B. Zhang, C. Li, J. Ma, Int. J. Commun. Syst. 28(11), 1789–1804 (2015)
17. J.W. Kim, J.S. In, K. Hur, J.W. Kim, Consumer Electron. IEEE Trans. 56(4), 2310–2316 (2010)
18. H. Pavithra, R. Shivashankar, G.R. Poornima, in Proceedings of the 2016 IEEE International Conference on Recent Trends in Electronics Information Communication Technology (IEEE, Piscataway, 2016), pp. 151–155
19. W. Wang, H.S. Shi, P.Y. Huang, B.J. Gao, J.P. Niu, J. Wang, J. Northwestern Polytechnical Univ. 34(6), 1016–1021 (2016)
20. Z.B. Yu, X.X. Kong, J.J. Pei, Transducer Micro-system Technol. 35(11), 44–46 (2016)
21. A. Somasundara, A. Ramamoorthy, M. Srivastava, IEEE Trans. Mob. Comput. 6(4), 395–410 (2007)
22. B. Suh, S. Berber, Electron. Lett. 52(2), 167–169 (2015)
23. F. Tashtarian, M. Hossein, Y. Moghaddam, K. Sohraby, S. Effati, IEEE Trans. Veh. Technol. 64(7), 3177–3189 (2015)
24. A.W. Khan, A.H. Abdullah, M.H. Anisi, J.I. Bangash, Sensors 14, 2510–2548 (2014)
25. A. Kaswan, P.K. Jana, M. Azharuddin, In proceedings of advances in computing, communications and informatics, ICACCI, 2017 International Conference on IEEE (2017), pp. 168–173
26. L. Danpu, Z. Kailin, D. Jie, China Commun. 10(3), 114–123 (2013)
27. R. Logambigai, A. Kannan, Wirel. Netw 22, 945–957 (2016)
28. M. Dorigo, L.M. Gambardella, IEEE Trans. Evol. Comput. 1(1), 53–66 (1997)
29. L.M. Gambardella, É.D. Taillard, M. Dorigo, J. Oper. Res. Soc. 50(2), 167–176 (1999)
30. J.W. Lee, B.S. Choi, J.J. Lee, IEEE Trans. Ind. Inf. 7(3), 419–427 (2011)
31. M.R. Majma, S. Almassi, H. Shokrzadeh, Int. J. Commun. Syst. 29, 959 (2016)
32. L. Shi, B. Zhang, K. Huang, J. Ma, Int. J. Commun. Syst. 26(10), 1341–1355 (2012)

Similarity-aware data aggregation using fuzzy c-means approach for wireless sensor networks

Runze Wan[1], Naixue Xiong[2*], Qinghui Hu[3], Haijun Wang[1] and Jun Shang[1]

Abstract

For resource-constrained IoT systems, data collection is one of the fundamental operations to reduce the energy dissipation of sensor nodes and improve the network lifetime. However, an anomaly or deviation will exert a great influence on the quality of data collected, especially for a data aggregation scheme. By taking into account data-aware clustering and detection of anomalous events, a similarity-aware data aggregation using a fuzzy c-means approach for wireless sensor networks is proposed. Firstly, by using a fuzzy c-means approach, the clustering process can be performed to organize sensors into clusters based on data similarity. Next, an effective support degree function is defined for further outlier diagnosis. Afterwards, the appropriate weight of valid data can be obtained by taking advantage of the probability distribution characteristics of normal samples within a certain period. Finally, the aggregation result in the cluster can be estimated. Practical database-based simulations have confirmed that the proposed data aggregation method can achieve better performance than traditional methods in terms of data outlier detection accuracy and relative recovery error.

Keywords: Fuzzy c-means, Data similarity, Aggregation, Wireless sensor networks

1 Introduction

Wireless sensor networks (WSNs) are typically composed of many small and low-cost sensor nodes with resource constraints, such as low memory capacity, less computational complexity, low communication bandwidth, and limited power. This new type of network demonstrates the characteristics of low cost, wide distribution, small volume, and flexible self-organizing [1]. With the rapid development, it has been successfully applied in the consumer electronics market and more and more widely used in the fields of target tracking, intelligent transportation, health prognosis, industrial automation, and so on. However, due to the WSN's imperfect nature, the sensor nodes need to be deployed densely to compensate for the quality of data collected [2, 3]. Nonetheless, for process-monitoring applications, high frequent sensing and the transmission of readings result in a large number of redundant samples, which may lead

to the waste of the node's energy and bandwidth resource as well as the reduction of the network lifetime. Therefore, how to employ spatiotemporal correlation of the readings between sensor nodes and develop efficient data redundancy reduction for saving the energy of the sensors are urgent problems.

Data aggregation is an effective method to solve the above problems [4]. The basic idea is to aggregate the samples of multi-sensors with a certain degree of redundancy rather than transmit raw data. It means that some nodes will act as aggregator to eliminate redundant data received from other sensor nodes and achieve desirable results for data accuracy. In practical application, the monitoring indicators, such as temperature, humidity, flow rate, or pressure, will demonstrate smooth and steady change in the majority of cases [5]. Once a sudden event occurs, the surrounding sensor nodes are generally able to detect the situation and obtain the readings synchronously. Therefore, the samples with large deviations from individual nodes may have a greater impact on the overall fusion results and influence the quality of data collected [6]. In this paper, we

* Correspondence: xiongnaixue@gmail.com
[2]Department of Mathematics and Computer Science, Northeastern State University, Tahlequah, OK, USA
Full list of author information is available at the end of the article

focus on spatio-temporal correlation of the readings in cluster-based WSNs. In particular, we cope with data-aware clustering and detection of anomalous events, and we use fuzzy c-means approach to organize sensors into clusters based on data similarity.

2 Methods

This study originates from the need for detecting spatial outliers in terms of the spatial correlations among neighboring sensor reading, which can get more accurate fusion results. Our approach uses the spatial temporal correlations of sensor's samples to detect outliers locally.

Compared with previous works, our contributions are presented as follow:

- We propose a novel similarity-aware data aggregation using fuzzy c-means approach for wireless sensor networks.
- We propose a theoretical analysis to determine the optimization of cluster formation.
- We conduct extensive simulations to demonstrate the performance of the algorithms. Simulation results show that our proposed method can achieve better performance than traditional methods in terms of data outlier detection accuracy and relative recovery error.

3 Related work

The traditional methods of data aggregation can be classified into two major categories: random theory-based and artificial intelligence-based approaches [7, 8]. The former includes the weighted average method, least square method, the Bayesian estimation, D-S evidence theory, and so on. The latter uses artificial neural network, fuzzy reasoning, or rough set to eliminate the anomalous data.

Izadi et al. [9] presented a fuzzy-based data fusion approach for WSNs to mitigate redundant data and reduce energy consumption. The authors utilized a fuzzy logic controller to obtain the confidence factor, and then the true value is distinguished and transmitted to the cluster head (CH) for multi-sensor data fusion. Fu [10] proposed double CHs model for secure and accurate data fusion, in which each cluster maintains dual CHs according to the reputation evaluation. All CHs make data fusion and transmit the results to the base station (BS), and the dissimilarity coefficient can be obtained by BS according to the fusion results. If the dissimilarity coefficient exceeds the threshold, the CH will be put into the blacklist and rotate the CH selection immediately. Xiang et al. [11] proposed a data aggregation method based on the compressive sensing theory. Particularly, they adopted diffusion wavelets to make the raw sensor data

sparse to decrease the communication overload as well as the computational complexity.

Furthermore, there are several strategies proposed in order to mitigate the energy hole problem. Sun et al. [12] proposed a data aggregation method of wireless sensor networks using artificial neural networks. The data fusion tree is established to reduce the packets flow and can update the leaf nodes dynamically. Aikaraki et al. [13] introduced a joint design of data aggregation with the routing technology, and presented a grid-based routing and aggregator selection scheme to achieve low energy dissipation and low latency without sacrificing quality. By investigating data fusion with communication constraint between the fusion center and each sensor, Xu et al. [14] presented a data fusion mechanism for target tracking in wireless sensor networks based on quantized innovations and Kalman filtering. By adding some delay time, all the data collected by relay node can be fused at one time so as to reduce the energy consumption. Aiming to ensure the data quality, Li et al. [15] proposed various metrics for QoS (quality of service) in the process of data aggregation, including lifetime, data delay, and retransmission rate. Also, the approach is discussed to ensure above QoS metrics in details.

Moreover, data outliers give rise to a very important impact on the correctness of data fusion results and the efficiency of IoT systems. In order to ensure the correctness of fusion results, the data outliers caused by such as software defects, occasionally failed communication, low battery, or malfunction on hardware should be excluded to avoid impact on the aggregation results. Actually, most of the monitoring targets or the occurrence of external events usually will be random and unexpected. With regard to the data outliers from anomalous events, the readings should be identified exactly. Krishnamachari et al. [16] proposed a distributed algorithm for fault-tolerant event region detection in wireless sensor networks, which can determine whether a node is abnormal. Besides, by exploiting the anomaly probability from adjacent nodes, only a few bit messages are sufficient to achieve fault-tolerant localization as events occurred. Tan et al. [17] presented a prediction model of data flow based on linear autoregressive analysis and further proposed a real-time detection algorithm for outliers identification and compression processing. Fernandes et al. [18] propose an autonomous profile-based anomaly detection system using principal component analysis and flow analysis to mitigate the impact of false data injection. By making inference of end-to-end measurements collected by relay nodes, Zheng et al. [19] proposed a trust-assisted framework for detecting and localizing network anomalies in a hierarchical sensor network, which also can obtain a flexible tradeoff between inference accuracy and probing overhead. Hu et al. [20]

presented outlier detection methods based on a neural
network for WSNs, which exploited historical data to
train the neural network to determine whether the actual measured value into the prediction interval so as to
distinguish the data anomalies.

4 Network model and cluster formation

4.1 Network model

We consider a cluster-based architecture for a wireless
sensor network, where all sensor nodes can monitor the
given condition and periodically send its collected data
to its CH. Most researches demonstrate that clustering
is considered as an efficient topology control method in
WSN to improve the scalability and lifetime of the whole
system [21]. By dividing the network, sensor nodes will
be grouped into different clusters based on certain rules
and each cluster has a cluster head [22]. CH is responsible for managing the cluster and receiving the set of
collected data from its member node during a certain
period. Also, in order to improve the efficiency of data
fusion, CH should have the ability to employ statistical
detection based on the sensor readings. It can detect
spatial outliers that deviate from normal data, thus ensuring the accuracy of data fusion.

In addition, the following assumptions about the network's topology are suggested:

1. At each period, the sensor nodes acquire the
monitoring readings at a fixed sampling rate with m
measures.
2. The original attribute information collected from the
sensor nodes can be fuzzified into a set of membership
functions.
3. In each cluster, the member nodes collect data in a
periodic manner. Subsequently, all member nodes will
send their data to the appropriate CH for data aggregation
at the end of a round.

4.2 Cluster formation algorithm

In this section, we discuss the cluster formation based
on data similarity by using fuzzy c-means approach.
Compared to other topologies, cluster-based network
topology is recently considered to be more effective for
aggregating data packets separately. In addition, most of
the existing data aggregation techniques based on clustering topology are dedicated to an event-driven data
model. Many hierarchical cluster formation algorithms
focus on the distance between nodes, residual energy,
geographic coverage, and so forth. In contrast, the main
purpose of our proposed method is to clear and amelior-
ate the collected data and provide the best information
to end users [23]. From a statistical point of view of the
correlation, the perceived data of same time slot can
demonstrate spatial-temporal correlation in the adjacent

monitoring region. If the monitoring indicators of perceptual physical objects in the region do not show great
fluctuation, there will be minimal deviation of the data
collected by the sensor nodes with close geographical location [24]. Therefore, cluster formation algorithm can
make use of the spatial-temporal correlated environmental data and partition the adjacent sensor nodes with
similar data instances into one cluster and different to
objects in other groups.

The fuzzy c-means (FCM) algorithm was proposed by
Bezdek [25] and has been used in cluster analysis, pattern recognition, image processing, and so forth. FCM is
a clustering method derived from unsupervised learning,
which uses fuzzy theory to divide a set of data points
into a set of fuzzy clusters according to certain partitioning criteria [26]. Suppose a WSN that consist of N-sen-
sor nodes randomly distributed over an area of $S \times S$
meters. By using of the sensor's respective geographical
location and collected data initially, the BS computes the
cluster centers and allocates sensor nodes to the clusters
by applying the FCM algorithm.

Each node is assigned a degree of membership u_{ij} to a
cluster C_k rather than completely being a member of
other clusters. According to [27], to achieve adequate
coverage rate, the optimal number of K clusters should
be determined by

$$K = \left\lceil \frac{\ln(1-\delta)}{\ln\left(1-3\sqrt{3}R^2/2S^2\right)} \right\rceil \tag{1}$$

where S represents the side length of square region, R
represents the sensor's communication radius, and δ denotes the coverage rate to be assured.

Assuming that sensor node s_i and s_j locate in the same
cluster, X_i and X_j represent their collected data sets during a fixed period. x_{ij} denotes the measure generated by
the sensor s_i at the time slot j, and m is the number of
samples in the fixed period. In the periodic data collection model, in order to minimize data redundancy and
still guarantee the accuracy of fusion results, studying
the variance between measurements is an analytical way
to choose appropriate nodes to form clusters. According
to information entropy theory, the entropy value of all
samples at time j can be obtained by

$$e_j = -\frac{1}{\ln m} \sum_{i=1}^{m} x_{ij} \ln x_{ij} \tag{2}$$

where $0 \le e_j \le 1$.

Since the utility value of the index is proportional to
its impact on fusion results, the weight value of variable
j can be defined as

$$\omega_j = \frac{1-e_j}{\sum_{j=1}^{m}\left(1-e_j\right)} \tag{3}$$

Next, the weighted Euclidean distance between the reading of node s_i and centroid point v_k of cluster C_k can be expressed as

$$\text{Dis}(s_i, v_k) = \sqrt{\sum_{j=1}^{m}\omega_j\left(x_{ij}-v_{kj}\right)^2} \tag{4}$$

where v_{kj} indicates the reference value of the centroid point at time j in cluster C_k.

Next, the objective function should be proposed to enhance the quality of the clusters and allocate sensor nodes into their most appropriate one [28]. By using FCM, the objective function, which will operate with iterative procedures, can be formulated as follows:

$$\min J(X, U, C_1, C_2, \cdots, C_K) = \sum_{k=1}^{K}\sum_{i=1}^{N}u_{ki}^{v_0}[\text{Dis}(s_i, v_k)]^2 \tag{5}$$

$$\text{s.t.} \quad \sum_{k=1}^{K}u_{ki} = 1, 1\leq i\leq N; \quad u_{ki}\in[0,1], \quad 1\leq k\leq K, \quad 1\leq i\leq N;$$
$$\sum_{i=1}^{N}u_{ki}\in[0,N], \quad 1\leq k\leq K.$$

where u_{ki} denotes the degree of membership being assigned to node s_i to join in the cluster C_k. U denotes the membership matrix of u_{ki}. v_0 is a weighting exponent on each fuzzy membership that determines the amount of fuzziness of the resulting classification, and is set to 2.

By using the Lagrange's multiplier to optimize formula (18), the problem is equivalent to find the minimum value of the Eq. (6).

$$F(U,\lambda) = \sum_{k=1}^{K}\sum_{i=1}^{N}w_i u_{ki}^{v_0}[Dis(s_i, v_k)]^2 + \sum_{i=1}^{N}\lambda\left(\sum_{k=1}^{K}u_{ki}-1\right) \tag{6}$$

where w_i indicates the weight value in process of data aggregation. It can be set as $1/N$ initially, and be updated by formula (22) at the end of each sampling period.

Note that the first order partial derivative should be equal to 0, i.e., $\partial F/\partial\lambda = 0$ and $\partial F/\partial u_{ki} = 0$. Then, we have

$$\partial F/\partial\lambda = 1-\sum_{k=1}^{K}u_{ki} = 0 \tag{7}$$

$$u_{ki} = \frac{\lambda}{w_i[\text{Dis}(s_i, v_k)]^2} \tag{8}$$

Thus, u_{ki} can be determined as

$$u_{ki} = \frac{w_i[Dis(s_i, v_k)]^2}{\sum_{k=1}^{K}w_i[Dis(s_i, v_k)]^2} \tag{9}$$

Similarly, suppose that $\partial F/\partial v_k = 0$, the reference vector of the centroid point in cluster C_k can be given as

$$v_k = \frac{\sum_{i=1}^{N}w_i u_{ki}^2 X_i}{\sum_{i=1}^{N}w_i u_{ki}^2} \tag{10}$$

5 Data aggregation

5.1 Data outlier detection

Outliers are often known as anomaly or deviation, which can even mislead systems into unsafe conditions. Whether the quality of data collected by WSNs is reliable and accurate or not will influence the performance of the whole system [29]. Therefore, data outliers should be detected and isolated in time so as to ensure the validity of data aggregation result and fusion efficiency. For clustered WSN, it is impossible for CH to determine the validity of the data sent by its members. However, the geographical relationship between the readings of sensor nodes within a certain physical spatial range or cluster may be an effective means to identify outliers through credible tests of the masses. In this sub-section, an effective support degree function is defined for further outlier diagnosis is introduced, which is based on a standard statistical distribution model and makes use of the measures between neighboring nodes.

As mentioned above, due to the spatial-temporal correlation in the adjacent monitoring region, the measurements between node s_i and s_j at the same sampling period will show relatively small differences. Hence, the support degree from node s_j to s_i can be expressed as consistency between the samples X_i and X_j.

First, to eliminate the influence of the measurement scale, the normalization processing of raw data is introduced to make a relatively objective comparison between measurement sets collected by different sensor node. The original attribute of raw data may belong to positive index or negative index. First, linear transformation of raw data can be given by

$$\begin{cases} x'_{ij} = x_{ij}/\hat{x}_j, & \text{if } x_j \text{ is positive index} \\ x'_{ij} = \hat{x}_j/x_{ij}, & \text{if } x_j \text{ is negative index} \end{cases} \tag{11}$$

where $\hat{x}_j$ denotes the ideal value for the samples at the time slot j, and x'_{ij} is the proximity of x_{ij} to ideal values.

Then, the normalized value of sample x_{ij} can be represented as

$$y_{ij} = x'_{ij} / \sum_{i=1}^{n} x'_{ij} \tag{12}$$

In order to characterize that mutual support relationship between sensor nodes, the support degree can be defined as

$$p_{ij} = \frac{\sum_{k=1}^{m} (y_{ik} - \hat{y}_k)(y_{jk} - \hat{y}_k)}{\sqrt{\sum_{k=1}^{m} (y_{ik} - \hat{y}_k)^2 \sum_{k=1}^{m} (y_{jk} - \hat{y}_k)^2}} \tag{13}$$

where $\hat{y}_k$ represents the mean value.

According to the above function, the support degree matrix P of the measures from all nodes in a cluster can be obtained

$$P = \begin{bmatrix} 1 & p_{12} & \cdots & p_{1n} \\ p_{21} & 1 & \cdots & p_{2n} \\ \vdots & \vdots & \ddots & \vdots \\ p_{n1} & p_{n2} & \cdots & 1 \end{bmatrix} \tag{14}$$

where p_{i1}, p_{i2}, $\cdots$, p_{in} are support degree of the samples collected by s_i from the member nodes in the same cluster, and n indicates the number of member nodes. By integrating the evaluation of all neighboring nodes, the comprehensive support degree of X_i can be calculated as

$$T_i = \frac{1}{n} \sum_{j=1}^{n} p_{ij} \tag{15}$$

Since T_i is relative to the amount of support degree from other member nodes or its nearest local neighbors, it indicates normal level compared to the majority of sensor readings. When a sensor sends abnormal data due to noise errors or malicious attacks, the readings will obviously deviate from the measures of other sensors. As a result, its comprehensive support is very small. Unless a large area of intra-cluster nodes fail simultaneously, the probability of that exceptional case will be very low and can be neglected.

Suppose that the comprehensive support of data X_i is T_i, if $T_i \geq \zeta$, X_i is determined as normal data. Otherwise, the data will be regarded as outliers. Among them, the parameter ζ is set as the availability threshold value. When the value of T_i is less than the threshold value ζ, the corresponding readings X_i will be processed to mitigate the influence on the aggregation result.

5.2 Data aggregation strategy

In this section, we present the data aggregation strategy to ensure the accuracy of the aggregation result. Before aggregation process, the data being collected from member nodes will be sent entirely to the cluster head, which can conduct outlier detection based on the centralized approach. If the data being received is valid, it will be put into data aggregation process. Otherwise, they should be rejected immediately. Therefore, the two types of memory buffers can be embedded in CH and corresponding parameter a and b is set to count the number of normal and outlier data uploaded by each member node. Under certain conditions, the probability distribution of normal and outlier data will be approximate to the posterior probability distribution with binomial model, which can obey the beta distribution. Therefore, the beta distribution characteristics can be employed to evaluate the data validity.

Set χ as the posterior probability of a random event, the probability of distribution can be obtained based on the Bayesian statistics as

$$\begin{aligned} P(\chi|a,b) &= \frac{P(\chi,a,b)}{P(a,b)} \\ &= \frac{\binom{a+b}{a} \chi^a (1-\chi)^b}{\int_0^1 \binom{a+b}{a} \chi^a (1-\chi)^b d\chi} \\ &= \frac{\chi^a (1-\chi)^b}{\int_0^1 \chi^a (1-\chi)^b d\chi} \end{aligned} \tag{16}$$

where $0 \leq \chi \leq 1$, $a > 0$, $b > 0$.

The probability density function of the parameter (a, b) can be expressed as

$$f(\chi|a,b) = \frac{\chi^{a-1}(1-\chi)^{b-1}}{\int_0^1 u^{a-1}(1-u)^{b-1} du} = \frac{\Gamma(a+b)}{\Gamma(a)\Gamma(b)} \chi^{a-1}(1-\chi)^{b-1} \tag{17}$$

Thus, we have

$$P(\chi|a,b) = \frac{\chi^a (1-\chi)^b}{f(a+1,b+1)} \tag{18}$$

As a result, the reliability of the monitored samples of the determined node will obey the beta distribution with parameters $a+1$ and $b+1$.

$$P(\chi|a+1,b+1) = \begin{cases} \dfrac{\chi^a(1-\chi)^b}{B(a+1,b+1)}, & 0 < \chi < 1 \\ 0, & \text{otherwise} \end{cases} \tag{19}$$

Therefore, the mathematical expectation $E(\chi)$ of beta distribution can be given as

$$E(\chi) = \frac{a+1}{a+b+2} \tag{20}$$

Considering the effect of data retransmission caused by channel quality, the uncertainty of the readings being collected by the determined node can be attenuated by introducing time attenuation factor. Accordingly, the time attenuation factor should be defined as

$$\theta = Td_i / Td_{ave} \tag{21}$$

where Td_i denotes the packet transmission delay of node s_i, and Td_{ave} denotes the average packet transmission delay of all member nodes.

Thus, formula (8) can be modified to:

$$E(\chi) = \frac{a+1}{a+\theta b+2} \tag{22}$$

The revised mathematical expectation can be defined as the weight value in process of data aggregation, and $w_i = E_i(\chi)$ will be allocated to member nodes.

Finally, the aggregation result in the cluster can be estimated by

$$X^* = \sum_{i=1}^{n} w_i X_i \tag{23}$$

6 Experiments

In this section, practical database-based simulations have been conducted to evaluate the performance of our method. Firstly, the datasets are derived from the real sensed data collected from 54 Mica2Dot sensors deployed in the Intel Berkeley Research Lab between February 28 and April 5, 2004 [30]. The sensed data included humidity, temperature, light, and voltage values collected. In the experiments, we first selected some measurements of temperature from the sensor nodes 36 until 43, for the time period from March 18, 2004, to March 20, 2004, corresponding to 2000 log rows. We do not take into account the other features (humidity, light, and voltage). The quantity of data is about 2.3 million readings; it was collected using the TinyDB in-network query processing system, built on the TinyOS platform. Based on the dataset, we add a given mass of outliers to simulate the occurrence of events, which can make the data fluctuate to a certain extent.

To evaluate the performance of our approach, we use the TOSSIM tool [31]. TOSSIM is a TinyOS simulation tool which simulates WSN physical and link layer features accurately. This allows validating the solution under realistic WSN deployment conditions. In each test, we repeat the simulation for 30 times and compute the mean of results. The key simulation parameters are summarized in Table 1.

In the experiment scenarios, outliers are simulated randomly, and 100 values of temperature are generated and then added to the dataset. In terms of the evaluation metrics, detection accuracy rate is defined as the ratio of outliers being detected to all outliers, and false alarm

Table 1 Simulation parameters

Parameter	Value
Area size	500×500 m
Number of sensor nodes N	81
Node's communication range	40 m
Transmission channel	Wireless channel
Propagation model	Normal path loss model
Data packet size	32 bytes
Bandwidth	200 kilobytes per second
Radio layer	CC2420 radio layer
Queue size	50 packets
δ	0.6~0.9
ζ	0.1~0.9
Outlier probability	5%, 10%, 15%
u_0	2
m	50

rate represents the ratio of normal data mistakenly detected as outliers.

First, our objective is to study the detection of abnormal data in accordance to our proposed method. Figure 1 shows the detection accuracy rate when varying δ and ζ. Since the support degree can differentiate the normal data and outliers, it can effectively guarantee the data accuracy when handling the aggregation process. This truth is clearly shown in Fig. 1 when the sensor node applies the aggregation phase and when δ and ζ increases. In addition, it can be noticed that the detection accuracy rate can be significantly improved as ζ increases. The reason is that once the node's support degree cannot be satisfied with the threshold's requirement, its readings will not be preserved and submitted to CH for data fusion. It is beneficial to the accuracy of the final fusion results.

Figure 2 shows the false alarm rate when varying δ and ζ. Indeed, we can find that the overall trend of false alarm rate is opposite to that of the detection accuracy rate, especially when δ is larger, the fluctuation is more obvious. The higher value of δ means that more clusters will be distributed in the monitoring region. For fixed number of total sensor nodes, it will cause the reduction of the number of members in a single cluster and make the determination of outliers more stringent. In addition, the increase of threshold ζ leads to higher level for support degree, and the samples with large deviation can easily be judged to be invalid. In the case of a monitoring indicator that suddenly changes dramatically and only a few nodes perceived it, their readings will be treated as anomalous data owing to low support degree. Thus, the detection accuracy rate will be increased.

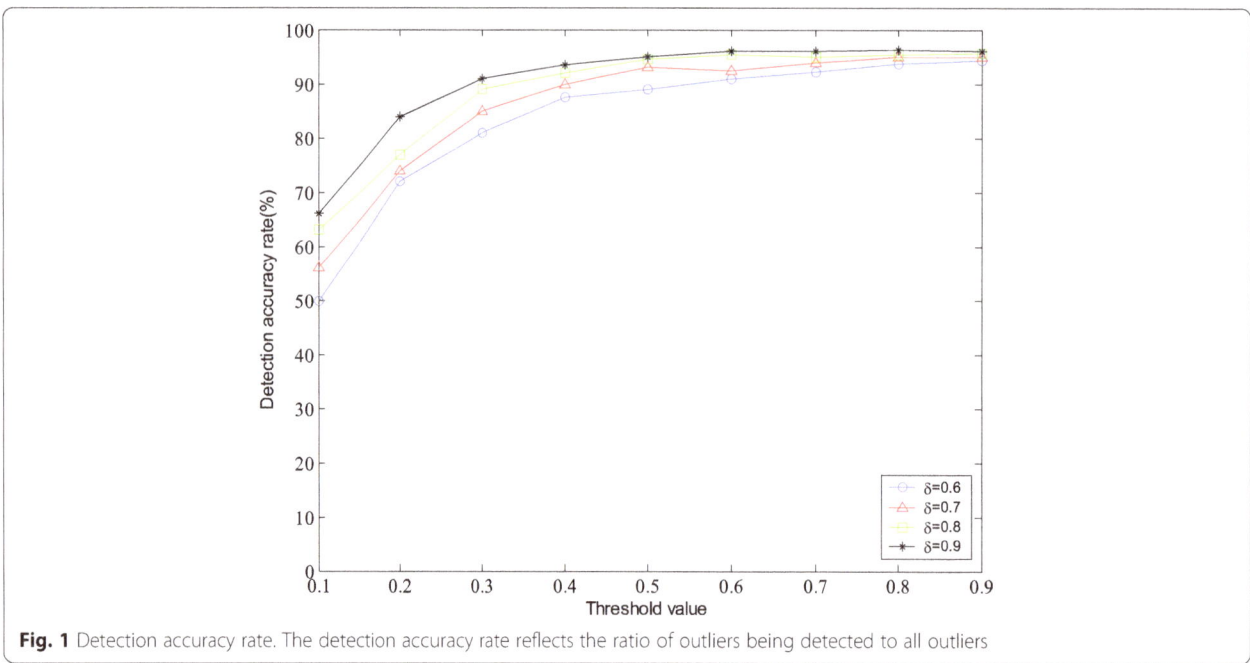

Fig. 1 Detection accuracy rate. The detection accuracy rate reflects the ratio of outliers being detected to all outliers

Figure 3 shows the deviation degree of aggregation result as a function of threshold ζ. Based on the obtained results, we can notice that ζ moves from the extreme values of 0.1 and 0.9 towards the optimal value 0.6. When the threshold is either too low or too high, the final fusion result is not ideal. This is due to the high accuracy and low false alarm rate of differentiating normal data and outliers in terms of appropriate availability threshold. Low threshold setting may result in low detection rate and thus affecting the accuracy of the final fusion result. Conversely, excessive value of specified threshold will make the criteria more rigorous, and the normal measurement will be identified as outlier. Moreover, if such a situation occurs in a continuous time, it will inevitably lead to a sharp decline in the weight of these valid data in data fusion processing and thus affecting the data accuracy.

Next, we further study the impact of different outlier probability on relative recovery error and make a comparison with KPFF [32] and DSADC [33]. Recovery accuracy

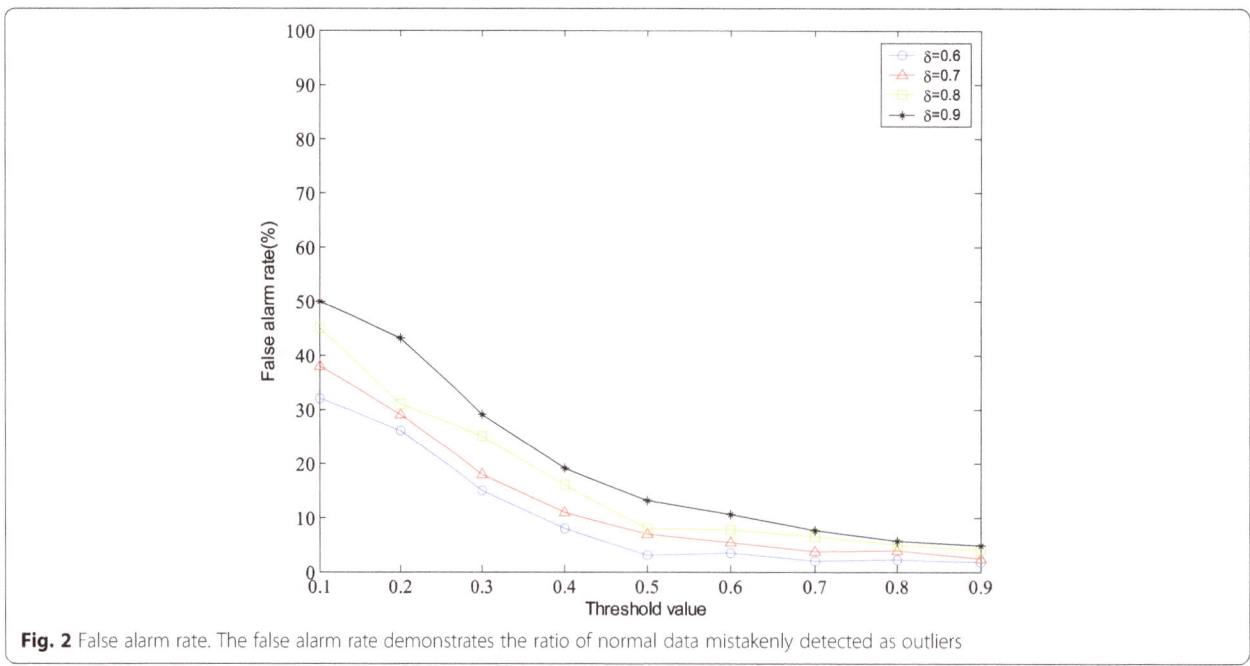

Fig. 2 False alarm rate. The false alarm rate demonstrates the ratio of normal data mistakenly detected as outliers

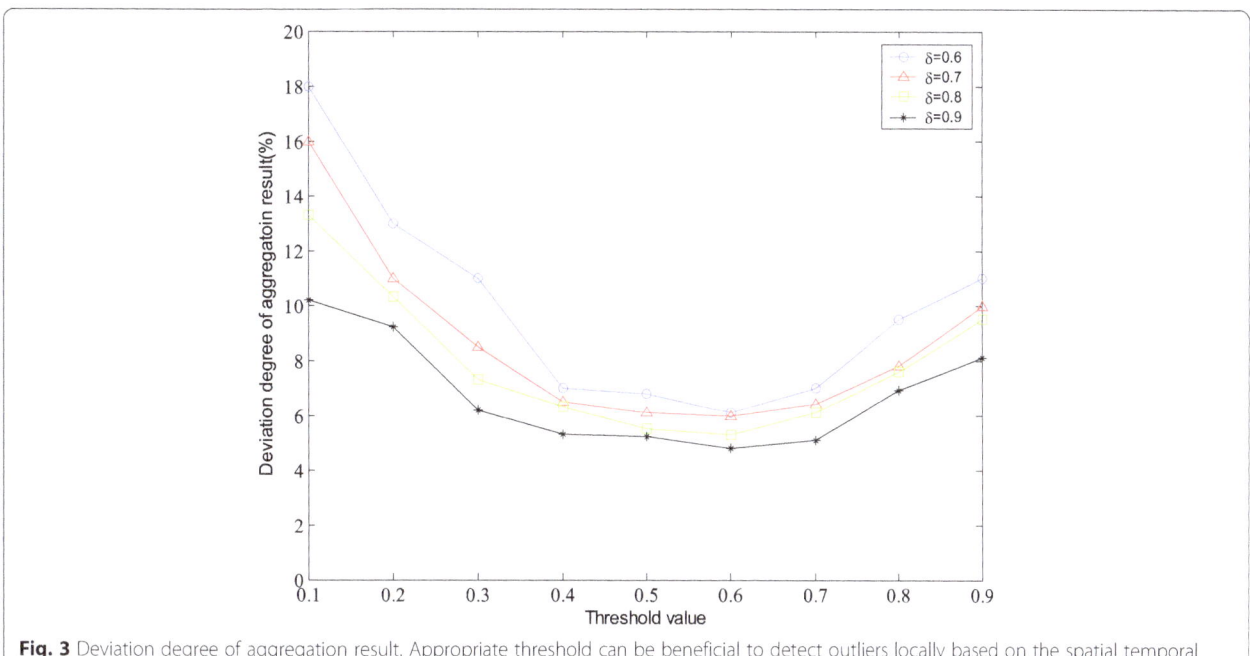

Fig. 3 Deviation degree of aggregation result. Appropriate threshold can be beneficial to detect outliers locally based on the spatial temporal correlations of the sensor's samples

is a normally used metric to evaluate the quality of data aggregation algorithms [34]. In this paper, recovery accuracy is mathematically defined by relative recovery error (RRE), which is the relative difference between original and recovered data matrices. Figures 4, 5 and 6 demonstrate the instant RRE along the timeline.

The obtained results show clearly that applied support degree in such a way is very effective. It also maintains adaptability with different outlier probability. From the

experiment results, it can be seen that the RRE curves of both KPFF and DSADC algorithms fluctuate dramatically. But we can still observe that nearly 90% RRE values of similarity-aware data aggregation using fuzzy c-means approach (SDAF) are below those of DSADC. The error of the fusion results obtained by SDAF is smaller than other methods especially as outlier probability increases. In the process of data aggregation, outlier samples can be identified effectively by diagnosis mechanism in

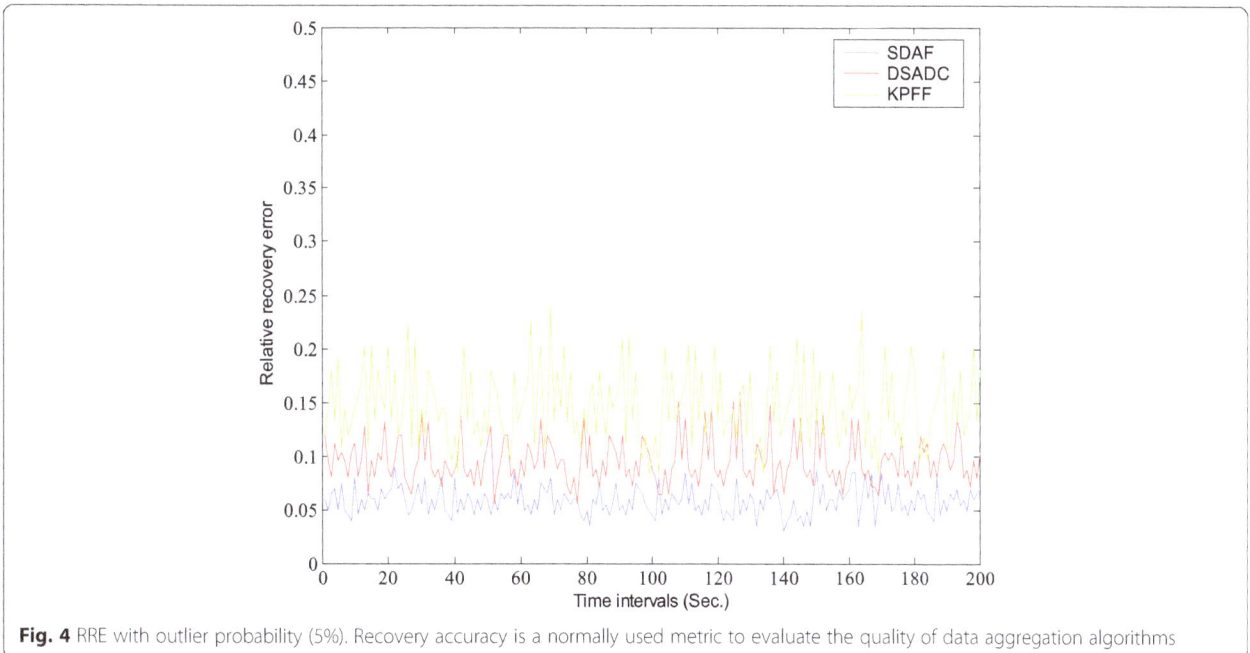

Fig. 4 RRE with outlier probability (5%). Recovery accuracy is a normally used metric to evaluate the quality of data aggregation algorithms

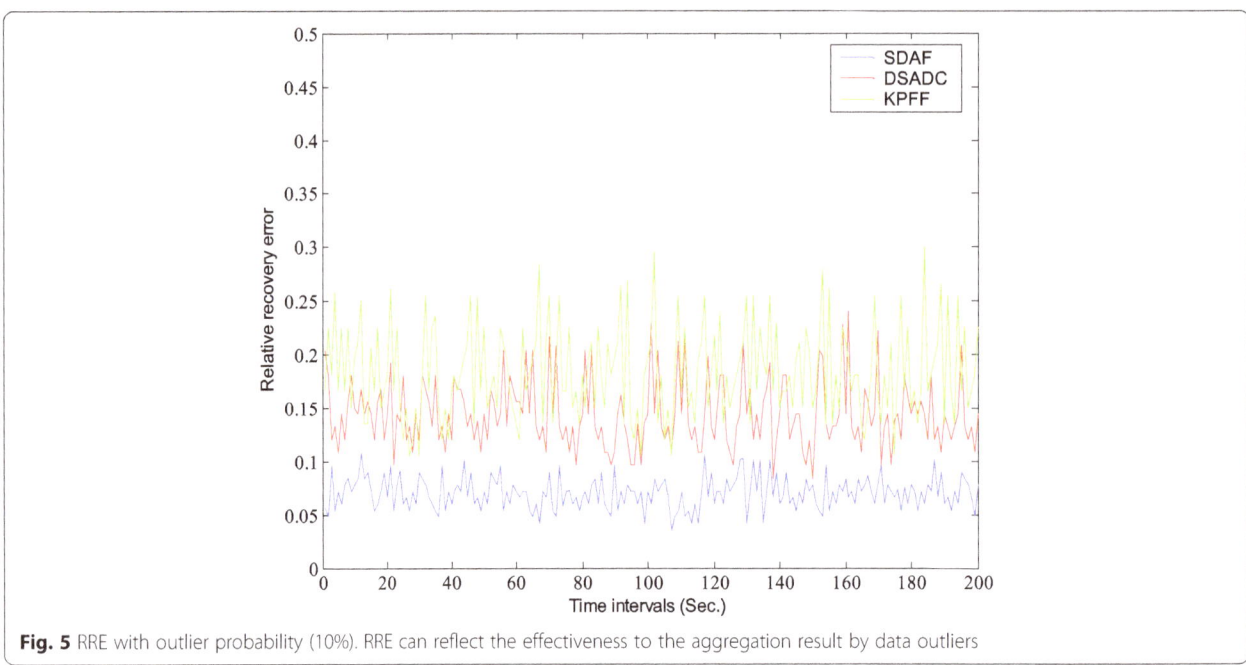

Fig. 5 RRE with outlier probability (10%). RRE can reflect the effectiveness to the aggregation result by data outliers

SDAF, and the outlier-free readings are further aggregated and transmitted to the CH. Therefore, it can reduce the effectiveness to the aggregation result by data outliers and avoid the possibility of misleading systems into unsafe conditions.

7 Conclusions

To minimize the energy consumption by redundant data and reduce the expense of transmissions to the sink, data aggregation technology is very essential for WSNs.

Data anomaly or deviation will exert a great influence on the quality of aggregated results. In this paper, we have proposed a similarity-aware data aggregation using a fuzzy c-means approach in clustered WSNs. By investigating the spatio-temporal correlations of sensor data and local detection of anomalous events, we presented a cluster formation algorithm based on fuzzy c-means approach. Then, we define an effective support degree function for further outlier diagnosis. Finally, based on statistical analysis of the outlier or outlier-free sensor

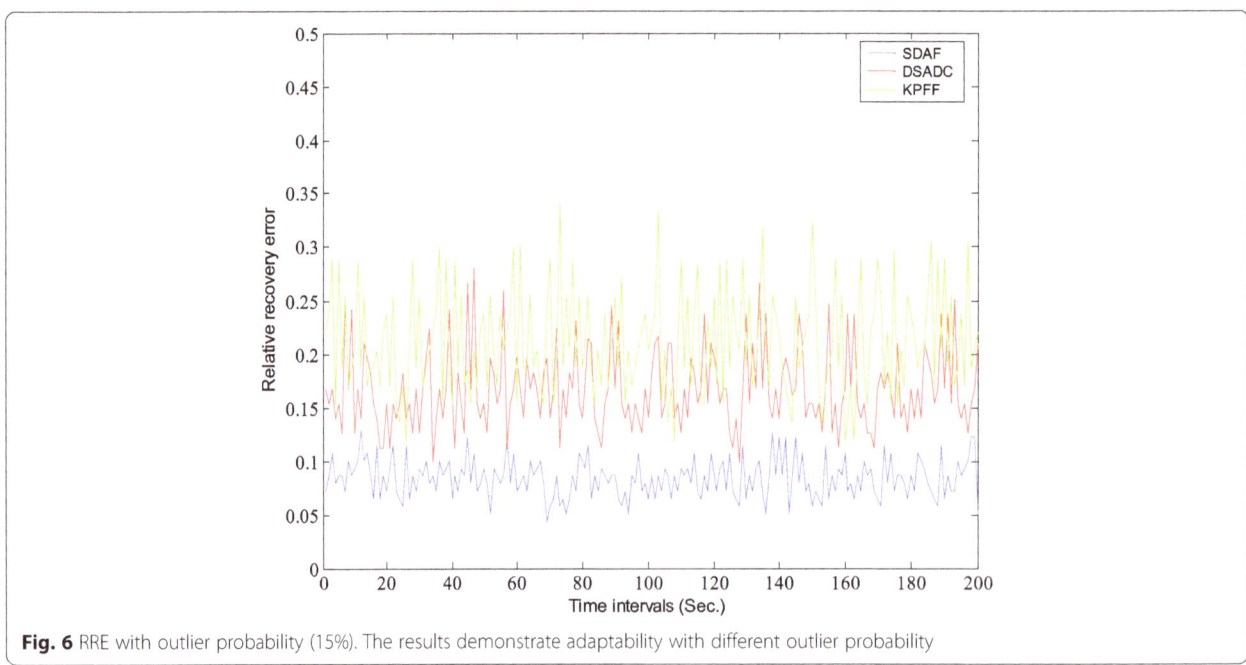

Fig. 6 RRE with outlier probability (15%). The results demonstrate adaptability with different outlier probability

data, the readings aggregation is conducted. Overall, the simulation results show that the proposed method can achieve better performance than traditional methods in terms of data outlier detection accuracy and relative recovery error.

In our future work, we plan to conduct the research on the analysis of outlier detection in terms of characteristics like the multi-dimension, detection mode, architectural structure, and correlation extraction.

Abbreviations
BS: Base station; CH: Cluster head; FCM: Fuzzy c-means; QoS: Quality of service; RRE: Relative recovery error; SDAF: Similarity-aware data aggregation using fuzzy c-means approach; WSNs: Wireless sensor networks

Acknowledgements
The authors acknowledged the anonymous reviewers and editors for their efforts in valuable comments and suggestions.

Funding
This research was supported in part by the Hubei Provincial Educational Science Program (Grant No. 2018GB073) and the Guangxi Nature Science Fund (Grant No. 2016GXNSFAA380226).

Authors' contributions
WR proposes the innovation ideas and theoretical analysis, and XN carries out experiments and data analysis. HQ also wrote parts of the manuscript. SJ and WH participated in the coordination of the study and reviewed the manuscript. All authors read and approved the final manuscript.

Competing interests
The authors declare that they have no competing interests.

Author details
[1]Hubei Co-Innovation Center of Information Technology Service for Elementary Education, Hubei University of Education, Wuhan, China. [2]Department of Mathematics and Computer Science, Northeastern State University, Tahlequah, OK, USA. [3]School of Computer Science & Engineering, Guilin University of Aerospace Technology, Guilin, China.

References
1. P. Zhong, Y.T. Li, W.R. Liu, Joint mobile data collection and wireless energy transfer in wireless rechargeable sensor networks. Sensors. 17(8), 1–23 (2017).
2. Y.Y. Zeng, C.J. Sreenan, L. Sitanayah, An emergency-adaptive routing scheme for wireless sensor networks for building fire hazard monitoring. Sensors. 11(3), 2899–2919 (2011).
3. C. Lin, N. Xiong, J.H. Park, T. Kim, Dynamic power management in new architecture of wireless sensor networks. Int. J. Commun. Syst. 22(6), 671–693 (2009).
4. H.J. Cheng, Y.Z. Chen, N.X. Xiong, Layer-based data aggregation and performance analysis in wireless sensor networks. J. Applied Mathematics. 2013, 1–12 (2013).
5. C. Zhou, S. Huang, N. Xiong, S.H. Yang, Design and analysis of multimodel-based anomaly intrusion detection systems in industrial process automation. IEEE Transactions on Systems, Man, and Cybernetics: Systems. 45(10), 1345–1360 (2015).
6. L. Shu, Y. Zhang, Z. Yu, L.T. Yang, M. Hauswirth, Context-aware cross-layer optimized video streaming in wireless multimedia sensor networks. J. Supercomput. 54(1), 94–121 (2010).
7. H.J. Cheng, D.Y. Feng, X.B. Shi, Data quality analysis and cleaning strategy for wireless sensor networks. EURASIP J. Wirel. Commun. Netw. 61, 1–11 (2018).
8. K. Lin, J.P.C. Rodrigues, N. Xiong, H.W. Ge, Energy efficiency QoS assurance routing in wireless multimedia sensor networks. IEEE System Journal. 5(4), 495–505 (2011).
9. I. Davood, J.H. Abawajy, S. Ghanavati, A data fusion method in wireless sensor networks. Sensors. 15(2), 2964–2979 (2015).
10. J.S. Fu, Y. Liu, Double cluster heads model for secure and accurate data fusion in wireless sensor networks. Sensors. 15(1), 2021–2040 (2015).
11. L. Xiang, J. Luo, C. Rosenberg, Compressed data aggregation: energy-efficient and high fidelity data collection. IEEE/ACM Trans. Netw. 21(6), 1722–1735 (2013).
12. L.Y. Sun, X.X. Huang, W. Cai, Data aggregation of wireless sensor networks using artificial neural networks. Chinese Journal of Sensors and Actuators. 24(1), 122–127 (2011).
13. J.N. Aikaraki, R. Uimustafa, A.E. Kamal, Data aggregation and routing in wireless sensor networks: optimal and heuristic algorithms. Comput. Netw. 53(7), 945–960 (2009).
14. J. Xu, J.X. Li, S. Xu, Data fusion for target tracking in wireless sensor networks using quantized innovations and Kalman filtering. Science China: Information science edition. 55(3), 530–544 (2012).
15. H. Li, H.Y. Yu, Research on data aggregation supporting QoS in wireless sensor networks. Application Research of Computers. 25(1), 64–67 (2008).
16. B. Krishnamachari, S. Iyengar, Distributed Bayesian algorithms for fault-tolerant event region detection in wireless sensor networks. IEEE Trans. Comput. 53(3), 241–250 (2004).
17. Y.H. Tan, Y.P. Lin, T. Dong, Real-time detection algorithm for anomaly data in sensor networks. Journal of System Simulation. 19(18), 4335–4341 (2007).
18. G.J. Fernandes, J.P.C. Rodrigues, M.L. Proença, Autonomous profile-based anomaly detection system using principal component analysis and flow analysis. Appl. Soft Comput. 34(9), 513–525 (2015).
19. S. Zheng, J.S. Baras, in 8th IEEE Communications Society Conference on Sensor, Mesh and ad hoc Communications and Networks(SECON). Trust-assisted anomaly detection and localization in wireless sensor networks (2011), pp. 386–394.
20. S. Hu, G.H. Li, W.W. Lu, Outlier detection methods based on neural network in wireless sensor networks. Computer Science. 41(11), 208–211 (2014).
21. R. Kumar, N. Singh, A survey on data aggregation and clustering schemes in underwater sensor networks. Int. J. Grid Distrib. Comput. 7(6), 29–52 (2014).
22. W. Guo, N. Xiong, A.V. Vasilakos, G. Chen, H. Cheng, Multi-source temporal data aggregation in wireless sensor networks. Wirel. Pers. Commun. 56(3), 359–370 (2011).
23. V. Chatzigiannakis, S. Papavassiliou, Diagnosing anomalies and identifying faulty nodes in sensor networks. IEEE Sensors J. 7(5), 637–645 (2007).
24. X. Wang, Q. Li, N. Xiong, Y. Pan, in International Conference on Wireless Algorithms, Systems, and Applications (WASA 2018). Ant colony optimization-based location-aware routing for wireless sensor networks (2018), pp. 109–120.
25. F. Herrera, Genetic fuzzy systems: status, critical considerations and future directions. Int. J. Comput. Intell. Res. 5, 59–67 (2005).
26. N. Goyal, M. Dave, A.K. Verma, in Int. Conf. Electron. Commun. Syst. (ICECS). Fuzzy based clustering and aggregation technique for under water wireless sensor networks (2014), pp. 1–5.
27. W.B. Heinzelman, A.P. Chandrakasan, H. Balakrishnan, An application-specific protocol architecture for wireless micro-sensor networks. IEEE Trans. Wirel. Commun. 1(4), 660–670 (2009).
28. J.S. Lee, W.L. Cheng, Fuzzy-logic-based clustering approach for wireless sensor networks using energy predication. IEEE Sensors J. 12, 2891–2897 (2012).
29. S.M. Reda, M. Abdelhamid, S. Hadj, A. Amar, Performance evaluation of network lifetime spatial–temporal distribution for WSN routing protocols. J. Netw. Comput. Appl. 35(4), 1317–1328 (2012).
30. Intel lab data home page. http://db.lcs.mit.edu/labdata/labdata.html. March 20, 2014.
31. P. Levis, N. Lee, M. Welsh, D. Culler, in Proc. of the 1st International Conference on Embedded Networked Sensor Systems. TOSSIM: accurate and scalable simulation of entire TinyOS applications (ACM Digital Library, Los Angeles, California, 2003), pp. 126–137.
32. H. Harb, A. Makhoul, D. Laiymani, in Proc. of the 10th IEEE International Conference on Wireless and Mobile Computing, Networking and Communications (WiMob). K-means based clustering approach for data aggregation in periodic sensor networks (2014), pp. 434–441.
33. G. Fernando, J. Gentian, N. Michele, S. Aldri, Data similarity aware dynamic node clustering in wireless sensor networks. Ad Hoc Networks. 24, 29–45 (2015).
34. Y. Sang, H. Shen, Y. Tan, N. Xiong, in Proc. of International Conference on Information and Communications Security. Efficient protocols for privacy preserving matching against distributed datasets (2006), pp. 210–227.

12

An adaptive retransmit mechanism for delay differentiated services in industrial WSNs

Ye Chen[1], Wei Liu[2], Tian Wang[3], Qingyong Deng[4,5], Anfeng Liu[1]* ⓘ and Houbing Song[6]

Abstract

The Internet of Things (IoT) is the latest Internet development, with billions of Internet-connected devices and a wide range of industrial applications. Wireless sensor networks are an important part of the Internet of Things. It has received extensive attention from researchers due to its large-scale, self-organizing, and dynamic characteristics and has been widely used in industry, traffic information, military, environmental monitoring, and so on. With the development of microprocessor technology, sensor nodes are becoming more and more powerful, which enables the same wireless sensor networks (WSNs) platform to meet the different quality of service (QoS) requirements of many applications. Applications for industrial wireless sensor networks range from lower physical layers to higher application layers. The same wireless sensor network sometimes needs to process information from different layers. Traditional protocols lack differentiated services and cannot make full use of network resources. In this paper, an Adaptive Retransmit Mechanism for Delay Differentiated Services (ARM-DDS) scheme is proposed to meet different levels of delays of applications. Firstly, we analyze the impact of different retransmit mechanisms and parameter optimization on delays and energy consumption. Based on the results of the analysis, in ARM-DDS scheme, for routes with transmission delay tolerance, energy-saving retransmission mechanisms are used, and low-latency retransmission mechanisms are used for latency-sensitive routes. In this way, the data routing delays of different applications are guaranteed within bound and the energy consumption of the network is reduced. What is more, ARM-DDS scheme makes full use of the residual energy of the network and uses a small delay routing retransmit mechanism in the far-sink area to reduce end-to-end delay. Both theoretical analysis and simulation experiments show that under the premise of the same reliability requirements, ARM-DDS scheme reduces data transmission delay 12.1% and improves network energy utilization 28%. Given that the reliability requirements of the data stream are different, the scheme can also extend the network lifetime.

Keywords: Industrial wireless sensor networks, Annular routing, Compressed sensing, Clustering, Energy utilization

1 Introduction

Wireless sensor networks (WSNs) are an important part of the Internet of Things (IoT). It has received extensive attention from researchers due to its large-scale, self-organizing, and dynamic characteristics and has been widely used in industry, traffic information, military, environmental monitoring, and so on. The development of modern industry is getting faster and faster, and the application research of industrial wireless sensor networks

is of great significance. With the development of microprocessor technology, sensing-based devices are becoming more and more powerful. The computing and storage capacity of smart sensing devices has been more than 10 years ago for personal computers [1–3]. The enrichment of sensing devices makes WSNs gradually evolve into a common sensing platform, which can provide data-aware services for a variety of applications [4–6]. Numerous applications can work together with the same WSNs as a platform to accomplish tasks at the same time, which greatly reduces the time and cost of application redeployment and operation of WSNs, and greatly promotes the development of WSNs [7–9]. For

* Correspondence: afengliu@mail.csu.edu.cn
[1]School of Computer Science and Engineering, Central South University, Changsha 410083, China
Full list of author information is available at the end of the article

example, sensing nodes deployed on the roadside can monitor traffic flow, collect temperature data, and events can also be monitored, so that multiple different applications tasks can be performed on one platform [10, 11]. This situation becomes more common as the processing capacity of sensor nodes increases, thus laying the foundation for the development of smart farming, smart transportation, smart industry, and smart cities [12–15].

Due to the diversity and complexity of applications running over WSNs [16], different applications may have different quality of service (QoS) requirements, especially for delay [13, 16–21]. For example, in the industrial automation production control line, the event data must be routed to the sink as soon as possible in the monitoring of fire and other dangerous events of important facilities and key equipment, otherwise, it will cause great loss of personnel and property. On the other hand, the humidity information monitoring is usually only needed to collect 3–4 times a day for the perception of environmental conditions such as the temperature of household crops, so the delay of data packet routing is tolerable and acceptable within a few minutes.

Another important issue in WSNs is the energy issue [11, 15, 20–24]. Because wireless sensor network nodes are powered by batteries and cannot be replaced in most scenarios, their energy is strictly limited [19, 25]. Therefore, designing an energy-efficient routing strategy is an important issue in WSNs [16, 18, 20, 26, 27]. Another issue is to ensure the reliability of data transmission [28]. Because the wireless sensor network is based on wireless communication and data transmission to the sink often passes through multi-hop routing, its communication reliability is far lower than that of the wired network. According to relevant research, the link packet loss rate of the wireless sensor network is often as high as 10%, in some cases even as high as 20–30% [28]. However, most applications have certain requirements for the reliability of data arriving at the sink. For example, in wireless communication with 90% reliability requirement, the reliability of data packet transmission after 10 hops routing is 34.9%, while after 20 hops, the reliability is only 12.2%. Therefore, in wireless communication, it is often necessary to take some reliability safeguard measures to make the reliability of data to the sink meet the threshold δ. At the same time, different applications have different reliability thresholds, and some applications also tolerate the loss of some data packets, thus δ is usually a constant less than 1, such as 90% [28]. For example, in large-scale monitoring of crops, usually missing part of the data packets does not affect the effective monitoring of crops, because only a certain proportion of data can make the correct decision [28].

At present, most routing strategies are mainly aimed at a single application. Because the same application has simple QoS requirements, so a fixed QoS routing strategy can meet the requirements. However, with the enhancement of the function of sensing nodes, many applications can run on the same WSNs platform. At this time, it is very urgent to provide a routing strategy with differentiated services with QoS [16, 29]. In general, the strategy that makes routing delays smaller will consume more energy. Conversely, the routing strategies that consume less energy will have a high end-to-end delay [21, 26]. Similarly, the reliability requirements of data transmission will also have an impact on energy consumption and delays. For example, wireless sensor networks usually use retransmitting mechanisms to ensure the reliability, so the higher the reliability requirements in applications [28], the greater the maximum number of retransmissions, the greater the delays caused by multiple retransmissions, and the more energy consumed, so the energy consumption and delays of applications with high-reliability requirements will be larger. Therefore, if applications with different QoS requirements are running on the same WSNs platform, if the same routing strategy is still adopted, that is, to select a routing strategy to satisfy the highest QoS requirements. Although this strategy can satisfy all applications' QoS requirements, the cost of the system is very high, which makes the overall lifetime of the system and the efficiency very low. Therefore, a strategy to provide differentiated services is urgently needed for the current WSNs.

There are already some routing strategies providing QoS differentiated services and studies on delay differentiated services also exist. Zhang et al. [16] proposed integrity and delay differentiated routing (IDDR) scheme to provide QoS differentiated services at a WSNs platform. In some applications, they are sensitive to data delays, while in some applications, they are sensitive to data integration requirements. Therefore, in the IDDR scheme, the delay-sensitive packets are routed to sink along the shortest path, thus minimizing delays. For packets sensitive to lose but not sensitive to delays, another routing strategy is adopted to avoid possible dropping on the hotspots in which with large data volume and high conflict and congestion. So, the IDDR scheme can guarantee two applications with different QoS requirements running on the same WSNs platform at the same time.

Obviously, for delay-sensitive data, although shortest routing can reduce delay, in the application of ultra-high delay requirement, the delay requirement cannot be satisfied even if the delay is reduced. That is to say, differentiated services are a best-effort strategy based on existing strategies, which has not been redesigned from the routing mechanism, so it can only satisfy a small range of delay differentiated services, so this mechanism is not applicable for applications that are particularly sensitive to delay requirements. On the other hand, in previous strategies, data that are not sensitive to delay will be bypassed. In fact, it is not advisable to increase the path length, because it will not bring any benefits to the system, but will only increase the delay and energy consumption, especially in data-sparse networks. Given the above problems, an Adaptive Retransmit Mechanism for Delay Differentiated Services (ARM-DDS) scheme is proposed to meet different levels of delays of applications; the main innovations of the scheme are as follows:

(1) An Adaptive Retransmit Mechanism for Delay Differentiated Services (ARM-DDS) scheme is proposed to meet the needs of applications with delay differentiated services. Firstly, we analyze the retransmit mechanism in depth and get the influence of different to retransmit modes and retransmit times on delay and energy consumption, as well as the optimal selection of retransmitting parameters on the premise of ensuring certain reliability of data transmission. In the ARM-DDS scheme, for routes with transmission delay tolerance, energy-saving retransmission mechanisms are used, and for latency-sensitive routes, low-latency retransmission mechanisms are used. In this way, the data routing delays of different applications are guaranteed within the bound and the energy consumption of the network is reduced, so delay differentiated services are realized very well.

(2) What is more, the ARM-DDS scheme makes full use of the residual energy of the network and uses a small delay routing retransmit mechanism in the far-sink area to reduce end-to-end delay. Therefore, the ARM-DDS scheme can further reduce the delay based on improving energy utilization.

(3) Both theoretical analysis and simulation results show that, compared with previous studies, the ARM-DDS scheme has the following advantages: (a) ARM-DDS can meet the delay requirements in a wide range. In previous studies, the delay range allowed by the delay differentiated services scheme is relatively narrow and cannot meet the highly sensitive requirements of different applications for delay, while the ARM-DDS scheme can reduce the delay to one third of the existing strategies, so it has a strong practical significance and a wide range of adaptability. (b) Compared with the highest QoS strategy, the ARM-DDS scheme can greatly reduce energy consumption and prolong network lifetime. (c) Compared with the general stop-wait automatic retransmit protocol, the ARM-DDS scheme reduces

data transmission delay by 12.1% and improves network energy utilization by 28%.

The rest of this paper is organized as follows: Section 2 shows the methods. Section 3 reviews related works. Section 4 describes the network model and problem statements. In Section 5, the detailed design of the ARM-DDS scheme is presented. The results of the theoretical performance analysis are given in Section 6. The experiment results are given in Section 7. The conclusion is presented in Section 8. The list of abbreviations is in the "Abbreviations" section.

2 Methods

In WSN, transmission of data with different delay requirements and reliability requirements is common, so it is important to differentiate services. Therefore, a method of differentiated services to improve the utilization of network resources is proposed. Differentiated services can effectively improve energy-efficiency. Because traditional routing algorithms cannot perform differentiated services for different requirements of data, transmission methods that use the highest latency requirements and reliability requirements for all data waste energy. The use of differentiated services for data with different delay requirements can reduce the amount of data sent by low-latency sensitivity and reduce the energy consumption of the network. Some data have high-latency-sensitivity requirements and can utilize the residual energy of the non-hotspot region to employ a lower latency retransmission mechanism. Therefore, such a differentiation method can meet the delay requirements of all data and improve energy utilization without affecting the life of the network. On the other hand, differentiated services for data streams with different reliability requirements can extend network lifetime, because packet reliability is guaranteed by the retransmission mechanism. Therefore, the data packet of different reliability requirements is differentiated, and the number of retransmissions of the data packet required by the low reliability is reduced, that is, the number of data retransmissions of the network is reduced, and the network life is prolonged. Therefore, the differentiated service method can meet the data transmission requirements of higher delay sensitivity, improve energy utilization, and extend network lifetime. The background and related work on this method can be found in Section 3.

3 Background and related work

More and more wireless sensing devices are currently being applied in a variety of fields [30, 31], such as military, industrial, agricultural, and daily life [32, 33]. And the number of these sensing devices is huge. According to [1], the number of devices currently connected to the Internet has exceeded that of humans, thus greatly promoting the development of the Internet of Things (IoT) [1, 3, 5, 11, 32, 34]. These sensing devices have huge computing and storage resources, which have caused major changes in the current network structure [1, 11, 32, 34, 35], mainly reflected in edge computing. Fog computing has become the focus of current research [1, 11, 32, 34, 35]. These new network structures are combined with the emerging artificial intelligence technology [36, 37], which brings unprecedented opportunities for the Internet of Things (IoT) and greatly promotes network development [34, 35]. As wireless sensor networks (WSNs) is the most important component of IoT, more and more applications are deployed in the same WSNs, which leads the research of WSNs to a new stage [17, 19], from single research to hybrid and differentiated services.

The most important task of WSNs is to perceive and acquire information in the surrounding environment and pass the perceived data to the sink for corresponding processing [13, 17, 20]. Since different applications have different delay requirements, the range involved in the delay is very wide, involving from the MAC layer, the network layer, to the application layer [13, 17–21, 26]. And delay is closely related to other related performance indicators, such as a retransmission mechanism designed due to the unreliability of network transmission. The retransmission mechanism will increase the delay, and also increase the energy consumption of the node. The reliability of the communication link is related to the transmission power of the node. Increasing the transmission power of the node can improve the reliability of the communication link, thereby reducing the retransmission of the data packet, reducing the delay, and reducing the extra energy consumption caused by the retransmission. However, increasing the transmission power of a node itself increases the energy consumption of the node. These, in turn, interact with the duty cycle working mode adopted by the node, routing mechanisms such as multi-path routing, and coding methods of the application layer. Therefore, the optimization of wireless sensor networks is a very complicated challenging issue. This section will discuss the related work.

3.1 Retransmission protocols and mechanisms
The retransmission protocol is an important mechanism to ensure the reliability of wireless transmission and is one of the most important research contents in WSNs. As mentioned earlier, the communication quality of WSNs is often very different due to the communication distance between nodes, transmission power, and the surrounding environment. In some networks, the packet

loss rate is as high as 20–23%, so some reliability measures are needed to ensure the reliability of packet transmission. At the same time, the strategy of guaranteeing the reliability of data transmission often affects the delay of data transmission. Therefore, the strategy of ensuring the reliability of data transmission is considered simultaneously with the delay optimization strategy.

(1) Send-and-wait automatic repeat-request (SW-ARQ) protocols. The SW-ARQ protocol is a basic data transmission protocol that guarantees the reliability of wireless communication [38]. It mainly guarantees the reliability of data transmission through the retransmission mechanism. Its main operating mechanism is as follows: Sender starts to send packets after establishing a connection with the receiver, and the receiver returns an acknowledge (ACK) to the sender after receiving the packet. After receiving the ACK, sender knows that the data packet has been successfully received by the receiver, so it can start the transmission of the next packet. If the packet sent by the sender is lost during the transmission, the receiver will not return an ACK back because of not having received the packet [38]. Thus, after the sender waits for a timeout period and does not receive the ACK, it considers that the transmitted data packet has been lost, and thereby retransmits the data packet. The sending and receiving process is the same as that of the first sending of the data packet. From the above process, the reliability of the SW-ARQ protocol data transmission is guaranteed by retransmitting the data packet multiple times. Let the probability of successful data transmission be p_i^s, then n data retransmissions can probabilistically guarantee the probability of successful transmission is $1 - (1-p_i^s)^n$. It can be seen that multiple retransmission mechanisms can significantly improve the reliability of data transmission but will increase the delay of data transmission. In data communication, the time it takes for the sender to send a packet to receive an ACK is called round-trip time (RTT), and the sender takes longer to perform a retransmission. It needs to wait for a time to out (RTO) time before starting a retransmission, while the RTO duration is longer than the RTT, because delay is the accumulation of time taken by multiple retransmissions. Thus, the more data retransmissions, the greater the delay [38]. In order to control the delay within the allowable range, it is often necessary to set a maximum number of retransmissions m. That is, when the maximum number of retransmissions of the same packet reaches m, the packet is not retransmitted, but discarded. It is easy to get the maximum number of retransmissions: Since the reliability of data transmission for m times is required to be $1 - (1-p_i^s)^m > \delta$, the reliability of data transmission can be guaranteed to meet the requirement when m

$= \lceil \lg(1-\delta)/ \lg(1-p_i^s)\rceil$. Setting the maximum number of retransmissions prevents extreme situations during a packet transmission: always unsuccessful transmission. Of course, in general, the packet transmission success rate p_i^s is relatively high, which is 80% or more. Therefore, in practice, the value of the data transmission success rate p_i^s is relatively high. Therefore, although the maximum number of retransmissions m is set, the average number of successful data transmissions is not large in the case where δ is less than 1, much smaller than m [38].

The above retransmission mechanism is mainly aimed at the reliability transmission of hop to hop. According to source node routing and multi-hop routing of the data packet, the above retransmission mechanism also has control strategies divided into end to end (E2E) and hop by hop (HBH) [38]. The control mode of E2E is that the source node sends the data packet to the sink; then, the latter returns the ACK after receiving the data packet. In this control mode, once the source node receives the ACK, it knows that the data packet has been successfully received by the sink, and thus begins to send the next packet. However, in this mode, the data packet has to go through multiple hops to reach the sink. Assuming that the hop count of the data packet from the source node to the sink is h, the success rate of the packet to the sink through k hops' routing is p^h. Obviously, if the success rate of one hop of the ACK when it returns is β, the probability that the ACK successfully returns to the source node is β^h. Therefore, the probability that the source node successfully receives the ACK returned is $p^h \cdot \beta^h$. If $\rho = 90\%$, $\beta = 95\%$, and $h = 10$, the probability that the source node can receive the ACK after sending the data packet $< 21\%$. To make the reliability rate of the data successfully reaching the sink, $\delta = 90\%$, the source node needs to send 10 times on average to guarantee the rate. For each retransmission, the packet needs to go through 10 hops to reach the sink, and the ACK returned by the sink needs to go through 10 hops to return to the source node, so the elapsed time (delay) is very long, and the RTO time of a retransmission will be longer. This shows that in the E2E control mode, the number of retransmissions of the source node is large, and the delay is large too [38]. However, this control method has one advantage that has not been noticed in previous studies: In an ideal wireless sensor network with high reliability of data transmission (the transmission reliability rate can be considered as 100%), since sink is the end point of data transmission of all nodes, the nodes that are close to the sink area bear a large amount of data, which is called hotspots. Due to the hotspots, the energy consumption of nodes near the sink area is much higher than in other areas, which may lead to early network death. This situation seriously damages the lifetime of the network. In the reliability control mode of E2E, since part of the data packets sent by the source node will gradually lose during their transmission to the sink, which will reduce the impact of hotspots. Hop by hop (HBP) reliability control is a hop-by-hop reliability control method [38]. In this way, if the reliability of receiving the data packet by the sink is δ, and the data packet needs to go through h hops to reach the sink, and when the data transmission reliability of each hop is $\sqrt[h]{\delta}$, the reliability after the h-hop routing of the data packet is guaranteed to be δ. In the HBP mode, to require the reliability of each hop to be $\sqrt[h]{\delta}$, that is, to require this process, the sender sends the data packets to the receiver, then receiver returns the ACK to the sender after having received, but if the sender does not receive the ACK, timeout retransmission is implemented. If the sender sets its maximum number of retransmissions $m = \lceil \lg(1-\sqrt[h]{\delta})/ \lg(1-p_i^s)\rceil$, the end to end reliability of data transmission will be guaranteed to be δ. In this HBP control mode, the control mode adopted by each node is single-hop control, and thus its end to end delay is the delay accumulation of each hop. Relatively speaking, the nodes of the HBP mode bear less data.

It can be seen from the above SW-ARQ mechanism that the biggest feature of this retransmission mechanism is that the protocol is simple and easy to implement [38]. Moreover, in this way, the extra energy consumption of the node is due to the energy consumption by transmitting the ACK, and the length of the ACK is relatively smaller than the length of the data packet, so the additional energy consumption is relatively small, and thus the energy consumption of the protocol is relatively small compared to the mechanism introduced later. However, the biggest shortcoming of this method is that each retransmission generates a timeout period, and the data packets may be needed to be retransmitted multiple times, resulting in a larger delay for this protocol.

In order to reduce delay and the large deficit of the SW-ARQ protocol, the researchers proposed an improved protocol for SW-ARQ. Go-Back-N protocol (GBN) [39] is just one of the improved protocols. The main operation of the GBN protocol is the following: The sender is no longer a serial mode that it sends a packet at a time and waits for an ACK to return like the SW-ARQ protocol [39]. In the GBN protocol, the sender continuously sends multiple data packets, and the data packet contains the sequence number of itself. If the receiver finds that the sequence number of the received packet is not continuous, which indicates that there is a packet loss during the transmission, therefore, the receiver sends a NACK to the sender, indicating the smallest sequence number of the packet in the unreceived packets. After the sender receives the NACK, it retransmits from the sequence number of the packet indicated in the NACK. The GBN protocol speeds up the transmission of data packets, thereby reducing delay [39].

Selective repeat protocol (SR) [40] is an improvement to the GBN protocol. In this protocol, the sender sends the packet in a similar manner to the GBN protocol. However, when the receiver returns the NACK of the unreceived packets, the SR protocol only retransmits the packets that were not successfully transmitted, thus reducing the system overhead.

The delay of the retransmission protocol mainly comes from the retransmission of the data packet, and the retransmission of the data packet will inevitably go through a TTO time, so the time taken is long. Multiple retransmissions and multi-hop routing will make the end to end delay very large. Therefore, if using the mechanism that does not return ACK, the delay can be reduced. Therefore, some researchers have proposed the mechanism of s packet rerouting. In this mechanism, the sender repeatedly sends the same data packet m times, and the s data packets all have a certain probability of reaching the sink. Therefore, after repeated transmissions for m times, the value of data transmission reliability can be ensured to meet the application requirements δ. Compared with the SW-ARQ protocol, since the receiver does not need to return ACK in this protocol, the delay is the least, but the node has the highest energy consumption. As previously analyzed, in the SW-ARQ protocol, the average expected number of times that a sender sends a packet is much smaller than the maximum number of retransmissions m. However, in the mechanism without feedback, the number of times the data packet is sent by each node is the maximum number of retransmissions m. Therefore, the energy consumption of this protocol is very large and rarely used in wireless sensor networks.

3.2 Routing and delay optimization work

(1) Multipath routing mechanism. The proposed multipath routing mechanism is not proposed for communication links and unreliability but it is similar to the above non-feedback communication protocol. The multi-path routing mechanism is that the sender routes the data packets from multiple routes to the sink at the same time. Therefore, as long as the data packets of one route can be successfully routed, the data packet transmission can be successful [41]. Multi-path routing mechanism is often proposed to defend against security attacks, so that if a route is attacked, there are still other routes that can successfully reach the sink [42]. Corresponding to this is the Security and Energy-efficient Disjoint

Route (SEDR) scheme. SEDR scheme using (T, M)-threshold secret-sharing mechanism divides the data into M shares, which route to the sink along different routes. If sink receives the T shares of the M shares, the sink can get the information of the entire packet [43]. And the attacker cannot get the packet information if the number of shares it gets is less than T. (1) Although this method is designed for security, it plays a role in improving the reliability of data transmission.

Another strategy to improve the reliability of data transmission is the packet reproduction method [44]. In such a routing strategy, since some data packets are gradually lost in the routing process, in the packet reproduction routing method, the packet is reproduced once every certain number of hops (such as i hops) [44], that is, to copy the source packet to M shares, each of which is transmitted forward along a different routing path. Some of the M packets in the process of routing will disappear. In order to ensure that there are packets that can reach the sink in the end, when each data packet is routed i hops, reproducing M shares for rerouting from different routes to make up for those lost packets in the routing process. Thus, the data packet can be guaranteed to reach the sink with high reliability. This packet reproduction method has a lower cost than the multi-path routing, and the reliability of the data routing is high [44].

(2) Opportunistic routing. Opportunistic routing is also an effective routing method for reducing delay. The essence of this method is to make full use of the broadcast characteristics of wireless communication, so that the delay of routing can be made smaller on the basis of ensuring the reliability of data transmission. Specifically, the method used by opportunistic routing is the following: Due to the unreliability of data transmission, the sender will select multiple receivers for transmission in opportunistic routing in order to improve the reliability of data routing. Since multiple receivers are selected for one data transmission, assuming the transmission reliability is p, when n receivers are selected in the opportunistic routing, the probability that the sender successfully transmits the data after the calculation is $1 - (1 - p)^n$. It can be seen that a data transmission once of opportunistic routing is equivalent to c times of transmissions in the SW-ARQ protocol, so that the delay can be effectively reduced. But in this method, you need to select n receivers at the same time. Such wireless sensor nodes can easily select n receivers in a non-duty cycle network, however, which is difficult in the duty cycle

based WSNs, due to the periodic awake/sleep of the nodes.

(3) Broadcast routing method. To overcome the packet loss of data transmission and reduce delay, Joo et al. [45] proposed a broadcast data transmission method in data fusion network. In some data fusions, n multiple data packets can be merged into one data packet. For example, in a wireless sensor network, the average temperature and average humidity in the surrounding environment are monitored. When multiple monitoring data meet, the average value can be calculated. The node only needs to transmit the average value when transmitting the data value. In the network of packets loss, the data packet has a certain loss ratio in the process of routing to the sink. Therefore, the method adopted by Joo et al. [46] is that each node that receives the data routes to the sink in a broadcast manner. By adopting such a method, the reliability of the data routing can be significantly improved. Because, after a node broadcasts, the data packet received by nodes that are closer to the sink than the broadcast node, is merged with its original data, which continues to be broadcasted forward. Therefore, if there are m nodes in the broadcast range of the original node, it can broadcast forward again, the probability of successful data transmission is $1 - (1-p_i^s)^m$. Each node that successfully receives the data packet will broadcast forward in the same way, so the probability of success of each route is $1 - (1-p_i^s)^m$. This greatly increases the probability of data transmission to the sink. Also, in this way, the delay is relatively small. Because only the packets that have been successfully transmitted survive and are continuously routed forward, this is equivalent to a multi-path routing strategy, and the time used by the packet first arriving in the sink is the transmission delay of this packet.

3.3 Transmit power and delay optimization work

In a network with high reliability of the communication link, the data packet does not need to be retransmitted, so its delay is small. While in a network with low reliability of the communication link, the data needs to be retransmitted multiple times, and the time consumed by one retransmission far exceeds the time of the first data transmission. Therefore, the method of improving the reliability of the communication link can be used to reduce delay and improve data transmission reliability. Thus, there are also some methods to reduce delay and improve data transmission reliability by improving the quality of communication links [46]. One of the more

effective methods is to improve the data transmission power. The central idea of this approach is that the communication success rate between nodes in a wireless router network is not only related to the distance between the nodes but also related to the transmit power of the sender nodes. The higher the transmit power of the sender nodes, the higher the signal-to-noise ratio (SNR) of the received signal of the receiver and the higher the success rate of receiving data [46]. It can be seen from the foregoing discussion that if the data transmission success rate is higher, the number of retransmissions required by the node is smaller, so that the delay caused by the retransmission can be effectively reduced. For example, in a network with a data transmission success rate of 98%, the probability of successfully reaching the sink after 10 hops is 81.7%. While in a network with a data transmission success rate of 80%, the success rate of the packet reaching the sink after a 10-hop routing is only 10.7%. It can be seen that increasing the success rate of data transmission can significantly reduce the number of retransmissions thus reducing delay [46]. Based on the above ideas, some researchers have proposed the strategy of increasing the transmission power of the sender to improve the reliability of data transmission, thus to reduce the delay. But the real strategy of improving the node's transmit power is more complicated. Because increasing the transmit power of a node requires more energy consumption, but the energy of the node is very limited. Thus, to increase the transmit power of the node, it should minimize the energy consumption of data per bit successfully transmitted [46]. The sender's transmit power is a nonlinear relationship with the successful transmission of the packet (see Fig. 1). If the communication distance between the nodes is fixed, increasing the transmit power of the sender can increase the receiver's reception rate. However, after the receiver's receiving rate reaches a certain value, it rises very little if the sender's power continues to be increased. Therefore, when the transmit power of the sender can be adjusted to the optimized value, the energy consumption for successfully transmitting each bit of data can be minimized. In terms of reducing the delay, increasing the sending power of the sender helps to reduce the number of retransmissions, thereby reducing delay [46].

3.4 Differentiated service-related work

Due to the development of current microprocessors, multiple applications can run in the same WSNs at the same time. Because different applications require different QoS indicators, research on differentiated services has only just begun, but there have been some researches about this.

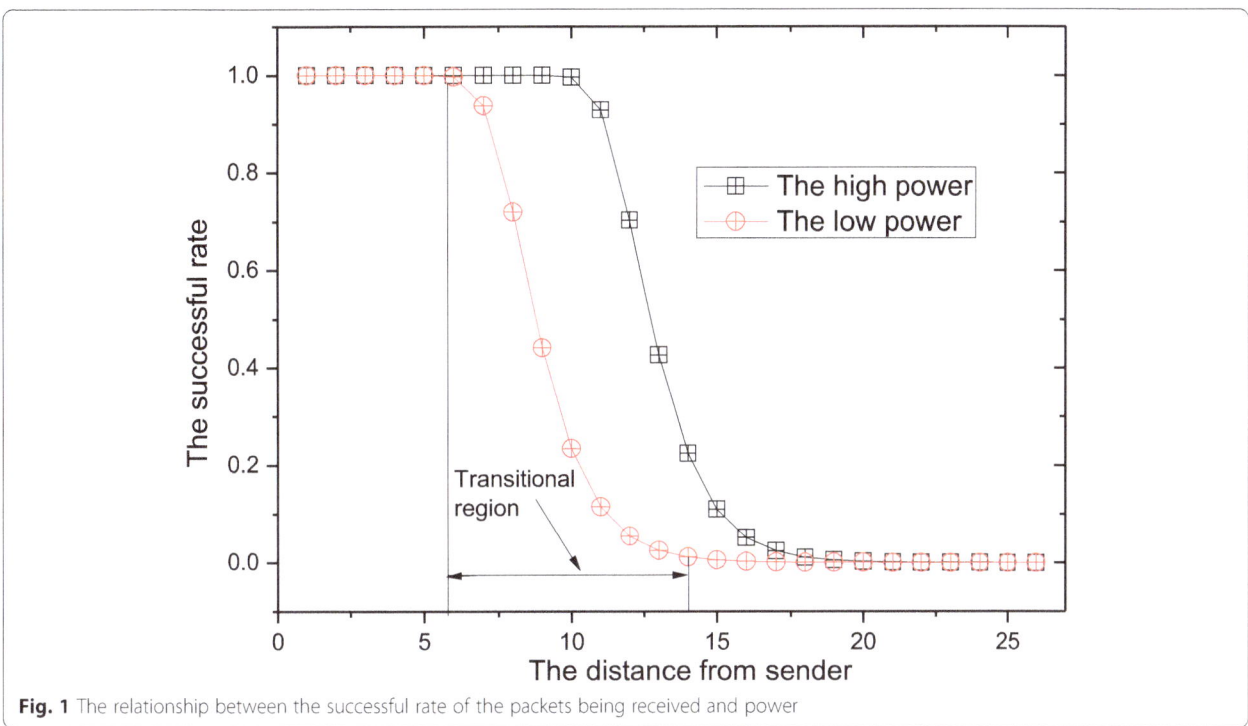

Fig. 1 The relationship between the successful rate of the packets being received and power

Zhang et al. [16] proposed an integrity and delay differentiated routing (IDDR) scheme, which is such a differentiating service strategy. In IDDR scheme, there are two different types of network applications. One is data integrity, which is an application of packet loss sensitivity, requiring an extremely low packet loss ratio. The other type is an application of delay sensitivity, which has high requirements for delay. In WSNs, the shortest routing method is generally adopted, which causes hotspots near the sink area. The hotspot area bears a large amount of data, which on the one hand makes the energy consumption of nodes in this area large. On the other hand, congestion occurs in the hotspots area, due to the large amount of data transmitted by the nodes in this area. The result of congestion is high packet loss rate and large data routing delay. Therefore, the solution adopted by the IDDR scheme is using the shortest routing method for delay-sensitive packets to reduce their delay. For data packets with high data integrity requirements, a detour is used. By bypassing the hotspots area, although the length of the route is increased, the congestion is small, thus meeting the requirements of data integrity. Therefore, IDDR can differentiate services.

Huang et al. [47] proposed a very effective delay in differentiated services (DDS) for data fusion networks. In a data fusion network, if a packet waits for a period of time for more packets on the intermediate node of the route, the amount of data reduced will be larger after more data packets are merged, thereby reducing the energy consumption of the route and improving the

network life. Therefore, from the perspective of reducing energy consumption, the longer the data packet stays on the node, the more data fusion can be performed, which can reduce energy consumption and thus help to improve network life. However, the longer the packet stays on the node, the greater the delay of the packet reaching the sink, which affects its usability. Thus, the method adopted by Huang et al. [47] is the following: For data with high delay requirements, set a smaller dwell time threshold to speed up data routing and reduce delay, and for data with less demanding delay, set a larger dwell time threshold to make more data packets meet for data fusion thus reducing energy consumption and improving network life.

4 The system model and problem statement

4.1 The network model

In this paper, a system model of a classical planar wireless sensor network is used. The system model is referenced to [48]. The system model structure is as follows:

(1) The research environment of this paper is a planar wireless sensor network, which is a circular network model with R as the radius of the network and r as the radius of each hop. There are many sensor nodes and one sink node in the planar network. Among them, the sensor nodes have the same initial energy. The sensor node cannot be replaced in the middle of use. For the sink node, there is no energy limitation [17, 18, 26]. The sensor nodes are

evenly distributed according to the density ρ in the network centered on the sink node. The power of the node can be adjusted according to the distance of the transmitted data [49]. In a transmission cycle, each sensor node collects data and generates a packet. The packet will be transmitted to the sink via the set multi-hop route.

(2) In the wireless sensor network studied in this paper, the sensor node position is fixed. And when the sensor node transmits data, its transmission path does not change. Data routing uses the shortest path by the policy. Since the probability distribution of network events, in reality, is relatively sparse, this paper does not consider the congestion phenomenon on the nodes, that is, the queue delay.

(3) When the node performs data transmission, the success rate of node i to transmit data to the next hop is p_i^s, and the success rate of the ACK returned after the next hop receives the data is q_i^a. If the transmit failed, it will be retransmitted.

4.2 System parameters

The local energy model used in this paper is referred to [50]. Let ϖ_s represent the energy consumption of data transmission, as shown in Eq. (1).The energy consumption ϖ_{rec} of the received data is given by Eq. (2).

$$\begin{cases} \varpi_s = \alpha \ell_s s^2 + \alpha E_{\mathrm{elec}} & \text{if } s < s_{td} \\ \varpi_s = \alpha \ell_m s^4 + \alpha E_{\mathrm{elec}} & \text{if } s > s_{td} \end{cases} \tag{1}$$

$$\varpi_{\mathrm{rec}} = \alpha E_{\mathrm{elec}} \tag{2}$$

Among them, the transmission circuit loss is represented by E_{elec}. The energy consumption of data transmission uses the free space (s^2 power loss) and multipath fading (s^4 power loss) channel model according to whether the distance exceeds the threshold s_{td}. When the power amplifier loss is based on the free space model, ℓ_s is used instead; when the transmission distance is greater than or equal to the threshold s_{td}, ℓ_m is selected as the power amplifier loss of the multi-antenna model. α is the number of bits in the data packet. The above parameter settings are shown in Table 1 [50]. The system parameters adopted in this paper are listed in Table 2.

Table 1 Network parameters

Parameter	Value
Threshold distance (r_0) (m)	87
Sensing range (r) (m)	≤ 80
Eelec (nJ/bit)	50
ε_{fs} (pJ/bit/m^2)	10
ε_{amp} (pJ/bit/m^4)	0.0013
Initial energy (E_{init}) (J)	0.5

4.3 Problem statement

The main goal of this paper is to design a communication approach to reduce the transmission delay of WSNs by improving the network energy utilization, and to ensure that the increase of Adaptive Retransmit Mechanism for Delay Differentiated Services (ARM-DDS) protocol.

The problem of the ARM-DDS protocol can be described as differentiated services for different transmission delay requirements, to meet the broad delay requirements. Data packets with different delay requirements and reliability requirements are processed differently to optimize the delay and energy utilization of the entire network. WSN data collection routing can be characterized by several performance indicators, as described below:

(1) End-to-end delay (denoted as $\mathcal{Y}_{\mathrm{ETE}}$) refers to the time from when the source node sends a packet to the sink. After the source node sends the data packet, it will reach the sink through several relay nodes. Each time a packet is forwarded via a relay node, it takes time. Therefore, the end-to-end delay is the sum of the transmission delays per hop. The end-to-end delay minimization can be expressed as follows:

$$\min(\mathcal{Y}_{\mathrm{ETE}}) = \min\left(\sum y_i\right) \tag{3}$$

(2) The effective energy utilization rate (expressed as η) refers to the ratio of the energy used in the network to the total energy of the network. Considering that o_i is the energy consumption of node i in the network, $\mathcal{O}_i$ is the initial energy of the node, so the maximum effective energy utilization is as follows:

$$\max(\eta) = \max\left(\frac{\sum o_i}{\sum \mathcal{O}_i}\right) \tag{4}$$

(3) Definition of network lifetime (expressed as $\mathcal{L}$) [1, 12, 14]. Network lifetime is defined as the death time of the first node in the network. $\mathcal{O}_i$ is the energy consumption of node i, and $\mathcal{O}_i$ is the initial energy of each node. Therefore, the maximum network lifetime is as follows:

Table 2 System parameters

Parameter	Value	Description
p_i^s	60%-95%	Probability of data being successfully sent to the subsequent node
q_i^a	80%-95%	Probability of ACK successful transmission
δ	90%-100%	End-to-end reliability of data packets
τ_{round}	50ms	round-trip time
τ_{rtto}	60ms	round-trip time to out
s	10ms	The minimum interval time of next data packets being sent
b	10ms	The minimum interval time of next ACK being sent
μ	10ms	Timeout interval of retransmission

$$\max(\mathcal{L}) = \max(\mathcal{O}_i/o_i) \tag{5}$$

In summary, the research questions in this paper can be summarized as follows. And the parameters related to the calculation are shown in Table 3.

$$\begin{cases} \min(\mathcal{Y}_{\text{ETE}}) = \min\left(\sum y_i\right) \\ \max(\eta) = \max\left(\dfrac{\sum o_i}{\sum \mathcal{O}_i}\right) \\ \max(\mathcal{L}) = \max\left(\dfrac{O_i}{o_i}\right) \\ s.t.\delta_i^{e2e} = \prod_{k \in \text{path to sink of node } i} \lambda^k \geq \delta \end{cases} \tag{6}$$

5 ARM-DDS strategy

5.1 Research motivation

The Adaptive Retransmit Mechanism for Delay Differentiated Services (ARM-DDS) scheme is mainly proposed for the research problem in which no effective methods work on the distinction of delay differentiated service. The main

goal of ARM-DDS is to reduce the energy consumption of nodes as much as possible. Under the premise of ensuring the reliability of data transmission to meet the application requirements, it can provide a wide range of differentiated services for the delay. The research motivation of this paper mainly comes from the following points.

(1) This paper provides a wide range of adaptive retransmission mechanism for delay differentiated services. The previous SW-ARQ protocol, GBN protocol, and SR protocol are all best-effort protocols, that is, the protocol takes into the optimized operation of both energy consumption and delay. Therefore, this kind of protocol uses the same service mechanism for the applications, so it cannot work on the situation of delay differentiated services in the development of wireless sensor networks. We have found that in the original retransmission mechanism, the same rules of sending/confirming data are working on the same packets. For example, in SW-ARQ, after sending a data packet, it waits for the ACK confirmation message (see Fig. 1). However, in the GBN protocol, the retransmission

Table 3 Parameters related to the calculation

Notation	Description
c	Persistent transmission times for the first time
n	Number of total nodes in network
$\gamma_h(\delta)$	The maximum number of retransmissions of hop h
e_p^t, e_p^r	Energy consumption of node for receiving and sending a data packet respectively
e_A^t, e_A^r	Energy consumption of node for receiving and sending an ACK respectively
ω_i	Data E2E delay of node i in ARM-DDS protocol
$D_{\xi,i}^{k,t}, X_{\xi,i}^{k,t}$	Number of node i for sending and receiving data packets in SW-ARQ protocol respectively
W_i^t, W_i^r	Network delay
S_h^{send}, S_h^{rec}	Number of hop h for sending and receiving data packets in ARM-DDS protocol respectively
S_h^{AS}, S_h^{AR}	Number of node i for sending and receiving data packets in ARM-DDS protocol respectively
k	Data stream number, indicating different data streams
Δ	A constant less than the radius

orders depend on the NACK whether to confirm the message receiving after sending data packet. In the SR protocol, the loss packet can be directly re-transmitted after receiving the NACK. To further reduce the delay, we use the methods in this paper: for the perfect application system of delay, when the first packet is sent, it is repeatedly sent c times (see Fig. 2). When $c = 1$, it is the same as the re-transmission mechanism currently used. When $c > 1$, the delay can be reduced. The reason is this, when $c > 1$, since the continuously sent c times, the probability that the port receives the data packet is greatly improved, that is, the original p_i^s is reduced to $1-(1-p_i^s)^c$. Therefore, the subsequent retransmission can be effectively reduced. The probability of being able to effectively reduce the delay.

The delay distinguishing service can be obtained by adjusting the parameter c. When $c = 1$, it is the retransmission mechanism that has been proposed. When c selects different values, the delay of data transmission will be different. Generally, when the c value is small, increasing the value of c will reduce the delay. However, taking the largest possible c value does not achieve the goal of reducing the delay. When the packet is sent with a minimum interval of s, it takes at least $(c-1)\,s$ to send c packets. If the value of c is too large, the delay will also increase. Therefore, the delay of the repeated transmission c is large, and the delay is limited. On the one hand, the data packet is sent c times at a time, and its energy consumption increases linearly with the increase of c. Therefore, the tradeoff between delay and energy consumption is needed to

obtain an optimized c value. As can be seen from the above analysis, The retransmission mechanism proposed in this paper can obtain different delay levels by selecting the difference of the parameter c. Overcoming the previous strategy, it only provides a delay level, which makes the ARM-DDS proposed in this paper have a wide delay service.

(2) Based on the first point, this paper proposes a service-differentiating architecture. This architecture is: For applications with high delay require-ments, a large retransmission mechanism under c value is used to make the data transmission delay small. For applications with low delay requirements, a retransmission mechanism with a smaller c value is used. Since the energy consumption of the node is proportional to the value of c, the delay is in-versely proportional to the value of c. Thus, to re-duce the delay system requires more energy consumption. However, the ARM-DDS strategy provides better overall performance than the previ-ous strategy because of the differentiated services provided. The main reasons are as follows: (a) The SW-ARQ protocol lacks differentiated services. In order to meet QoS requirements of all the data when using the SW-ARQ protocol, only the strat-egy that meets the highest QoS requirements in the network can be adopted. Therefore, comprehensive energy consumption is much higher than the strat-egy proposed in this paper. (b) There is no differen-tiation strategy proposed in this paper. The previous strategy is a kind of pseudo-differentiated service, rather than truly differentiated services, be-cause, in those strategies, the strategy can only reach one service level. The retransmission

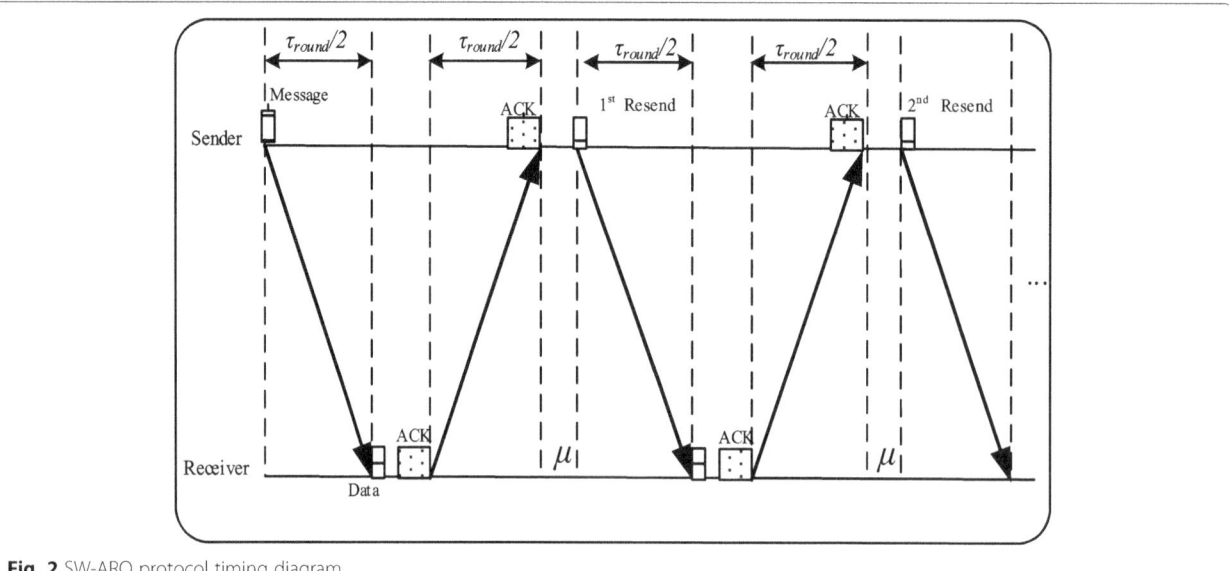

Fig. 2 SW-ARQ protocol timing diagram

mechanism of this paper is a real differentiated service, capable of offering a variety of levels of service on a retransmission mechanism.

(3) The ARM-DDS scheme proposed in this paper has a different strategy than the previous one: We analyzed that the near-sink area in the wireless sensor network bears a large amount of data, so its energy consumption is large, which is hotspots. The energy of the hotspots area node has a residual, so in the ARM-DDS scheme, we fully increase the residual energy of these far-sink areas to appropriately increase the value of c in the retransmission protocol and use a smaller c value in the near-sink area. By making full use of the remaining energy of the network and improving energy utilization, it can increase the network lifetime under the same conditions as the previous strategy or further reduce the delay when the network life is the same.

Based on the above three points, the ARM-DDS protocol uses differentiated services for different requirements data to reduce transmission delay. When transmitting data streams with different reliability requirements, set different maximum retransmission times for different required data streams and the maximum number of retransmissions near the sink, improving network life, meeting higher reliability requirements, and improving energy utilization at far-sink nodes.

5.2 General design of ARM-DDS

In this section, the algorithm of the ARM-DDS scheme differentiated service is given by analyzing the data transmission mode.

Figures 2 and 3 represent the difference and improvement between the ARM-DDS strategy and the traditional SW-ARQ protocol. The data transmission of the SW-ARQ protocol is shown in Fig. 2. The sender sends a packet every time it transmits. Sending a packet to the receiver requires $\frac{1}{2}\tau_{round}$ time. The success rate is p_i^s. Receiver immediately sends an ACK to the sender when a packet is received. The ACK also needs $\frac{1}{2}\tau_{round}$ time to the sender. The rate is q_i^a. When sender receives the ACK, it can be determined that the data transmission is successful. After sender sends the data packet, if the ACK is not received after waiting for $\tau_{round} + \mu$ time, it is considered that the sending of the data packet failed, and then sender needs to send the data packet again. Therefore, under the SW-ARQ protocol, additional $\tau_{round} + \mu$ time is required for each data retransmission. The sender will resend the data. Therefore, the SW-ARQ protocol requires $\tau_{round} + \mu$ time for each data to be retransmitted.

The ARM-DDS scheme proposed in this paper can continuously send multiple sets of data. When using ARM-DDS scheme transmission, first determine the c value according to the requirements of the application. As shown in Fig. 3, the time when the data packet is sent to the receiver is $\frac{1}{2}\tau_{round}$, which is the same as the traditional protocol. When sending c packets in succession, the interval of sending each packet is s. Therefore, when the packets of group c arrive at receiver, the time is $\frac{1}{2}\tau_{round} + (s-1)c$. The algorithm of receiver and the SW-ARQ protocol is consistent, and when a packet is received, an ACK is returned. The minimum interval for sending ACK is b. As long as one ACK is received by

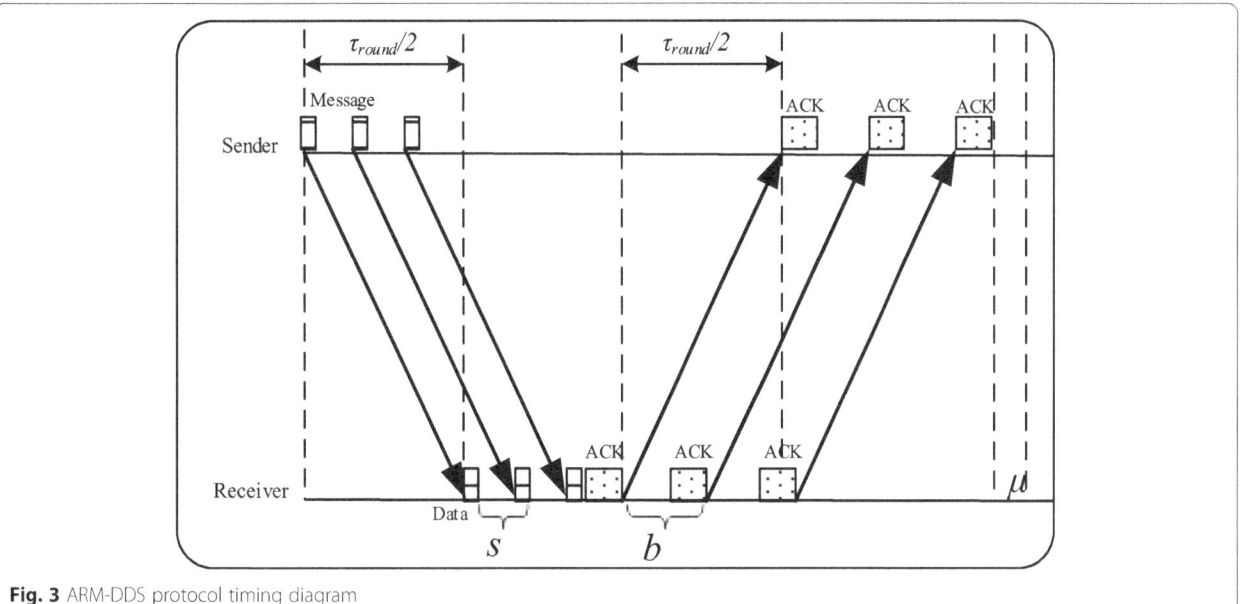

Fig. 3 ARM-DDS protocol timing diagram

sender, it is regarded as successful. The transmission of the c-group data packets using the transmission strategy of the ARM-DDS scheme requires $\tau_{\text{round}} + (s + b - 2)c$ delay. If the traditional protocol is used, this delay becomes $(\tau_{\text{round}} + \mu)c$. In the network, the data transmission time τ_{round} is much larger than the minimum interval between data and ACK transmission $(s + b - 2)$. Therefore, according to the above analysis, it can be obtained. The following conclusions are drawn: (1) When the number of retransmissions is small, the time required for the ARM-DDS protocol to transmit data with the SW-ARQ protocol is similar.

(2) When multiple retransmissions are required, the transmission delay of ARM-DDS will increase $(s + b - 2)c$. The transmission delay of the SW-ARQ protocol is increased by $(\tau_{\text{round}} + \mu)c$. At this point, ARM-DDS has an advantage in latency. That is to say, the ARM-DDS scheme has a greater advantage in delay when it is required to retransmit data multiple times.

The algorithm of the ARM-DDS scheme is as follows: In a planar wireless sensor network, all nodes are awake and all nodes know their next hop node. When transmitting data, depending on the timeliness of the data, the data has different delay requirements. Time-sensitive data needs to arrive as quickly as possible, so the node must transmit the data as much as possible to ensure that its latency is low. Data with relatively low timeliness does not require more energy. Distinguish data with latency requirements so that each data can be transmitted within its required delay.

Algorithm 1: ARM-DDS sender algorithm
1: Find node i, which is at the h hop.
2: The delay requirement of the data w^k reliability requirement δ^k and the ratio ϱ_k.
3: According to Theorem 1, the maximum number of retransmissions $\gamma_h^k(\delta^k)$ is obtained.
4: Use algorithm 2 to select the correct c_k value.
5: Send c_k group data and listen for ACK.
6: **If** No ACK was received within $\tau_{round} + \mu$
7:     **Then** k=c_k
8:         **While**(k+c_k<$\gamma_h^k(\delta^k)$)
9:             send packet c_k
10:            **If** node i received an ACK
11:                **Then**   **End while**
12:            **End if**
13:            **Else**
14:                k=k+c_k
15:        **End while**
15:        **If** k<$\gamma_h^k(\delta^k)$
16:            send packet $(\gamma_h^k(\delta^k) - k)$
17   **End if**
18: **End if;**

The algorithm of the receiving node is as follows: In the ARM-DDS protocol, the receiving node waits to receive a data packet, and if a group of data is received, it sends an ACK, otherwise, it continues to listen to the data packet.

The preceding discusses the transmission and reception methods of the ARM-DDS protocol. Now, the optimal value of the number of packets c transmitted each time under the differentiated service is discussed. Therefore, the calculation of the optimal value has the following algorithm:

Algorithm 2: ARM-DDS scheme c value algorithm
1:It is known that the transmission single-hop delay requirement is w^k, and the current hop count of the node is h.
2:Using Theorem 1, the maximum retransmission number $\gamma_h^k(\delta^k)$ is obtained.
3: **For** $c = 1$ to S_h^{send} **Do**
4:    **If** $h \mathrel{!}=1$ **then**
5:        $d_j \leftarrow 0$ // ARM-DDS scheme single hop delay
6:            **For** $m = 1$ to c **Do**
7:                $d_j \leftarrow$ Delay of successful transmission
8:            **End for**
9:            **For** $m = c + 1$ to $\gamma_h^k(\delta^k)$ **Do**
10:               $d_j \leftarrow$ Expected delay
11:           **End for**
12:       **If** $d_j < w^k$ **then**
13:           $c \leftarrow$ m
14:       **End if**
15:   **End if**
16:   **Else** c $\leftarrow$ 1
17: **End for**

Algorithm 2 assigns the appropriate c value to applications according to different delay requirements. The ARM-DDS protocol does not send multiple sets of data at one time to hotspots in the near-sink area. Because hotspots increase the number of sending groups, the energy consumption increases, the node death rate increases, and the network life decreases, which is undesirable for the research. Therefore, for a node with a hop count of 1, it is consistent with the SW-ARQ protocol, that is, $c = 1$. In this way, it is possible to differentiate the data stream without reducing the network life and reduce the purpose of delay.

Figure 4 is an E2E delay comparison of the two protocols. The experimental environment is a planar network with a radius of 400 m centered on the sink node. Its emission radius is 80 m, reliability $\delta=0.8, p_i^s = 0.8, q_i^a = 0$ .8. As can be seen from Fig. 3, the ARM-DDS protocol delay is less than the SW-ARQ delay. And the further away from the sink, the greater the delay is reduced. When applications require different delays, the ARM-DDS policy can change the value of c to cause the data to be sent with the expected delay.

Regarding the loss of energy, the energy consumption of the ARM-DDS protocol proposed in this paper will be even larger. Because the ARM-DDS protocol increases the number of data transmissions compared to the SW-ARQ protocol, energy consumption is greater. Studies have shown that in a traditional flat network, when the lifetime of a network is exhausted, the energy

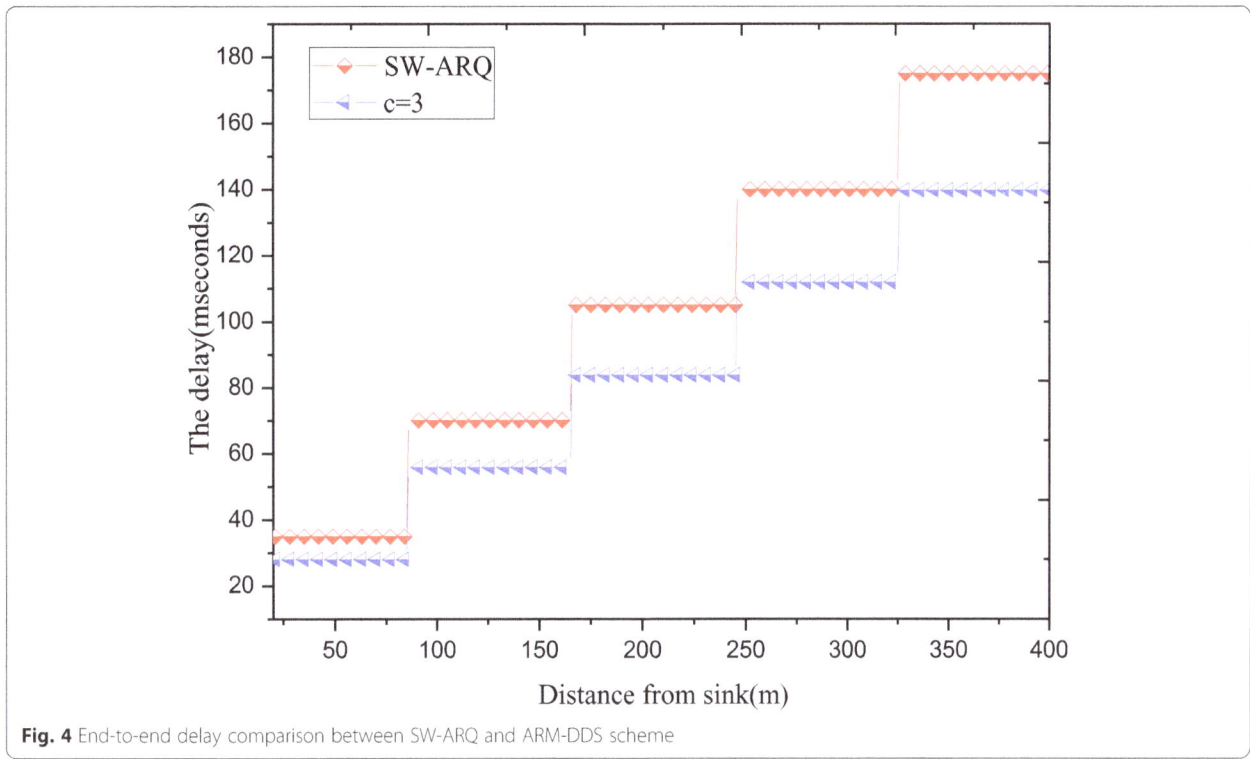

Fig. 4 End-to-end delay comparison between SW-ARQ and ARM-DDS scheme

remaining in the entire network is still as high as 90% [51]. Therefore, the ARM-DDS strategy can make full use of the remaining energy and increase the number of transmission groups at the far-sink without losing the network lifetime. Use this method to increase the energy utilization of the entire network [52, 53]. Therefore, adopting the above method does not affect the lifetime and can reduce the delay. Under the optimization of the differentiated service, the energy utilization of the entire network can also be increased.

5.3 Parameter optimization of ARM-DDS protocol

In the previous section, we mainly studied the transmission algorithm and c value algorithm of ARM-DDS scheme. This section mainly discusses two other important parameters of the differentiated service: the data volume ξ_i and the maximum number of retransmissions $\gamma_h^{ARM}(\delta)$. Among them, $\gamma_h^{ARM}(\delta)$ is the main parameter affecting network delay, and the data quantity ξ_i is the main parameter affecting energy consumption.

When studying the difference in data volume between the ARM-DDS scheme and the traditional protocol, first analyze the data situation of the nodes in the network under the assumption that the data between the nodes is not lost, to provide a calculation for the case where the data loss rate between nodes is > 0, the basics.

According to the reference [16], when the node transmits data with the transmission radius r, if the data

transmission between the nodes is not lost, the distance of the node i from the sink is $d, d = hr + \Delta$ (where Δ is a less than r Constant), then the amount of data borne by the node ξ_i is shown as equation (7).

$$\xi_i = \frac{x(x+1)}{2d}r + x + 1 \tag{7}$$

$$x = \left\lfloor \frac{R-\Delta}{r} \right\rfloor \tag{8}$$

In a planar wireless sensor network, all data must have certain reliability requirements when it reaches the sink [54]. The statistical reliability required for transmission to the sink is represented by δ. The maximum number of retransmissions of the ARM-DDS scheme $\gamma_h^{ARM}(\delta)$ is different from the maximum number of retransmissions in the legacy protocol. In the SW-ARQ protocol, when applications with different reliability δ requirements are simultaneously transmitted since the service cannot be distinguished, all data can be transmitted only according to the maximum δ. This will cause the data that has already met δ to be retransmitted, the excess energy is wasted, and the delay is added. ARM-DDS scheme was proposed to optimize the above situation. The protocol can reasonably distinguish between applications with different δ requirements and set the appropriate number of retransmissions for each application.

For any node i in the planar network, the success rate of the node per-hop transmission is p_i^s. According to the reference, in the planar wireless sensor network, the SW-hop-by-hop ARQ protocol is used, and the node with the hop count h is used. When the maximum number of retransmissions is as follows, the reliability is guaranteed to be δ:

$$\gamma_h(\delta) = \left\lceil \frac{\log\left(1-\delta^{\frac{1}{h+1}}\right)}{\log\left(1-p_i^s\right)} \right\rceil \tag{9}$$

The biggest difference between the ARM-DDS protocol and the SW-ARQ protocol is the maximum number of retransmissions. Under the differentiated service, the maximum number of retransmissions of the ARM-DDS protocol must be less than or equal to the SW-ARQ protocol. Theorem 1 gives the maximum number of retransmissions $\gamma_h^{\mathrm{ARM}}(\delta)$ of the ARM-DDS protocol.

Theorem 1: In the ARM-DDS protocol, it is assumed that there are m kinds of data packets with different reliability requirements δ^k in the data stream to be transmitted, each ratio $\varrho_k\%$ (where k is the number of different data streams, $0 \le k \le m$), the maximum number of retransmissions under different data reliability requirements is $\gamma_h^k(\delta^k)$, so the maximum number of retransmissions under the

$$\gamma_h^{\mathrm{ARM}}(\delta) = \left\lceil \sum_{k=1}^m \gamma_h^k(\delta^k)\varrho_k \right\rceil \tag{10}$$

Proof In the SW-hop-by-hop ARQ transmission mode, the maximum number of retransmissions of a hop is $\gamma_h(\delta)$. Based on this, the ARM-DDS protocol differentiates the data streams with different reliability requirements. When the reliability requirement is δ^k, $\gamma_h^k(\delta^k)$ times need to be retransmitted, and the data stream is the proportion of the overall data stream which is $\varrho_k\%$, so we can conclude that the maximum retransmission number of the data stream with the reliability requirement of δ^k is $\gamma_h^k(\delta^k)\varrho_k$. Similarly, all the different reliability requirements data will be obtained. The flow is weighted and calculated:

$$\gamma_h^{\mathrm{ARM}}(\delta) = \left\lceil \sum_{k=1}^m \gamma_h^k(\delta^k)\varrho_k \right\rceil$$

In the previous, the retransmission times and data amount parameters of the ARM-DDS protocol are proposed. To get the influence of these parameters on the whole network, it is necessary to calculate the amount of data and the amount of ACK when transmitting data and analyze the difference between the ARM-DDS scheme and the SW-ARQ protocol data transmission.

In the SW-ARQ protocol, the number of data packets and the amount of information that each node bears are as follows:

Let the distance of the node i from the sink be d; then, we know that $d = r \times h + \Delta$ if the transmission and reception (SW-ARQ) mode are used. Considering that the reliability required for transmission is δ, the data commitment amount $D_{\xi,i}^{h,t}$ of the node i, and the ACK amount $X_{\xi,i}^{h,t}$ are

$$D_{\xi,i}^{h,t} = Z_h^h(\delta) + Z_{h+1}^h(\delta)\left(1+\frac{r}{d}\right) + Z_{h+2}^h(\delta)$$
$$\times \left(1+\frac{2r}{d}\right)\ldots + Z_x^h(\delta)\left(1+\frac{(x-h)r}{d}\right) \tag{11}$$

$$X_{\xi,i}^{h,t} = D_{\xi,i}^{h-1,t}p_i^s = W_h^h(\delta) + W_{h+1}^h(\delta)\left(1+\frac{r}{d}\right)$$
$$+ W_{h+2}^h(\delta)\left(1+\frac{2r}{d}\right)$$
$$+ \ldots W_x^h(\delta)\left(1+\frac{(x-h)r}{d}\right) \tag{12}$$

$$D_{\xi,i}^{h,r} = D_{\xi,i}^{h-1,t} - Z_h^h(\delta) \tag{13}$$

$$X_{\xi,i}^{h,r} = X_{\xi,i}^{h,r} = D_{\xi,i}^{h,t}p_i^s q_i^a \tag{14}$$

$$Z_{h+j}^h(\delta) = \frac{1-\left(1-p_i^s q_i^a\right)^{\gamma_{h+j}(\delta)}}{p_i^s q_i^a} \tag{15}$$

$$W_h^h(\delta) = 0 \tag{16}$$

$$W_{h+j}^h(\delta) = Z_{h+j}^{h+1}(\delta)p_i^s \tag{17}$$

Theorem 2: When multiple sets of data of ARM-DDS protocol are transmitted simultaneously, the amount of data is given as follows: According to Theorem 1, it is assumed that the hop count of node i is h, considering the reliability requirement is δ^k, and the maximum number of retransmissions is $\gamma_h^k(\delta^k)$. Therefore, the expected number of transmissions per hop S_h^{send}, the amount of reception S_h^{rec}, and the amount of ACK S_h^{AR} and the number of transmission S_h^{AS} are as follows:

$$S_h^{\mathrm{send}} = E_{h+0}^h(\delta^k)p_i^{s0} + \left(1+\frac{r}{d}\right)E_{h+1}^h(\delta^k)p_i^{s1}$$
$$+ \left(1+\frac{2r}{d}\right)E_{h+2}^h(\delta^k)p_i^{s2}$$
$$+ \ldots\left(1+\frac{(x-h)r}{d}\right)E_x^h(\delta^k)p_i^{sx-h} \tag{18}$$

$$S_h^{AS} = S_h^{\mathrm{rec}} = p_i^s\left(\left(1+\frac{r}{d}\right).E_{h+1}^h(\delta^k)p_i^{s1}+\right.$$

$$\left(1+\frac{2r}{d}\right)E_{h+2}^h(\delta^k)p_i^{s2}$$
$$+ \ldots\left(1+\frac{(x-h)r}{d}\right)E_x^h(\delta^k)p_i^{sx-h} \tag{19}$$

$$S_h^{AR} = S_h^{\text{send}} p_i^s q_i^a \tag{20}$$

Among them,

$$E_j^h(\delta^k) = c_h + \sum_{v=1}^{\lfloor \gamma_h^k(\delta^k)/c_h \rfloor} v p_i^s q_i^a \left(1 - \left(1 - p_i^s q_i^a\right)^{c_h}\right)\left(1 - p_i^s q_i^a\right)^{(v-1)c_h}$$
$$+ \gamma_j^k(\delta^k)\left(1 - p_i^s q_i^a\right)^{\lfloor \gamma_h^k(\delta^k)/c_h \rfloor} \tag{21}$$

Proof Because the ARM-DDS strategy is used, the node's data needs to be sent at least c_h times. Therefore, the probability that the transmitted data has at least c_h times is 1. If at least one of the transmitted data is successfully transmitted, the data transmission will stop. Only when this c_h data transmission fails will c_h packets be sent again. The probability of successful transmission within $2c_h$ times is $\left(1 - \left(1 - p_i^s q_i^a\right)^{c_h}\right)\left(1 - p_i^s q_i^a\right)^{c_h}$. This means the first set of data transfers failed, but the second set of c_h data transfers was successful. And sender also successfully received the ACK from the next hop. The probability that the second group c_h transmission is successful is the probability that all the first c_h group data transmission failures $\left(1 - p_i^s q_i^a\right)^{c_h}$ are multiplied by the probability that the second group data is successfully sent to receive. And the probability that sender successfully receives the ACK is $1 - \left(1 - p_i^s q_i^a\right)^{c_h}$. By analogy, the probability of successful transmission in the first vc_h is ($1 - \left(1 - p_i^s q_i^a\right)^{c_h})\left(1 - p_i^s q_i^a\right)^{(v-1)c_h}$. Until $(v+1)c_h$ is greater than $\gamma_h^k(\delta^k)$, the $\gamma_h^k(\delta^k) - \lfloor \gamma_h^k(\delta^k)/c_h \rfloor$ group data is continuously transmitted. Therefore, the expected value sent by the node is

$$E_j^h(\delta^k) = c_h + \sum_{v=1}^{\lfloor \gamma_h^k(\delta^k)/c_h \rfloor} v p_i^s q_i^a \left(1 - \left(1 - p_i^s q_i^a\right)^{c_h}\right)\left(1 - p_i^s q_i^a\right)^{(v-1)c_h}$$
$$+ \gamma_j^k(\delta^k)\left(1 - p_i^s q_i^a\right)^{\lfloor \gamma_h^k(\delta^k)/c_h \rfloor}$$

When a packet from an $h+j$ hop node is sent to a node with a hop count of h, the amount of data that the node needs to bear is: $\frac{(d+jr)}{d}$. Therefore, the number of packets sent from the node of $h+j$ hop to h hop is $E_{h+j}^h(\delta^k)\frac{(d+jr)}{d}$. The probability of sending to h hop is p_i^{sj}. Therefore, the number of packets sent by the node with the hop count is

$$S_h^{\text{send}} = E_{h+0}^h(\delta^k)p_i^{s0} + \left(1 + \frac{r}{d}\right)E_{h+1}^h(\delta^k)p_i^{s1}$$
$$+ \left(1 + \frac{2r}{d}\right)E_{h+2}^h(\delta^k)p_i^{s2}$$
$$+ \dots \left(1 + \frac{(x-h)r}{d}\right)E_x^h(\delta^k)p_i^{sx-h}$$

The number of ACKs that need to be sent is as follows: Each node receives an ACK and sends an ACK. Therefore, the number of ACKs sent is equal to the

number of received packets. The number of received packets is the number of data sent by the previous node multiplied by p_i^s (because there is a packet loss). Therefore, there is the following formula:

$$S_h^{AS} = S_h^{\text{rec}} = p_i^s\left(\left(1 + \frac{r}{d}\right)E_{h+1}^h(\delta^k)p_i^s + \left(1 + \frac{2r}{d}\right)E_{h+2}^h(\delta^k)p_i^{s2}\right.$$
$$\left. + \dots\left(1 + \frac{(x-h)r}{d}\right)E_x^h(\delta^k)p_i^{sx-h}\right)$$

Considering that the number of packets sent by the node i with the hop count h to the next hop is S_h^{send}, the packets arriving at the next hop are p_i^s. Each packet is received by the receive and returns an ACK. The probability that the return ACK is successfully received by sender is q_i^a. Therefore, the number of ACKs received by the node is

$$S_h^{AR} = S_h^{\text{send}} p_i^s q_i^a$$

As can be seen from Section 5.2, the larger the c value, the smaller the transmission delay. In the ARM-DDS scheme, if the c value is greater than 1, the transmission delay compared to the SW-ARQ protocol will be reduced. In the ARM-DDS protocol, since the non-hotspots area has more energy remaining, we can set the non-hotspots area to have a larger c value. This reduces transmission delays, ensures network lifetime, and increases energy utilization.

In Section 5.3, it can be concluded from Theorem 1, that the number of retransmissions of the ARM-DDS scheme is less than or equal to the traditional SW-ARQ protocol. In the performance of the data amount, according to Algorithm 2 and Theorem 2, it can be concluded that since the ARM-DDS strategy uses a larger c value in the non-hotspots area, the amount of transmitted and received data is higher than the SW-ARQ protocol. In the hotspots area, the number of retransmissions of the ARM-DDS policy is smaller than that of the SW-ARQ protocol, and the number of data transmitted each time is equal to the conventional protocol. Therefore, the data amount of the ARM-DDS policy at this time is smaller than the SW-ARQ protocol.

6 The experimental results and analysis

This section mainly analyzes the two most important factors of the ARM-DDS protocol for the network: the impact of transmission delay and energy consumption. Discuss the performance of the ARM-DDS protocol in these two aspects under differentiated services.

6.1 Transmission delay

Theorem 3: In the ARM-DDS protocol, the node i of any h hop and its distance to the sink is d, $d = hr + \Delta$.

The end-to-end delay ω_h is as follows when transmitting data with different delay requirements:

$$\omega_h = \sum_{j=1}^{h} w_{h2j} \tag{22}$$

$$
\begin{aligned}
w_{h2j} = &\sum_{v=1}^{c_j} \left(p_i^s \prod_{m=1}^{v} (1-p_i^s) \left(s(m-1) + \frac{1}{2}\tau_{\text{round}} \right) \right) \\
&+ \sum_{v=c_j+1}^{2c_j} \left(p_i^s \prod_{m=1}^{v} (1-p_i^s) \left(s(m-1) + \frac{1}{2}\tau_{\text{round}} \right) \right) \\
&+ \ldots \sum_{v=(k-1)c_j+1}^{\lfloor \gamma_h^k(\delta^k)/c_j \rfloor} \left(p_i^s \prod_{m=1}^{v} (1-p_i^s) \left(s(m-1) + \frac{1}{2}\tau_{\text{round}} \right) \right) \\
&+ \sum_{v=\lfloor \gamma_h^k(\delta^k)/c_j \rfloor +1}^{\gamma_j(\delta)} \left(p_i^s \prod_{m=1}^{v} (1-p_i^s) \left(\frac{1}{2}\tau_{\text{round}} + \mu \right. \right. \\
&\left. \left. + (c_j-1)(s+b) + \tau_{\text{rtto}}(v-c_j-1) \right) \right)
\end{aligned} \tag{23}
$$

Proof According to Theorem 1, under the delay required by the application, the maximum number of retransmissions of the node of the h hop is $\gamma_h^k(\delta^k)$. The transmission delay can be expressed as $w_h = \sum_{v=0}^{\gamma_h^k(\delta^k)} p_v P_v$, where p_v is the probability of successfully transmitting a packet for the v time, and P_v is the delay of the vth successful transmission. When the node sends data to the next hop, all packets need to arrive: $s(c_h-1) + \frac{1}{2}\tau_{\text{round}}$. Where c_h is the number of consecutively transmitted packets for the h hop, and s is the data sent each time interval between. If the first set of c_h data transmission fails, the sender will transmit the second set of data of c_h until the maximum number of retransmissions is less than c_h times.

Therefore, we can conclude that the probability of successful arrival at the vth ($v \le \gamma_h(\delta)$) transmission is $p_i^s \prod_{m=1}^{v}(1-p_i^s)$. The maximum expected delay after receive processing the data packet is $\frac{1}{2}\tau_{\text{round}} + \mu + (c_h-1)(s+b) + \tau_{\text{rtto}}(v-c_h-1)$.

In summary, the delay for h hop to jump to j is as follows:

$$
\begin{aligned}
w_{h2j} = &\sum_{v=1}^{c_h} \left(p_i^s \prod_{m=1}^{v} (1-p_i^s) \left(s(m-1) + \frac{1}{2}\tau_{\text{round}} \right) \right) \\
&+ \sum_{v=c_j+1}^{2c_h} \left(p_i^s \prod_{m=1}^{v} (1-p_i^s) \left(s(m-1) + \frac{1}{2}\tau_{\text{round}} \right) \right) \\
&+ \ldots \sum_{v=(k-1)c_i+1}^{kc_h} \left(p_i^s \prod_{m=1}^{v} (1-p_i^s) \left(s(m-1) + \frac{1}{2}\tau_{\text{round}} \right) \right) \\
&+ \sum_{v=kc_j+1}^{\gamma_j(\delta)} \left(p_i^s \prod_{m=1}^{v} (1-p_i^s) \left(\frac{1}{2}\tau_{\text{round}} + \mu \right. \right. \\
&\left. \left. + (c_h-1)(s+b) + \tau_{\text{rtto}}(v-c_h-1) \right) \right)
\end{aligned}
$$

The end-to-end delay of the h-th hop is

$$\omega_h = \sum_{j=1}^{h} w_{h2j}$$

Theorem 4: In a planar wireless sensor network with radius R, the known transmission radius is r and the node distribution density is ρ. The average end-to-end delay per hop is ω_h. The average weighted delay for the entire network is as follows:

$$\omega = \sum_{h=1}^{h=x} \frac{\left[(hr)^2 - ((h-1)r)^2 \right] \omega_h}{R^2} \tag{24}$$

Proof Suppose the radius of the planar network is R, the radius per hop is r, the density of nodes is ρ, and the end-to-end delay per hop is ω_h. Since each hop of the planar network can be approximated as a ring shape, the area of the hth hop can be found as $\pi(hr)^2 - \pi((h-1)r)^2$. Therefore, the node of the hth jump has a total of $\rho[\pi(hr)^2 - \pi((h-1)r)^2]$, and the average weighted delay of the entire planar network is

$$\sum_{h=1}^{h=x} \frac{\rho \left[(hr)^2 - ((h-1)r)^2 \right] \omega_h}{\pi R^2 \rho}$$

which is

$$\sum_{h=1}^{h=x} \frac{\left[(hr)^2 - ((h-1)r)^2 \right] \omega_h}{R^2}$$

This paper uses the above formula theorem to calculate the performance of the ARM-DDS protocol in terms of network transmission delay.

6.2 Energy consumption

In WSNs, nodes in the non-hotspots area still have a lot of energy left when the network life expires, causing waste. Therefore, the remaining energy of the non-hotspots region node can be fully utilized to set a larger c value and reduce the transmission delay. The nodes of hotspots need not lose excess energy. Therefore, setting $c = 1$ does not reduce the network lifetime. The following is the theorem about network energy consumption.

Theorem 5: In the ARM-DDS protocol, considering that the energy consumption of nodes transmitting and receiving data packets is e_p^t and e_p^r, respectively, the energy consumption of transmitting and receiving ACKs by nodes is e_A^t and e_A^t, respectively. Then the energy consumption of the h-th hop node is as follows:

$$E_h^i = S^{\text{send}} e_p^t + S^{\text{rec}} e_p^r + S^{AS} e_A^t + S^{AR} e_A^r$$

Proof From Theorem 2, the transmission received data packet S_h^{send} and the reception amount S_h^{rec} and the ACK bearer S_h^{AR} and the transmission amount S_h^{AS} are obtained.

So the energy consumption of a node is the product of the number of transmissions and the energy required to transmit each packet:

$$o_1^i = S^{send} e_p^t + S^{rec} e_p^r + S^{AS} e_A^t + S^{AR} e_A^r$$

Theorem 5 is given to provide a theoretical basis for network lifetime. Investigating the energy consumption of hotspot nodes helps determine network lifetime.

This article does not consider low probability burst situations. Assume that all nodes consume the same amount of energy per round. After deriving the energy consumption of all nodes in the network, the network lifetime can be given.

Theorem 6: Assuming that the lifetime of a planar network is $\mathcal{L}$, the total energy of the nodes in the hotspots region $\mathcal{O}_1^{all}$ is related to the energy consumption o_1^i

. We can get the following formula:

$$\mathcal{L} = \frac{\mathcal{O}_1^{all}}{o_1^i} = \frac{\mathcal{O}_1^{all}}{S_1^{send} e_p^t + S_1^{rec} e_p^r + S_1^{AS} e_A^t + S_1^{AR} e_A^r} \quad (25)$$

The subscript refers to the hotspot area node closest to the sink. The average lifetime of these nodes determines the lifetime of the entire network.

Proof Because in a flat network, we can know from the energy consumption formula that the energy consumption of the near-sink node is the largest. Once the network has energy-saving exhaustion, it means that the planar network is exhausted. Assume that in one cycle, the energy consumption of the near-sink node is $o_1^i = S_1^{send} e_p^t + S_1^{rec} e_p^r + S_1^{AS} e_A^t + S_1^{AR} e_A^r$, the total energy of the node is $\mathcal{O}_1^{all}$. Therefore, in the case of constant physical conditions, the node near the sink can run the $\mathcal{L} = \frac{\mathcal{O}_1^{all}}{o_1^i}$ period, after which the node will run out of energy, so define it as the lifetime of the entire network.

This paper uses the above formula theorem to calculate the performance of the ARM-DDS protocol in terms of network energy consumption.

7 Simulation and analysis results

In a planar wireless sensor network, the one-hop delay of a packet refers to the time it takes from being sent by the current node to being received by the next hop node. The end-to-end delay of a packet is the time it takes to be transmitted from the source node to the sink. In this section, the experimental results of the ARM-DDS protocol and the SW-ARQ protocol are given.

The experimental parameters of this paper are set as follows: $R = 500, r = 80, p_i^s = 0.8, q_i^a = 0.8, \tau_{round} = 50$ ms. The number of data bits is 100 b, and the number of ACK bits is 10 b. The latency and reliability requirements of the data stream are as follows:

When $\delta = 0.8$, the one-hop delay requirement is 25 ms and accounts for 30% of the data stream.

When $\delta = 0.9$, the one-hop delay requirement is 30 ms and accounts for 50% of the data stream;

When $\delta = 0.95$, the one-hop delay requirement is 35 ms and accounts for 20% of the data stream.

7.1 Delay comparison

The comparison between the traditional SW-ARQ protocol and the ARM-DDS protocol in terms of delay is given in this section. When analyzing the transmission performance of the two protocols, the c value of the ARM-DDS protocol should be determined first.

Algorithm 2 gives the value of c in different network environments, as shown in Fig. 4. The end-to-end delay at different c values is given in Fig. 5. Combined with Figs. 4 and 5, the following conclusions are given: (1) The lower the data transmission delay, the larger the value of c. The reason is that the larger the number of simultaneous transmission groups c, the smaller the expected time required for successful transmission. Therefore, when transmitting time-sensitive data, the value of c will be larger. (2) The reduced delay of the ARM-DDS protocol is limited. This is because if the c-group data is sent at the same time, and the transmission time is increased by $(c - 1)s$. When the value of c is too large, an increase in the transmission time causes an increase in delay.

Therefore, setting an appropriate c value is especially important.

Figures 6 and 7 show the one-hop delay of the ARM-DDS protocol and the SW-ARQ protocol under different data streams with different delay requirements. From Fig. 6, the following conclusions can be drawn:

(1) In the SW-ARQ protocol, considering that the reliability of the data reaching the sink is δ, the number of retransmissions of the far-sink nodes is greater than the number of retransmissions of the near-sink nodes. The reason is that the data sent by the node of the far-sink area has to go through more hops before it reaches the sink. To ensure data reliability, it must be retransmitted more times. Therefore, the delay of far-sink nodes is higher than the near-sink node.

(2) In terms of one-hop delay of the near-sink node, both protocols use "one transmission and one reception" data transmission method, so the one-hop delay of the two is the same.

(3) When the number of transmission hops is greater than 1, the ARM-DDS protocol will send multiple sets of data each time, reducing the amount of time it takes to listen to the ACK (as shown in Fig. 3). Therefore the ARM-DDS protocol has a lower single-hop delay. When the SW-ARQ protocol can meet the delay requirement, the maximum number

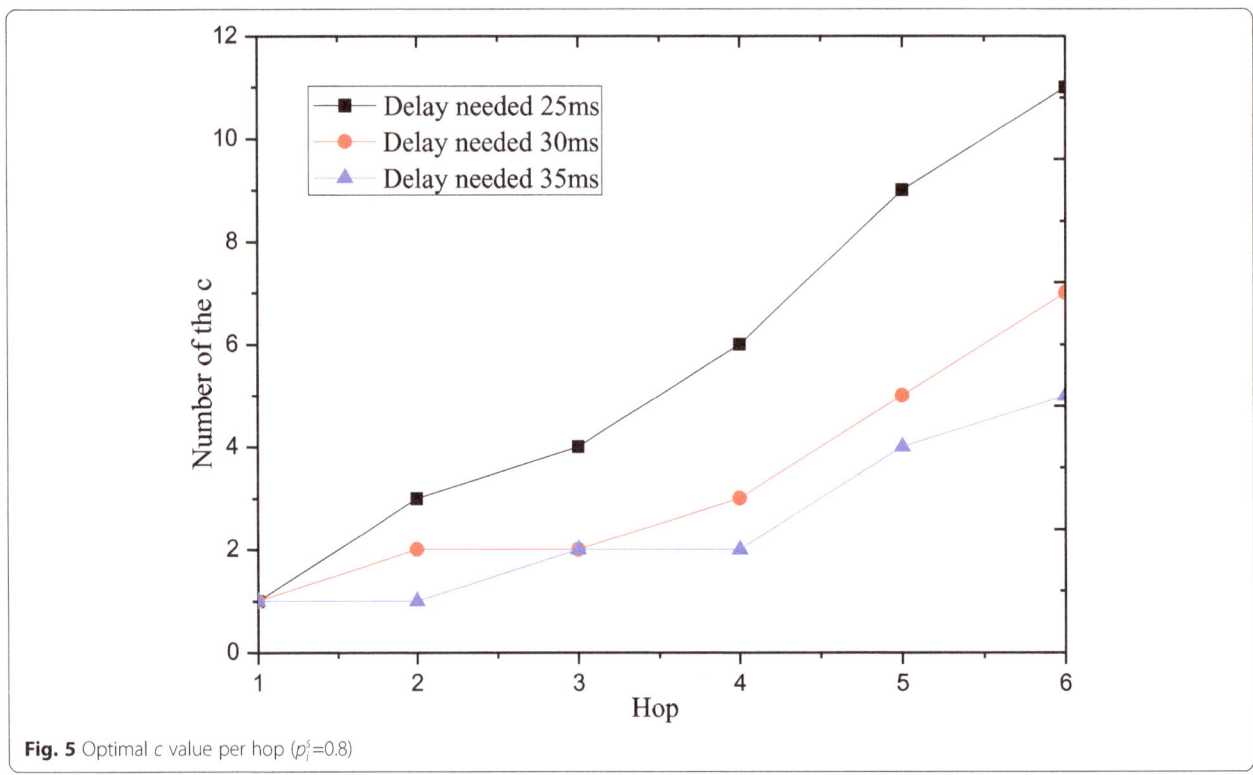

Fig. 5 Optimal c value per hop (p_i^s=0.8)

of retransmissions of the data can only be set according to the highest reliability requirement in the data stream.

(4) The delay of the DMDT protocol fluctuates because of the network environment setting and the integer

value of c. Finally, as the value of c becomes larger, the one-hop delay will be significantly reduced.

(5) Therefore, the SW-ARQ protocol delay will still be greater than the ARM-DDS protocol. Therefore, the one-hop delay of the ARM-DDS protocol is less

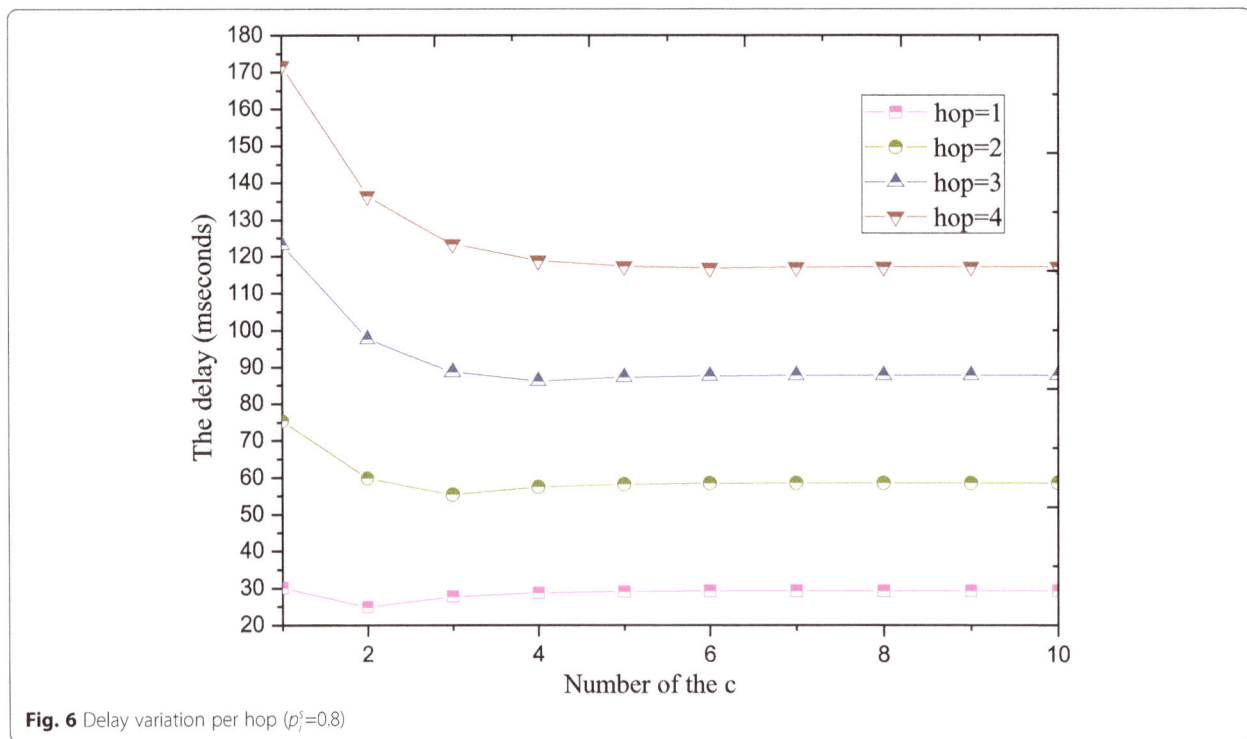

Fig. 6 Delay variation per hop (p_i^s=0.8)

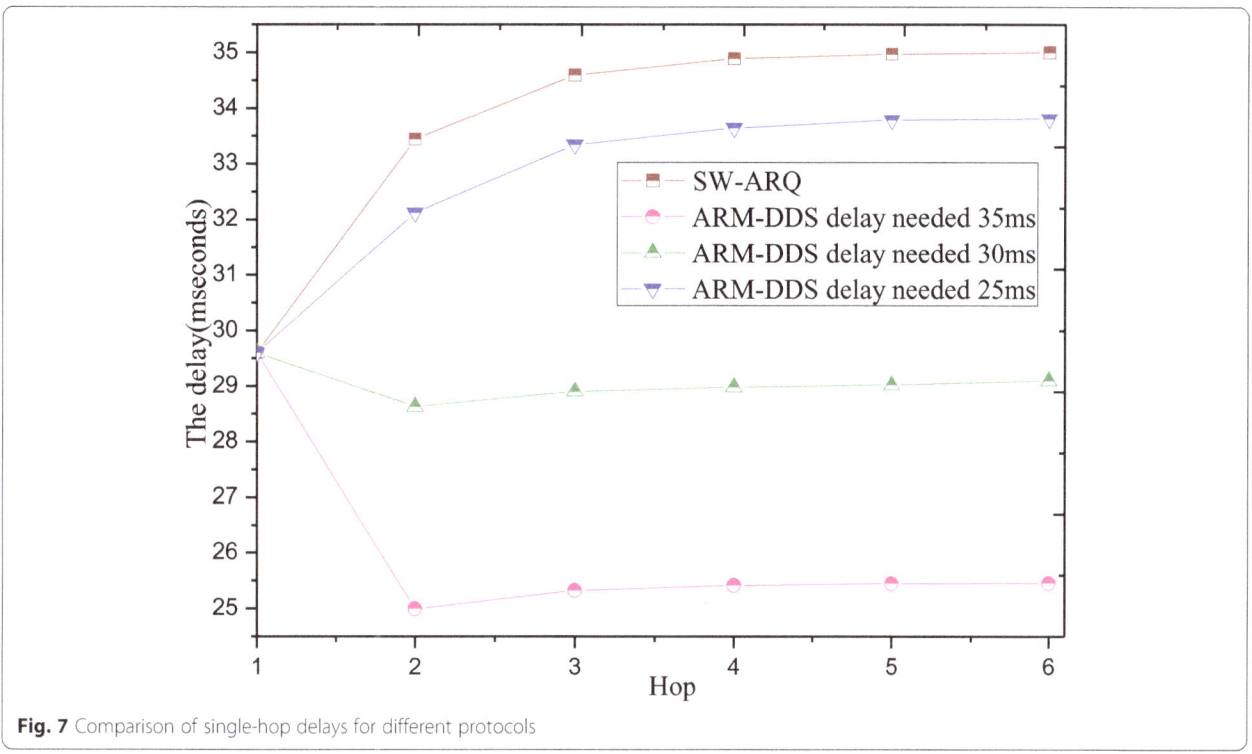

Fig. 7 Comparison of single-hop delays for different protocols

than or equal to the SW-ARQ protocol, which indicates that the ARM-DDS protocol can effectively reduce the transmission delay compared to the SW-ARQ protocol.

Figures 8 and 9 are end-to-end (E2E) delays at different success rates p. The following can be seen from the figures:

(1) Nodes near the sink are insensitive to the change of the transmission success rate p, because nodes near the sink only need to transmit data once. Therefore, when the transmission success rate p changes, the influence on the first hop is minimal.
(2) It can be seen from Fig. 8 that as the value of p becomes smaller, the maximum number of retransmissions per hop becomes larger. Because of the lower transmission success rate, the probability of data transmission failure increases, and the probability of retransmission increases. As shown in the figures, the end-to-end delays of both protocols increase as p becomes smaller.
(3) As the value of p becomes smaller, the ARM-DDS protocol reduces delay greater than the SW-ARQ protocol.

Therefore, the worse the network environment the ARM-DDS protocol works in, the more the transmission delay is reduced.

Figure 10 is the E2E delay of the ARM-DDS protocol and the SW-ARQ protocol under the comprehensive reliability change of the data stream. The following can be seen from Figs. 10 and 11:

(1) The differentiated data stream which has the different reliability requirements that cannot be met with the SW-ARQ protocol. Therefore, the transmission delay of the protocol does not change by following when the overall reliability of data stream changes.
(2) The overall delay time of the ARM-DDS protocol will be more largely under the higher reliability requirements comprehensively. The reason is that the increase of transmission delay is caused by the need of the node for more retransmission times to ensure the data arrival rate accurately when the comprehensive reliability of the data stream becomes higher.
(3) As can be seen from Fig. 10, the ARM-DDS protocol can reduce the transmission delay time significantly, besides the obvious reduction of the delay was procured by the larger hop count of the node.
(4) The reduction of the ARM-DDS protocol's delay is more obvious while the reliability requirement is high, and when the reliability requirement is low, the reduction of the near-sink node's delay is more obvious. Because the maximum number of

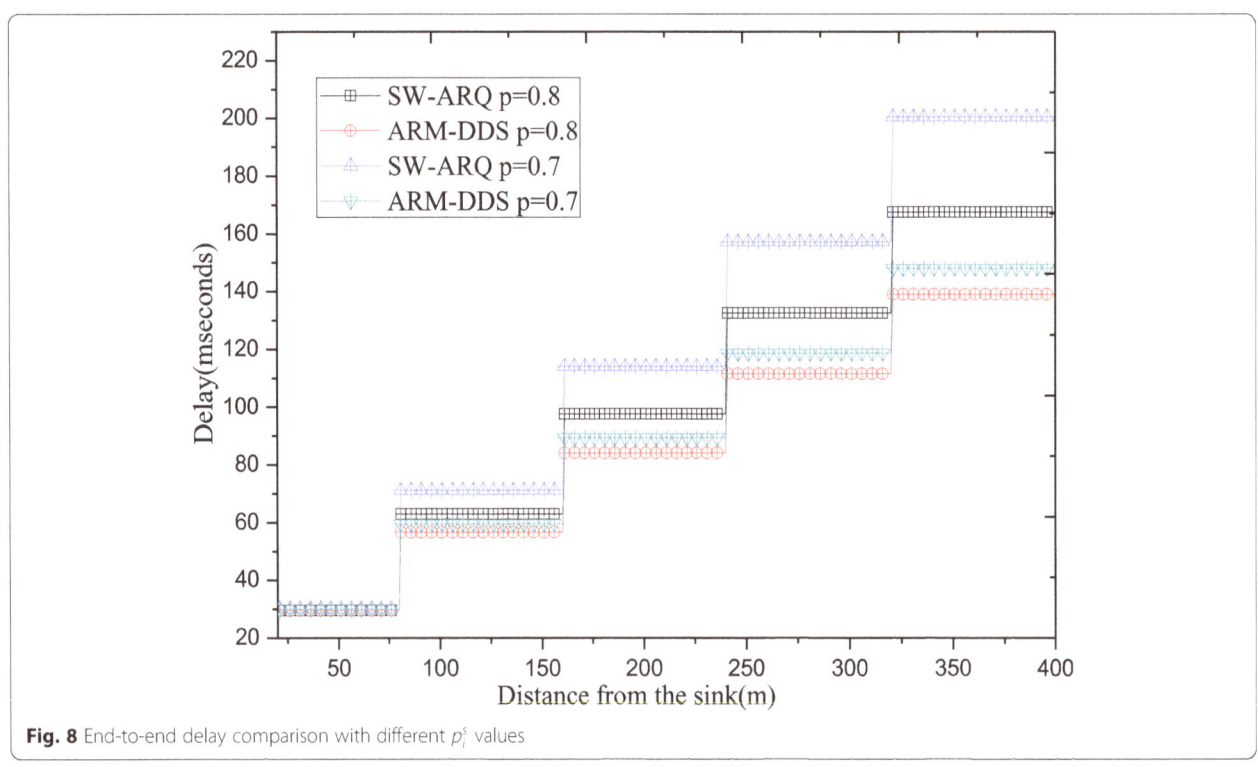

Fig. 8 End-to-end delay comparison with different p_i^s values

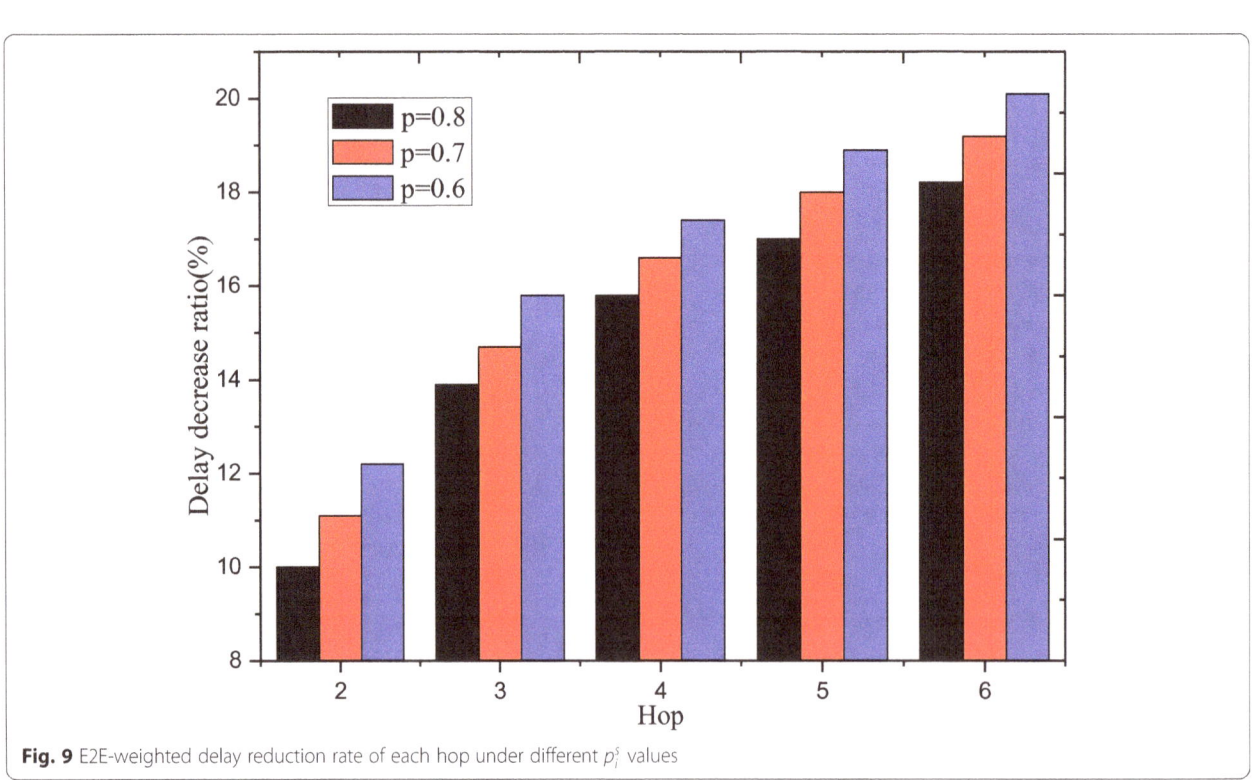

Fig. 9 E2E-weighted delay reduction rate of each hop under different p_i^s values

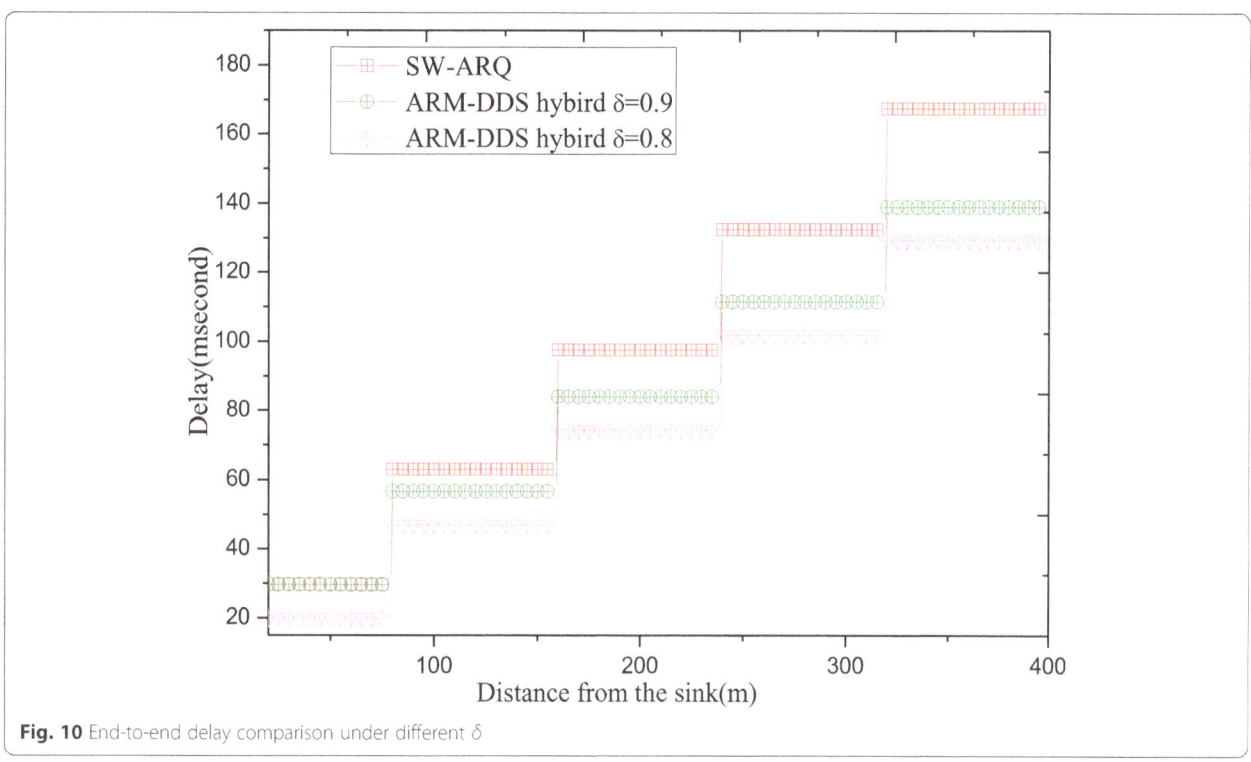

Fig. 10 End-to-end delay comparison under different δ

retransmissions times which near the sink nodes is small, at the same time, ARM-DDS protocol transmits c packets when it nears the sink, which makes it easier to meet the data arrival rate under high-reliability requirements.

However, when the reliability requirement is low, the c value near the sink is close to 1, and the delay reduction rate becomes smaller.

Figure 12 illustrates the average weighted delay of the ARM-DDS protocol and the SW-ARQ protocol for

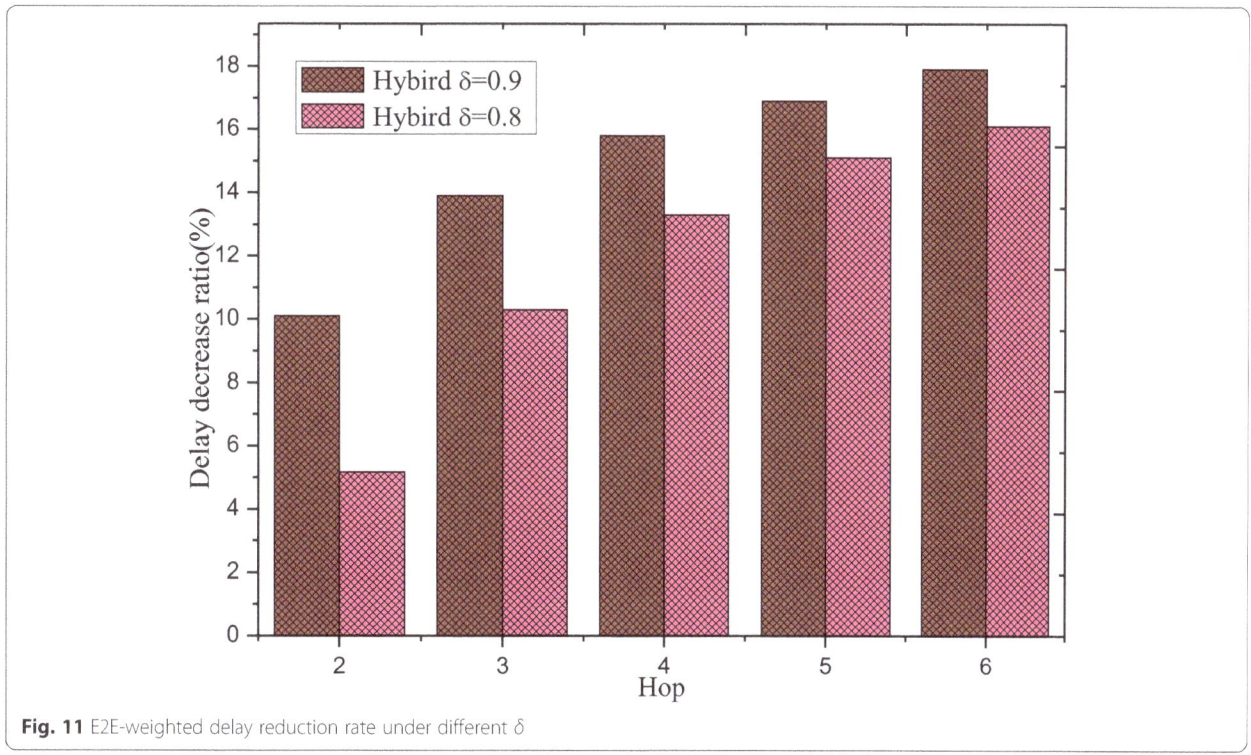

Fig. 11 E2E-weighted delay reduction rate under different δ

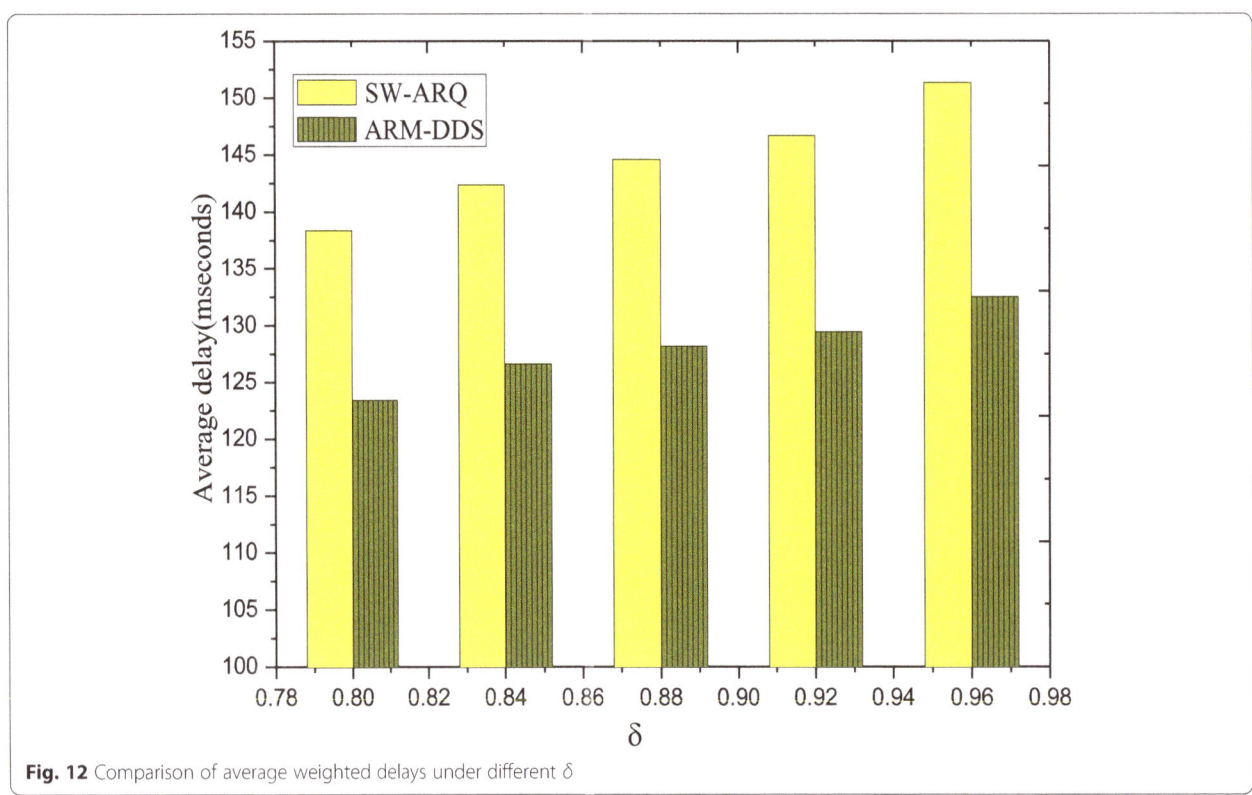

Fig. 12 Comparison of average weighted delays under different δ

different reliability requirements. It can be seen from Fig. 12:

(1) With the following of reliability increasing, the average weighted delay of both protocols reveals the upward trends. Furthermore, in order to meet the higher reliability, the number of retransmissions which expected by the node is increased, and it obviously causes the increasing of the average weighted delay.

(2) In terms of average weighted delay, the traditional SW-ARQ protocol is superlative than the ARM-DDS protocol because of the SW-ARQ protocol delay speed of almost 145 ms, while the ARM-DDS can reduce the delay to 127 ms. There are some explanations about the situations of both protocols. The ARM-DDS protocol uses c-group data to transmit at the same time; it only needs to return an ACK once, which greatly reduces the delay of ACK return. In the case that the traditional SW-ARQ protocol costs many times to return the ACKs under the lower transmission success rate situations. But as for the ARM-DDS protocol, it can reduce the delay which is caused by saving as part of return time. Although the ARM-DDS protocol can reduce latency in this case, its advantages are not obviously in the case of good network conditions in comparison with the SW-ARQ protocol.

Figure 13 shows the delay reduction rate of the ARM-DDS protocol and the SW-ARQ protocol under the different reliability requirements δ. It is obviously to optimize the transmission delay when the reliability δ is between 0.80 and 0.96, the ARM-DDS protocol can reduce the delay by 12–14% in compare with the traditional SW-ARQ protocol. As the reliability requirements increase, the rate of delay reduction increases. Explain that the ARM-DDS strategy performs better under high-reliability requirements.

Thus, the ARM-DDS protocol can be used to transport data streams that require more timeliness.

Figures 14 and 15 show the delay comparison of the ARM-DDS protocol with the traditional SW-ARQ protocol under different τ_{round}. The following conclusions can be drawn:

(1) τ_{round} has a great influence on a data transmission delay. When τ_{round} increases, the end-to-end delay of each node will increase significantly.

(2) The larger the τ_{round}, the more significant the effect on the ARM-DDS protocol to reduce the delay. Because in the traditional SW-ARQ protocol, the periods during the node which send a data packet and prepare to send the second data packet named as τ_{round}. According to Fig. 3. The node under the ARM-DDS protocol continuously transmits c data packets; it takes $c(k-1) + \frac{1}{2}\tau_{round}$ time

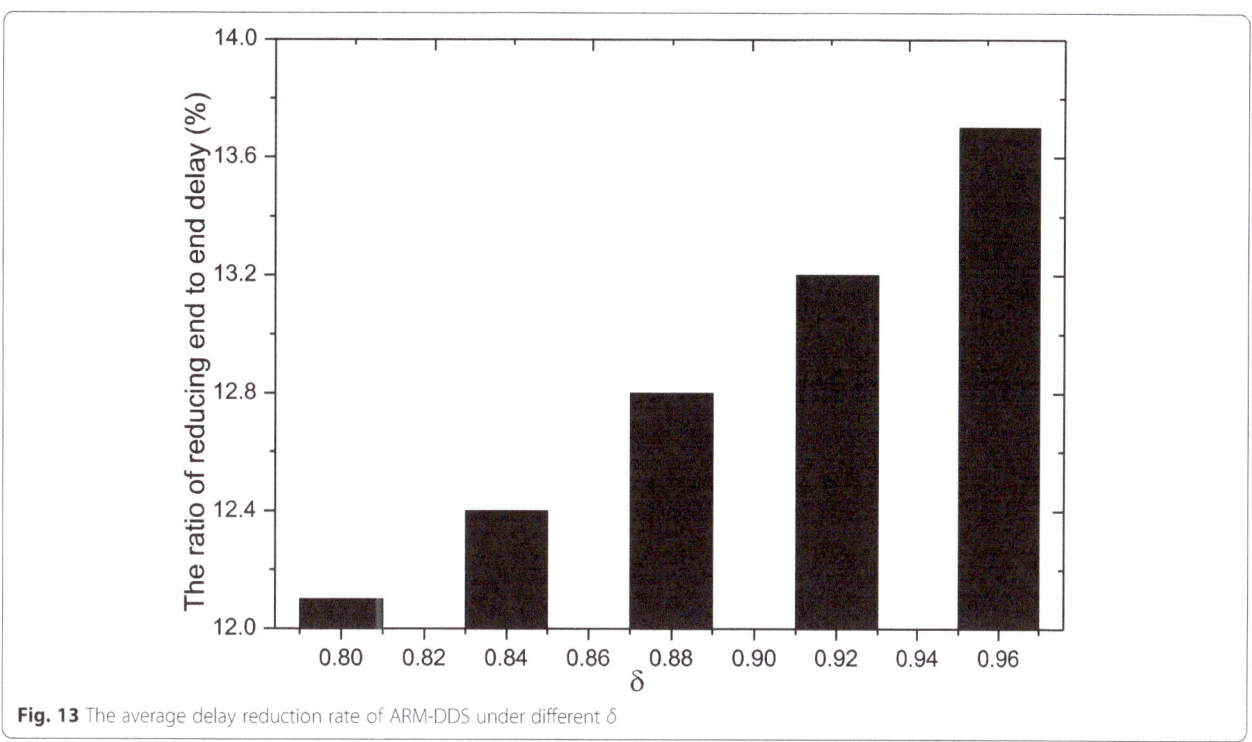

Fig. 13 The average delay reduction rate of ARM-DDS under different δ

and only cost the $c(k-1) + \tau_{\text{round}}$ which adds the extra time of receiving an ACK.

However, if SW-ARQ sends c packets, it will consume the time associated with $c\tau_{\text{round}}$. So when τ_{round} increases, the delay of SW-ARQ is bigger than that of ARM-DDS.

Figure 16 is intended to examine the delay reduction of the ARM-DDS protocol that is influenced by the different transmission radii r. The conclusion is given by the information from Fig. 16: With the transmission radius changes, it causes the number of hops of the node changes as following. But the delay did not change

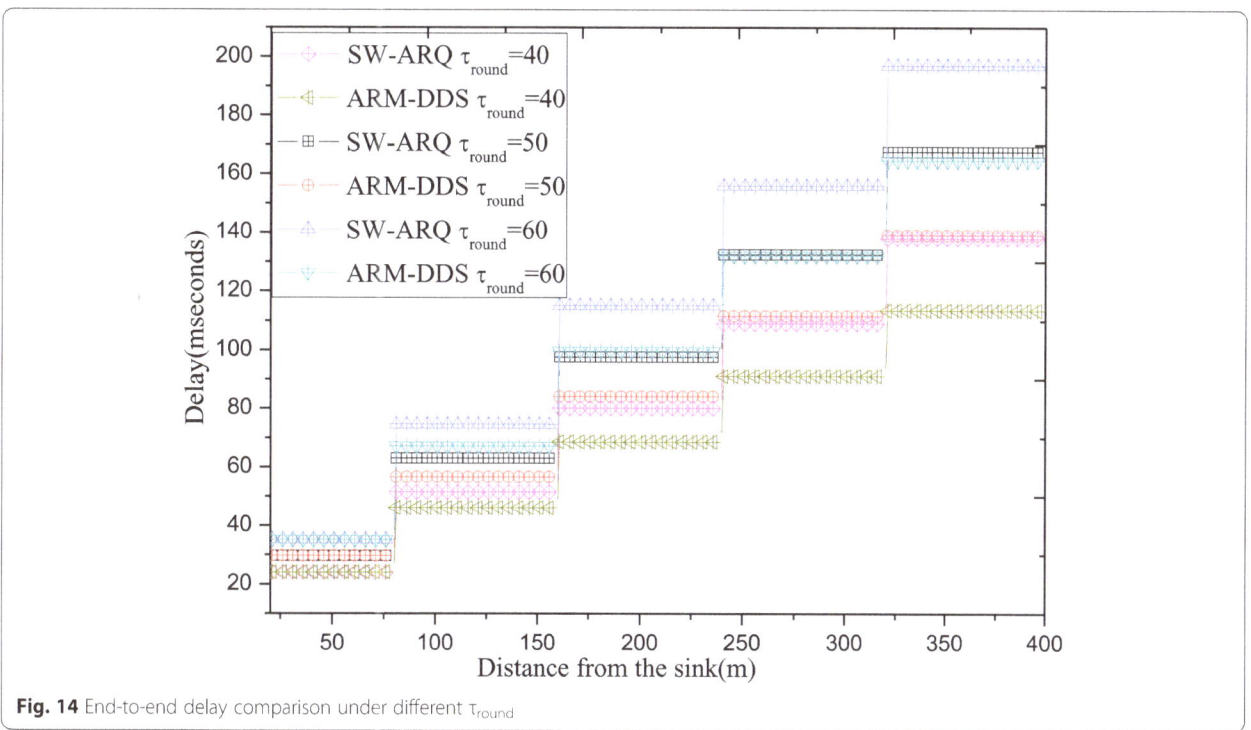

Fig. 14 End-to-end delay comparison under different τ_round

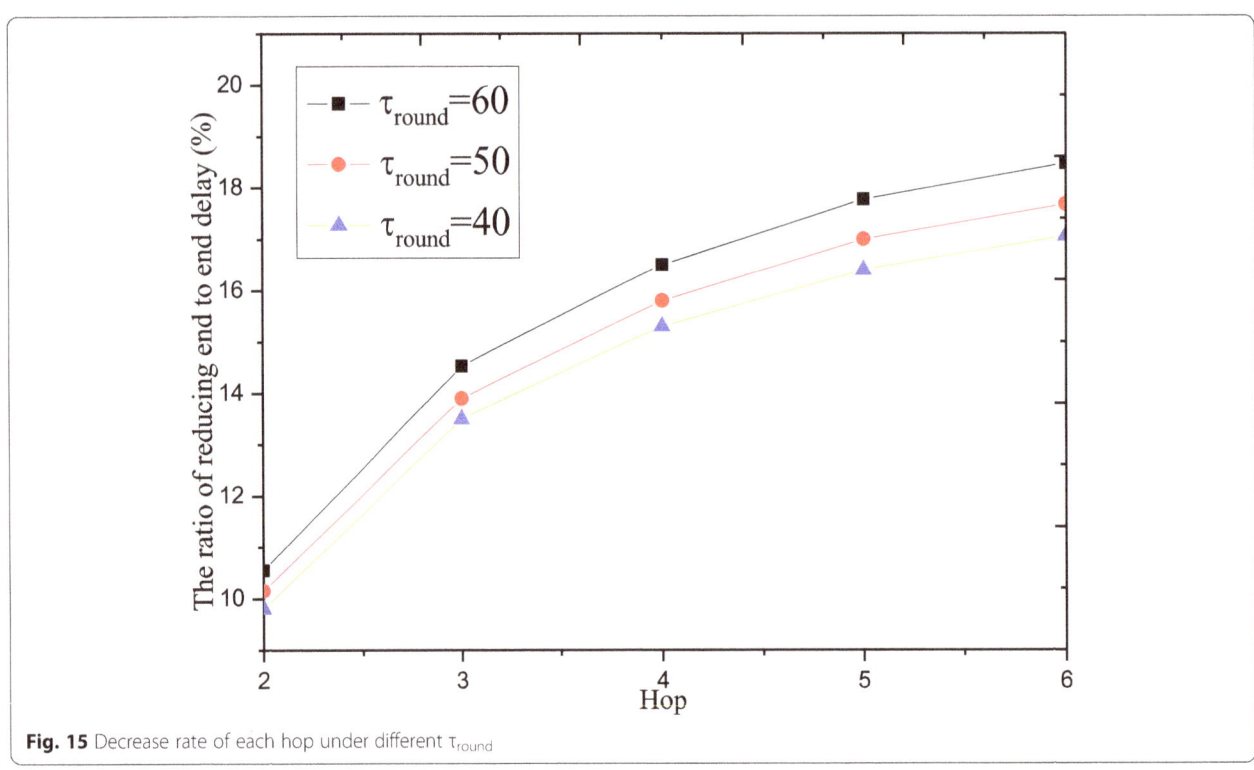

Fig. 15 Decrease rate of each hop under different τ_{round}

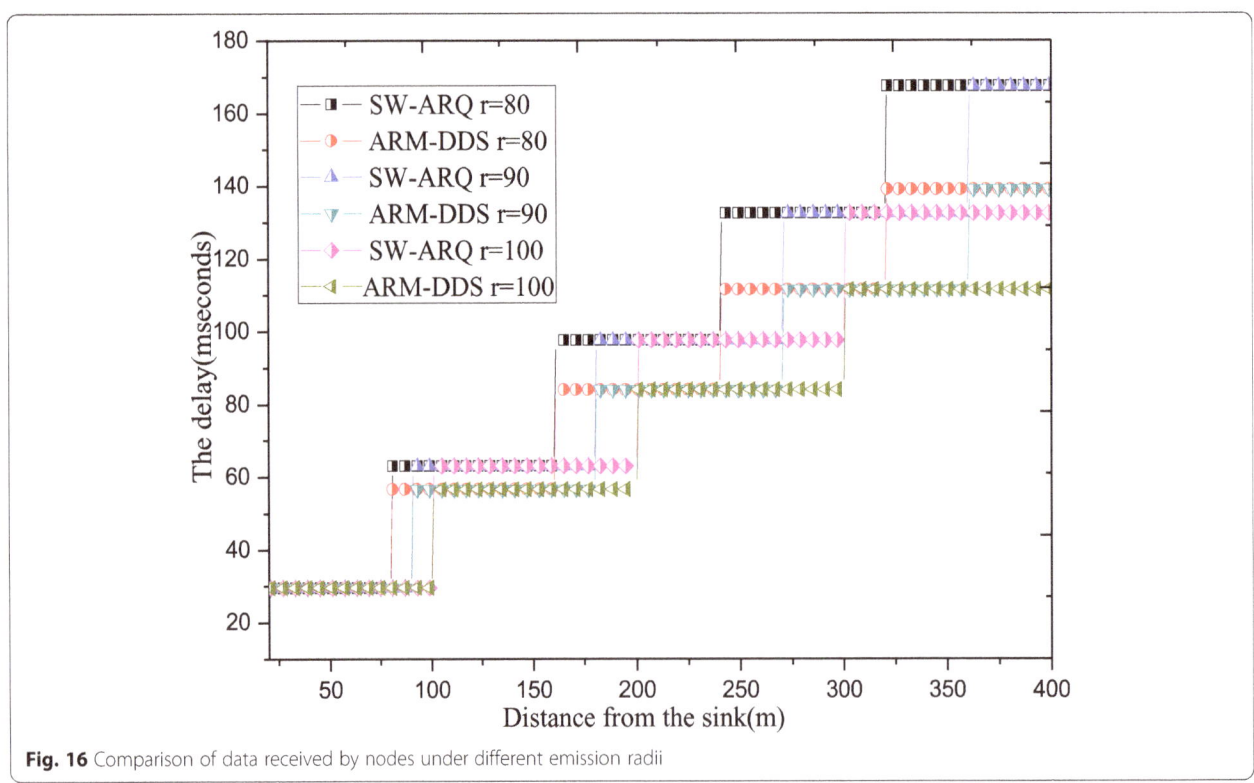

Fig. 16 Comparison of data received by nodes under different emission radii

significantly. So we conclude that the variation of the transmission radius does not significantly affect the ARM-DDS reduction delay.

7.2 Energy consumption comparison

Figures 17, 18, and 19 show the results of the comparison of data reception and transmission amount between the legacy protocol and the ARM-DDS protocol. The information can be obtained from Figs. 17, 18, and 19:

(1) When the ARM-DDS protocol which is near the sink node has been adopted, its transmission data amount is smaller than the SW-ARQ protocol. Because in the differentiated service of the ARM-DDS protocol, each data stream sets the number of transmissions depends on its reliability requirements. However, the SW-ARQ protocol selects the data stream with the highest reliability requirement to set the number of retransmissions of all data packets. At that moment, the amount of transmitted data near the sink of the ARM-DDS protocol will be smaller than the SW-ARQ protocol which is caused by the number of retransmissions of the ARM-DDS protocol is smaller than the conventional protocol.

(2) To make the comparison of receiving data volume of the two protocols, the data reception amount of the ARM-DDS protocol is larger than the SW-ARQ protocol. Because the data is transmitted using the ARM-DDS protocol at the far-sink nodes, the amount of reception will increase due to the number of times the node sends will be greater than the maximum number of retransmissions.

(3) As long distances between a node to sink, the amount of data transmitted and received by the ARM-DDS protocol is dramatically improving in comparison with the SW-ARQ protocol.

Hence, it is found that the energy consumption of ARM-DDS which is far from the sink protocol is larger than the energy consumption of the SW-ARQ protocol, and the ARM-DDS protocol could effectively improve the energy usage rate of the network.

Figure 20 emphasis that the maximum number of c value retransmissions times of the ARM-DDS protocol interactive with the SW-ARQ protocol under different p_i^s. As shown in Fig. 20: With the ARM-DDS protocol reducing the delay as much as possible, its c value is under the maximum number of retransmissions. The ARM-DDS protocol does not affect the network lifetime due to the value of c selected by the near-sink node that does not exceed the maximum number of retransmissions. On the other hand, the ARM-DDS protocol can increase the node energy consumption in the far-sink area. Thus, the ARM-DDS protocol can increase the energy utilization of the entire network without reducing network lifetime.

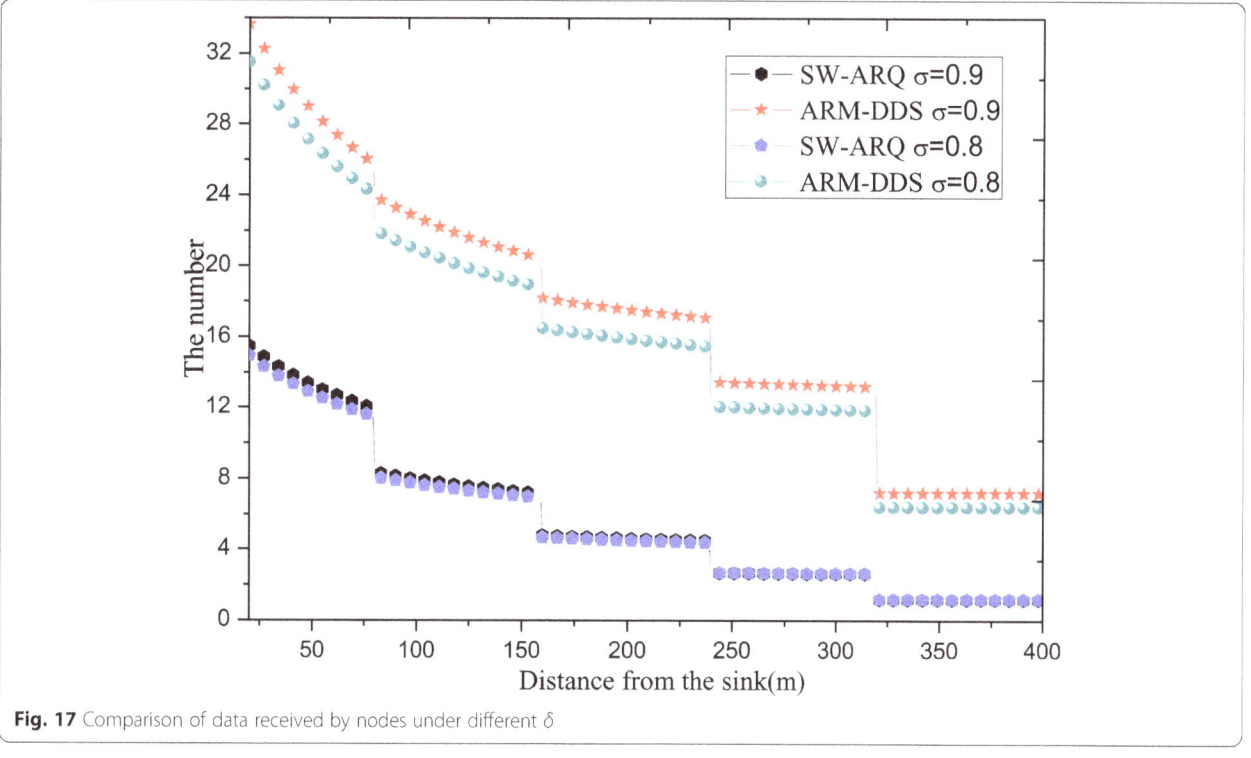

Fig. 17 Comparison of data received by nodes under different δ

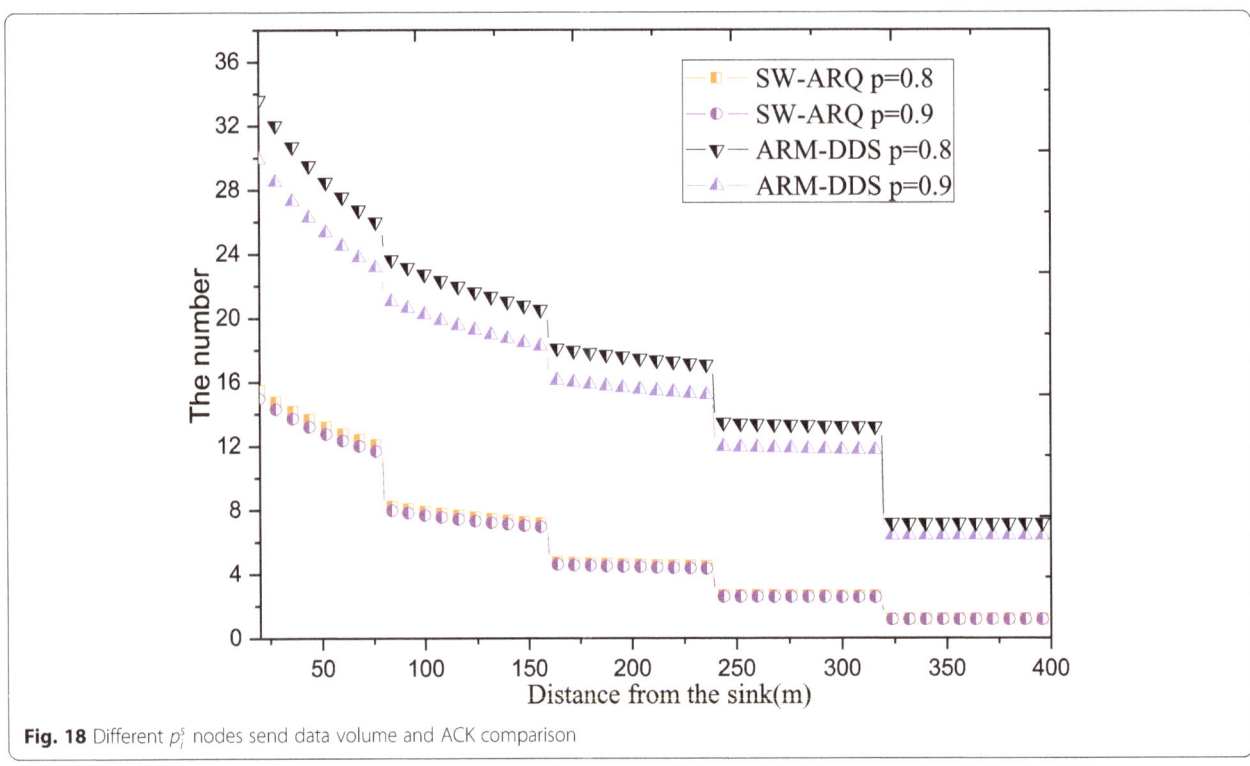

Fig. 18 Different p_i^s nodes send data volume and ACK comparison

From the system model given in Section 4, the network lifetime is defined as the death time of the first node in the network. In a flat network, the nodes near the sink not only need to undertake the energy of data transmitting that was created by themselves but also the amount of data transferred from the far-sink nodes due to the "many to one" data collective models throughout the centralization of the sink. Hence, the node near the sink will have the largest energy consumption in the entire network. So the first dead node in the network is

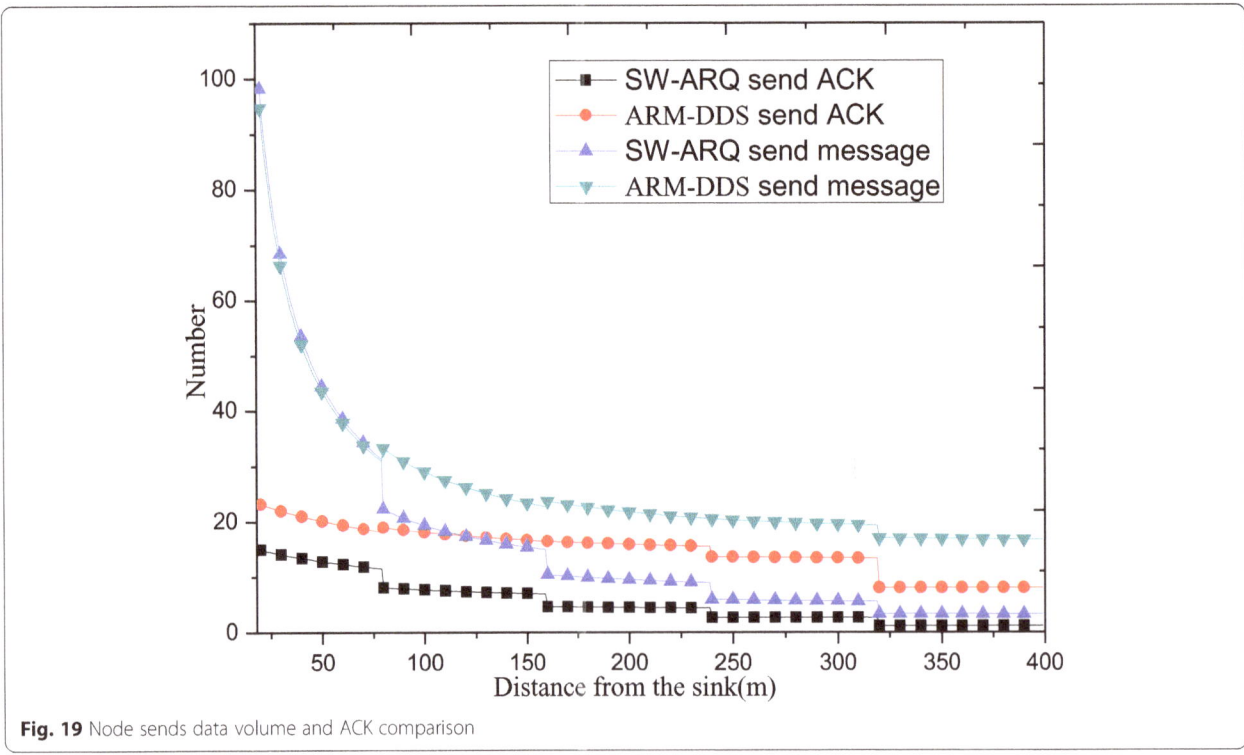

Fig. 19 Node sends data volume and ACK comparison

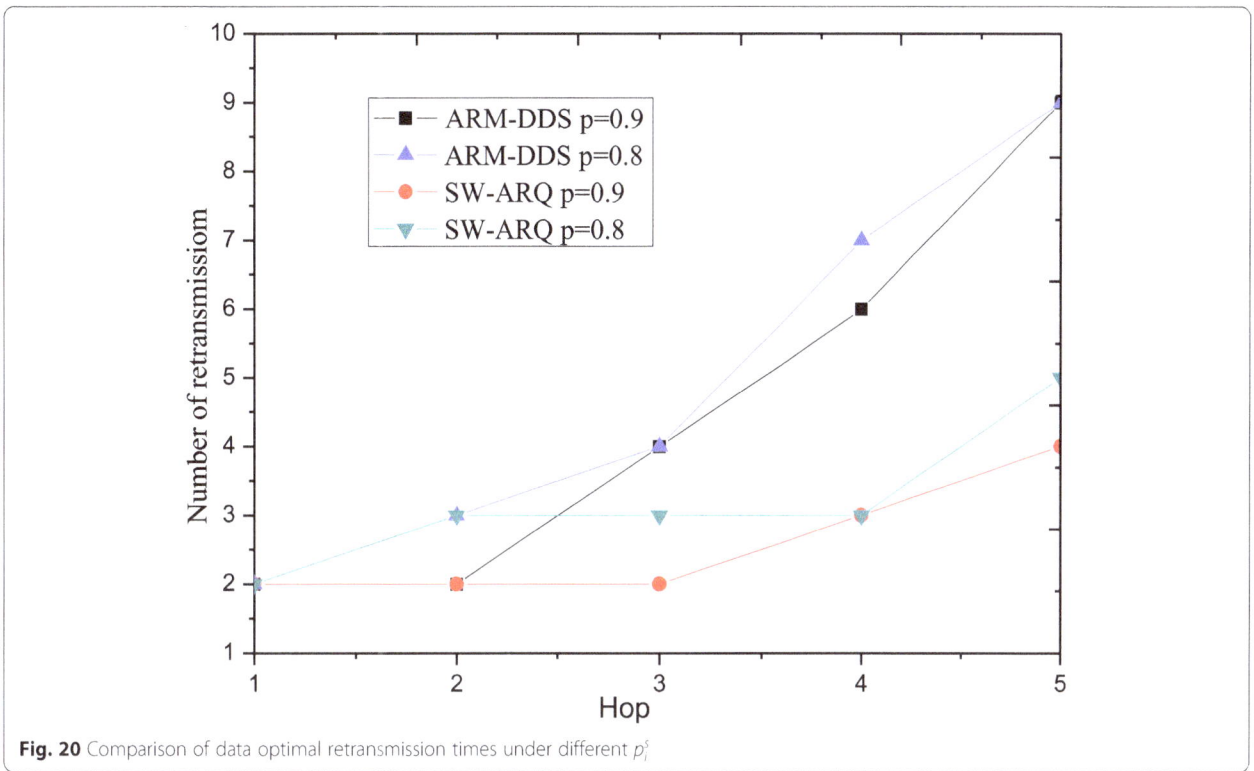

Fig. 20 Comparison of data optimal retransmission times under different p_j^s

usually one of the nodes close to sink. Figures 21 and 22 show the total energy consumption of a node in one cycle according to Theorem 2. Figures 21 is a comparison diagram of the SW-ARQ protocol and ARM-DDS energy consumption. The conclusions are as follows:

(1) The node farther away from the sink which can cost the small energy consumption. This phenomenon also applies to both protocols. Because according to Eq. (5), the farther away from the node, the smaller the amount of data he needs

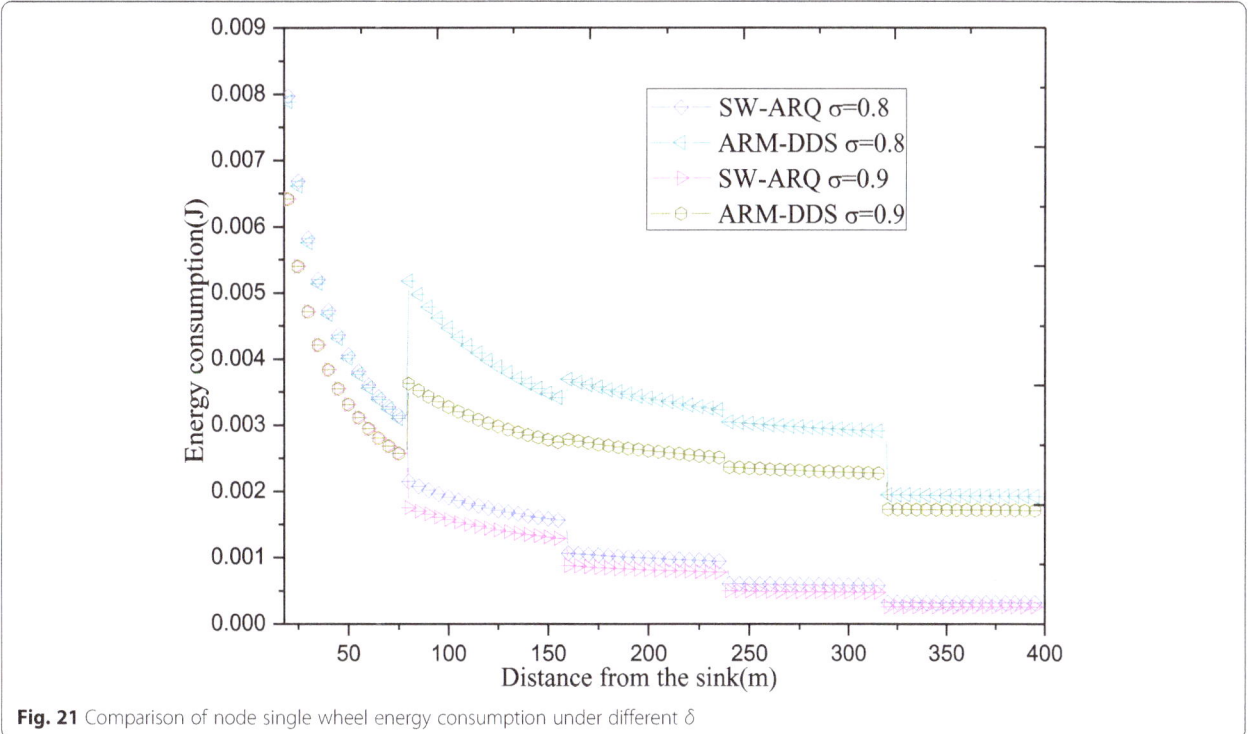

Fig. 21 Comparison of node single wheel energy consumption under different δ

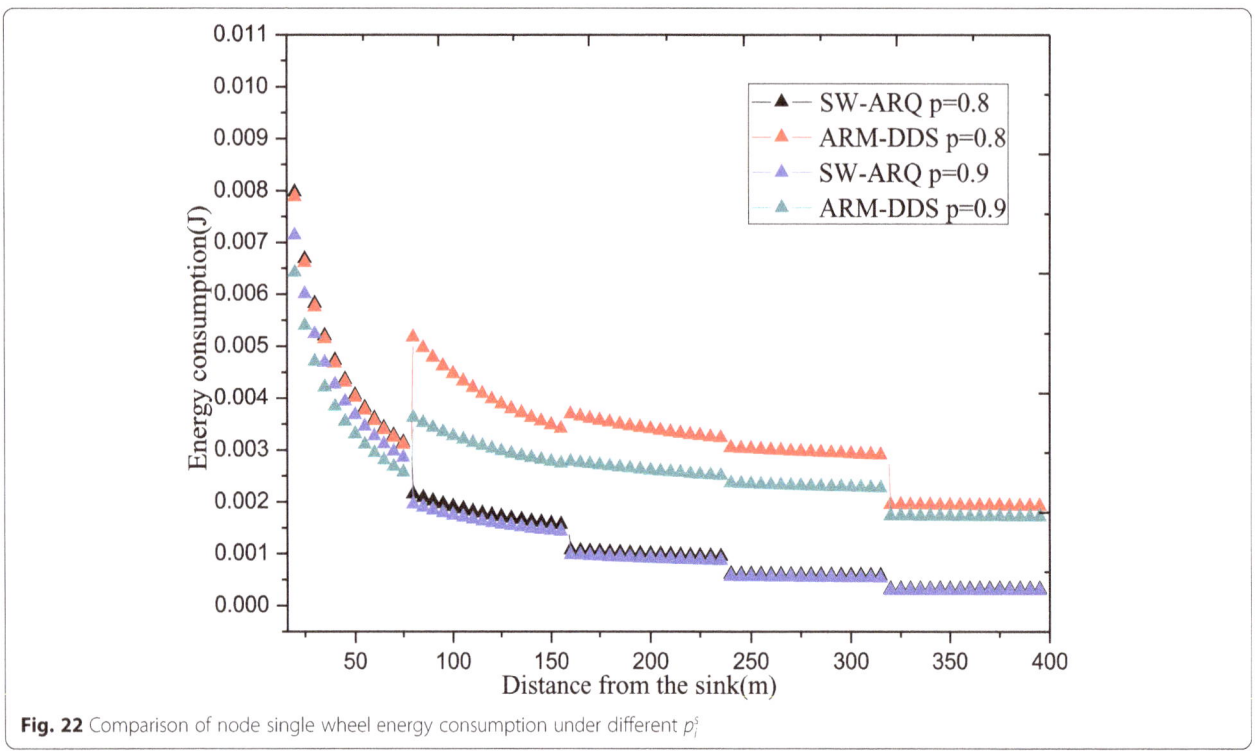

Fig. 22 Comparison of node single wheel energy consumption under different p_i^s

to transmit. So the farther away from the sink, the less energy the node consumes.

(2) The ARM-DDS protocol consumes more energy than the SW-ARQ protocol when studying the energy consumption of nodes at the far-sink. This is because the ARM-DDS transmission scheme continuously sends c packets of data every hop. In contrast, the SW-ARQ protocol only sends one packet at a time. Considering the far-sink node has a lot of residual energy, the c value set by the ARM-DDS protocol will exceed the maximum number of retransmissions. Therefore, the total energy consumption of the nodes of the ARM-DDS protocol is generally higher than the SW-ARQ protocol.

(3) A higher data transmission success rate will make the energy consumption of the node smaller. Because Eq. (7) shows that the larger the transmission success rate p_i^s, the smaller the maximum number of retransmissions of the node. According to Algorithm 2, the value of c of the ARM-DDS protocol will also become smaller.

(4) The energy consumption of the near-sink nodes of the SW-ARQ protocol will be higher. The reason is that the ARM-DDS protocol does not require all data streams to be retransmitted according to the maximum reliability requirements.

Therefore, the SW-ARQ protocol at the near-sink nodes expects the number of retransmissions to be

higher than the ARM-DDS protocol. If the network transmission environment deteriorates, the transmission success rate will become lower. As can be seen from the figure, the energy consumption of the ARM-DDS protocol is smaller than the SW-ARQ protocol. And it is more obvious when it is higher than the p_i^s value.

Figure 23 illustrates the energy utilization rate under different strategies. Energy utilization rate refers to the ratio of the remaining energy of the network to the initial total energy of the network when the network dies. From the experimental results of Fig. 23, the energy utilization rate of ARM-DDS is about 65%, and the energy utilization rate of SW-ARQ is about 37%. It can be seen that the ARM-DDS proposed in this paper significantly improves network energy utilization.

From Fig. 24, it can be shown that: ARM-DDS increases almost 28% of energy utilization in compare with the SW-ARQ protocol. Moreover, energy utilization has a minimal effect on both the transmission and reliability requirements of the network. All of these indicate that the ARM-DDS protocol proposed in this paper has a positive influence on energy utilization compared to the SW-ARQ protocol.

8 Network lifetime and reliability optimization

In terms of network life optimization, the ARM-DDS scheme can increase $\frac{\overline{\delta}}{\delta_{max}}$ % relative to the SW-ARQ protocol, where $\overline{\delta}$ is the average weighted reliability re-

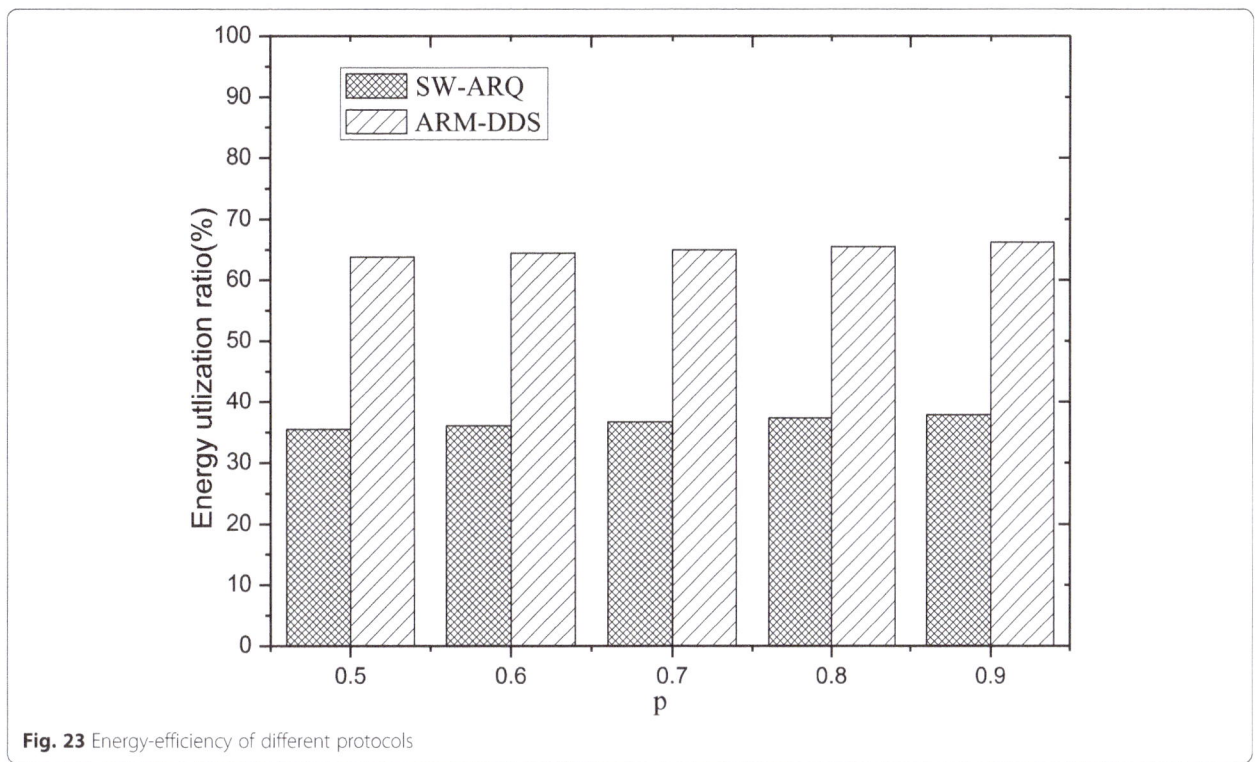

Fig. 23 Energy-efficiency of different protocols

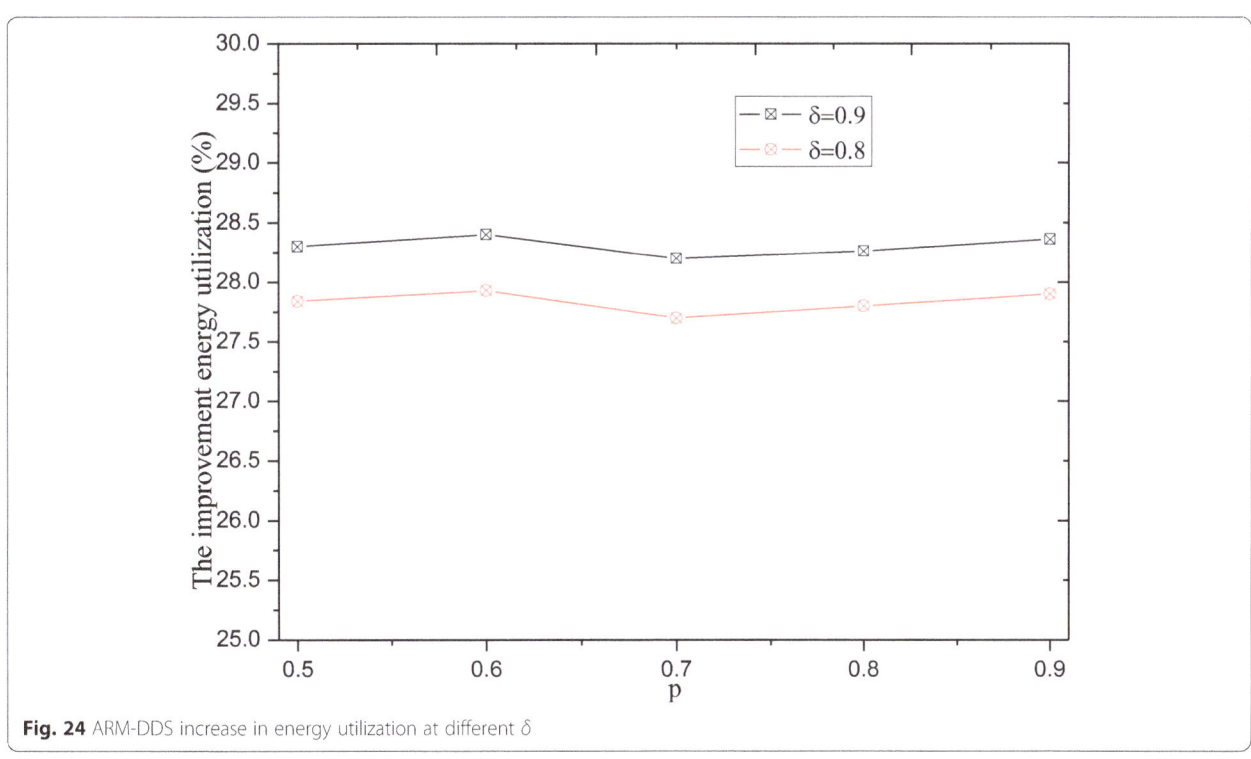

Fig. 24 ARM-DDS increase in energy utilization at different δ

quirement for all data streams, and δ_{max} is the maximum reliability requirement for all data streams. Figure 25 shows that the situation of network lifetimes for the two protocols under different reliability. The ARM-DDS protocol has a longer lifetime than the SW-ARQ protocol network, because the ARM-DDS protocol adopts the methods of differentiated services. Each data stream sets the value of c according to the needs of reliability. However, the SW-ARQ protocol is ruled to select the highest reliability requirements and to transfer all of the information in the data stream.

Therefore, the amount of data received and transmitted by the ARM-DDS protocol which near-sink nodes is smaller than the SW-ARQ protocol, which optimizes the lifetime of the network.

Figures 26 and 27 are the maximum numbers of retransmissions and the average number of retransmissions of the SW-ARQ protocol at the hop = 3 in different network environments. In the ARM-DDS protocol, the c does not exceed when the data transmission with high timeliness requiring the maximum number of retransmissions. Because the delay is guaranteed to be as small as possible, the set of c is not too large. Otherwise, when multiple sets of data are continuously transmitted, the time of data transmission will be accumulated and it still affects the delay. However, the value of c will exceed the maximum number of retransmissions when the data stream transmissions among the rules of timeliness are without high requirement.

Figure 28 shows the transmission reliability under the ARM-DDS protocol and the SW-ARQ protocol. The following can be seen from the figure:

(1) The reliability of the ARM-DDS protocol at the far-sink nodes is higher than that of the SW-ARQ protocol. Because the c value of the ARM-DDS protocol at the far-sink nodes is greater than the maximum number of retransmissions of the SW-ARQ protocol, therefore, the data reliability of the ARM-DDS protocol at the far-sink node is higher than that under the SW-ARQ protocol. This ensures a high data arrival rate even when the transmission conditions are not very good. In terms of transmission reliability of near-sink nodes, since the ARM-DDS protocol uses a differentiated service algorithm, the weighted reliability is lower than the SW-ARQ protocol. However, the ARM-DDS protocol can still meet the requirements of the system while reducing the overall reliability requirements.

(2) When transmitting the high-reliability requirement data, the SW-ARQ protocol waits for an ACK when sending a data packet, which results in a very small increase in transmission reliability at the far-sink nodes. When it is necessary to transmit very reliable data, the SW-ARQ protocol often fails to meet the requirements. The ARM-DDS protocol only starts to listen for ACKs after continuously transmitting multiple sets of data packets. Such a transmission

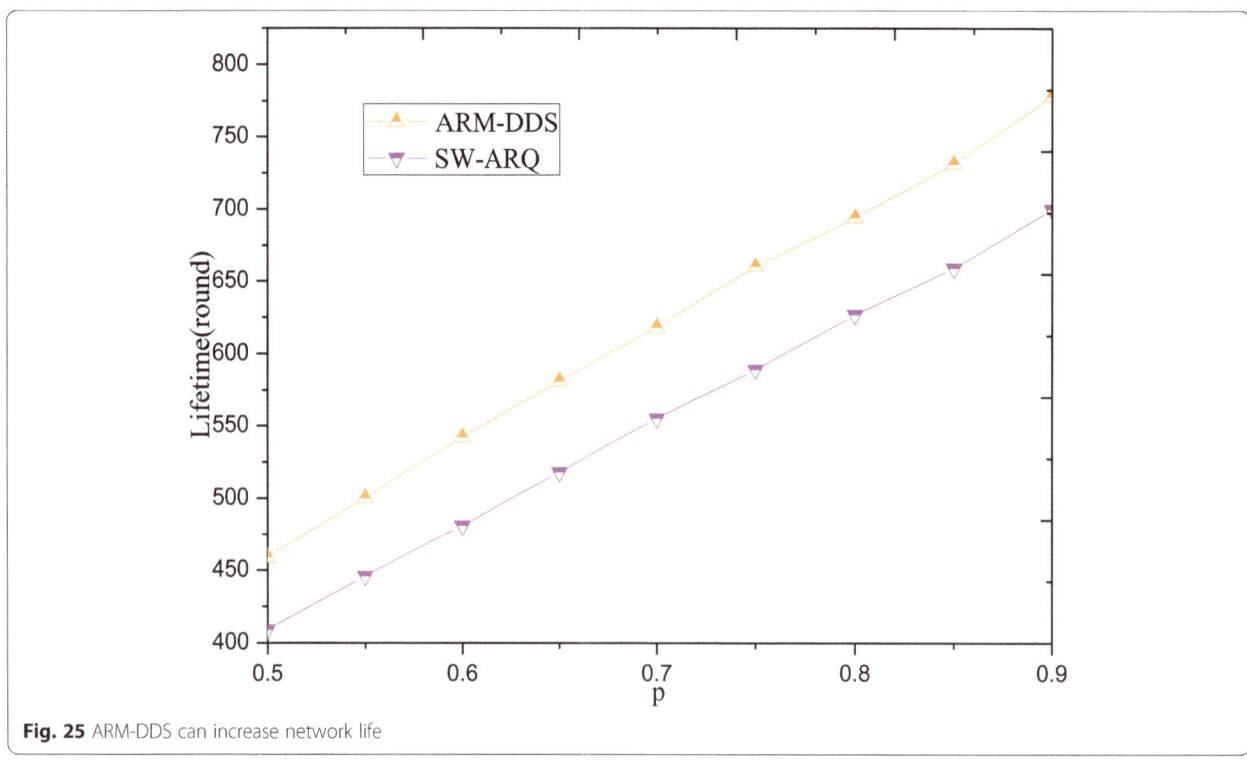

Fig. 25 ARM-DDS can increase network life

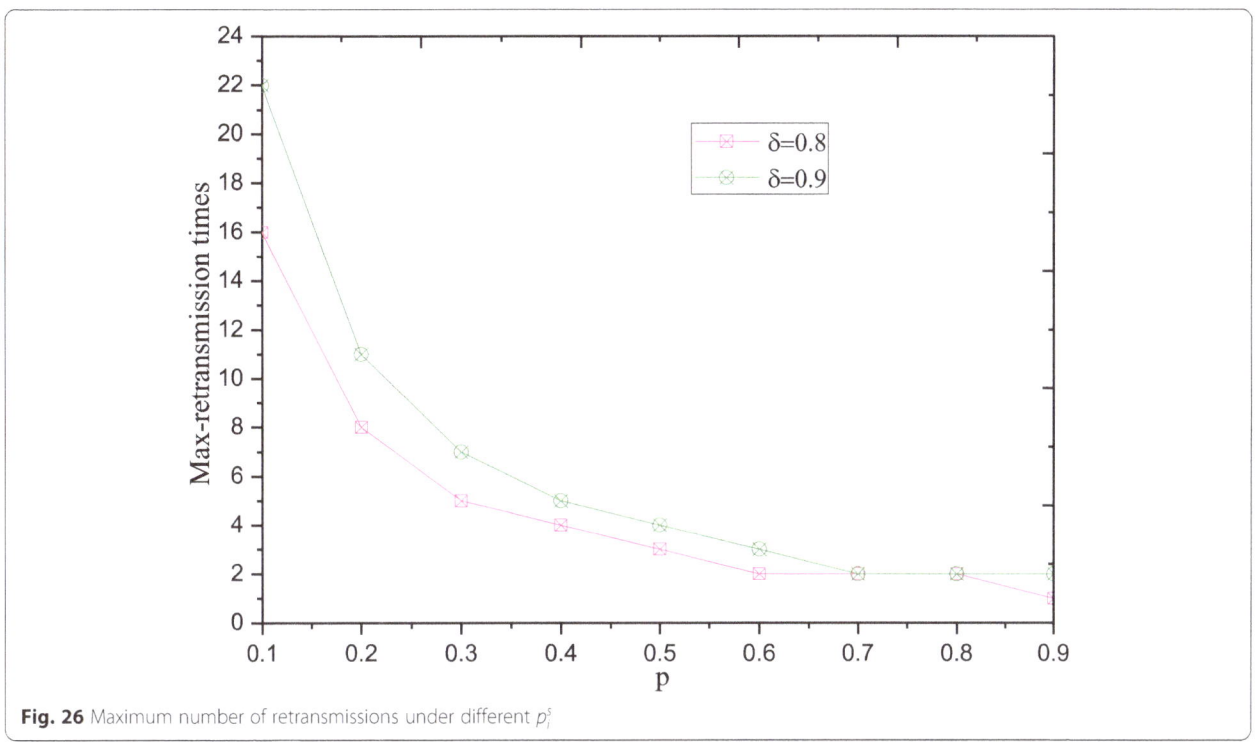

Fig. 26 Maximum number of retransmissions under different p_j^s

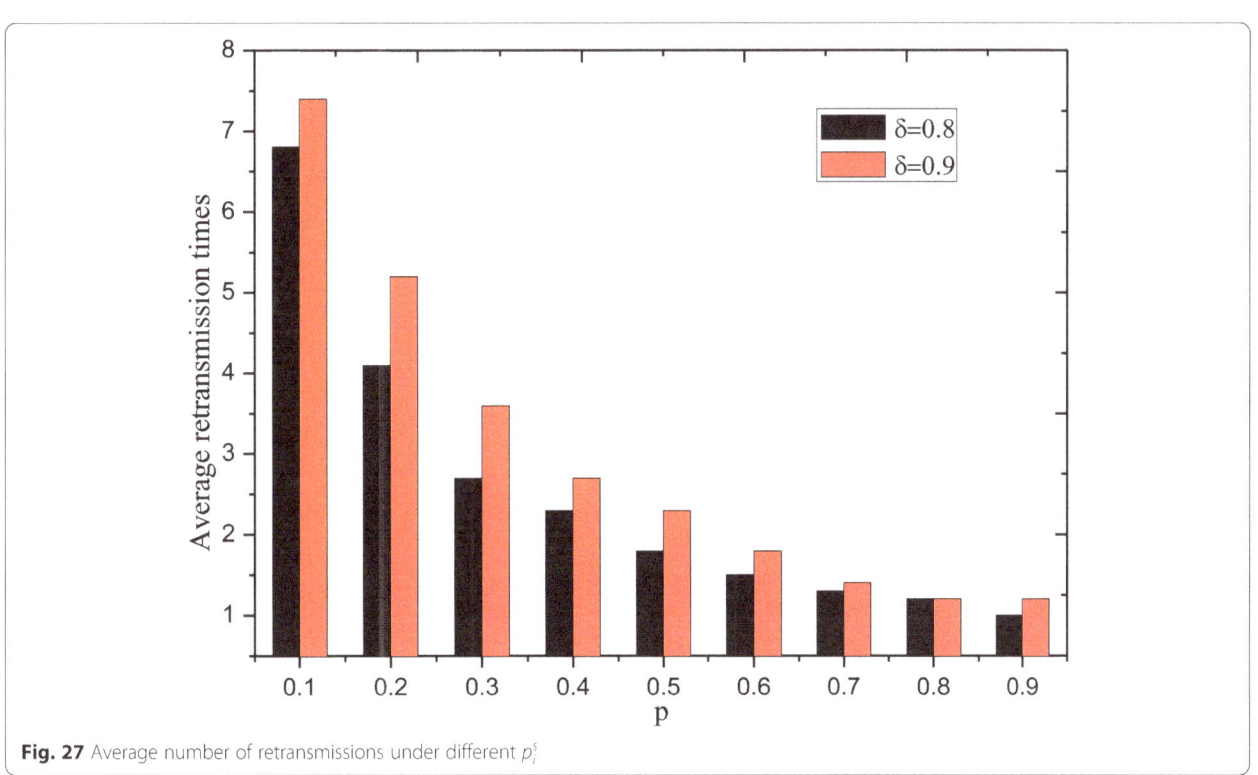

Fig. 27 Average number of retransmissions under different p_j^s

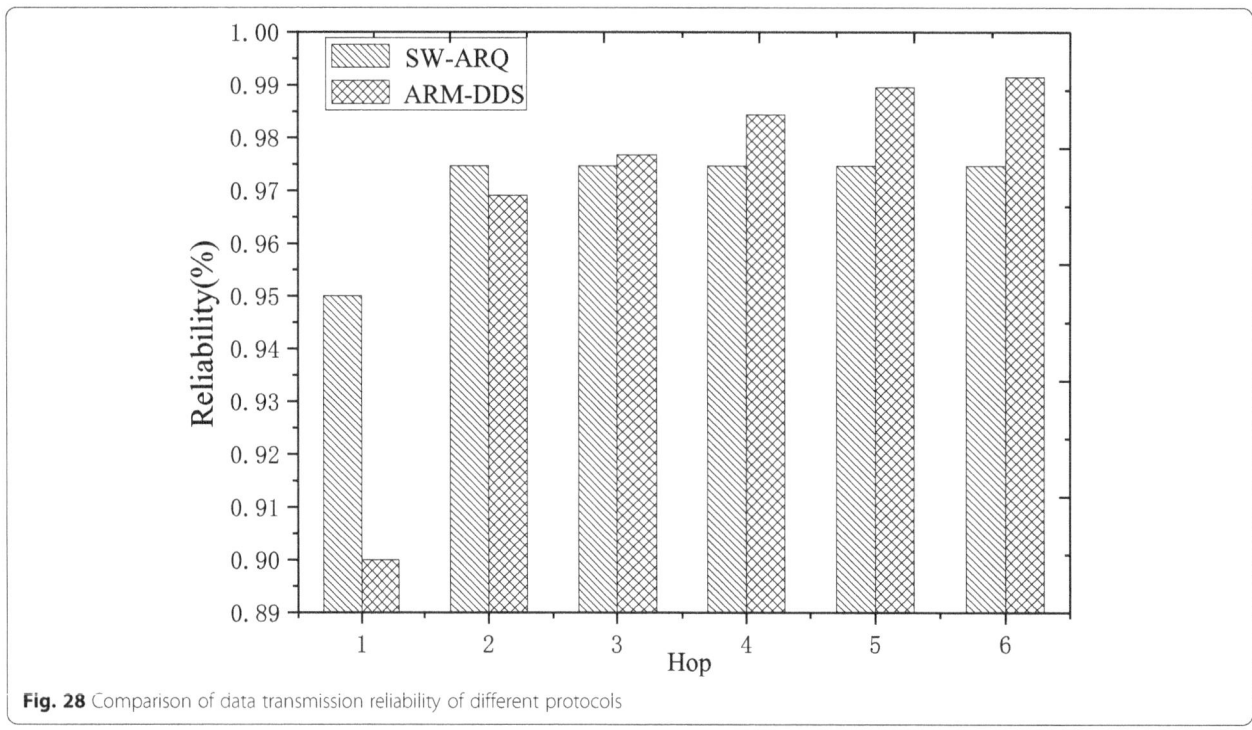

Fig. 28 Comparison of data transmission reliability of different protocols

method greatly increases the upper limit of the number of data packets sent by the system in one transmission period. Therefore, the ARM-DDS protocol can meet the data transmission requirements of higher reliability. Finally, it can be concluded that the ARM-DDS protocol can reduce the number of overall data retransmissions through differentiated services. The ARM-DDS protocol has advantages over the SW-ARQ protocol when transmitting data with high-reliability requirements.

9 Conclusion

The most important task in wireless sensor networks is to sense and capture information about the surrounding environment and pass the information back to the sink. Since different applications have different delay and reliability requirements for data, how to effectively differentiate services is important for improving network performance. This paper proposes an Adaptive Retransmit Mechanism for Delay Differentiated Services (ARM-DDS) scheme mainly for the research that there is no better delay differentiated service. In the ARM-DDS protocol, the application allocation is faced with different requirements and different transmission methods. Thus, it can meet all application requirements while reducing delays and increasing network lifetime. At the same time, the remaining energy in the network is utilized to increase the energy utilization of the network. This paper tests the performance of the ARM-DDS protocol through theoretical analysis and simulation

experiments. The experimental results show that ARM-DDS can effectively differentiate the data according to the traditional protocol. In the delay aspect, as long as the E2E delay requirement of the application is not lower than 87.9% of the SW-ARQ protocol, the protocol can be used for transmission. In terms of energy consumption, the ARM-DDS protocol can improve energy utilization by 28%. Finally, according to the reliability requirements, it increases the network life.

Abbreviations
ACK: Acknowledge; ARM-DDS: Adaptive Retransmit Mechanism for Delay Differentiated Services; E2E: End to end; GBN: Go-Back-N protocol; HBH: Hop by hop; IDDR: Integrity and delay differentiated routing; IoT: Internet of Things; MAC: Media Access Control; QoS: Quality of service; RTO: Time to out; RTT: Round-trip time; SEDR: Security and Energy-efficient Disjoint Route; SNR: Signal-to-noise ratio; SR: Selective repeat protocol; SW-ARQ: Send-and-wait automatic repeat-request; WSNs: Wireless sensor networks

Acknowledgement
The authors thank the person who provided meticulous and valuable suggestions for improving the paper.

Authors' contributions
YC performed the experiment, analyzed the experiment results, and wrote the manuscript. WL, TW, QD, and HS commented on the manuscript. AL conceived of the work and wrote part of the manuscript. All authors read and approved the final manuscript.

Funding
This research was funded by the National Natural Science Foundation of China (No. 61772554, No. 61602398, No. 61672447) and the Hunan Provincial National Natural Science Foundation of China (No. 2017JJ3316, No. 2019JJ50592).

Competing interests

The authors declare that they have no competing interests.

Author details

[1]School of Computer Science and Engineering, Central South University, Changsha 410083, China. [2]School of Informatics, Hunan University of Chinese Medicine, Changsha 410208, China. [3]Department of Computer Science and Technology, Huaqiao University, Xiamen 361021, China. [4]The Key Laboratory of Hunan Province for Internet of Things and Information Security, Xiangtan University, Xiangtan 411105, China. [5]College of Information Engineering, Xiangtan University, Xiangtan 411105, China. [6]Department of Electrical, Computer, Software, and Systems Engineering, Embry-Riddle Aeronautical University, Daytona Beach, FL 32114, USA.

References

1. S. Sarkar, S. Chatterjee, S. Misra, Assessment of the suitability of fog computing in the context of Internet of Things. IEEE Transactions on Cloud Computing 6(1), 46–59 (2018)
2. Y. Deng, H. Hu, N. Xiong, W. Xiong, L. Liu, A general hybrid model for chaos robust synchronization and degradation reduction. Inf. Sci. 305(39), 146–164 (2015)
3. X. Xiang, W. Liu, T. Wang, M. Xie, X. Li, H. Song, A. Liu, G. Zhang. Delay and energy efficient data collection scheme based matrix filling theory for dynamic traffic IoT. EURASIP Journal on Wireless Communications and Networking, 168 (2019). DOI: 10.1186/s13638-019-1490-5.
4. L. Hu, A. Liu, M Xie, T Wang. UAVs joint vehicles as data mules for fast codes dissemination for edge networking in smart city. Peer-to-Peer Networking and Applications, (2019). DOI: https://doi.org/10.1007/s12083-019-00752-0.
5. Y. Liu, A. Liu, T. Wang, X. Liu, N. Xiong, An intelligent incentive mechanism for coverage of data collection in cognitive Internet of Things. Futur. Gener. Comput. Syst. 100, 701–714 (2019)
6. X. Sun, W. Liu, T. Wang, Q. Deng, A. Liu, N. Xiong, S. Zhang, Two-hop neighborhood information joint double broadcast radius for effective code dissemination in WSNs. IEEE Access 7(1), 88547–88569 (2019)
7. Z. Wang, T. Li, N. Xiong, Y Pan. A novel dynamic network data replication scheme based on historical access record and proactive deletion. The Journal of Supercomputing, 62 (1), 227-250 (2012).
8. X. Xiang, W. Liu, A. Liu, N. Xiong, Z. Zeng, Z. Cai. Adaptive duty cycle control based opportunistic routing scheme to reduce delay in cyber physical systems. International Journal of Distributed Sensor Networks. 15 (4), Doi: https://doi.org/10.1177/1550147719841870 2019.
9. T. Shu, W. Liu, T. Wang, M. Zhao, N. Xiong, M. Ma, X. Li, A. Liu, Broadcast based code dissemination scheme for duty cycle based wireless sensor networks. IEEE Access 7(1), 105258–105286 (2019)
10. Y. Liu, M. Ma, X. Liu, N. Xiong, A. Liu, Y. Zhu. Design and analysis of probing route to defense sink-hole attacks for Internet of Things security. IEEE Transactions on Network Science and Engineering, (2018). DOI: https://doi.org/10.1109/TNSE.2018.2881152.
11. T. Wang, L. Qiu, G. Xu, A. Sangaiah, A. Liu. Energy-efficient and trustworthy data collection protocol based on mobile fog computing in Internet of Things. IEEE Transactions on Industrial Informatics, (2019). DOI: https://doi.org/10.1109/TII.2019.2920277.
12. W. Shi, W. Liu, T. Wang, Z. Zeng, G. Zhi. Adding duty cycle only in connected dominating sets for energy efficient and fast data collection, IEEE Access, (2019) Doi: https://doi.org/10.1109/ACCESS.2019.293/626.
13. M. Huang, W. Liu, T. Wang, Q. Deng, A. Liu, M. Xie, M. Ma, G. Zhang. A game-based economic model for price decision making in cyber-physical-social systems. IEEE Access, (2019) Doi: https://doi.org/10.1109/ACCESS.2019.2934515.
14. X. Deng, G. Li, M. Dong, K Ota. Finding overlapping communities based on Markov chain and link clustering. Peer-to-Peer Networking and Applications, 10(2), 411-420 (2017).
15. J. Ren, Y. Zhang, N. Zhang, D. Zhang, X. Shen, Dynamic channel access to improve energy efficiency in cognitive radio sensor networks. IEEE Trans. Wirel. Commun. 15(5), 3143–3156 (2016)
16. J. Zhang, F. Ren, S. Gao, H. Yang, et al., Dynamic routing for data integrity and delay differentiated services in wireless sensor networks. IEEE Trans. Mob. Comput. 14(2), 328–343 (2014)
17. Q. Wang, W. Liu, T. Wang, M. Zhao, X. Li, M. Xie, M. Ma, G. Zhang, A. Liu,

18. Reducing delay and maximizing lifetime for wireless sensor networks with dynamic traffic patterns. IEEE Access 7(1), 70212–70236 (2019)
18. Q. Li, A. Liu, T. Wang, M. Xie, N. Xiong, Pipeline slot based fast rerouting scheme for delay optimization in duty cycle based M2M communications. Peer-to-Peer Networking and Applications. (2019). https://doi.org/10.1007/s12083-019-00753-z
19. Y. Yuan, W. Liu, T. Wang, Q. Deng, A. Liu, H. Song. Compressive sensing based clustering joint annular routing data gathering scheme for wireless sensor networks. IEEE Access, (2019) Doi: https://doi.org/10.1109/ACCESS.2019.2935462.
20. C. Lin, YX He, N Xiong. An energy-efficient dynamic power management in wireless sensor networks. Fifth International Symposium on Parallel and Distributed Computing , 37, (2006).
21. F. Wang, W. Liu, T. Wang, M. Zhao, M. Xie, H. Song, X. Li, A. Liu, To reduce delay, energy consumption and collision through optimization duty-cycle and size of forwarding node set in WSNs. IEEE Access 7(1), 55983–56015 (2019)
22. S. Sarang, M. Drieberg, A. Awang, R. Ahmad, A QoS MAC protocol for prioritized data in energy harvesting wireless sensor networks. Comput. Netw. 144, 141–153 (2018)
23. H. Cheng, N. Xiong, LT Yang, YS Jeong. Distributed scheduling algorithms for channel access in TDMA wireless mesh networks. The Journal of Supercomputing, 63(2), 407-430 (2013).
24. L. Yin, J. Gui, Z. Zeng. Improving energy efficiency of multimedia content dissemination by adaptive clustering and D2D multicast. Mobile Information Systems, 2019, 5298508 (2019). DOI: https://doi.org/10.1155/2019/5298508.
25. Y. Bi, G. Han, C. Lin, X. Wang, Q. Zhang, Effective packet loss elimination in IP mobility support for vehicular networks. IEEE Netw. (2019). https://doi.org/10.1109/MNET.2019.1900093
26. N. Xiong, AV Vasilakos, J Wu, YR Yang, A Rindos, Y Zhou, WZ Song. A self-tuning failure detection scheme for cloud computing service. 2012 IEEE 26th International Parallel and Distributed Processing Symposium. 40, (2012).
27. X. Liu, M. Zhao, A. Liu, K. Wong, Adjusting forwarder nodes and duty cycle using packet aggregation routing for Body Sensor Networks. Information Fusion 53, 183–195 (2020)
28. Y. Liu, A. Liu, Y. Li, Z. Li, Y. Choi, H. Sekiya, J. Li, APMD: A fast data transmission protocol with reliability guarantee for pervasive sensing data communication. Pervasive and Mobile Computing 41, 413–435 (2017)
29. Y. Yang, N. Xiong, NY Chong, X Défago. A decentralized and adaptive flocking algorithm for autonomous mobile robots. The 3rd International Conference on Grid and Pervasive Computing, (2008).
30. X. Deng, T. He, L. He, J. Gui, Q. Peng, Performance analysis for IEEE 802.11 s wireless mesh network in smart grid. Wirel. Pers. Commun. 96(1), 1537–1555 (2017)
31. Y. Bi, G. Han, C. Lin, M. Guizani, X. Wang, Mobility Management for intro/inter domain handover in software defined networks. IEEE Journal on Selected Areas in Communications. 37(8), 1739–1754 (2019)
32. X. Peng, J. Ren, L. She, D. Zhang, J. Li, Y. Zhang, BOAT: A block-streaming app execution scheme for lightweight IoT devices. IEEE Internet-of-Things Journal 5(3), 1816–1829 (2018)
33. H. Cheng, D. Feng, X. Shi, et al., Data quality analysis and cleaning strategy for wireless sensor networks. EURASIP J. Wirel. Commun. Netw. 2018, 61 (2018). https://doi.org/10.1186/s13638-018-1069-6
34. Y. Liu, A. Liu, N. Xiong, T. Wang, W. Gui. Content propagation for content-centric networking from location-based social networks. IEEE Transactions on Systems Man Cybernetics-Systems, (2019). DOI: https://doi.org/10.1109/TSMC.2019.2898982.
35. G. Zhang, T. Wang, G. Wang, A. Liu, W. Jia. Detection of hidden data attacks combined fog computing and trust evaluation method in sensor-cloud system. Concurrency and Computation: Practice and Experience, (2018). DOI: https://doi.org/10.1002/cpe.5109.
36. H. Cheng, Z. Su, N. Xiong, Y. Xiao, Energy-efficient node scheduling algorithms for wireless sensor networks using Markov Random Field model. Inf. Sci. 329, 461–477 (2016)
37. C. Zhang, R. Chen, L. Zhu, A. Liu, Y. Lin, F. Huang. Hierarchical information quadtree: efficient spatial temporal image search for multimedia stream. Multimedia Tools and Applications, (2018). DOI: https://doi.org/10.1007/s11042-018-6284-y.
38. W. Liu, P. Zhuang, H. Liang, J. Peng, Z. Huang, Distributed economic dispatch in microgrids based on cooperative reinforcement learning. IEEE Transactions on Neural Networks and Learning 29(6), 2192–2203 (2018)

39. Y. Liu, A. Liu, Z. Chen, Analysis and Improvement of send-and-wait automatic repeat-reQuest protocols for wireless sensor networks. Wirel. Pers. Commun. **81**(3), 923–959 (2015)

40. W. Turin, Throughput analysis of the Go-Back-N protocol in fading radio channels. IEEE Journal on Selected Areas in Communications **17**(5), 881–887 (1999)

41. E. Weldon, An improved selective-repeat ARQ strategy. IEEE Trans. Commun. **30**(3), 480–486 (1982)

42. A. Laouid, A. Dahmani, A. Bounceur, R. Euler, F. Lalem, A. Tari, A distributed multi-path routing algorithm to balance energy consumption in wireless sensor networks. Ad Hoc Netw. **64**, 53–64 (2017)

43. Y. Sang, H. Shen, Y. Tan, N. Xiong. Efficient protocols for privacy preserving matching against distributed datasets. International Conference on Information and Communications Security, 210-227, (2006).

44. A. Liu, Z. Zheng, C. Zhang, Z. Chen, et al., Secure and energy-efficient disjoint multipath routing for WSNs. IEEE Trans. Veh. Technol. **61**(7), 3255–3265 (2012)

45. J. Zhang, A. Liu, P. Hu, J Long. A fuzzy-rule-based packet reproduction routing for sensor networks. International Journal of Distributed Sensor Networks, 14(4), (2018). DOI: 1550147718774016.

46. C. Joo, N.B. Shroff, On the delay performance of in-network aggregation in lossy wireless sensor networks. IEEE/ACM Transactions on Networking (TON) **22**(2), 662–673 (2014)

47. M. Huang, A. Liu, T. Wang, C. Huang. Green data gathering under delay differentiated services constraint for Internet of Things. Wireless Communications and Mobile Computing, 2018, 9715428 (2018). DOI: https://doi.org/10.1155/2018/9715428.

48. Q. Chen, S.S. Kanhere, M. Hassan, Analysis of per-node traffic load in multi-hop wireless sensor networks. IEEE transaction on wireless communications **8**(2), 958 967 (2009)

49. A.F. Liu, P.H. Zhang, Z.G. Chen, Theoretical analysis of the lifetime and energy hole in cluster based Wireless. Journal of Parallel and Distributed Computing **71**(10), 1327–1355 (2011)

50. Z. Rosberg, R.P. Liu, T.L. Dinh, Y.F. Dong, S. Jha, Statistical reliability for energy efficient data transport in wireless sensor networks. Wirel. Netw **16**, 1913–1927 (2010)

51. H. Zhou, J. Wu, H. Zhao, S. Tang, C. Chen, J. Chen, Incentive-driven and freshness-aware content dissemination in selfish opportunistic mobile networks. IEEE Transactions on Parallel and Distributed Systems **26**(9), 2493–2505 (2015)

52. T. Wang, H. Luo, W. Jia, A. Liu, M Xie. MTES: An intelligent trust evaluation scheme in sensor-cloud enabled industrial Internet of Things, IEEE Transactions on Industrial Informatics, (2019) DoI: https://doi.org/10.1109/TII.2019.2930286.

53. Y. Liu, A. Liu, X. Liu, M. Ma, A trust-based active detection for cyber-physical security in industrial environments. IEEE Transactions on Industrial Informatics (2019). https://doi.org/10.1109/TII.2019.2931394

54. H. Teng, K. Zhang, M. Dong, K. Ota, A. Liu, M. Zhao, T Wang. Adaptive transmission range based topology control scheme for fast and reliable data collection. Wireless Communications and Mobile Computing, 2018, 4172049 (2018). DOI: https://doi.org/10.1155/2018/4172049.

13

A bi-population QUasi-Affine TRansformation Evolution algorithm for global optimization and its application to dynamic deployment in wireless sensor networks

Nengxian Liu[1] (iD), Jeng-Shyang Pan[1,2,3]* (iD) and Trong-The Nguyen[2,4] (iD)

Abstract

In this paper, we propose a new Bi-Population QUasi-Affine TRansformation Evolution (BP-QUATRE) algorithm for global optimization. The proposed BP-QUATRE algorithm divides the population into two subpopulations with sort strategy, and each subpopulation adopts a different mutation strategy to keep the balance between the fast convergence and population diversity. What is more, the proposed BP-QUATRE algorithm dynamically adjusts scale factor with a linear decrease strategy to make a good balance between exploration and exploitation capability. We compare the proposed algorithm with two QUATRE variants, PSO-IW, and DE algorithms on the CEC2013 test suite. The experimental results demonstrate that the proposed BP-QUATRE algorithm outperforms the competing algorithms. We also apply the proposed algorithm to dynamic deployment in wireless sensor networks. The simulation results show that the proposed BP-QUATRE algorithm has better coverage rate than the other competing algorithms.

Keywords: Differential evolution, Particle swarm optimization, Bi-population, QUATRE algorithm, Global optimization, Dynamic deployment, Wireless sensor networks

1 Introduction

In the last few decades, there have been many optimization demands arising not only from the scientific community but also from various real-world applications. Generally, the approach to solving these optimization problems often begins with designing the objective function which can model the objectives of optimization problems [1]. Many optimization approaches have been proposed to meet these demands aiming at finding optimal solutions. Some Swarm-based intelligence optimization algorithms, such as particle swarm optimization (PSO) [2], ant colony optimization (ACO) [3], differential evolution (DE) [1],

artificial bee colony (ABC) optimization [4], and QUasi-Affine TRansformation Evolution (QUATRE) algorithm [5], and so on, have been developed to tackle these complex optimization problems.

QUATRE algorithm was first presented by Meng et al. in [5] that discussed the relationship between QUATRE algorithm and the other two swarm-based intelligence algorithms PSO and DE. In 1995, Kennedy and Eberhart firstly introduced the PSO algorithm [2]. As PSO is simple, powerful, and straightforward to implement, many researchers have studied this technique and developed various improved variants [6–8]. DE was introduced by Storn and Price [1] in 1995, which was arguably one of the most powerful optimization algorithms. As well, many DE variants have been proposed to enhance the performance of DE algorithm [9–11], and QUATRE algorithm is one of them proposed to conquer representational or positional bias of DE

* Correspondence: jspan@cc.kuas.edu.tw; lylnx@fzu.edu.cn
[1] College of Mathematics and Computer Science, Fuzhou University, Fuzhou 350108, China
[2] Fujian Provincial Key Laboratory of Big Data Mining and Applications, Fujian University of Technology, Fuzhou 350118, China
Full list of author information is available at the end of the article

algorithm [12]. QUATRE's conventional notation is "QUATRE/x/y" which denotes types of QUATRE variants. It is worth noting that the notation of QUATRE is more general than the DE's notation "DE/x/y/z" [13].

The canonical QUATRE algorithm and its variants can be found in literatures [12–17]. The QUATRE has the advantages of simplicity, few control parameters to set, and convenient to be used, but it has some weaknesses as the DE algorithm such as it will be premature convergence, and will search stagnation and may be easily trapped into local optima. Population diversity plays important role in alleviating these weaknesses. Therefore, it is important to keep the balance between diversity and convergence. In [16], S-QUATRE has been proposed which uses sort strategy to improve the performance of QUATRE algorithm. And S-QUATRE divides the population into the better and the worse groups and evolves the individuals in the worse group. The other algorithms which partition population into two groups or several subpopulations to maintain population diversity and to enhance the performance of algorithms, such as CMA-ES, PSO, DE and CSO, can be found in previous literature [18–22]. On the other hand, both mutation strategies and control parameter scale factor F have significant effects on the performance of QUATRE variants. Different mutation strategies in QUATRE algorithm have different performance over various optimization problems [13] because different mutation strategy has different search ability and convergence rate. Usually, similar to the DE algorithm, adopting larger F value in QUATRE algorithm means the algorithm is more focused on exploration, while a smaller F value means more exploitation [23]. Therefore, in this paper, in order to improve the performance of QUATRE algorithm, we propose a novel Bi-Population QUATRE algorithm with a sort strategy and a linear decrease scale factor F (BP-QUATRE), and each subpopulation has a different mutation strategy.

The remainder of the paper is arranged as follows. In Section 2, we briefly review the QUATRE algorithm. Our proposed Bi-Population QUasi-Affine TRansformation Evolution (BP-QUATRE) algorithm is given in Section 3. In Section 4, we apply the proposed algorithm to dynamic deployment in wireless sensor networks. What is more, the experimental analysis of our proposed algorithm under CEC2013 test suite and simulation results in wireless sensor networks are presented in Section 4. Finally, Section 6 gives the conclusion.

2 Canonical QUATRE algorithm

Meng et al. have proposed the QUATRE algorithm for solving optimization problems [5]. QUATRE is an abbreviation of QUasi-Affine TRansformation Evolution, and the reason the authors naming the algorithm QUATRE is that individuals in QUATRE algorithm evolve by using an affine transformation-like equation. The detailed evolution equation of QUATRE is as follows.

$$X \leftarrow M \otimes X + \overline{M} \otimes B \tag{1}$$

where M is an evolution matrix and $\overline{M}$ is a binary inverted matrix of M. The elements of them are either 0 or 1. The binary invert operation means to invert the values of the matrix. The reverse values of zero elements are ones, while the reverse values of one elements are zeros. Equation 2 shows an example of binary inverse operation.

$$M = \begin{bmatrix} 1 & & & \\ 1 & 1 & & \\ & & \cdots & \\ 1 & 1 & \cdots & 1 \end{bmatrix}, \overline{M} = \begin{bmatrix} 0 & 1 & 1 & 1 \\ 0 & 0 & 1 & 1 \\ & & \cdots & 1 \\ 0 & 0 & \cdots & 0 \end{bmatrix} \tag{2}$$

M is transformed from an initial matrix M_{ini} which is initialized by a lower triangular matrix with the elements equaling to ones. Transforming from M_{ini} to M contains two consecutive steps. In the first step, every element in each row vector of M_{ini} is randomly permuted. In the second step, the sequence of the row vectors is randomly permuted with all elements of each row vector unchanged. An example of the transformation with ps = D is given in Eq. 3. Usually, the population size ps is larger than the dimension, while the matrix M_{ini} needs to be extended according to population size ps. Equation 4 shows an example of ps = 2D + 2. Generally, when ps % D = k, the first k rows of the D × D lower triangular matrix are included in M_{ini} and adaptively change M according to M_{ini} [12].

$$
M_{ini} = \begin{bmatrix} 1 & & & \\ 1 & 1 & & \\ & & \cdots & \\ 1 & 1 & \cdots & 1 \end{bmatrix} \sim \begin{bmatrix} & & & 1 & \\ & & \cdots & & \\ 1 & 1 & \cdots & 1 \\ & 1 & 1 & \end{bmatrix}
$$
$$= M \tag{3}$$

$$
M_{ini} = \begin{bmatrix} 1 & & & \\ 1 & 1 & & \\ & & \cdots & \\ 1 & 1 & \cdots & 1 \\ 1 & & & \\ 1 & 1 & & \\ & & \cdots & \\ 1 & 1 & \cdots & 1 \\ & & \vdots & \\ 1 & & & \\ 1 & 1 & & \end{bmatrix} \sim \begin{bmatrix} 1 & & \cdots & 1 \\ & & \cdots & 1 \\ & & \cdots & \\ 1 & 1 & & \\ & & 1 & \\ 1 & 1 & \cdots & 1 \\ & & 1 & \\ 1 & & & 1 \\ & & \vdots & \\ 1 & 1 & \cdots & 1 \\ & 1 & & \end{bmatrix}
$$
$$= M \tag{4}$$

$X = [X_{1,\ G}, X_{2,\ G}, \ldots, X_{i,\ G}, \ldots, X_{ps,\ G}]^{T}$ is the population matrix with ps individuals. $X_{i,\ G} = [x_{i1}, x_{i2}, \ldots, x_{iD}]$ is the ith row vector of the matrix X, which denotes the location of ith individual of the Gth generation, and each individual $X_{i,\ G}$ is a candidate solution for an optimization problem, and D denotes the dimension number of objective function, where $i \in \{1, 2, \ldots, ps\}$. The operation "$\otimes$" stands for component-wise multiplication of the elements in each matrix, which is the same as ".*" operation in Matlab software. $\mathbf{B} = [B_{1,\ G}, B_{2,\ G}, \ldots, B_{i,\ G}, \ldots, B_{ps,\ G}]^T$ is the donor matrix, and it has several different calculation schemes (mutation strategies) which are listed in Eqs. (5)–(8) [7].

$$\text{QUATRE/best/1}: B = X_{gbest,G} + F \cdot (X_{r1,G}\text{-}X_{r2,G}) \tag{5}$$

$$\text{QUATRE/ rand/1}: B = X_{r0,G} + F \cdot (X_{r1,G}\text{-}X_{r2,G}) \tag{6}$$

$$\text{QUATRE/target/1}: B = X + F \cdot (X_{r1,G}\text{-}X_{r2,G}) \tag{7}$$

$$\begin{aligned}\text{QUATRE/target-to-best/1B}\\ = X + F \cdot (X_{gbest,G}\text{-}X) + F \cdot (X_{r1,G}\text{-}X_{r2,G})\end{aligned} \tag{8}$$

$X_{gbest,\ G} = [X_{gbest,\ G}, X_{gbest,\ G}, \ldots, X_{gbest,\ G}]^T X_{gbest,\ G} = [X_{gbest,\ G}, X_{gbest,\ G}, \ldots, X_{gbest,\ G}]^T$ denotes a row vector-duplicated matrix with each row vector equaling to the global best individual $X_{gbest,\ G}$ of the Gth generation. F can be considered as amplification factor, whose value region is (0, 1] for most optimization problems. $X_{r1,\ G}, X_{r2,\ G}$ and $X_{r3,\ G}$ are a set of random matrices which are generated by randomly permutating the sequence of row vectors in the matrix $\mathbf{X}$ of the Gth generation.

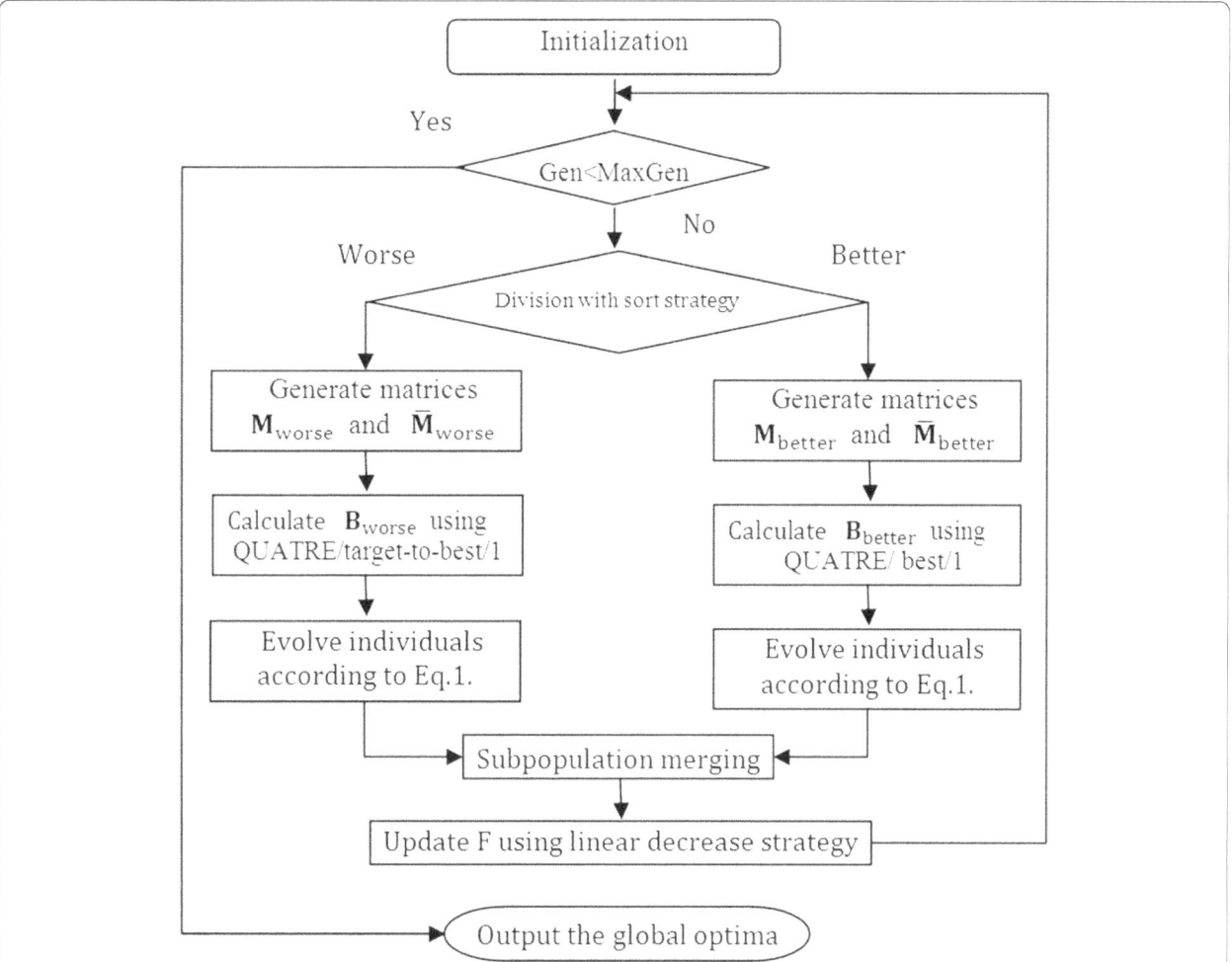

Fig. 1 The main framework of BP-QUATRE. The flowchart of BPQUATRE algorithm consists of population initialization, population division, subpopulation evolution, subpopulation merging and approach of updating parameter scale factor F. Gen is the current generation number and MaxGen is the maximum generation number. Better means the subpopulation with better fitness values (i.e., with the lower fitness values for a minimization problem). Worse means the subpopulation with worse fitness values. The better subpopulation evolves by adopting mutation strategy "QUATRE/best/1" to make good exploitation around the individuals with better fitness values and to have good convergence rate. The better subpopulation evolves by adopting mutation strategy "QUATRE/best/1" to make good exploitation around the individuals with better fitness values and to have good convergence rate. The worse subpopulation evolves by using mutation strategy "QUATRE/target-to-best/1" to make a good exploration around the individuals with worse fitness values and to preserve population diversity

3 Bi-population QUATRE algorithm with sort strategy and linear decrease scale factor F (BP-QUATRE)

In this section, we describe the main idea of our proposed algorithm BP-QUATRE. As mentioned above, because easily trapping into local optima and premature convergence is the weakness of QUATRE algorithm. In order to alleviate the above weaknesses, BP-QUATRE consisting of population initialization, population division, subpopulation evolution, subpopulation merging, and an approach to update the parameter scale factor, F, is proposed in this paper. The main framework of BP-QUATRE is shown in Fig. 1.

3.1 Bi-population division and mutation strategies

Usually, as conventional evolutionary algorithms, the QUATRE algorithm suffers from the problem of premature convergence, i.e., the population is too early to lose diversity and fall into local optima. Multi-population approach helps to increase population diversity and alleviate premature convergence [20]. Inspired by this, we use a Bi-population approach to enhance the diversity of the population. In our proposed algorithm, the individuals in the population are firstly sorted after initialization according to the fitness values, and then the entire population is equally divided into two subpopulations based on the sorted sequence, say pop_{better} and pop_{worse}, respectively. As we know, different mutation strategy of QUATRE algorithm has different search abilities. Mutation strategy "QUATRE/best/1" uses the best individual to guide the population and has a fast convergence rate and good local search ability around the best individual. Therefore, the subpopulation pop_{better} evolves by adopting mutation strategy "QUATRE/best/1" to make good exploitation around the individuals with better fitness values and to have good convergence rate. On the other hand, mutation strategy "QUATRE/target-to-best/1" is a strong exploration-biased strategy, because this strategy generates donor individual using the best individual and two random selected individuals. Thus, the subpopulation pop_{worse} evolves by using mutation strategy "QUATRE/target-to-best/1" to make a good exploration around the individuals with worse fitness values and to preserve population diversity. Therefore, this bi-population division and different subpopulation having different mutation strategy approach can make a trade-off between the population diversity and convergence rate.

3.2 Linear decrease scale factor

Scale factor plays an essential role in balancing exploration and exploitation ability of QUATRE algorithm during the search phases. In [5], the authors illustrate the effect of different scaling factor values on the performance of the QUATRE algorithm. And there is no fixed parameter setting which can achieve the best performance for all kinds of problems. It is significant to find a good method to dynamically adjust the

scaling factor value. According to [6, 24] for most population-based optimization algorithm, it is a good idea for the algorithm to have more exploration ability in the early stages of the search in order to sample diverse zones of the search space. In the later stages of the search, the algorithm should possess more exploitation ability to search the relatively small space where the potential global optimum lies in. Namely, at the beginning of the search, the scale factor of the algorithm should be larger. While with the increment of generations, the scale factor of algorithm should be decreased to increase the exploitation ability. Hence, we use the linear decrease strategy proposed in [6] to dynamically adjust the value of scale factor which can be described as fellow.

$$F = F_{max} - (F_{max} - F_{min}) \cdot Gen/\text{MaxGen} \qquad (9)$$

where F_{max} and F_{min} are the predetermined maximum and minimum values of scale factor F. Gen is the current generation number, and MaxGen is the maximum generation number.

The pseudo code of BP-QUATRE algorithm is shown in Algorithm 1.

Algorithm 1 BP-QUATRE Algorithm

Initialization:
Initialize the searching space V, dimension D, Gen=1, the benchmark function f(X), randomly Initialize the population X with ps individuals, and evaluate the fitness values of all individuals.

Iteration:
1: **while** Gen < MaxGen || !stopCriterion do
2:     Sort individuals based on fitness values and then divide them into pop_{better} and pop_{worse} according to sort order
3:     Generate matrices $\mathbf{M}_{better}$ and $\mathbf{M}_{worse}$ using Eq. 4. Generate matrices $\overline{\mathbf{M}}_{better}$ and $\overline{\mathbf{M}}_{worse}$ using Eq. 4.
4:     Calculate donor matrix $\mathbf{B}_{better}$ using Eq. 5. Calculate donor matrix $\mathbf{B}_{worse}$ using Eq. 8.
5:     Evolve individuals in each sub-population using Eq. 1.
6:     Evaluate fitness values of all individuals.
7:     for i = 1 to ps do
8:         if $f(X_i) \leq f(X_{pbest,i})$ then
9:             $X_{pbest,i} = X_i$
10:        end if
11:     end for
12:     $\mathbf{X} = \mathbf{X}_{pbest}$, $X_{gbest} = opt\{\mathbf{X}_{pbest}\}$.
13:     Update scale factor F according to Eq.9.
14:     $Gen = Gen + 1$
15: **end while**

Output:
The global optimum X_{gbest}, global best fitness value $f(X_{pbest})$.

---

4 Apply the proposed BP-QUATRE algorithm to dynamic deployment in wireless sensor networks

In this section, we apply the proposed BP-QUATRE algorithm to dynamic deployment in wireless sensor networks (WSN). The WSN becomes a popular research field [25–29] due to its great value in real-world applications such as environment monitoring, healthcare applications, and forest fire detection. The WSN is composed of a large number of battery-powered, multifunctional, and resources-constrained sensor nodes. The performance of the whole WSN depends on the position of the sensors which affect the coverage, connectivity, energy efficiency, and network lifetime. In some applications, the locations of the sensors are predetermined by static ways. However, in some cases such as battlefield, underwater, and disaster-affected regions where is difficult to predetermine the locations by static ways, only dynamic deployment strategies can be adopted. In dynamic deployment, the sensors are first randomly placed within the area of interest and then sensors can relocate their locations by using information from other sensor nodes. But random initial deployment may not ensure effective coverage. In order to enhance the coverage rate of the whole WSN, a number of algorithms have been developed for dynamic node deployment, including virtual force [30], Voronoi diagram [31], and swarm intelligence algorithms [33–36]. Many swarm intelligence algorithms are employed in sensor deployment, such as the particle swarm optimization (PSO) [32, 33], artificial bee colony algorithm (ABC) [34], differential evolution (DE) [35], and so forth. In this study, the proposed QUATRE algorithm is first applied to dynamic deployment in WSN with the aim of improving the coverage rate. The proposed algorithm is compared with PSO-based and DE-based dynamic deployment algorithm.

4.1 Sensor detection model

Without losing generality, this paper assumes that each sensor node can move and has the same sensor radius and communication range. There are two sensor detection models in wireless sensor networks: binary detection model and probability detection model [36]. In the binary detection model, the detected possibility of the event concerned is 1 within the sensing radius. Otherwise, the probability is 0. This model can be expressed by the Eq. 9 [37].

$$C_{xy}(P, s_i) = \begin{cases} 1, & d(P, s_i) < r \\ 0, & \text{otherwise} \end{cases} \quad (10)$$

where r represents sensor radius and $d(P, s_i)$ denotes the Euclidean distance between point P and the sensor node s_i. Although the binary sensor model is relatively simple, the uncertainties in measurement are not taken into account. Generally, sensor detections are imprecise in practical, so the detection probability $C_{xy}(P, s_i)$ needs to be presented in probabilistic terms. Therefore, we use the probabilistic detection model in the paper, which can be expressed by the Eq. 10 [38].

$$C_{x,y}(P, s_i) = \begin{cases} 1, & d(P, s_i) \le r - r_e \\ e^{\left(-\alpha_1 \lambda_1^{\beta_1} / \lambda_2^{\beta_2} + \alpha_2\right)}, & r - r_e < d(P, s_i) \le r + r_e \\ 0, & d(P, s_i) > r + r_e \end{cases} \quad (11)$$

where $r_e (0 < r_e < r)$ is the measure of uncertainty. α_1, α_2, β_1, and β_2 are detection parameters related to the characteristics of sensors. $\lambda_1 = r_e - r + d(P, s_i)$ and $\lambda_2 = r_e + r - d(P, s_i)$ are the input parameters. In general, the detection probability covered by sensor node may be less than 1. This means that it is necessary to overlap the sensor detection area to compensate for the potential low detection probability [39]. And we assume that sensors observe independently. Considering a point P (x, y) in the overlap region of a set of sensors S, the joint detection probability of point P can be calculated by the Eq. 11.

$$C_{x,y}(S) = 1 - \prod_{s_i \in S} \left(1 - C_{x,y}(P, s_i)\right) \quad (12)$$

Let C_{th} is the threshold of predefined effective detection probability. This implies that the point P (x, y) can be effectively covered if

$$\min_{x,y} \left\{ C_{x,y}(S) \right\} \ge C_{th} \quad (13)$$

4.2 Dynamic deployment based on BP-QUATRE algorithm

The purpose of sensor deployment algorithm is to determine an optimal sensor distribution in the region of interest, which is similar to the swarm intelligence algorithm for solving complex optimization problems. Therefore, it is possible to apply BP-QUATRE algorithm to the dynamic deployment problem of WSN.

In the BP-QUATRE algorithm, the individual is composed of the coordinate representing its position in the solution space. In dynamic deployment, the individual represents the deployment of the sensors in the sensed area. Supposing the number of sensors is N, the dimension of the individual is set to $2N$ and the individual encoding is expressed as $X_i = [x_{i1}^1, x_{i1}^2, x_{i2}^1, x_{i2}^2, ..., x_{iN}^1, x_{iN}^2]$. The elements represent the x and y coordinates of sensors from 1 to N in turn.

The fitness function of the BP-QUATRE corresponds to the coverage rate of the network. Coverage rate is an

Table 1 Parameters settings

Algorithm	Parameters settings
BP-QUATRE	Fmax = 0.9, Fmin = 0.4
QUATRE variants	F = 0.7
PSO-IW	Wmax = 0.9, Wmin = 0.4, c1 = 2, c2 = 2
DE	F = 0.7, Cr = 0.1

Table 2 Performance for BP-QUATRE, QUATRE/target-to-best/1, and QUATRE/best/1

30D	QUATRE/target-to-best/1			QUATRE/best/1			BP-QUATRE		
	Best	Mean	Std	Best	Mean	Std	Best	Mean	Std
1	0.0000E+00	4.5475E-15	3.2155E-14	0.0000E+00	2.2737E-14	6.8905E-14	0.0000E+00	5.9117E-14	1.0075E-13
2	3.4918E+04	2.0307E+05	9.1401E+04	6.5779E+04	3.2526E+05	1.7428E+05	8.0342E+04	3.2499E+05	1.8932E+05
3	3.3344E-08	1.2152E+04	4.2407E+04	6.4313E-02	1.1505E+06	3.0040E+06	4.0302E-05	6.8721E+05	2.9931E+06
4	1.3712E+00	9.6559E+00	7.2335E+00	4.1346E+00	2.1215E+01	1.5656E+01	6.7090E+00	3.8516E+01	2.5066E+01
5	0.0000E+00	1.0914E-13	2.2504E-14	0.0000E+00	1.0914E-13	2.2504E-14	0.0000E+00	1.1369E-13	2.2968E-14
6	2.8398E-09	1.8153E+00	6.2792E+00	2.9877E-04	7.3092E+00	1.0870E+01	4.5328E-03	1.0260E+01	8.0606E+00
7	1.9057E-01	3.8445E+00	4.0489E+00	1.1364E+00	2.1349E+01	1.7591E+01	7.2738E-02	4.9636E+00	3.9591E+00
8	2.0749E+01	2.0933E+01	5.1788E-02	2.0849E+01	2.1004E+01	5.8828E-02	2.0827E+01	2.0953E+01	4.7373E-02
9	1.0130E+01	2.6120E+01	6.0410E+00	5.8076E+00	1.6079E+01	5.5873E+00	9.7462E+00	2.2635E+01	5.7088E+00
10	5.6843E-14	2.7047E-02	1.5235E-02	0.0000E+00	2.3154E-02	1.5496E-02	7.3960E-03	2.0791E-02	1.1579E-02
11	2.3082E+01	2.7925E+01	2.5361E+00	1.3929E+01	2.6411E+01	8.3326E+00	1.7053E-13	4.4972E+00	1.8006E+00
12	9.0630E+01	1.1724E+02	1.3886E+01	3.9359E+01	7.5733E+01	1.8919E+01	1.9899E+01	5.3234E+01	1.5281E+01
13	1.0132E+02	1.3263E+02	1.5255E+01	5.2197E+01	1.1252E+02	3.2964E+01	5.8213E+01	1.0634E+02	2.4064E+01
14	9.9796E+02	1.3914E+03	1.9533E+02	1.0081E+02	8.1082E+02	2.6067E+02	1.0878E+01	1.9290E+02	1.2680E+02
15	5.3901E+03	6.3005E+03	3.4954E+02	3.3512E+03	5.1268E+03	7.7891E+02	2.7376E+03	4.0668E+03	5.3740E+02
16	1.8510E+00	2.3565E+00	2.4224E-01	1.3247E+00	2.4246E+00	4.5226E-01	9.4808E-01	1.8071E+00	4.0821E-01
17	5.5094E+01	6.1017E+01	2.5301E+00	2.1370E+01	5.5165E+01	1.1812E+01	1.7609E+00	3.0571E+01	7.3126E+00
18	1.6116E+02	1.8968E+02	9.6887E+00	1.0816E+02	1.6029E+02	2.5169E+01	5.2189E+01	1.0584E+02	2.3166E+01
19	2.7404E+00	4.6348E+00	4.7267E-01	1.6863E+00	3.6567E+00	7.7921E-01	1.0161E+00	1.7223E+00	3.4349E-01
20	1.0903E+01	1.1871E+01	3.6616E-01	1.0342E+01	1.2062E+01	6.4629E-01	1.0035E+01	1.1254E+01	4.9221E-01
21	2.0000E+02	3.0384E+02	7.7553E+01	2.0000E+02	3.1932E+02	8.3201E+01	2.0000E+02	3.0271E+02	8.2769E+01
22	1.0740E+03	1.4937E+03	1.9286E+02	4.5570E+02	8.3787E+02	2.5055E+02	6.2475E+01	2.4699E+02	1.0913E+02
23	4.5981E+03	6.0880E+03	5.3171E+02	3.7200E+03	5.3213E+03	8.4120E+02	2.6820E+03	4.0320E+03	7.2841E+02
24	2.0017E+02	2.1733E+02	1.4867E+01	2.1116E+02	2.3769E+02	1.1610E+01	2.0020E+02	2.3468E+02	1.6071E+01
25	2.3653E+02	2.5235E+02	8.2988E+00	2.4178E+02	2.5758E+02	8.0331E+00	2.4144E+02	2.5694E+02	1.2474E+01
26	2.0001E+02	2.4606E+02	6.6115E+01	2.0001E+02	2.4558E+02	6.4221E+01	2.0001E+02	2.1684E+02	4.6111E+01
27	3.2346E+02	6.1723E+02	1.7389E+02	5.5559E+02	6.9637E+02	9.3585E+01	4.1966E+02	7.5219E+02	1.5375E+02
28	1.0000E+02	3.5818E+02	2.5128E+02	1.0000E+02	3.7948E+02	2.8873E+02	3.0000E+02	3.0000E+02	2.9792E-13
win	9	10	14	3	1	4	12	16	9
lose	15	17	13	21	26	23	13	12	19
draw	4	1	1	4	1	1	3	0	0

important aspect to measure the performance of WSN. Let each sensor can cover an area C_i and A is the total size of the region of interest. Then, the coverage rate CR is calculated by the Eq. 13.

$$CR = \frac{\cup C_i}{A} \quad i \in N \tag{14}$$

However, it is too complicated to calculate the coverage rate of randomly deployed sensor networks by Eq. 13. Therefore, this paper uses the grid scanning method [37] to evaluate the coverage rate. According to [37], CR is evaluated as the Eq. 14.

$$CR = \frac{m}{n} \tag{15}$$

5 Experimental results and discussion

A set of experiments was conducted to evaluate the performance of the proposed algorithm BP-QUATRE and its application to dynamic deployment in WSN.

5.1 Experimental results for BP-QUATRE

In this subsection, we evaluate the performance of the proposed BP-QUATRE algorithm on CEC2013 [40] test

Table 3 Performance for BP-QUATRE, PSO-IW, and DE algorithms

30D	PSO-IW			DE/best/1/bin			BP-QUATRE		
	Best	Mean	Std	Best	Mean	Std	Best	Mean	Std
1	1.8516E-02	2.0880E+03	2.3796E+03	2.2737E-13	2.2737E-13	0.0000E+00	0.0000E+00	5.9117E-14	1.0075E-13
2	9.2004E+05	2.9077E+07	4.8633E+07	1.4886E+07	2.6850E+07	7.1628E+06	8.0342E+04	3.2499E+05	1.8932E+05
3	7.5454E+08	9.3345E+10	1.8679E+11	1.8850E+08	6.7124E+08	2.5438E+08	4.0302E-05	6.8721E+05	2.9931E+06
4	1.4853E+03	5.8973E+03	5.1338E+03	2.3767E+04	3.7255E+04	6.3675E+03	6.7090E+00	3.8516E+01	2.5066E+01
5	8.1621E-01	1.2112E+03	1.2376E+03	1.1369E-13	1.3642E-13	4.5936E-14	0.0000E+00	1.1369E-13	2.2968E-14
6	3.8139E+01	2.4618E+02	2.6320E+02	1.6157E+01	2.4642E+01	1.1567E+01	4.5328E-03	1.0260E+01	8.0606E+00
7	7.4531E+01	2.3039E+02	1.4010E+02	4.5255E+01	5.8459E+01	7.7884E+00	7.2738E-02	4.9636E+00	3.9591E+00
8	2.0758E+01	2.0918E+01	5.8419E-02	2.0778E+01	2.0945E+01	4.7712E-02	2.0827E+01	2.0953E+01	4.7373E-02
9	2.2437E+01	2.9920E+01	3.3769E+00	2.3097E+01	2.9119E+01	1.8981E+00	9.7462E+00	2.2635E+01	5.7088E+00
10	1.4411E+00	5.7765E+02	4.3368E+02	6.7846E+00	1.5520E+01	4.3221E+00	7.3960E-03	2.0791E-02	1.1579E-02
11	7.2539E+01	1.6044E+02	3.8346E+01	5.6843E-14	6.9647E-01	1.0101E+00	1.7053E-13	4.4972E+00	1.8006E+00
12	9.8398E+01	1.8126E+02	6.7573E+01	1.1538E+02	1.5072E+02	1.5277E+01	1.9899E+01	5.3234E+01	1.5281E+01
13	1.5549E+02	2.4254E+02	4.7152E+01	1.2579E+02	1.6918E+02	1.3208E+01	5.8213E+01	1.0634E+02	2.4064E+01
14	2.1910E+03	3.6743E+03	8.1801E+02	1.3049E+00	2.6590E+01	5.0098E+01	1.0878E+01	1.9290E+02	1.2680E+02
15	3.0712E+03	4.6861E+03	8.3189E+02	5.1246E+03	6.1285E+03	3.9430E+02	2.7376E+03	4.0668E+03	5.3740E+02
16	5.2554E-01	1.2755E+00	3.6444E-01	1.5328E+00	2.3124E+00	3.2555E-01	9.4808E-01	1.8071E+00	4.0821E-01
17	1.1866E+02	1.7987E+02	3.7182E+01	2.6033E+01	3.1227E+01	9.1319E-01	1.7609E+00	3.0571E+01	7.3126E+00
18	1.1506E+02	1.6382E+02	2.8799E+01	1.9757E+02	2.2046E+02	1.1380E+01	5.2189E+01	1.0584E+02	2.3166E+01
19	6.9242E+00	2.8457E+03	6.4680E+03	2.7883E+00	3.8670E+00	3.9209E-01	1.0161E+00	1.7223E+00	3.4349E-01
20	1.0638E+01	1.2595E+01	7.4314E-01	1.2114E+01	1.2813E+01	2.5585E-01	1.0035E+01	1.1254E+01	4.9221E-01
21	2.0438E+02	3.9378E+02	1.9091E+02	2.0000E+02	2.9010E+02	7.6842E+01	2.0000E+02	3.0271E+02	8.2769E+01
22	2.4780E+03	3.8977E+03	7.9507E+02	1.1651E+02	2.5046E+02	1.8276E+02	6.2475E+01	2.4699E+02	1.0913E+02
23	2.9470E+03	5.0143E+03	9.3015E+02	5.5175E+03	6.5088E+03	4.1567E+02	2.6820E+03	4.0320E+03	7.2841E+02
24	2.8274E+02	2.9936E+02	1.0622E+01	2.5730E+02	2.7222E+02	6.3487E+00	2.0020E+02	2.3468E+02	1.6071E+01
25	2.8413E+02	3.0698E+02	9.4788E+00	2.7937E+02	2.8866E+02	4.6272E+00	2.4144E+02	2.5694E+02	1.2474E+01
26	2.0007E+02	3.6095E+02	5.3876E+01	2.0112E+02	2.0202E+02	5.0229E-01	2.0001E+02	2.1684E+02	4.6111E+01
27	1.0119E+03	1.1676E+03	7.9394E+01	9.4887E+02	1.0521E+03	4.0959E+01	4.1966E+02	7.5219E+02	1.5375E+02
28	1.1640E+02	1.9266E+03	7.1657E+02	3.0000E+02	3.2379E+02	1.6825E+02	3.0000E+02	3.0000E+02	2.9792E-13
win	3	2	0	2	4	17	22	22	11
lose	25	26	28	25	24	11	5	6	17
draw	0	0	0	1	0	0	1	0	0

suite for real-parameter optimization, which includes unimodal functions (f1-f5), multimodal functions (f6-f20), and composition functions (f21-f28). The names and search ranges of this 28 benchmark functions can be found in [40], and they are shifted to the same global best location $O\{o_1, o_2, \ldots, o_d\}$.

Firstly, we compare the BP-QUATRE with the two QUATRE variants "QUATRE/target-to-best/1" and "QUATRE/best/1" as BP-QUATRE employs these two mutation strategies. Then, we compare the BP-QUATRE with inertia weight PSO and standard DE due to the relationship among them as described in ref [5]. The parameter settings of the algorithms are shown in Table 1. The dimensions of all functions are set to 30. The population size ps is set to 100 for each algorithm, and the maximal number of function evaluation (NFE) is 3,000,000. We run each algorithm on each benchmark function 50 times independently. The best, mean, and standard deviation of the function error are collected in Table 2 and Table 3. The simulation results of some benchmark functions are shown in Fig. 2.

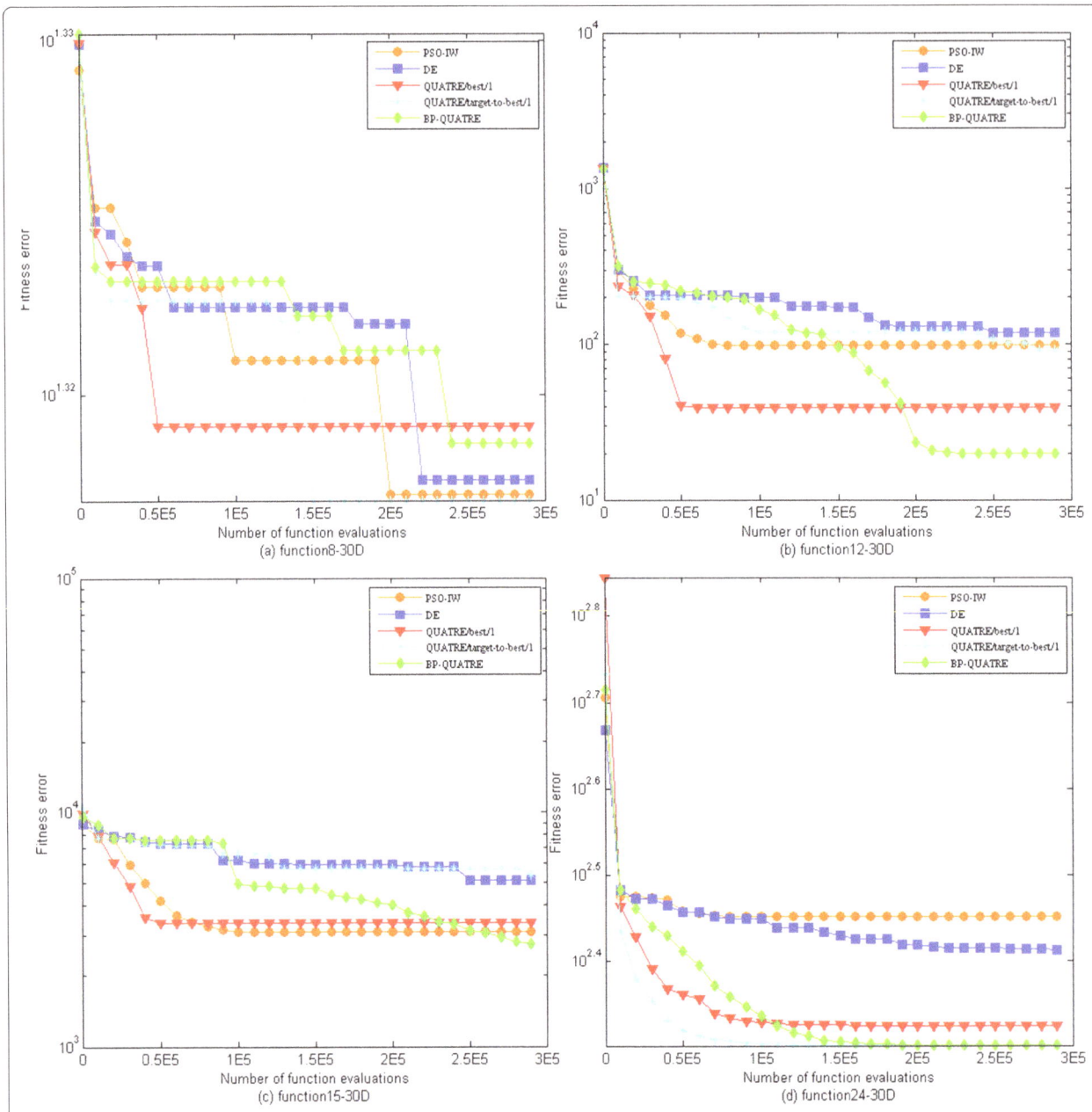

Fig. 2 Fitness errors vs. number of function evaluations of functions f8, f12, f15, and f24. The figure presents the fitness error and the convergence speed comparison by employing the best value of 50 runs obtained by each competing algorithm on 30-D optimization. The functions f8, f12, f15, and f24 figures are presented here. NFE means number of function evaluations

From Table 2, we can see that BP-QUATRE has significantly better performance than the other two QUATRE variants over 28 benchmark functions. The BP-QUATRE finds 12 best values and 16 mean values of CEC2013 benchmark functions in comparison QUATRE variants. This is because the BP-QUATRE can take advantage of different mutation strategies to maintain population diversity, and its linear decrease scale factor

control strategy is helpful to balance exploration and exploitation ability. For the standard deviation, the "QUATRE/target-to-best/1" has better performance than "QUATRE/best/1" and BP-QUATRE algorithms, and BP-QUATRE algorithm has better performance than "QUA-TRE/best/1." In addition, we can observe that QUATRE variants with different mutation strategies have different performance. The "QUATRE/target-to-best/1" performs

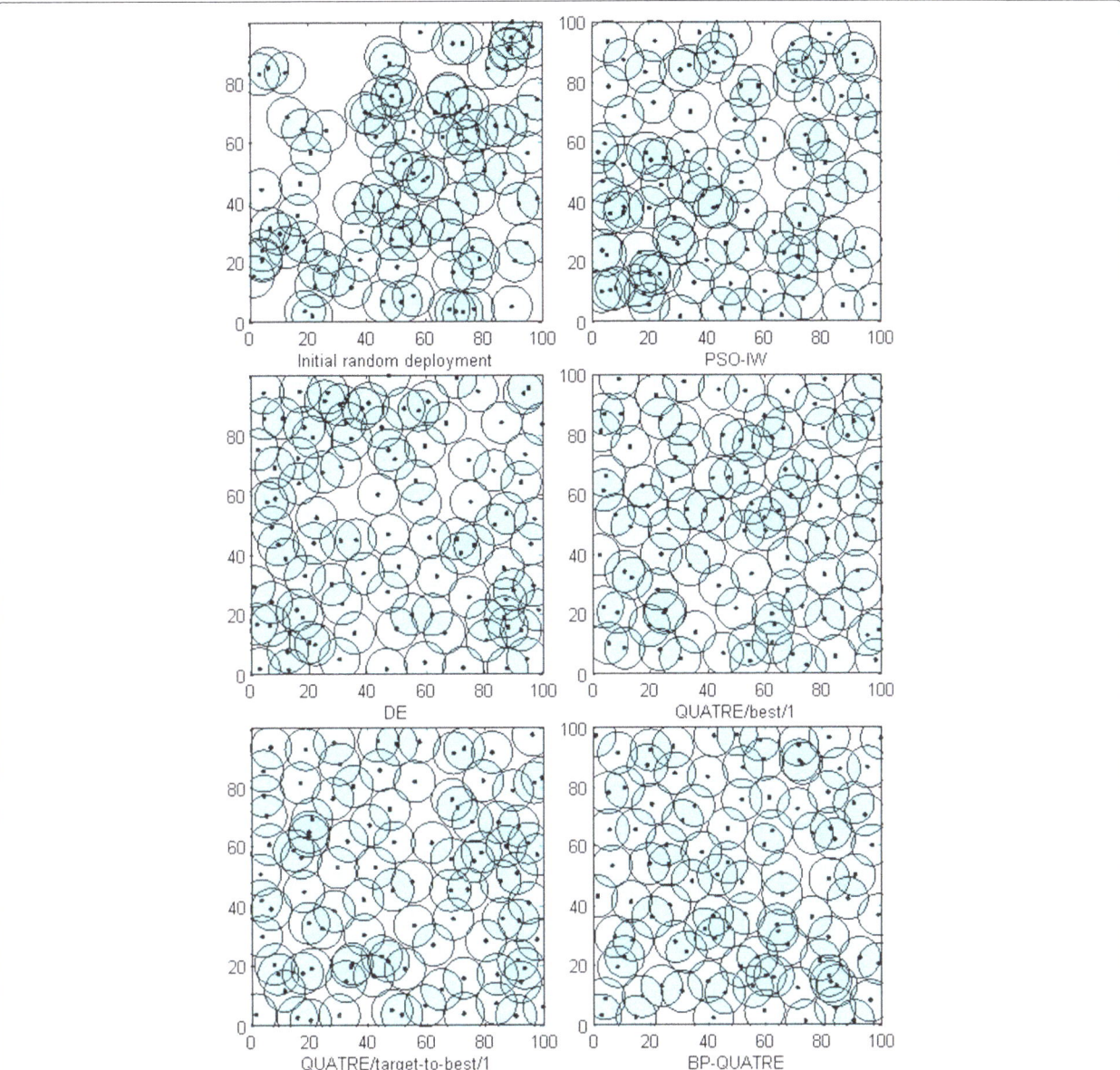

Fig. 3 Initial random deployment and final sensor positions based on each algorithm. The figure presents Initial random deployment and the final deployment of 100 sensor nodes in the monitored target area based on each competing algorithm (i.e., PSO-IW, DE, QUATRE/best/1, QUATRE/target-to-best/1, and BP-QUATRE). The black dots represent the final positions of the sensor nodes, and the circles represent the relative monitoring areas of the sensor nodes. The coverage rates obtained by PSO-IW, DE, QUATRE/best/1, QUATRE/target-to-best/1, and BP-QUATRE are 88.73%, 91.83%, 93.67%, 93.05%, and 93.89%, respectively

better on unimodal and composition functions than "QUATRE/best/1,", while the "QUATRE/best/1" performs better on multimodal functions than "QUATRE/target-to-best/1."

From Table 3, we can see that, for the best value, the PSO-IW algorithm finds 3 minimum values of 28 benchmark functions. The DE algorithm finds 2 minimum values of 28 benchmark functions. While our proposed BP-QUATRE algorithm finds 22 minimum values

of 28 benchmark functions in comparison with PSO-IW and DE algorithms, and thus, it has overall better performance than the contrasted algorithms. For the mean, our proposed algorithm also has significantly better performance than the competing algorithms. For the standard deviation, the DE algorithm has better performance than PSO-IW and BP-QUATRE algorithms, and BP-QUATRE algorithm has better performance than PSO-IW algorithm. Overall, our proposed

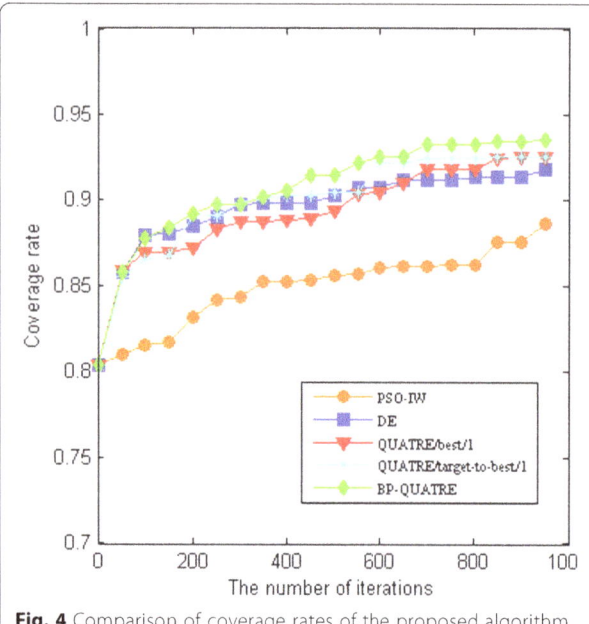

Fig. 4 Comparison of coverage rates of the proposed algorithm with the other algorithms, e.g., PSO-IW, DE, QUATRE/best/1 and QUATRE/targetto-best/1

Section 5.1 except that the acceleration coefficients c1 and c2 of the PSO are set to 1. The population size ps is set to 40, and the number of iterations is 1000. We run each algorithm 10 times independently with the same initialization.

One of initial deployments and the final best deployments of WSN after executing all competing algorithms based on the probabilistic detection model are shown in Fig. 3. The best convergences of each algorithm are shown in Fig. 4 by coverage rate for the number of iterations. The best, mean, and standard deviation of the coverage rates for the mentioned algorithms are given in Table 4. Obviously, it can be seen that our proposed BP-QUATRE has better performance than other two QUATRE variants and all QUATRE algorithms have better performance than PSO-IW and DE algorithm. In other words, BP-QUATRE has better coverage rate than the other four competing algorithms in the dynamic deployment of WSN. This is certainly related to the more powerful exploration and exploitation capability of the BP-QUATRE algorithm.

6 Conclusion

This paper proposes a novel BP-QUATRE algorithm for optimization problems. In BP-QUATRE, the population is divided into two subpopulations with sort strategy, and each subpopulation employs a different mutation strategy to balance between the diversity and convergence rate. In addition, adjusting scale factor with linear decrease strategy is adopted in BP-QUATRE algorithm to balance between exploration and exploitation ability. The proposed algorithm is verified under CEC2013 test suite. The experimental results demonstrate that the proposed BP-QUATRE algorithm not only has better performance than QUATRE variants "QUATRE/target-to-best/1" and "QUATRE/best/1," but also has better performance than the PSO-IW algorithm and DE algorithm. We also apply the proposed BP-QUATRE algorithm to dynamic deployment in WSN. The simulation results demonstrate that the proposed BP-QUATRE algorithm has better coverage rate than the other competing algorithms. In the future work, we will apply BL-QUATRE algorithm to classify music genre [41].

BP-QUATRE algorithm has better performance than the other two competing algorithms.

5.2 Simulation results for dynamic deployment in WSN

Simulations are conducted to evaluate the performance of BP-QUATRE algorithm in the dynamic deployment of WSN. The simulation results of the proposed algorithm are compared with the results of the PSO-IW, DE, and two QUATRE variants.

To make the simulation results more reliable, the parameter settings such as the target area, the number of sensors, and their detection radius are according to [34]. The monitored target area is a square region with a size of $100\,m \times 100\,m$, and 100 sensor nodes are scattered randomly on this target region. The parameter settings for the probabilistic detection model are $\alpha_1 = 1$, $\alpha_2 = 0$, $\beta_1 = 1$, and $\beta_2 = 1.5$. And the detection radius of each sensor node is 7 m, the uncertainty parameter of measurement r_e is 3.5 m, and the communication radius r_c is 21 m. The effective detection threshold c_{th} is 0.7. The control parameters of each algorithm are the same as in

Table 4 Performance for competing algorithms in the dynamic deployment of WSN

	PSO-IW	DE	QUATRE/best/1	QUATRE/target-to-best/1	BP-QUATRE
Best	0.8873	0.9183	0.9367	0.9305	0.9389
Mean	0.8629	0.9143	0.9212	0.9187	0.9362
Worse	0.8443	0.9084	0.9149	0.9123	0.9331
Std	0.0117	0.0030	0.0067	0.0053	0.0021

Abbreviations
BP-QUATRE: Bi-population QUATRE; DE: Differential evolution; PSO: Particle swarm optimization; PSO-IW: Inertia weight PSO; QUATRE: QUasi-Affine TRansformation Evolution algorithm; Std: Standard deviation; WSN: Wireless sensor networks

Acknowledgements
The authors would like to thank Prof. Zhenyu Meng for providing the code of QUATRE.

Authors' contributions
NL proposed the idea of the BP-QUATRE algorithm. He also carried out the simulations and drafted the paper. JP supervised the work and introduced the idea of applying the proposed BP-QUATRE algorithm into dynamic deployment in WSN. TN gave some suggestions for the paper and revised the manuscript. All authors read and approved the final manuscript.

Authors' information
Nengxian Liu is a Ph.D. candidate in the College of Mathematics and Computer Science, Fuzhou University. His research interest includes computing intelligence and sensor networks.
Jeng-Shyang Pan received the Ph.D. degree in Electrical Engineering from the University of Edinburgh, UK, in 1996. Now, he is the Dean in the College of Information Science and Engineering, Fujian University of Technology. His current research interests include soft computing, sensor networks and signal processing.
Trong-The Nguyen received his Ph.D. degree in Communication Engineering from the National Kaohsiung University of Applied Sciences, Taiwan, in 2016. He is currently a lecturer in the College of Information Science and Engineering, Fujian University of Technology. His current research interests include computational intelligence and sensor networks.

Funding
No funding.

Competing interests
The authors declare that there are no competing interests.

Author details
[1]College of Mathematics and Computer Science, Fuzhou University, Fuzhou 350108, China. [2]Fujian Provincial Key Laboratory of Big Data Mining and Applications, Fujian University of Technology, Fuzhou 350118, China. [3]College of Computer Science and Engineering, Shandong University of Science and Technology, Qingdao 266510, China. [4]Department of Information Technology, University of Manage and Technology, Haiphong 180000, Vietnam.

References
1. R. Storn, K. Price, Differential evolution-a simple and efficient heuristic for global optimization over continuous spaces. J. Global Optim. **11**(4), 341–359 (1997)
2. J. Kennedy, R. Eberhart, *Particle Swarm Optimization* (Proceedings of IEEE International Conference on Neural Networks, Perth, 1995)
3. M. Dorigo, V. Maniezzo, A. Colorni, Ant system: Optimization by a colony of cooperating agents. IEEE Trans. Syst. Man Cybern. Part B Cybern. **26**(1), 29–41 (1996)
4. D. Karaboga, B. Basturk, A powerful and efficient algorithm for numerical function optimization: artificial bee colony (ABC) algorithm. J. Glob. Optim. **39**(3), 459–471 (2007)
5. Z.Y. Meng, J.S. Pan, X. HR, QUasi-Affine TRansformation Evolutionary (QUATRE) algorithm: A cooperative swarm based algorithm for global optimization. Knowl.-Based Syst. **109**, 104–121 (2016)
6. Y.H. Shi, R. Eberhart, *A Modified Particle Swarm Optimizer* (Proceedings of the IEEE Conference on Evolutionary Computation, Anchorage, 1998)
7. B. Liu, L. Wang, et al., Improved particle swarm optimization combined with chaos. Chaos Solison. Fract. **25**(5), 1261–1271 (2005)
8. R. Cheng, Y.C. Jin, A social learning particle swarm optimization algorithm for scalable optimization. Inf. Sci. **291**, 43–60 (2015)
9. J.H. Liu, J. Lampinen, A fuzzy adaptive differential evolution algorithm. Soft. Comput. **9**(6), 448–462 (2005)
10. J. Brest, S. Greiner, et al., Self-adapting control parameters in differential

evolution: A comparative study on numerical benchmark problems. IEEE Trans. Evol. Comput. **10**(6), 646–657 (2006)
11. S. Rahnamayan, H.R. Tizhoosh, M.M.A. Salama, Opposition-based differential evolution. IEEE Trans. Evol. Comput. **12**(1), 64–79 (2008)
12. Z.Y. Meng, J.S. Pan, *QUasi-affine TRansformation Evolutionary (QUATRE) algorithm: A parameter-reduced differential evolution algorithm for optimization problems* (IEEE Congress on Evolutionary Computation (CEC), Vancouver, 2016)
13. Z.Y. Meng, J.S. Pan, *QUasi-Affine TRansformation Evolutionary (QUATRE) Algorithm: The Framework Analysis for Global Optimization and Application in Hand Gesture Segmentation* (2016 IEEE 13th International Conference on Signal Processing, Chengdu, 2016)
14. Z.Y. Meng, J.S. Pan, Monkey king evolution: A new memetic evolutionary algorithm and its application in vehicle fuel consumption optimization. Knowl.-Based Syst. **97**, 144–157 (2016)
15. Z.Y. Meng, J.S. Pan, X.Q. Li, *The QUasi-Affine TRansformation Evolution (QUATRE) Algorithm: An Overview* (The Euro-China Conference on Intelligent Data Analysis and Applications, Málaga, 2017)
16. J.S. Pan, Z.Y. Meng, et al., *QUATRE algorithm with sort strategy for global optimization in comparison with DE and PSO variants* (The Euro-China Conference on Intelligent Data Analysis and Applications, Málaga, 2017)
17. Z.Y. Meng, J.S. Pan, QUasi-Affine TRansformation Evolution with External ARchive (QUATRE-EAR): An enhanced structure for differential evolution. Knowl.-Based Syst. **155**, 35–53 (2018)
18. I. Loshchilov, M. Schoenauer, M. Sebag, *Bi-Population CMA-ES Agorithms with Surrogate Models and Line Searches* (Proceedings of the 15th Annual Conference Companion on Genetic and Evolutionary Computation, Amsterdam, 2013), pp. 1177-1184
19. J.F. Chang, S.C. Chu, et al., A parallel particle swarm optimization algorithm with communication strategies. J. Inf. Sci. Eng. **21**(4), 809–818 (2005)
20. Y. WJ, J. Zhang, *Multi-population differential evolution with adaptive parameter control for global optimization* (Proceedings of the 13th Annual Genetic and Evolutionary Computation Conference, Dublin, 2011)
21. L.Z. Cui, G.H. Li, et al., Adaptive differential evolution algorithm with novel mutation strategies in multiple sub-populations. Comput. Oper. Res. **67**, 155–173 (2016)
22. P.W. Tsai, J.S. Pan, *Parallel Cat Swarm Optimization* (Proceedings of the Seventh International Conference on Machine Learning and Cybernetics, Kunming, 2008)
23. S. Das, P.N. Suganthan, Differential evolution: A survey of the state-of-the-art. IEEE Trans.Evol. Comput. **15**(1), 4–31 (2011)
24. S. Das, A. Konar, et al., *Two improved differential evolution schemes for faster global search* (Genetic and Evolutionary Computation Conference, Washington DC, 2005)
25. J.S. Pan, L.P. Kong, T.W. Sung, et al., α-Fraction first strategy for hierarchical model in wireless sensor networks. J Internet Technol. **19**(6), 1717–1726 (2018)
26. S. Abdollahzadeh, N.J. Navimipour, Deployment strategies in the wireless sensor network: A comprehensive review. Comput.Commun., **91**, 1–16 (2016)
27. J.S. Pan, L.P. Kong, T.W. Sung, et al., A clustering scheme for wireless sensor networks based on genetic algorithm and dominating set. J Internet Technol. **19**(6), 1111–1118 (2018)
28. H. YFa, C.H. Hsu, Energy efficiency of dynamically distributed clustering routing for naturally scattering wireless sensor networks. J Netw Intell **3**(1), 50–57 (2018)
29. X.C. Bao, C.Z. Deng, et al., Topology optimization of fault tolerant target monitoring for energy harvesting wireless sensor network. J Netw Intell **2**(4), 310–321 (2017)
30. Y. Zou, K. Chakrabarty, *Sensor Deployment and Target Localization Based on Virtual Forces* (Proceedings of the 22nd Annual Joint Conference of the IEEE Computer and Communications Societies, San Francisco, 2003)
31. G.L. Wang, G.H. Cao, Movement-assisted sensor deployment. IEEE T Mobile Comput. **5**(6), 640–652 (2006)
32. X.L. Wu, L. Shu, et al., *Swarm Based Sensor Deployment Optimization In Ad-Hoc Sensor Networks* (Proceedings of the 2nd International Conference on Embedded Software and Systems, Xi'an, 2005)
33. M. Rout, R. Roy, Optimal wireless sensor network information coverage using particle swarm optimization method. Int. J. Electron. Lett. **5**(4), 491–499 (2017)
34. X.Y. Yu, J.X. Zhang, et al., A faster convergence artificial bee colony algorithm in sensor deployment for wireless sensor networks. Int. J. Distrib. Sens. N **1-9**, 2013 (2013)

35. L.Z. Jin, J. Jia, et al., *Node Distribution Optimization in Mobile Sensor Network Based on Multi-Objective Differential Evolution Algorithm* (2010 Fourth International Conference on Genetic and Evolutionary Computing, Shenzhen, 2010)

36. Y. Zou, K. Chakrabarty, *Sensor Deployment and Target Localization Based on Virtual Forces* (IEEE INFOCOM 2003. Twenty-second Annual Joint Conference of the IEEE Computer and Communications Societies, San Francisco, 2003)

37. K. Chakrabarty, S.S. Iyengar, et al., Grid coverage for surveillance and target location in distributed sensor networks. IEEE. T. Comput. **51**(12), 1448–1453 (2002)

38. S.J. Li, C.F. Xu, et al., *Sensor Deployment Optimization for Detecting Maneuvering Targets* (Proceedings of the 7th international conference on information fusion, Philadelphia, 2005)

39. S.M.A. Salehizadeh, A. Dirafzoon, et al., *Coverage in Wireless Sensor Networks Based on Individual Particle Optimization* (Proceedings of the IEEE International Conference on Networking, Sensing and Control, Chicago, 2010)

40. J. Liang et al., *Problem Definitions and Evaluation Criteria for the CEC 2013 Special Session on Real-Parameter Optimization* (Computational Intelligence Laboratory, Zhengzhou University, Zhengzhou, China and Nanyang Technological University, Singapore, Technical report 2012, 2013)

41. Y.T. Chen, C.H. Chen, et al., A two-step approach for classifying music genre on the strength of AHP weighted musical features. Math **7**(1), 19 (2019)

14

Adaptive cascade single-shot detector on wireless sensor networks

Zhiyong Wei[*] and Fengling Wang

Abstract

The target detection model based on convolutional neural networks has recently achieved a series of exciting results in the target detection tasks of the PASCAL VOC and MS COCO data sets. However, limited by the data set for a particular scenario, some techniques or models applied to the actual environment are often not satisfactory. Based on cluster analysis and deep neural network, this paper proposed a new Statistic Experience-based Adaptive One-shot Network (SENet). The whole model solved the following practical problems. (1) By clustering the existing image classification dataset ImageNet, a common set of target detection datasets is formed, and a data set named ImageNet iLOC is formed to solve the object detection. The problem of single and insufficient quantities in the task. (2) We use cluster analysis on the size and shape of objects in each sample, which solves the problem of inaccurate manual selection of suggested areas during object detection. (3) In the multi-resolution training and prediction process, we reasonably allocate the size and shape of the suggested frame at different resolutions, greatly improve the utilization rate of the proposed frame, reduce the calculation amount of the model, and further improve the real-time performance of the model. The experimental results show that the model has a breakthrough in accuracy and speed (FPS reaches 54 in the case of a 3.4% increase in mAP).

Keywords: Cluster, Detection, Deep neural networks, Filter banks, Detection data sets

1 Introduction

Convolutional neural network (CNN) was widely used in the 1990s (such as model [1]), but with the rise of support vector machines in the field of computer vision, CNN entered a period of low tide. In 2012, the image classification model proposed by Krizhevsky et al. [2] demonstrated the revolutionary image classification accuracy in ILSVRC (ImageNet Large Scale Visual Recognition Challenge, ILSVRC) [3], rekindling people's interest in CNN. A series of image classification models based on CNN are continually proposed, and the image classification records of ILSVRC are also refreshed again and again. At the same time, the mean average precision (mAP) and detection speed of the target detection reference data set PASCAL VOC [4] and MS COCO [5] are also constantly increasing. Firstly, the success of selective search (SS) [6] and region proposal based volume and neural networks (R-CNN) [7] has driven advances in object detection. Although R-CNN was very time consuming to

propose, the cost of the detection model was greatly reduced by sharing the convolution between the proposed regions [2, 7]. Typical research results, such as Fast R-CNN, have achieved real-time target detection speeds without the time-consuming recommendations of regional recommendations, using target detection models with extremely deep convolutional neural networks [2]. However, the time-consuming problem of regional recommendations remains the performance bottleneck of the most advanced target detection systems. Next, region proposal networks (RPN) using CNN instead of selective lookups is proposed [7]. RPN shares a partial convolutional layer with the most advanced target detection network, and by testing the shared convolution, the cost of the calculation suggestion box is further reduced. However, the entire detection model (Faster R-CNN) needs to train a proposed network and a detection network [7], which is still too cumbersome, inefficient, and not easy to optimize relative to the detection model of a single network.

YOLO regards target detection as a single regression problem. It first extracts the input image through a

* Correspondence: ncvtweizhiyong@163.com
Nanning College for Vocational Technology, Nanning 530008, Guangxi, China

traditional convolutional neural network to form an S × S grid (e.g., 7 × 7). Each grid produces two bounding boxes of different sizes and shapes (7 × 7 × 2 = 98) relative to the original image, each bounding box representing the coordinates of a potential object and the probability of belonging to a certain category.

The SSD model is similar to YOLO. The main difference is that the SSD architecture combines multiple feature maps of different resolutions in the neural network for target prediction, naturally processing objects of different sizes, and improving detection quality [8]. However, the single dataset, poor integration, low speed and accuracy, and difficulty in optimization are still many of the problems faced by the object detection model. This chapter proposes an end-to-end, single neural network target detection model, which mainly has the following contributions:

By applying the clustering analysis algorithm to the existing ImageNet for image classification, a set of general methods for making image classification data sets into target detection data sets is formed, which solves the problem of single and insufficient samples in object detection tasks.

The paper uses cluster analysis to determine the size and shape of the bounding box in each sample, and obtains the prior knowledge of the size and shape of the bounding box of the object in the data set. Based on these priors, the design of the model's border frame is guided, and the object detection process is solved manually. Design suggestions for areas that are not accurate.

In the multi-resolution training and prediction process, this paper greatly improves the utilization rate of the proposed frame by reasonably assigning the size and shape of the suggestion frame of different resolution detection layers, and at the same time, reduces the calculation amount of the model and further improves the calculation. The real-time and accuracy of the model.

1.1 Related works

In this part, we first introduce the neural network-based target detection and recognition methods. On this basis, the plant anomaly detection technology was reviewed, and the research progress of false positive detection technology was reviewed.

1.2 Image-based object detection and feature extractors

In recent years, visual media gained through the Internet has proliferated. A large amount of data brings new opportunities and challenges to the application of neural networks. Since Alex Net [2] first applied the convolutional neural network (CNN) to image classification

tasks in the ImageNet Large Scale Visual Identity Competition (ILSVRC-2012) [3], it consists of eight layers. CNN demonstrated superior performance compared to traditional manual computer vision algorithms. Therefore, in recent years, several deep neural network structures have been proposed to improve the accuracy of the same task.

Object detection and recognition is an important issue in recent years. In the case of detecting specific categories, earlier applications focused on image classification from object-centric [9]. The goal is to categorize images that may contain objects. However, the new main paradigm is not only to classify and accurately locate objects in an image [10]. Therefore, current prior art object methods for object detection are primarily based on deep CNN [3]. They are divided into two phases and a one-stage approach. Two-stage methods are often associated with region-based convolutional neural networks, such as faster R-CNN [7], region-based fully connected networks (R-FCN) [11]. In these frameworks, the region proposal network (RPN) generates a set of candidate object locations in the first level, and the second level uses CNN to classify each candidate location into one of the classes or backgrounds. It uses a deep network to generate features for backward use by the RPN to extract recommendations. In addition to the system based on regional recommendations, a first-level framework for object detection has been proposed. Recent SSDs [8], Yolo [12] and Yolo v2 [13] have shown promising results, resulting in a real-time detector similar to the accuracy of a two-stage detector. In the past few years, it has also been demonstrated that deeper neural networks achieve higher performance than simple models in image classification tasks [3]. However, with significant performance improvements, the complexity of deep structures has also increased, such as VGG [14], ResNet [15], GoogleNet [16], ResNext [17], DenseNet [18], dual path network [19], and Senet [20]. As a result, deep artificial neural networks often have much more trainable model parameters than the number of samples they accept [21]. Although a large number of data sets are used, neural networks tend to over-fitting [1]. On the other hand, several strategies have been applied to improve performance in deep neural networks. For example, increasing the number of samples increases the data [22], weighting regularization to reduce model over-fitting [23], random discarding off-activation [24], and batch normalization [25]. While these strategies have proven to be effective in large networks, the lack of data or category imbalances remains a challenge for several applications. There is no specific way to understand the complexity of artificial neural networks for their application to any problem. Therefore, the importance of developing a strategy specifically designed for applications that include limited data and class imbalance issues. Moreover, depending on the complexity of the application, today's challenge is to design a deep learning approach that can perform complex tasks while maintaining lower computational costs.

1.3 Anomaly detection in plants

The problem of plant diseases is an important issue directly related to people's food safety and welfare. Diseases and pests affect food crops, which in turn cause significant losses in the peasant economy. The effects of disease on plants are becoming challenging in crop protection and healthy food production. Traditional methods for identifying and diagnosing plant diseases depend primarily on the visual analysis of experts within the area, or in the laboratory. These studies often require high expertise in the field, in addition to the probability of not successfully diagnosing a particular disease, thus leading to erroneous conclusions and treatment (proposed 2018). In these cases, in order to obtain quick and accurate decisions, automated systems will provide efficient support to identify diseases and pests of infected plants [26, 27]. Recent advances in computing technology, particularly graphics processing units (GPUs), have led to the development of new image-based technologies, such as efficient deep neural networks. The application of deep learning has also expanded into the field of precision agriculture, and it has also shown satisfactory results while solving complex problems. Some applications include disease identification for several crops such as Cole [26], apples [28], bananas [29], wheat [30], and cucumbers [31]. The CNN-based approach constitutes a powerful tool that is used as a feature extractor in multiple jobs. Mohanty et al. [27] compared two CNN architectures, Alexnet and Googlenet, to identify 14 crop species and 26 diseases using large disease databases and healthy plants. Their results show a system that effectively classifies images containing specific diseases into crops that use transfer learning. However, the disadvantage of this work is that its analysis is based only on images collected in the lab, not in actual field scenarios. Therefore, it does not cover all the changes contained there. Similarly, Sladjevic et al. [32] identified 13 healthy leaf plant diseases using the Alexnet CNN architecture. They used several strategies to avoid overfitting and improve classification accuracy, such as data enhancement techniques, to increase the size of the data set and improve efficiency while training CNN. The average accuracy of the system is 96.3%. Recently, Liu et al. [28] proposed a method for apple leaf disease identification based on a combination of Alexnet and GoogleNet architectures. The system was trained to identify four types of apple leaf disease using an image data set collected in the laboratory with a total accuracy of 97.62%. In [33], ferencinos evaluated various CNN models to detect and diagnose plant diseases using leaf images of healthy and infected plants. The system is capable of classifying 58 different plant/disease combinations from 25 different plants. In addition, the experimental results show interesting comparisons when using images collected in the laboratory compared to images collected in the field. The use of two types of images gives promising results with an optimum accuracy of 99.53% given by the VGG network. However, when images acquired in the field are used for testing rather than laboratory images, the success rate is significantly reduced. In fact, according to the author, this proves that image classification under real field conditions is more difficult and complicated than using images collected in the laboratory.

Although the above work has achieved good results in the identification of plant diseases, the challenges of complex field conditions, infection changes, various pathologies in the same image, and surrounding objects have not been studied. They mainly use images acquired in the lab, so they cannot handle all the situations that occur in real scenes. Moreover, they are all based on the method of disease classification. In contrast, Fuentes et al. [26] proposes a system that can successfully detect and locate nine Cole pests and diseases using images collected in the field, including actual cultivation conditions. This approach differs from other methods in that it generates a set of bounding boxes that contain the location, size, and category of the disease and/or pest in the image. This work examines different meta-architectures and CNN feature extractors to identify and locate suspicious regions in an image. The results show that the performance of the method reaches 83%. However, the system has some difficulties that make it impossible to achieve higher performance. They mentioned that due to the lack of samples, some highly variant classes are often confused with other classes, resulting in false positives or low precision. According to the idea in [26], our current work aims to solve the above problems and improve the results by focusing on false positives and class imbalances. On the other hand, our method has studied several techniques to make the system more robust to inter- and intra-species changes in Cole pests and diseases.

2 Our methods
2.1 Classified dataset to test dataset

The target detection data set is costly to produce (you need to label the number of objects in the image, the size, shape, location, and type of each object), and the public can obtain less (more famously, only PASCAL VOC and COCO data sets). The problem of single type of data set severely limits the accuracy and application range of the target detection model. On the other hand, because the image classification data set is relatively easy to make (only the type of the object needs to be labeled), there are many kinds of types. The publicly available data set is used to train the image classification model. More famous are MNIST [34], CIFAR [35], ImageNet [3], Youtube-8 M [36], Open Image [37], etc. These data sets contain nearly 10,000 kinds and hundreds of millions. At the same time, some well-known image classification models achieved very high accuracy (even 100%) on the relevant data sets. Do we have a way to automatically add location tags to these well-known classification datasets? In the field of machine learning, there are many unsupervised algorithms that are very suitable for finding the shape of randomly distributed data. Spectral clustering is one of them [38]. Spectral theory is the study of how the properties of a graph are described by several easily calculated quantities. Spectral clustering of a graph is an important tool for describing graphs. The usual method is to encode the graph into a matrix and then calculate the eigenvalues of the matrix (also called spectrum spectrum) [38]. In other words, spectral clustering uses a weighted adjacency matrix and its spectral map to analyze the data by graph segmentation [39]. Therefore, spectral clustering can achieve more powerful data representation in the feature space by using the main components in the data revealed by the spectrum, thus facilitating data clustering.

The object detection data set usually consists of the center coordinates of the bounding box of the object, the length and width, and the category label of the object, which are usually represented by $\{cx, cy, w, h, category\}$. This chapter uses spectral clustering algorithm and ResNet [17] image classification model to propose a method to make the classification data set into object detection data set. The algorithm first uses spectral clustering to obtain the contours (shapes) of the objects in the image, and calculates the position coordinates $\{cx, cy, w, h\}$ of each object; using ResNet to classify the objects in the image to obtain a confidence that it belongs to a certain category. Degree (confidence). The algorithm flow is divided into the following stages to complete the production of the test data set:

1) Obtain image left and lower edge feature coordinate vectors $= \{\vec{l}_1, \vec{l}_2, ..., \vec{l}_m\}; D = \{\vec{d}_1, \vec{d}_2, ..., \vec{d}_n\}$, respectively, of size $m \times n$.

2) Cluster analysis of the pixels of the image using spectral clustering to obtain a rough outline of each object $O = \{O_1, O_2 ... O_M\}$ in the image, and a pixel coordinate vector O_j of each object (cluster) $C^j = \{\vec{c}_1^j, \vec{c}_2^j ... \vec{c}_K^j\}$.

3) Calculate the object coordinate vector set C^j based on the formula (1–4), the four coordinates of the nearest and farthest distance of each coordinate vector to the edge set $\{L, D\}$ $T_j = \{\vec{t}_u^j, \vec{t}_d^j, \vec{t}_l^j, \vec{t}_r^j\}$, using coordinates $(x_1, y_1), (x_2, \Delta_y), (\Delta_x, y_3), (x_4, y_4)$ indicates.

$$\vec{t}_u^j = \max_{k \in K} \text{distance}\left(D, \vec{c}_k^j\right) \tag{1}$$

$$\vec{t}_d^j = \min_{k \in K} \text{distance}\left(D, \vec{c}_k^j\right) \tag{2}$$

$$\vec{t}_l^j = \min_{k \in K} \text{distance}\left(L, \vec{c}_k^j\right) \tag{3}$$

$$\vec{t}_r^j = \max_{k \in K} \text{distance}\left(L, \vec{c}_k^j\right) \tag{4}$$

4) After getting the four vertices of the object, we calculate the width, height, and center coordinates of the bounding box based on the formula (5–7):

$$w = x_4 - \Delta_x \tag{5}$$

$$h = y_1 - \Delta_y \tag{6}$$

$$(c_x, c_y) = (\Delta_x + w, \Delta_y + h) \tag{7}$$

5) Finally, we use ResNet to calculate the category confidence of the object in the rectangle. If the confidence of this object belongs to a category is greater than 0.85, then the correct label of the algorithm output sample (ground-truth label) $\{(c_x^j, c_y^j), w, h, c\}$. Otherwise, ignore this object. See Fig. 1 for the specific process and a schematic diagram of the algorithm.

Algorithm 1 Generating object detection dataset based on spectral clustering

---

Input: Classification sample X and the size of the dataset is n.

Output: Detection object D

    Function CLASSIFICATIONTODETECTION(X)

        Initialization $k_{cluster} = 6$

        for $i = 1$ to n **do**

        Compute the left and bottom coordinates of sample X_i

        Update the coordinate set $\{L, D\}$

        Spectral clustering for detection object O based on sample X_i

        for $j = 1$ to O_M **do**

            Compute the upper vertex t_u^j, bottom vertex t_d^j, left vertex t_l^j and right vertex t_r^j of object O_j

            Update the object coordinate $T_j = \{\tilde{t}_u^j, \tilde{t}_d^j, \tilde{t}_l^j, \tilde{t}_r^j\}$

            Compute the width w and height h of the anchor box

            Compute the center coordinate (c_x^j, c_y^j) of the anchor box

            Update the positioning information of O_j as $b_j = \{(c_x^j, c_y^j), w, h\}$

            Compute the parameter confidence of object O_j as $c = ResNet(b_j)$

            if $\max_c(c_i) > 0.85$ **then**

                Compute the ground truth box of object O_j as $gtb_j = \{(c_x^j, c_y^j), w, h, c\}$

                Update the detection dataset with gtb_j

            else

                continue

            end if

        end for

        end for

        return the new detection dataset D

    end function

2.2 Boundary box clustering (k-means++)

At present, the recommended boxes of classical object detection models are based on manual experience, and the size and shape of the boundary boxes are set manually. In reality, the manual designed boundary boxes are usually neither flexible nor robust [3, 7, 9]. For this reason, this paper uses k-means++ algorithm to cluster the shape of the object in the sample, so as to objectively obtain the shape distribution of the object in the sample. Standard k-means++ calculates the distance from each element to the cluster center based on the Euclidean distance. Because the distance between each element and the cluster center varies greatly among objects of different sizes, using absolute distance to calculate the shape of the object will make k-means++ unable to converge correctly. Therefore, this paper uses formula (8) instead of the standard Euclidean distance calculation method to complete the clustering analysis of object shape, in which box is the sample to be clustered (including the width and height of the object), centroid is the center of the current clustering, and the function of intersection over union (IoU) outputs the overlap ratio between the object and the clustering cluster. Figure 2a shows that when k-means++ chooses different K values, $k = 6$ is selected to balance the speed and the overlap rate of IoU. Figure 2b shows the distribution of objects of different sizes and shapes in the original image by

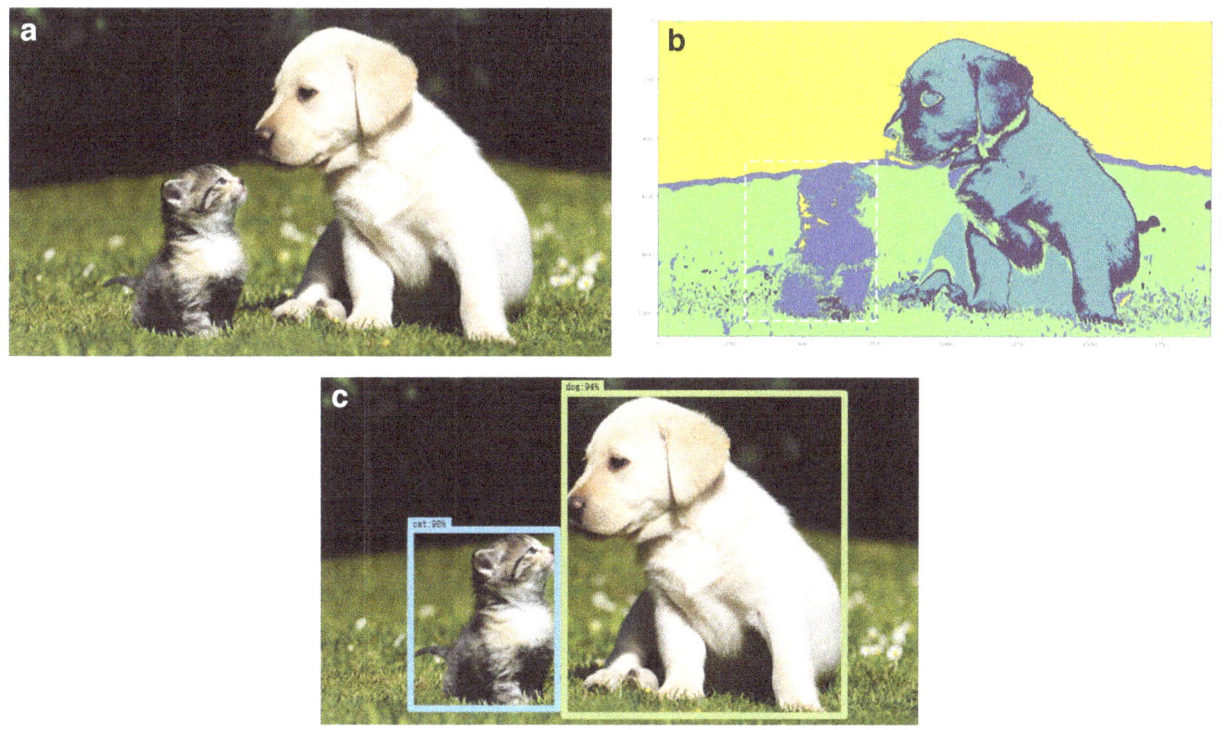

Fig. 1 A monitoring system with wireless sensor networks. Figure 1 shows a monitoring system with wireless sensor networks. **a** Original image. **b** Anchor box defining. **c** Object recognition

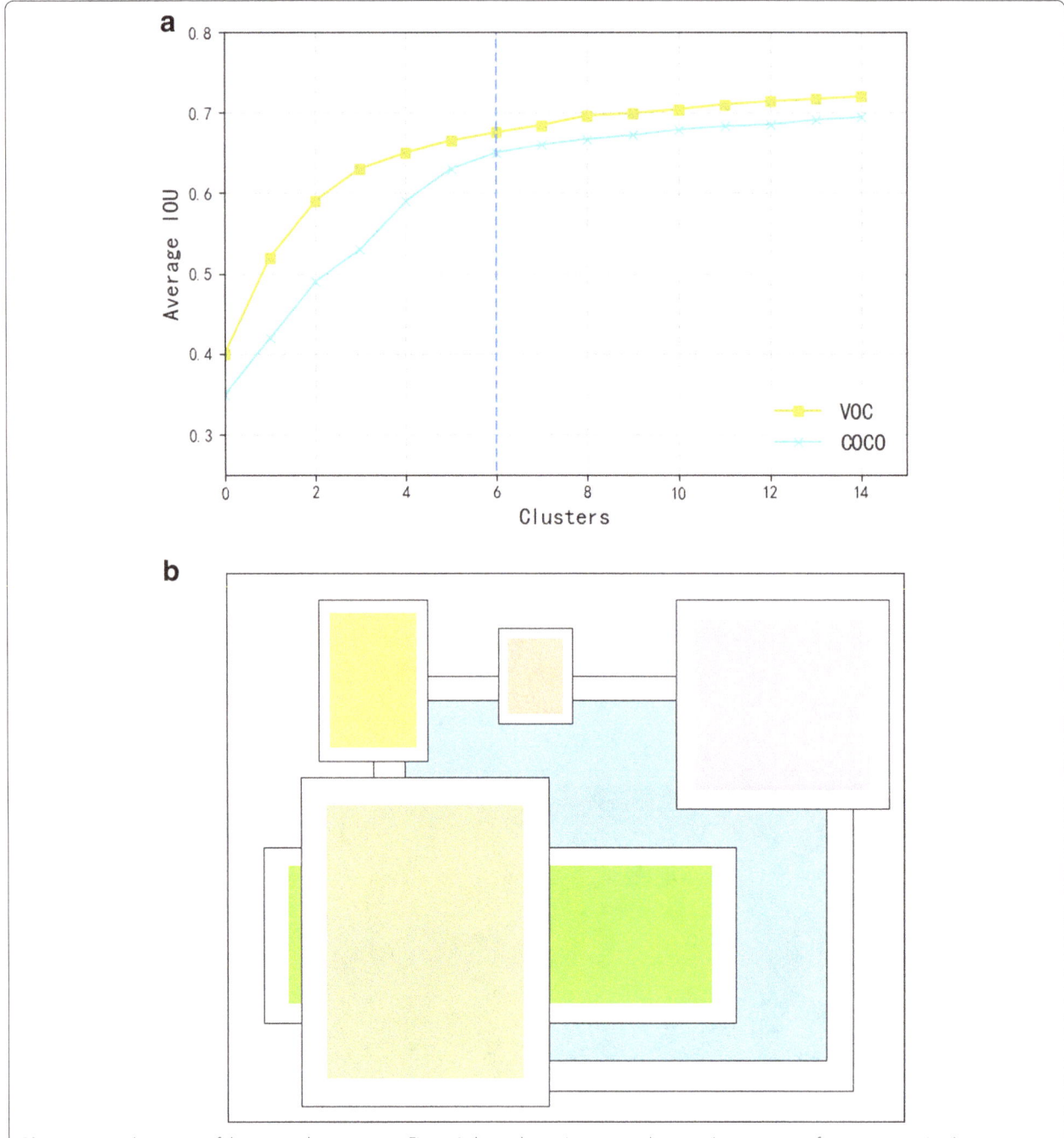

Fig. 2 A general overview of the approach we propose. Figure 2 shows the entire proposed system. Input images of any size are trained in our primary diagnostic unit, which generates boundary boxes and their location and category of infected areas in the image. The secondary diagnostic unit uses the bounding box as an input, and the secondary diagnostic unit independently trains the CNN filter bank to reduce the number of false positives generated by the primary unit. Both systems are further integrated into the level and location. K-means++ clusters the border shapes of training samples from PASCAL VOC and COCO datasets. **a** It shows that when k-means++ chooses different K values, $k = 6$ is chosen to balance the speed and the overlap rate of the IOU. **b** K-means + + clustering results show that thin and high boundaries account for the majority of the samples

clustering the ImageNet and COCO data sets. We find that the shape of objects is different from the default box set manually. The boundary boxes of high and thin shapes occupy the majority of shapes. Again, it proves the necessity of clustering the shape of samples to obtain the priori shape of objects.

$$d(\text{box}, \text{centroid}) = 1 - IoU(\text{box}, \text{centroid}) \qquad (8)$$

Figure 3a shows the method of setting default boxes manually. Due to lack of objective analysis, it can not effectively improve the phenomenon of overlapping areas

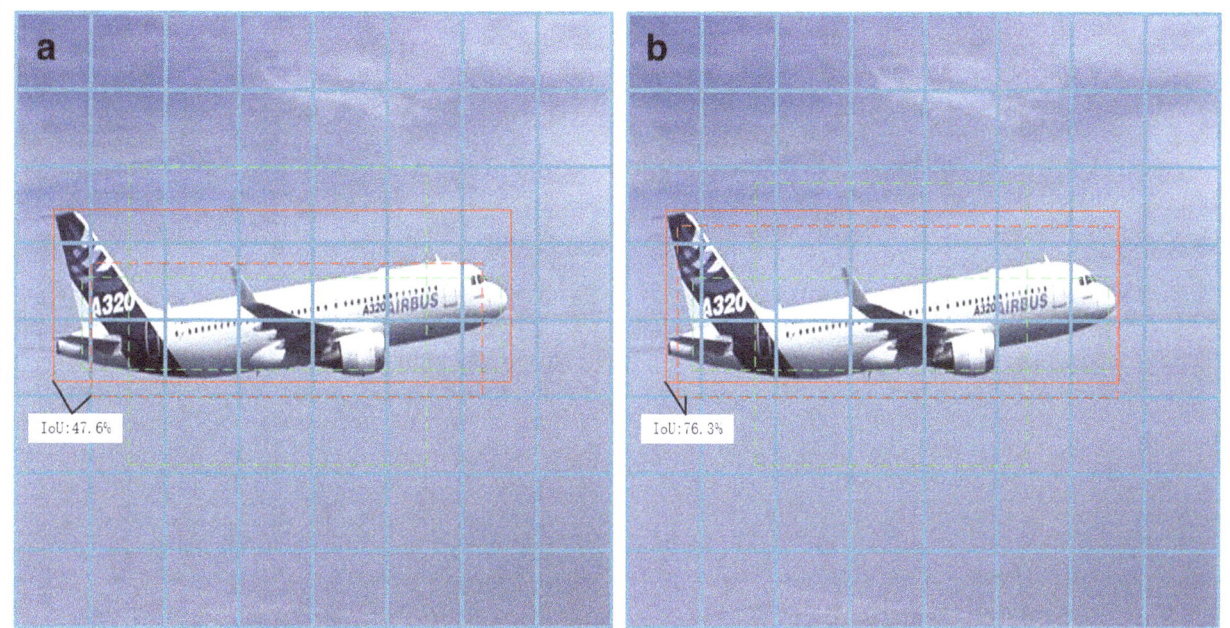

Fig. 3 Primary diagnosis unit for bounding box detection. Figure 3 shows the process by which the primary diagnostic unit detects suspicious areas containing disease and pests in the input image. It is similar to Fuentes et al. (2017b). Compared the manual setting of boundary boxes with the k-means++ clustering generation of boundary boxes, in which the red solid border is the correct border and the red dotted border is the highest border in IoU of different methods of generating boundary boxes. **a** The size and shape of the default boundary box set manually are usually not able to obtain high IoU, and **b** the prior knowledge of the boundary box is obtained from tens of millions of samples using k-means++ (thin, high borders account for the majority)

of interest. The red solid border is the correct border, and the red dotted border is the highest border of IoU for different methods of generating boundary boxes. The average IoU of default boxes manually designed is only 47.6%. Figure 3b shows that after getting the priori boundaries of object shape and size from a large number of samples by clustering analysis, we can reduce the square boundaries, increase the boundaries of thin and high edges appropriately, effectively improve the proportion of IoU, and further improve the convergence speed and detection accuracy of the model.

It is different from K-means clustering algorithm used in Yolo and DSSD. K-means++ has better stability and efficiency in clustering results. Firstly, in order to ensure the high detection speed of the model, the default box size and shape in the model are usually taken as parameters, which are set up when the model is initialized, and no changes are made. This requires that the clustering results on the training set must be as stable as possible. However, K-means clustering algorithm is very sensitive to the initialization of parameters of the algorithm. Each randomly selected K clustering centers will produce completely different clustering results, which confuses researchers in choosing the priori shape of the boundary box. In addition, if the selection of clustering centers is inappropriate, the clustering results will be quite different, and even some results can not reflect the characteristics

of the data set. Cluster centers of k-means++ are divided in turn from least to most. Each time the elements farthest from the current cluster centers are selected as new cluster centers, which reduces the uncertainty of random selection of k-means, improves the speed and accuracy of clustering, and provides a better basis for the selection of priori boxes for the detection model.

2.3 Model training

During training, the data input to the classifier includes the feature mapping of the nth detection layer, the category of the object in the correct border, and the probability that the object belongs to a certain category. The data input to the detector includes the feature mapping of the nth detection layer, the position coordinates of the correct and priori borders, and the offset errors of the output priori and the correct borders. The reality is that there are many differences between the IoU of each priori box and the correct border. We use two methods to judge a priori box as a positive sample: (1) the highest priori box overlapping the correct border box IoU, or (2) the IoU of a priori box and any correct border is greater than 0.7 (a correct border can be used as a label for multiple prioris). Although the second condition is sufficient to determine the positive sample, we still use the first condition, because in some rare cases, the second condition may not find the positive sample (for example, an object is too small or too large). If the IoU ratio of a non-positive sample to all correct borders is less than 0.3, then we consider some prior boxes as negative samples. Finally, we discard the priori boxes that do not contribute to training, neither positive samples nor negative samples. The training method of the model originates from the multi-task minimization objective function, but extends to the category that can recognize multiple objects. The whole objective loss function is the weighting of detection loss (reg) and classification loss (cls):

$$L(x, c, p, g) = \frac{1}{N}\left(L_{cls}(x, c) + \alpha L_{reg}(x, p, g)\right) \quad (9)$$

Where x is the feature mapping of the output of different detection layers, c is the category of objects in a priori box d, p is the coordinate of the prediction box, and g is the correct border coordinate value. N is the number of matching priori frames. If $N = 0$, the loss is set to 0, and alpha controls the weight of detection error. Similar to Faster R-CNN, our regression method calculates the center $(\hat{t}_i^x, \hat{t}_i^y)$, width $(\hat{t}_i^w)$, and height $(\hat{t}_i^h)$ of the prediction box (p) and the center (t_j^x, t_j^y), width (t_j^w), and height (t_j^h) of the prediction box (p), respectively. Our regression method calculates the offsets between the center; the detection loss is based on the coordinate vector of the prediction box (p) calculated by smooth L1 based on the offset between $\{\hat{t}_i^m\}$ and the coordinate vector t_j^m of ground truth box g; AMME can train the model from end to end.

$$L_{reg}(x, p, g) = \sum_{i \in Pos}^{N} \sum_{m \in \{c_x, c_y, w, h\}} x_{ij}^k \text{smooth}_{L1}\left(\hat{t}_i^m - t_j^m\right)$$

$$(10)$$

Where:

$$\hat{t}_i^{c_x} = \frac{p_i^{c_x} - d_i^{c_x}}{d_i^w}, \hat{t}_i^{c_y} = \frac{p_i^{c_y} - d_i^{c_y}}{d_i^h}$$

$$\hat{t}_i^w = \log\frac{p_i^w}{d_i^w}, \hat{t}_i^h = \log\frac{p_i^h}{d_i^h}$$

$$t_j^{c_x} = \frac{g_j^{c_x} - d_i^{c_x}}{d_i^w}, t_j^{c_y} = \frac{g_j^{c_y} - d_i^{c_y}}{d_i^h}$$

$$t_j^w = \log\frac{g_j^w}{d_i^w}, t_j^h = \log\frac{g_j^h}{d_i^h}$$

The classified loss function is as follows:

$$L_{cls}(x, c) = -\sum_{i \in Pos}^{N} x_{ij}^p \log(\hat{c}_i^p) - \sum_{i \in Neg} \log(\hat{c}_i^0) \quad (11)$$

Where:

$$\hat{c}_i^p = \frac{\exp(c_i^p)}{\sum_p \exp(c_i^p)}$$

The model can use the error function above to optimize the proposed regions generated by all the prior boxes, but this will be biased toward negative samples, because their number of samples dominates. Therefore, in the training process, we use the min-batch method to randomly select 128 recommended areas at a time, and force the proportion of positive and negative samples to be kept at 1:1. If there are less than 64 positive samples in an image, the negative samples are filled in small batches. We randomly initialize all the new multi-resolution detection layers so that their parameters obey the Gauss distribution with zero mean and 0.1 variance. The basic network layer is initialized by pre-training the ImageNet classification model VGG-16.

3 Results

The hardware of the experiment is accomplished on a Dell server and equipped with two GTX-1080Ti GPUs. The operating system is Ubuntu 16.464 bits, which runs the Tensor-flow deep learning framework, and uses Tensor-board to monitor the training process. All the experimental results are based on VGG16 and trained in advance on ILSVRC datasets. Target detection training set and test set are Passcal VOC 2007, 2012, COCO, and our ImageNet iLOC data set based on Section 3.1. We use the AMME optimizer proposed in chapter 4 to fine-tune the model. The default parameters are learning_rate = 0.001, beta 1 = 0.9, beta 2 = 0.999, and = 1e-08. The learning rate of different data sets is slightly different from the setting of parameters of beta 1 and beta 2, which will be described in detail later.

3.1 Experimental results in PASCAL VOC 2007

On this data set, our SENET method is compared with SSD, YOLO, and Faster R-CNN. The data set used in this section includes PASCAL VOC 2007, training set in 2012, and verification set in PASCAL VOC 2007 and 2012, totaling 16,551 images. The test set uses PASCAL VOC 2007 test, including 4952 images. In the first 40 K iterations of the model, AMME uses learning_rate = 0.001, beta 1 = 0.9, beta 2 = 0.999, Euro = 1e-08, then reduces learning_rate = 0.0007, beta 1 = 0.75, beta 2 = 0.777, Euro = 1e-08, and then iterates 20 K. Table 1 shows that SENET's accuracy when using 300*300 as input has exceeded that of SSD model with the same size as input. This again shows that in Section 5.2.1, we

Table 1 The results of PASCAL VOC 2007 test set

Method	mAP	Aero	Bike	Bird	Boat	Bottle	Bus	Car	Cat	Chair	Cow	Table	Dog	Horse	Mbike	Person	Plant	Sheep	Sofa	Train	TV
Faster R-CNN	73.2	76.2	79.0	70.9	65.5	52.1	83.1	84.7	86.4	52.0	81.9	65.7	84.8	84.6	77.5	76.7	38.8	73.6	73.9	83.0	72.6
YOLO V2	73.4	86.3	82.0	74.8	59.2	51.8	79.8	76.5	90.6	52.1	78.2	58.5	89.3	82.5	83.4	81.3	49.1	77.2	62.4	83.8	68.7
SSD300	74.3	75.5	80.2	72.3	66.3	47.6	83.0	84.2	86.1	54.7	78.3	73.9	84.5	85.3	82.6	76.2	48.6	73.9	76.0	83.4	74.0
SSD512	76.8	82.4	84.7	78.4	73.8	53.2	86.2	87.5	86.0	57.8	83.1	70.2	84.9	85.2	83.9	79.7	50.3	77.9	73.9	82.5	75.3
EAO300	74.9	75.7	80.1	74.3	66.6	53.6	82.0	83.6	85.7	58.6	78.2	75.9	83.7	83.3	82.7	77.2	49.9	73.9	75.3	82.6	74.6
EAO512	78.1	85.7	85.4	78.8	71.3	55.4	84.9	87.3	86.9	59.2	82.8	74.3	85.9	87.1	85.7	81.9	54.5	78.7	74.1	84.9	76.3

can better match the correct border by clustering the boundaries of the training set, so as to improve the accuracy of the model. When the image training model of 512×512 is input more, SENET's mAP easily surpasses Faster R-CNN (mAP reaches 78.1%, 5.9%, 1.3% higher than SSD512). Moreover, most of its high confidence tests are correct, and the recall rate is about 85–90%. Compared with the R-CNN step training using two stitching methods, our SENET model directly regresses the shape of the object and the category of the classified object, so it is easier to train and optimize, so it has less detection error.

Table 1 shows the results of PASCAL VOC2007 test set. Among them, the input image sizes of model SSD, YOLO, and Faster R-CNN are 512×512. Our SENET model uses two sizes (300×300) and (512×512) to compare with each baseline model.

To illustrate the performance of two SENET models with different input sizes in more detail, we use tensorboard detection and analysis tools from Google tensorflow to analyze the training process of the model. As can be seen from Fig. 4 because of our multi-resolution detection layer, using more efficient priori box generation method can get better IOU value, which is conducive to accelerating the convergence speed and improving the accuracy of the model. Compared with SSD (right side of Fig. 4), the convergence speed of the model is faster

(about 60 K iteration model has converged), and mAP is higher than SSD. At the same time, we can also see that SENET's convergence is more stable than SSD's.

Most detection models detect smaller objects with worse performance than larger objects, mainly because after multi-layer convolution, the feature mapping of the smallest objects at the top level may not have any information left. In Fig. 5 based on the clustering results, we use different shape and number of priori boxes in different detection layers, which makes the model less sensitive to the size of boundary boxes than SSD. The experimental results also show that our SENET model has better performance and stronger robustness than SSD when detecting smaller objects.

3.2 Experimental results of MS COCO

MS COCO 2015 has 91 classifications, each of which has about 10,000 samples, and each sample (image) has about one to six objects. To further validate the proposed method, we train SENET 300*300 and 512*512 models on MS COCO datasets. Because COCO data sets have many kinds of objects, many objects to be detected in a single sample and small objects to be detected, the gradient oscillation of the model is large when it starts training. First, the parameters of learning_rate = 0.001, $\beta_1 = 0.9$, $\beta_2 = 0.99$, $\epsilon = 10^{-8}$, batch-size = 128 are trained

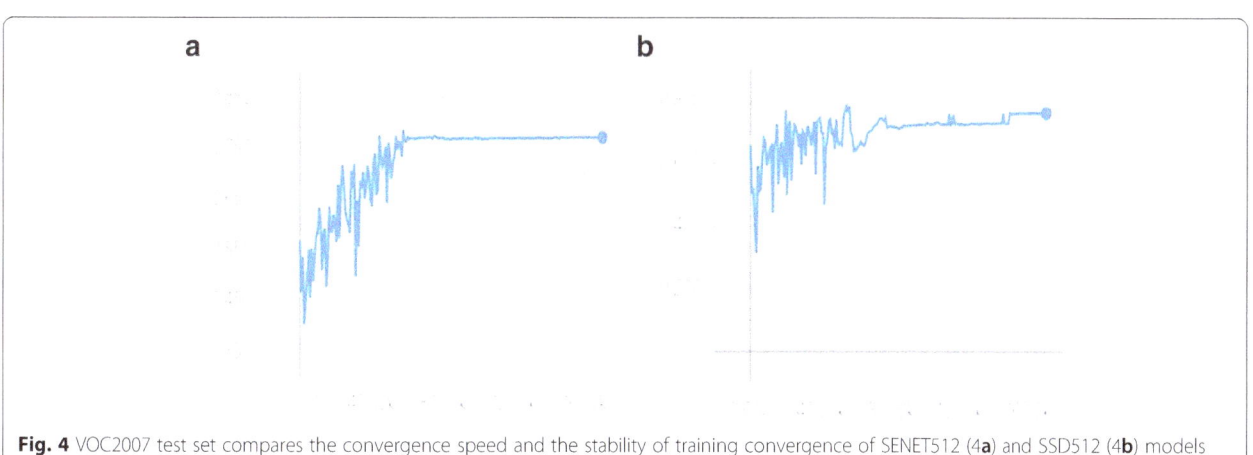

Fig. 4 VOC2007 test set compares the convergence speed and the stability of training convergence of SENET512 (**4a**) and SSD512 (**4b**) models

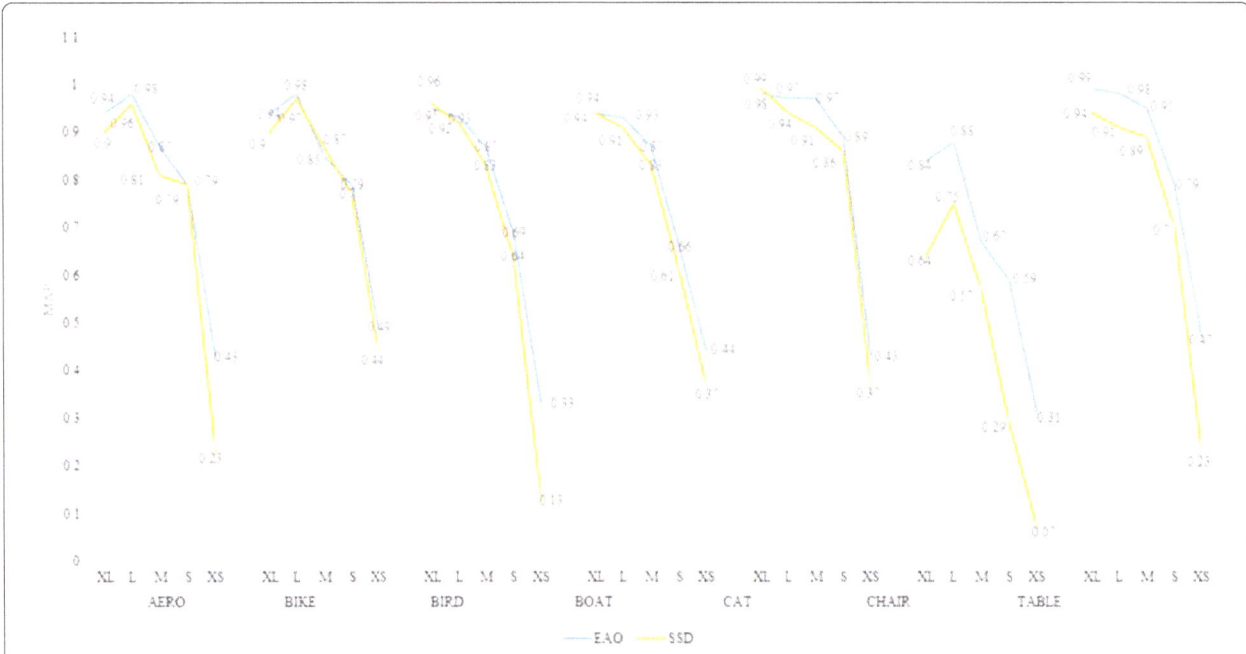

Fig. 5 The results of the bounding box detector demonstrate an imbalance between the classes. Each column represents a comparison of the number of borders for each class using different intersections within a 10% to 90% joint threshold range. SENET and SSD models are used to compare the accuracy of identifying objects of different sizes. The overall performance of SENET model is better than that of contrast model, and the recognition accuracy of chair with sparse structure is much higher than that of SSD

and iterated 140 K times. Then, the parameters of ADAM are adjusted to learning_rate = 0.0009, $\beta_1 = 0.9$, $\beta_2 = 0.999$, and $\epsilon = 10^{-8}$, batch-size = 64 continue to iterate 50 K times.

Following the strategy mentioned in Section 5.3.2, we cluster the size and shape of objects in COCO datasets and use smaller priori boxes for all detection layers. Then, based on the method of Section 5.3.3, the shape and number of cell priori frames in different detection layers are designed. We evaluated the mAP value with IOU < [0.5:0.05:0.95] (standard COCO measurement method, simply quoted as mAP@[.5,.95]) and mAP@0.5 (PASCAL VOC measurement method).

Table 2 shows the test results of each model on test-dev2015. Similar to what we observed on PASCAL VOC datasets, SENET 300*300 outperforms Faster R-CNN and YOLO in mAP@0.5 and mAP@0.95, and is very close to SSD512. However, whether SENET300 or SENET3512, its mAP@0.5 is significantly better than SSD and YOLO. We speculate that this is because the size of objects in MS COCO datasets is too small, and SSD and YOLO models are not good at locating many small objects accurately, which leads to model failure. The experimental results also show that by increasing the size of the input image to 512*512, SENET is more accurate than all baseline models in both test criteria. The experimental results also show that SENET512 model

is better than ION [162]. It is a multi-size version of Fast R-CNN, which uses cyclic neural network to explicitly simulate feature context. In Fig. 6 some detection results of MS COCO test-dev using 512*512 model and convergence process of SENET model boundary box are shown.

4 Discussion

Firstly, aiming at the problem of single sample and high production cost of current object detection data set, based on clustering analysis algorithm and image classification model, this paper proposes a method of integrating classification data into detection data set, and makes ImageNet iLOC detection data set from ImageNet classification data set. The experimental results show that the accuracy of the model can be improved by 4.3% by using the proposed ImageNet iLOC detection data set and

Table 2 Testing results of MS COCO show that SENET model performs well in small objects

Mode	Boxes	mAP@[0.5:0.95]	mAP@0.5	mAP@0.75
Faster R-CNN	RPN 300	21.9	42.5	21.9
ION	–	22.4	42.7	18.7
YOLO	97	21.4	43.8	19.6
SSD512	8735	25.5	44.9	22.7
SENET300	5965	21.7	45.1	24.6
SENET512	5965	26.9	47.6	28.2

Fig. 6 Some false positives produced in the primary diagnostic unit. **a** The ulcer sample was detected as plague. **b** The gray mold sample was detected as ulcer disease. **c** The low temperature sample was detected as ulcer disease. The detection effect of MS COCO dataset is shown, **a** the convergence process of boundary box location and object classification, and **b** the detection effect of SENET512 on COCO dataset

continuing to train the model. Then, aiming at the problem that the current popular object detection model can not design the default box manually and accurately, this chapter uses K-means++ clustering algorithm to cluster the shape of the object in the sample, and obtains the shape distribution of the object. Based on this priori knowledge, the designed priori box model and ground true box IOU are higher, which greatly improves the convergence speed of model training. Finally, according to the characteristics of different multi-resolution detection layers corresponding to different size regions of the original image, we carefully design the size and shape of the priori box of each detection layer to form an empirical adaptive single network detection model SENET. In the experimental part, I continue to train SENET model with ImageNet iLOC data set. The results show that the accuracy of the SENET model proposed in this chapter is improved by about 3–4% on average compared with other benchmark models. Future work will further improve the practicability of the model. For example, the model will start from the video of a specific scene (e.g., unmanned driving), real-time object detection in video content, and ensure high speed and accuracy.

Abbreviations

CNN: Convolutional neural network; fps: Frame per second; GPU: Graphics processing units; ILSVRC: ImageNet Large Scale Visual Recognition Challenge; IoU: Intersection over union; mAP: Mean average precision; R-CNN: Region proposal based CNN; RPN: Region proposal networks

Acknowledgements

Subsidies from the Training Program for 1000 Young and Middle-aged Key Teachers in Guangxi Colleges and Universities, Nanning Special Expert Project Funding.

Authors' contributions

AF designed the study, performed the experiments, and data analysis, and wrote the paper. DP and SY advised on the design of the system and analyzed to find the best method for efficient recognition of diseases and pests of Cole plants. JL provided the facilities for data collection and contributed with the information for the data annotation. All authors read and approved the final manuscript.

Authors' information

Zhiyong Wei received his B.Sc. degree in 2005 from Guilin Institute of Technology, received his M.S. degree in 2016 from Guangxi University, now he is Senior Engineer in Nanning College For Vocational Technology.His main research interest include image and video processing algorithms and data mining.
Fengling Wang received the B.S received M.S. degree in 2006 from BeiJing University Of Technology. now he is Professor in Nanning College For Vocational Technology. His main research interests include computer software theory, image processing algorithms and data mining.

Competing interests

The authors declare that they have no competing interests.

References

1. Pereyra G, Tucker G, Chorowski J, et al. Regularizing neural networks by penalizing confident output distributions[J]. arXiv preprint arXiv:1701.06548, 2017

2. A. Krizhevsky, I. Sutskever, G. Hinton, Imagenet classification with deep convolutional neural networks[J]. Adv. Neural Inf. Proces. Syst. 25(2), (2012)

3. O. Russakovsky, J. Deng, H. Su, et al., Imagenet large scale visual recognition challenge[J]. Int. J. Comput. Vis. 115(3), 211–252 (2015)

4. M. Everingham, L. Van Gool, C.K.I. Williams, et al., The pascal visual object classes (voc) challenge[J]. Int. J. Comput. Vis. 88(2), 303–338 (2010)

5. T.Y. Lin, M. Maire, S. Belongie, et al., in European Conference on Computer Vision. Microsoft coco: common objects in context[C] (Springer, Cham, 2014), pp. 740–755

6. J.R.R. Uijlings, K.E.A. Van De Sande, T. Gevers, et al., Selective search for object recognition[J]. Int. J. Comput. Vis. 104(2), 154–171 (2013)

7. S. Ren, K. He, R. Girshick, et al., Faster r-cnn: towards real-time object detection with region proposal networks[J]. IEEE Trans. Pattern Anal. Mach. Intell., 39(6), 1137–1149 (2017)

8. W. Liu, D. Anguelov, D. Erhan, et al., in European Conference on Computer Vision. Ssd: single shot multibox detector[C] (Springer, Cham, 2016), pp. 21–37

9. O. Russakovsky, Y. Lin, K. Yu, et al., in European Conference on Computer Vision. Object-centric spatial pooling for image classification[C] (Springer, Berlin, Heidelberg, 2012), pp. 1–15

10. C. Szegedy, A. Toshev, D. Erhan, Deep neural networks for object detection[C]. Adv. Neural Inf. Proces. Syst., 2553–2561 (2013)

11. J. Dai, Y. Li, K. He, et al., R-fcn: object detection via region-based fully convolutional networks[C]. Adv. Neural Inf. Proces. Syst., 379–387 (2016)

12. J. Redmon, S. Divvala, R. Girshick, et al., in Proceedings of the IEEE Conference on Computer Vision and Pattern Recognition. You only look once: unified, real-time object detection[C] (2016), pp. 779–788

13. J. Redmon, A. Farhadi, in Proceedings of the IEEE Conference on Computer Vision and Pattern Recognition. YOLO9000: better, faster, stronger[C] (2017), pp. 7263–7271

14. Simonyan K, Zisserman A. Very deep convolutional networks for large-scale image recognition[J]. arXiv preprint arXiv:1409.1556, 2014

15. K. He, X. Zhang, S. Ren, et al., in Proceedings of the IEEE Conference on Computer Vision and Pattern Recognition. Deep residual learning for image recognition[C] (2016), pp. 770–778

16. C. Szegedy, W. Liu, Y. Jia, et al., in Proceedings of the IEEE Conference on Computer Vision and Pattern Recognition. Going deeper with convolutions[C] (2015), pp. 1–9

17. S. Xie, R. Girshick, P. Dollár, et al., in Proceedings of the IEEE Conference on Computer Vision and Pattern Recognition. Aggregated residual transformations for deep neural networks[C] (2017), pp. 1492–1500

18. G. Huang, Z. Liu, L. Van Der Maaten, et al., in Proceedings of the IEEE Conference on Computer Vision and Pattern Recognition. Densely connected convolutional networks[C] (2017), pp. 4700–4708

19. Y. Chen, J. Li, H. Xiao, et al., Dual path networks[C]. Adv. Neural Inf. Proces. Syst., 4467–4475 (2017)

20. J. Hu, L. Shen, G. Sun, in Proceedings of the IEEE Conference on Computer Vision and Pattern Recognition. Squeeze-and-excitation networks[C] (2018), pp. 7132–7141

21. Zhang C, Bengio S, Hardt M, et al. Understanding deep learning requires rethinking generalization[J]. arXiv preprint arXiv:1611.03530, 2016

22. Bloice M D, Stocker C, Holzinger A. Augmentor: an image augmentation library for machine learning[J]. arXiv preprint arXiv:1708.04680, 2017

23. van Laarhoven T. L2 regularization versus batch and weight normalization[J]. arXiv preprint arXiv:1706.05350, 2017

24. N. Srivastava, G. Hinton, A. Krizhevsky, et al., Dropout: a simple way to prevent neural networks from overfitting[J]. J. Mach. Learn. Res 15(1), 1929–1958 (2014)

25. Ioffe S, Szegedy C. Batch normalization: accelerating deep network training by reducing internal covariate shift[J]. arXiv preprint arXiv:1502.03167, 2015

26. A. Fuentes, S. Yoon, S. Kim, et al., A robust deep-learning-based detector for real-time tomato plant diseases and pests recognition[J]. Sensors 17(9), 2022 (2017)

27. S.P. Mohanty, D.P. Hughes, M. Salathé, Using deep learning for image-based plant disease detection[J]. Front. Plant Sci. 7, 1419 (2016)

28. B. Liu, Y. Zhang, D.J. He, et al., Identification of apple leaf diseases based on deep convolutional neural networks[J]. Symmetry 10(1), 11 (2017)

29. J. Amara, B. Bouaziz, A. Algergawy, A deep learning-based approach for banana leaf diseases classification[C]. BTW (Workshops)., 79–88 (2017)

30. S. Sankaran, A. Mishra, R. Ehsani, et al., A review of advanced techniques for detecting plant diseases[J]. Comput. Electron. Agric. 72(1), 1–13 (2010)

31. Y. Kawasaki, H. Uga, S. Kagiwada, et al., *Basic study of automated diagnosis of viral plant diseases using convolutional neural networks[C]//International Symposium on Visual Computing* (Springer, Cham, 2015), pp. 638–645

32. S. Srdjan, A. Marko, A. Andras , et al., Deep neural networks based recognition of plant diseases by leaf image classification[J]. Comput. Intell. Neurosci. **2016**, 1–11 (2016)

33. K.P. Ferentinos, Deep learning models for plant disease detection and diagnosis[J]. Comput. Electron. Agric. **145**, 311–318 (2018)

34. Y. LeCun, C. Cortes, C.J.C. Burges, The MNIST database of handwritten digits, 1998[J]. URL http://yann.lecun.com/exdb/mnist. **10**, 34 (1998)

35. C. Wagner, *The sources of El cavallero Cifar[M]* (1903)

36. Abu-El-Haija S, Kothari N, Lee J, et al. Youtube-8m: a large-scale video classification benchmark[J]. arXiv preprint arXiv:1609.08675, 2016

37. A.W. Fitzgibbon, A. Zisserman, in *European Conference on Computer Vision.* Automatic camera recovery for closed or open image sequences[C] (Springer, Berlin, Heidelberg, 1998), pp. 311–326

38. A.Y. Ng, M.I. Jordan, Y. Weiss, On spectral clustering: analysis and an algorithm[C]. Adv. Neural Inf. Proces. Syst., 849–856 (2002)

39. P.F. Felzenszwalb, D.P. Huttenlocher, Efficient graph-based image segmentation[J]. Int. J. Comput. Vis. **59**(2), 167–181 (2004)

Trust evaluation model with entropy-based weight assignment for malicious node's detection in wireless sensor networks

Xueqiang Yin[1,2*] and Shining Li[1]

Abstract

Trust management is considered as an effective complementary mechanism to ensure the security of sensor networks. Based on historical behavior, the trust value can be evaluated and applied to estimate the reliability of the node. For the analysis of the possible attack behavior of malicious nodes, we proposed a trust evaluation model with entropy-based weight assignment for malicious node's detection in wireless sensor networks. To mitigate the malicious attacks such as packet dropping or packet modifications, multidimensional trust indicators are derived from communication between adjacent sensor nodes, and direct and indirect trust values will be estimated based on the corresponding behaviors of those sensor nodes. In order to improve the validity of trust quantification and ensure the objectivity of evaluation, the entropy weight method is applied to determine the proper value of the weight. Finally, the indirect trust value and direct trust value are synthesized to obtain the overall trust. Experimental results show that the proposed scheme performs well in terms of the identification of malicious node.

Keywords: Trust evaluation model, Entropy, Weight assignment, Security, Wireless sensor networks

1 Introduction

Nowadays, wireless sensor networks (WSNs) have become one of the most useful technologies and attracted more and more attention from researchers [1]. Owing to the capabilities of data acquisition, processing, and transmission, the sensor nodes can be deployed in many application scenarios, such as environmental monitoring, battlefield detection, industrial safety monitoring and health care, etc. However, due to the unmanned environment and the characteristics of energy-constrained, the sensors are vulnerable to various attacks. By capturing some normal nodes, the attackers can change their behavior and then insert false data or decisions to mislead the decision-making of the whole network. In addition, the sensor nodes may be problem-prone to non-malicious errors, such as inadequate residual energy and faults of wireless transceiver or components, and then, result in unreliable data generation [2]. Especially, to improve the energy efficiency, data aggregation in

sensor networks is needed. Once a node is captured, the errors or forged data sent by the fault node will impact the entire fusion result. Therefore, network security of WSNs is a crucial problem to be solved [3].

In the field of network security, asymmetric cryptography is widely used to deal with external attacks in the Internet, peer-to-peer, and ad hoc networks. However, due to the complexity and demand of huge computational memory, the encryption algorithm is not suitable for WSNs due to limited processing power and resource constraints [4]. In addition, the security mechanism based on encryption can only solve external security problems and cannot effectively deal with internal attacks. In WSNs, the particularity of nodes, different from other networks, can refuse to cooperate with service requesters to save energy, and those nodes are being called selfish nodes. Although they do not actively attack the network, a large number of selfish nodes may cause serious consequences. Obviously, the existing encryption mechanism is incapable to identify the risks caused by authenticated selfish nodes. Therefore, it is necessary to establish an effective security mechanism to solve those problems [5]. In recent years, trust management has

* Correspondence: yinxueqiang@mail.nwpu.edu.cn

[1]School of Computer Science, Northwestern Polytechnical University, Xi'an 710029, People's Republic of China

[2]The 15th Research Institute of China Electronic Technology Group Corporation, Beijing 100083, People's Republic of China

been regarded as an effective complementary mechanism to ensure the security of sensor networks. Based on historical behavior, the trust value of a node can be evaluated to estimate the reliability based on the performance of specific tasks. At present, many typical trust models have been proposed for WSNs, which derives from game theory [6], Bayesian estimation [7], D-S evidence [8], fuzzy logic [9], etc. The above models all identify malicious nodes through trust evaluation to a certain extent and provide a theoretical basis for further research. Under the open network environment, the trust between sensor nodes will vary dynamically with time and behavior The trust value obtained by the trust model should also change with the communication behavior between nodes, thus effectively restrain the abnormal increase or decrease of trust value and resist the influence of flattery or slander between nodes. How to define the trust relationship in the model as well as improve the efficiency of the model implementation becomes an important issue.

The rest of this paper is organized as follows: after the related works are summarized in Section 2, the trust evaluation model is presented in detail in Section 3. In Section 4, we present the steps of secure communication under the proposed model. We evaluate the performance of our trust evaluation model in Section 5. And finally, we conclude this paper in Section 6.

2 Related work

For internal malicious node attacks, most of the effective defense measures are often built on trust model management. Trust evaluation can be abstractly referred to as the estimation of the relevant evidence affecting the trust of the subject. Generally, trust can be measured in a way similar to information or knowledge and formulated as degree of trust. The trust value can be defined as the combination of direct trust and indirect trust, which can be given a certain weight according to the specific application requirements [10]. Behavior-based trust evaluation models can be divided into centralized and distributed trust evaluation model. Different working modes directly affect the data exchange mode between participants in trust assessment, as well as the data processing, trust calculation during the phase of trust assessment.

In the centralized trust evaluation model, the center obtains global information such as exchange records between sensor nodes or user's feedback and calculates trust according to a certain rule. Ganeriwal et al. [11] proposed a reputation-based framework for high integrity sensor networks, and they introduced beta function to calculate reputation and trust value. In [12], Probst et al. introduced a statistical method into a trust management model. Trust value can be estimated based on the

direct and indirect experience of nodes and confidence interval is applied to identify malicious behavior. Cheng et al. [13] presented a trust model based on D-S evidence theory, in which the comprehensive trust value can be obtained by D-S combination rules. According to historical behavior, trust fluctuation, and recommendation inconsistency in a certain period of time, Anita et al. [14] proposed a routing trust prediction model based on fuzzy theory. The model can predict the subsequent behavior of neighbor nodes, but it may lead to the loss of information. Aivaloglou et al. [15] proposed a hybrid trust and reputation management model based on the certificate-based method and behavior-based mechanism. The model utilizes the knowledge of network topology and data flow to support the highly diversified needs of node's roles. Combining fuzzy and gray theory, Wu et al. [16] proposed a trust model with incentive mechanism to evaluate the reliability of nodes. However, because of the complexity of model calculation, it is not suitable for sensor networks with limited processing capacity of nodes. Generally, the centralized trust evaluation model has the characteristics of a relatively simple structure and less difficulty to implement. However, due to overwhelmingly dependent on central nodes, load balancing and robustness become the bottleneck of further exploitation.

In contrast, in the distributed trust evaluation model, the trust degree does not depend on the support of the central entity. Through the direct interaction with the evaluated entity, the recommendation of the direct interaction from all entities can be synthesized to estimate the trust degree. In [17], Jiang et al. proposed an efficient distributed trust model according to the exchange messages from all sensor nodes, and the trust metrics include communication overhead, energy consumption, and data validity. In [18], Bao et al. proposed a hierarchical dynamic trust management protocol for clustered wireless sensor networks and develop a probability model using stochastic Petri net techniques to analyze the performance. Zhang et al. [19] proposed a multi-level trust management framework. In this framework, three levels of trust, namely subjective trust, objective reputation, and recommendation trust, are used to establish trust relationships among nodes. The shortcomings lie in the lack of trust sharing and update mechanism. To ensure the security of data forwarding and improve energy efficiency, Tang et al. [20] proposed a trust-based secure routing scheme using the trace back approach, in which the data and notification employ a dynamic probability of marking and logging during routing selection. Based on the hierarchical network structure, Liao et al. [21] proposed a weighted trust evaluation strategy, which updates the weighted trust value continuously by comparing the data collected by sensor nodes and the final data fusion results. The anomaly nodes detection and trusted data filtering mechanism can obtain good performance and scalability. To achieve the tradeoff between energy conservation and network security, Liao et al. [21] presented a mixed and continuous monitor-forward model based on game theory to mitigate the selective-forwarding attack, in which the monitoring node conducts a strategy continuously to determine the duration of behavior surveillance. However, the selfishness and rationality of sensor nodes are not thoroughly considered.

Taking into account of energy consumption and secure routing, many studies combine the construction of trust model with the clustering management mechanism of nodes. Shaikh et al. [22] proposed a group-based trust management mechanism and applied it to cluster-structured wireless sensor networks. The calculation of trust value is achieved by monitoring the communication behavior between neighbor nodes, including member node's trust, cluster head's trust, cluster trust, and base station trust. The trust model can effectively resist malicious node attacks and protect malicious nodes from defamation and defamation attacks as well as keep energy-efficiency. Zhou et al. [23] proposed a trust evaluation model based on the autonomous behavior of sensor nodes. Sensor nodes acquire direct or indirect trust values by monitoring the behavior of neighbor nodes. Cluster heads calculate comprehensive trust values according to D-S evidence theory. By trust evaluation, malicious nodes can be effectively identified and malicious nodes can be restricted to become cluster heads. Crosby et al. [24] designed a distributed trust-based cluster head election mechanism. The trust table was constructed by monitoring the transmission process of neighbor nodes, and the trust degree was calculated. Then, the reliable cluster head was elected according to the trust degree, which ensured the reliability of data fusion and network security. For secured data fusion, Fu et al. [25] introduced a cluster-based trust model with double cluster heads structure, in which the dissimilarity coefficient is defined to evaluate the data fusion results. If the fusion results exceed the threshold value, it demonstrates that the cluster head is possible to be compromised nodes and then to be added to the blacklist.

3 Trust model

3.1 Trust indicators

The purpose of trust evaluation is to provide support for trust decision-making to establish a reliable relationship between the entities. Combining with the implementation of security strategy, it can form a general trust management system. In WSNs, the sensor's authentication depends not only on the historical data of the node itself, but also on the adjacent nodes with spatio-temporal correlation. The characteristics of node behavior often vary with time, and the regularity has some statistical characteristics. Therefore, the behavior of nodes can be analyzed, and a quantitative evaluation model can be established through the history of interaction between nodes. Specifically, the sensor nodes in adjacent areas monitor each other and calculate their trust, which can effectively identify malicious nodes to resist network attacks.

The selection of trust factors is the premise and foundation of calculating node's direct trust, and the trust elements should conform to the characteristics of WSNS. Malicious attacks launched by nodes mainly include stealing, tampering with perceptual information, injecting a lot of error information, etc. Therefore, we can analyze the data repetition rate, the number of data packets, data correlation, and the volatility of data latency.

Definition 1: *Data repetition rate*. The data repetition rate of samples can reflect the node's abnormal behavior owing to repeat sending packets continuously.

$$\mathrm{DRR}_{i,j}(u, v, t) = \frac{S_{u,v}(t) - SP_{u,v}(t)}{S_{u,v}(t)} \qquad (1)$$

where $S_{u,\,v}(t)$ is the number of sent samples at time t, and $SP_{u,\,v}(t)$ is the number of the repeated samples.

Definition 2: *Packet size abnormality*. If the number of samples during the monitoring cycle is too large, it may be a denial of service attack. On the other hand, if the number is too small, the possibility of selfish behavior is high.

$$\mathrm{PSA}(u, v, t) = \frac{|\,S_{u,v}(t) - \Delta S(t)\,|}{S_{u,v}(t)} \qquad (2)$$

where $\Delta S(t)$ denotes the expected value for the number of samples.

Definition 3: *Data correlation*. The data collected by neighbor nodes have certain correlation, and the difference between normal nodes should be within a certain range.

$$\mathrm{DC}(u, v, t) = \alpha e^{-r[D_u(t) - D_v(t)]^2} \qquad (3)$$

where α is the attractiveness parameter, and r represents the distance between node s_u and s_v. $D_u(t)$ and $D_v(t)$ represent the measured value of node s_u and s_v, respectively.

Definition 4: *Volatility of transmission delay*. Due to signal interference and other factors in wireless communication, data transmission delay will occur in nodes. The neighboring nodes have temporal and spatial correlation. The transmission delay of networks should fluctuate within a certain range.

$$VTD(u,v,t) = \frac{\sum\limits_{k=1}^{h} RT(u,k)-ST(u,k)}{\sum\limits_{k=1}^{h} RT(v,k)-ST(v,k)} \qquad (4)$$

where h denotes the average number of hops between node s_u and s_v, and $RT(i,k)$, and $ST(i,k)$ represents the time of receipt and delivery of samples, respectively.

3.2 Clustering objective function

Quantification of trust relationship needs to meet the dynamic requirements of the environment, and it should also show the exact emphasis according to the impact of measurement indicators [26, 27]. Generally, under the condition of multiple monitoring indicators, the weight value has certain experience and subjectivity, which is not conducive to the validity of trust quantification and evaluation [28, 29].

In this paper, the trust evaluation model divides the nodes into categories of normal nodes, relay nodes, and base station in the perception layer. In the process of evaluating node behavior trust, only relay nodes generate recommended trust values among themselves, and it is assumed that the base station is fully trusted. Let s_1, s_2, $\cdots$, s_n denote n adjacent relay nodes of the evaluated target. According to the index mentioned above, the observation vector $(r_{i1}\ r_{i2}\ \cdots\ r_{im})$ is obtained at the ith relay node. The evaluation matrix $R_{n \times m}$ can be constructed, in which r_{ij} represents the evaluation result of jth indicator from ith relay node, $1 \le i \le n$, $1 \le j \le m$.

Generally, under the condition of multiple monitoring indicators, the establishment of weights has certain experience and subjectivity, which is disadvantageous to the validity of trust quantification and evaluation [30, 31]. In this paper, the weight of monitoring index will be solved based on the method of entropy weight.

First, the membership matrix $U_{m \times n}$ is defined, and the matrix element u_{ij} represents the degree of membership of r_{ij} with constraint of

$$\sum\limits_{j=1}^{m} u_{ij} = 1, 0 \le u_{ij} \le 1. \qquad (5)$$

Next, to indicate the difference between the recommendation entity and the expectation caused by the objective deviation, we define the recommended deviation Δ as:

$$\Delta_j = \frac{1}{n}\sqrt{\sum\limits_{i=1}^{n} u_{ij} \sum\limits_{j=1}^{m} (r_{ij}-\bar{r}_j)^2} \qquad (6)$$

where $\bar{r}_j$ represents the average value of jth indicator, and $\bar{r}_j = \frac{1}{m}\sum_{i=1}^{m} r_{ij}$

The objective of clustering is to find the optimal clustering vector so as to minimize the overall recommended deviation, and the objective function can be expressed as:

$$\min\{\Delta^2\} = \min\{\sum\limits_{j=1}^{m}\sum\limits_{i=1}^{n} u_{ij} \sum\limits_{i=1}^{m}\left[\frac{1}{n}(r_{ij}-\bar{r}_j)\right]^2\} \qquad (7)$$

According to the definition of membership matrix $U_{m \times n}$, u_{ij} can be regarded as the probability that the ith entity belongs to the jth monitoring index. Therefore, the information entropy of jth monitoring index for the relay node s_i can be calculated as:

$$H = -u_{ij}\ln(u_{ij}) \qquad (8)$$

Accordingly, the total information entropy of the matrix $U_{m \times n}$ can be expressed as:

$$H^* = -\sum\limits_{j=1}^{m}\sum\limits_{i=1}^{n} u_{ij}\ln(u_{ij}) \qquad (9)$$

In order to minimize the clustering function and optimize the overall information entropy, the optimization process can be described as follows:

$$\min\{-\sum\limits_{j=1}^{m}\sum\limits_{i=1}^{n} u_{ij}(\sum\limits_{i=1}^{m}\left[\frac{1}{n}(r_{ij}-\bar{r}_j)\right]^2) + \frac{1}{\rho}\sum\limits_{j=1}^{m}\sum\limits_{i=1}^{n} u_{ij}\ \ln(u_{ij})\} \qquad (10)$$

where ρ is the equilibrium factor of the equation.

By using the Lagrange multiplier method [32, 33], the constraint $\sum_{j=1}^{m} u_{ij} = 1$ can be introduced into the Lagrange multiplier λ, and the Eq. (9) can be transformed to

$$L(u_{ij},\lambda,t_j) = \sum\limits_{j=1}^{m}\sum\limits_{i=1}^{n} u_{ij}(\sum\limits_{i=1}^{m}\left[\frac{1}{n}(r_{ij}-\bar{r}_j)\right]^2)$$
$$+ \frac{1}{\rho}\sum\limits_{j=1}^{m}\sum\limits_{i=1}^{n} u_{ij}\ \ln(u_{ij})$$
$$+ \lambda\left|\sum\limits_{i=1}^{n} u_{ij}-1\right| \qquad (11)$$

Solving the objective function $L(u_{ij},\lambda,\bar{r}_j)$, u_{ij} can be derived as

$$u_{ij} = \frac{\exp(-\rho\sum\limits_{i=1}^{n}(r_{ij}-\bar{r}_j)^2)}{\sum\limits_{j=1}^{m}\exp(-\rho\sum\limits_{i=1}^{n}(r_{ij}-\bar{r}_j)^2)} \qquad (12)$$

According to the membership degree and recommendation deviation degree, the weight values of monitoring indicators with normalization can be obtained as

$$
\begin{cases}
CR_j = \bar{r}_j \times (1-k_j) \\
w_i = \dfrac{CR_j}{\sum\limits_{i=1}^{n} CR_j}
\end{cases}
\tag{13}
$$

$$
\omega_i = \dfrac{\sum\limits_{j=1}^{m} |r_{ij}-\bar{r}_{ij}|}{\sum\limits_{i\in\Omega}\sum\limits_{j=1}^{m} |r_{ij}-\bar{r}_{ij}|}
\tag{16}
$$

where CR represents the comprehensive recommendation for jth indicator from entities.

Finally, based on recommendation trust and corresponding weight value, the quantitative results of direct trust evaluation between relay node k and monitored node can be estimated as:

$$
DTrust_k = \sum_{j=1}^{m} w_j r_{kj}
\tag{14}
$$

Then, the indirect trust can be estimated as

$$
ITrust_k = \dfrac{\sum\limits_{i\in\Omega} \omega_i DTrust_i}{|\Omega|}
\tag{17}
$$

Definition 5: *Total trust.* By synthesizing indirect trust with direct trust, the total trust can be obtained as follows:

$$
TTrust = \theta DTrust + (1-\theta)ITrust
\tag{18}
$$

where $\theta \in [0,1]$ and indicate the trustworthiness degree to the trust value.

3.3 Indirect trust

The indirect trust can be regarded as the recommendation from the third party [34]. The indirect trust value of node s_u to node s_v is composed of the direct trust value of all recommendation nodes to node s_v, and the recommended nodes are referred to as the common neighbor nodes of node s_u and s_v. However, not all recommendation nodes are trustworthy, and unreliable recommendation will provide false information to evaluate the trustworthiness of the nodes, which will affect the trust of the sensor nodes that create and manipulate the data.

In order to calculate the indirect trust value accurately through recommendation nodes, it is necessary to select trusted neighbors as recommendation nodes. First, we define a specified trust threshold δ, and the nodes with direct trust higher than δ will be selected as the recommended neighbors set. As multiple nodes push trust values to a single node at the same time, it may bring opportunities to malicious nodes. Malicious nodes intentionally elevate or degrade the trust of a node by sending false or conflicting recommendation trust values. Therefore, multiple recommendation trust problems must be solved through trust merge rules. Suppose Ω denote represents the set of trusted neighbor nodes of the evaluated node and relay node s_k, and $|\Omega|$ represents the number of nodes in set Ω. Firstly, the average value of all evaluation result will be calculated as

$$
\bar{r}_{ij} = \dfrac{\sum\limits_{i\in\Omega} r_{ij}}{|\Omega|}
\tag{15}
$$

Next, the weight ω_i of the recommendation node can be obtained as:

4 Secure communication

Based on the proposed trust evaluation model, the secure communication is introduced based on AODV protocol. In the initial stage, the identity-based cryptography mechanism is applied to verify the legitimacy of nodes and establish a trusted network environment. Then, the specific flow of its secure communication is as follows:

Step 1: Initially, all nodes broadcast their own identity information in the network. All neighbor nodes in the communication range can calculate their shared keys according to the private keys and identification, which can be used to encrypt and decrypt exchange message between them.

Step 2: If the source node s_S prepares to communicate with the destination node s_D, it will query the local routing table whether exists a route to the destination. If not, the route to node s_D should be established by the following steps.

Step 3: s_S broadcasts query message RREQ to its neighbor s_k, and s_k will determine whether the same query information has been processed. If so, the current request message is discarded. Otherwise, the number of hops in the query message will plus 1.

Step 4: Then, s_S will retrieve the direct trust of neighbor s_k in its local storage module and broadcast a trust query message to its neighbor nodes. All the nodes receiving the source node trust query information check whether they have the trust value of the node s_k. If so, the trust response message encrypted with the shared secret key between itself and the source node s_S will be returned.

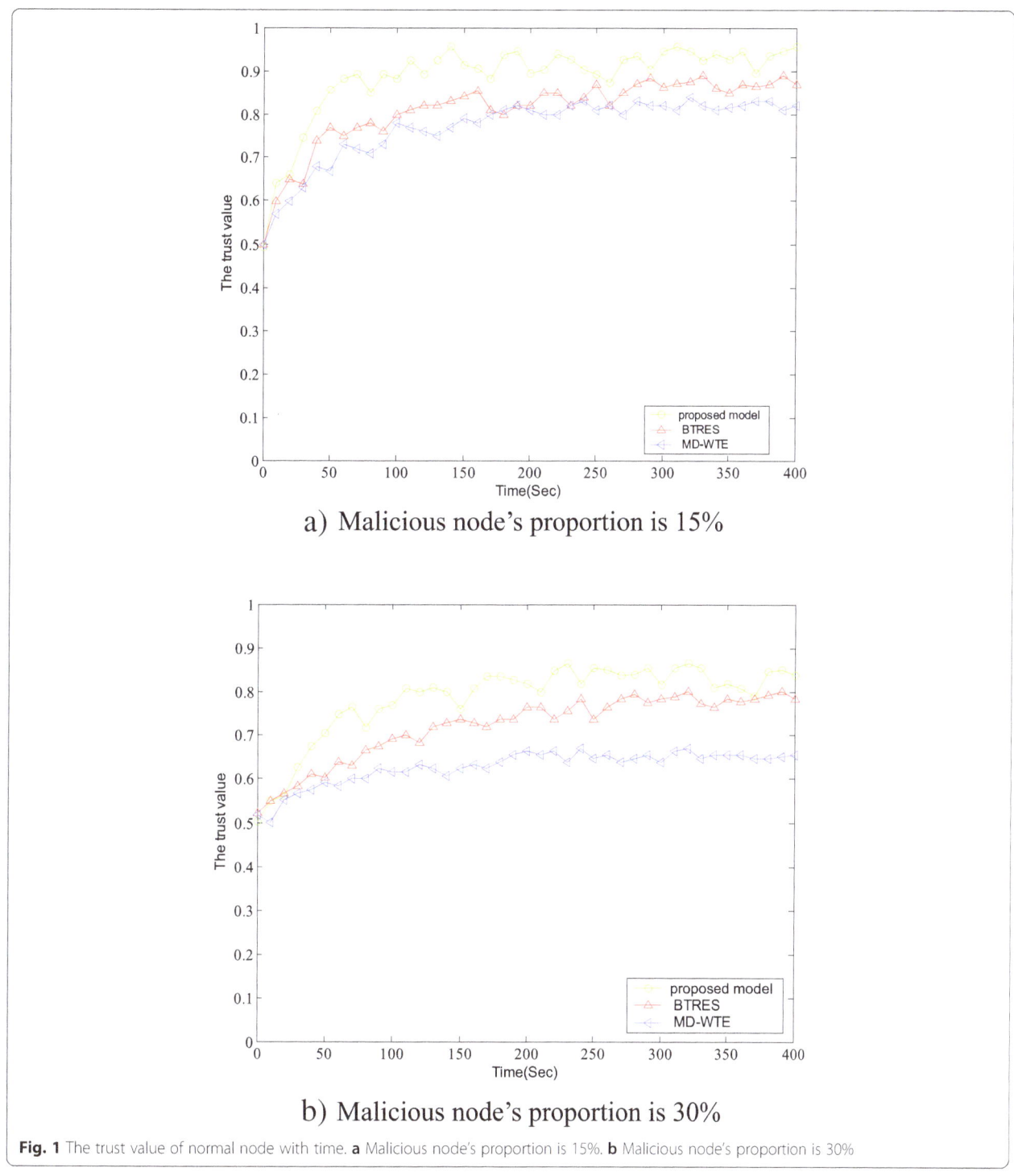

a) Malicious node's proportion is 15%

b) Malicious node's proportion is 30%

Fig. 1 The trust value of normal node with time. **a** Malicious node's proportion is 15%. **b** Malicious node's proportion is 30%

Step 5: Finally, s_S can obtain the comprehensive trust of the node s_k. If trustworthy, it will be regarded as the next hop node and continue to forward the routing query message RREQ. Otherwise, return to execute Step 2.

Step 6: Execute Step 3 repeatedly until a trusted route from s_S to destination s_D can be resolved.

5 Experimental results

In order to verify the validity of the proposed trust model for wireless sensor networks, simulation experiments are conducted. The size of the network is 100m × 100m. One hundred sensor nodes are randomly distributed in the region, and the base station is located in the center of the monitoring area. The perception

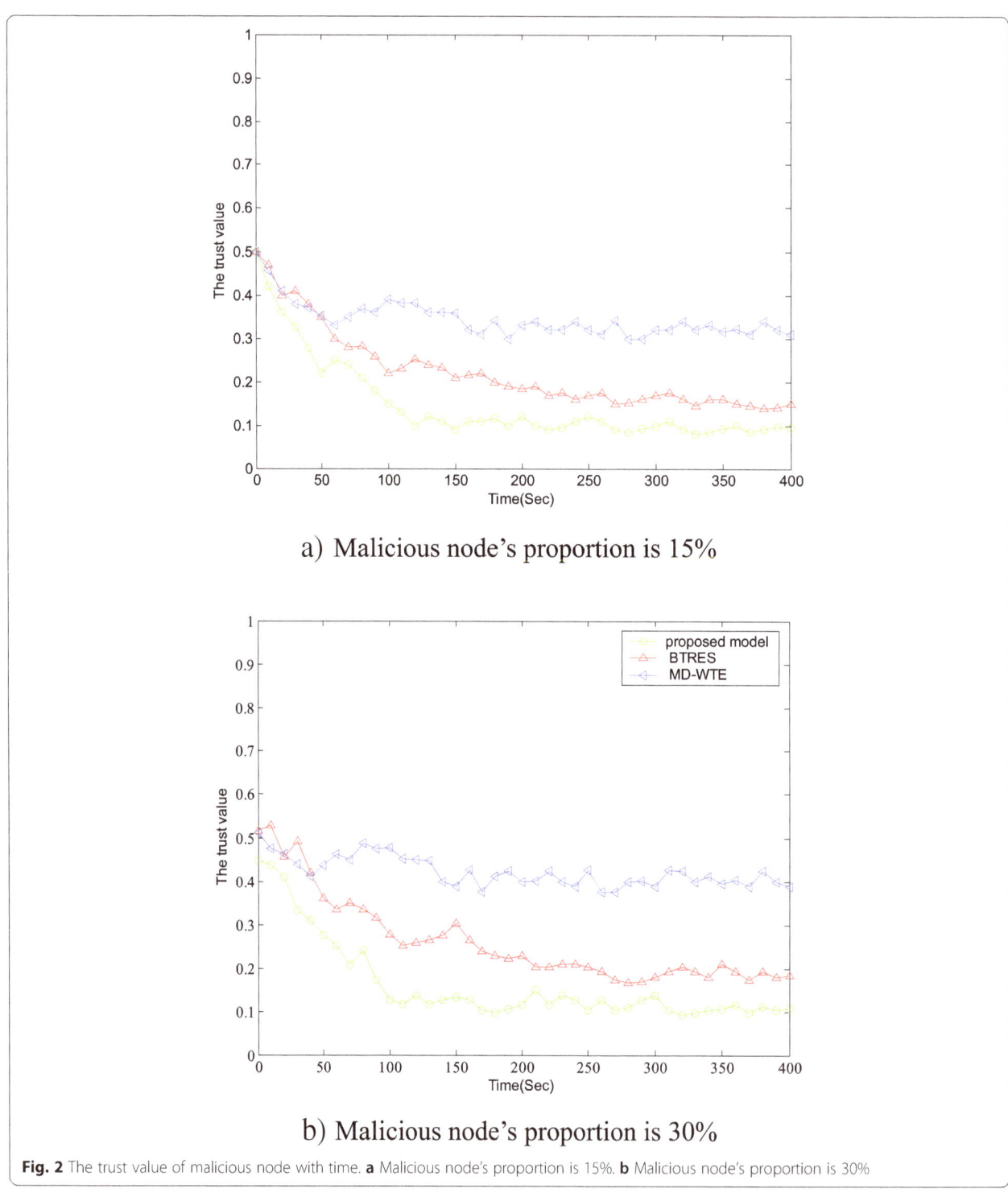

a) Malicious node's proportion is 15%

b) Malicious node's proportion is 30%

Fig. 2 The trust value of malicious node with time. **a** Malicious node's proportion is 15%. **b** Malicious node's proportion is 30%

radius of nodes is 20 m and the communication range is set to 40 m. In the simulation experiment, the proportion of malicious nodes is arranged from 0 to 40%, and they simulated by the several kinds of attacks, including selective forwarding attack, data forgery attack, DoS attack, and on/off attack. Each simulation time is 400 s, and the time period of trust update is equal to 10 s. The

assignment of other parameters is as $m = 4$, $\theta = 0.5$ and $\delta = 0.7$.

The performance of the proposed method is compared with that of MD-WTE [34] and BTRES [35]. Figures 1 and 2 show the trust value of the normal node and malicious node with time as malicious node proportion is 15% and 30%, respectively. The experimental results

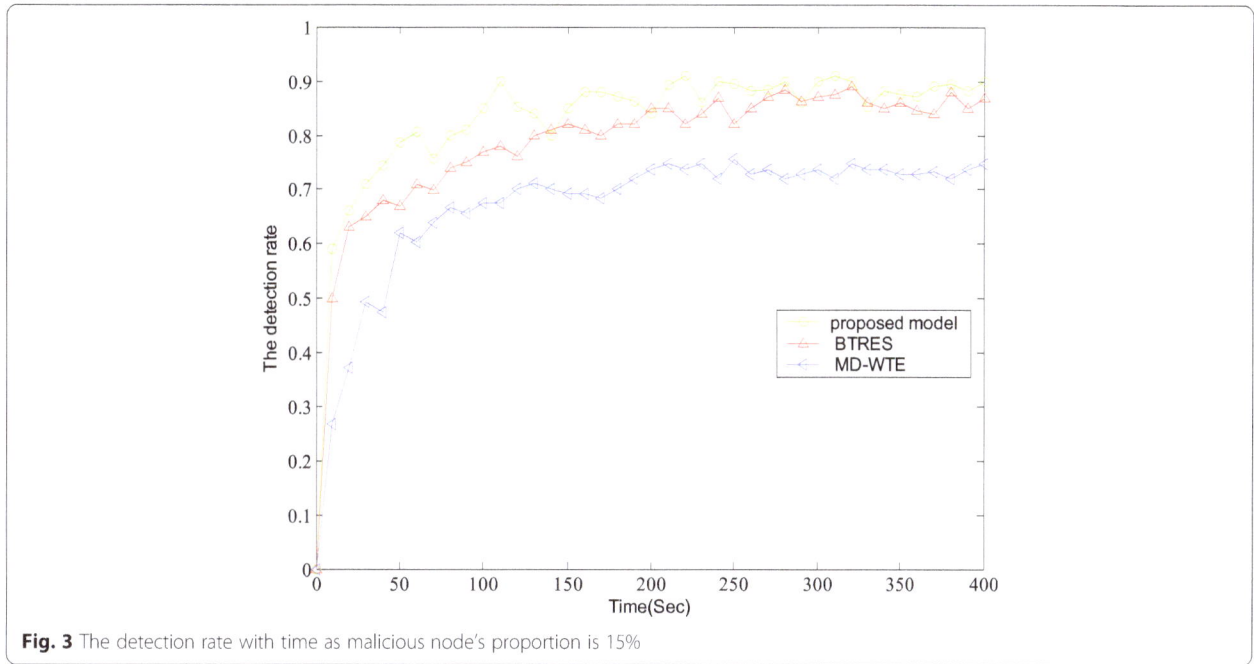

Fig. 3 The detection rate with time as malicious node's proportion is 15%

show that in MD-WTE, the average trust value of normal nodes is similar, but the trust value of abnormal nodes is obviously higher than that of other methods. As a result, the malicious nodes cannot be distinguished clearly. That is because only one trust value is applied in MD-WTE, and the malicious sensor nodes can hide malicious behavior of their sensing through trusted transfer function. The nodes can still maintain high reliability by masking malicious packet loss via trusted sensing behavior. In our proposed model and BTRES, the direct trust values are closer to the object trust values compared with the integrated trust values since the integrated trust values are more or less influenced by the malicious recommendations. However, they take communication behavior into account to calculate sensor nodes' trust value and improve the accuracy of recommendation trust against the selective forwarding attack and the data forgery attack.

As can be seen from Fig. 2, when the proportion of malicious nodes is about 30%, the trust value of normal

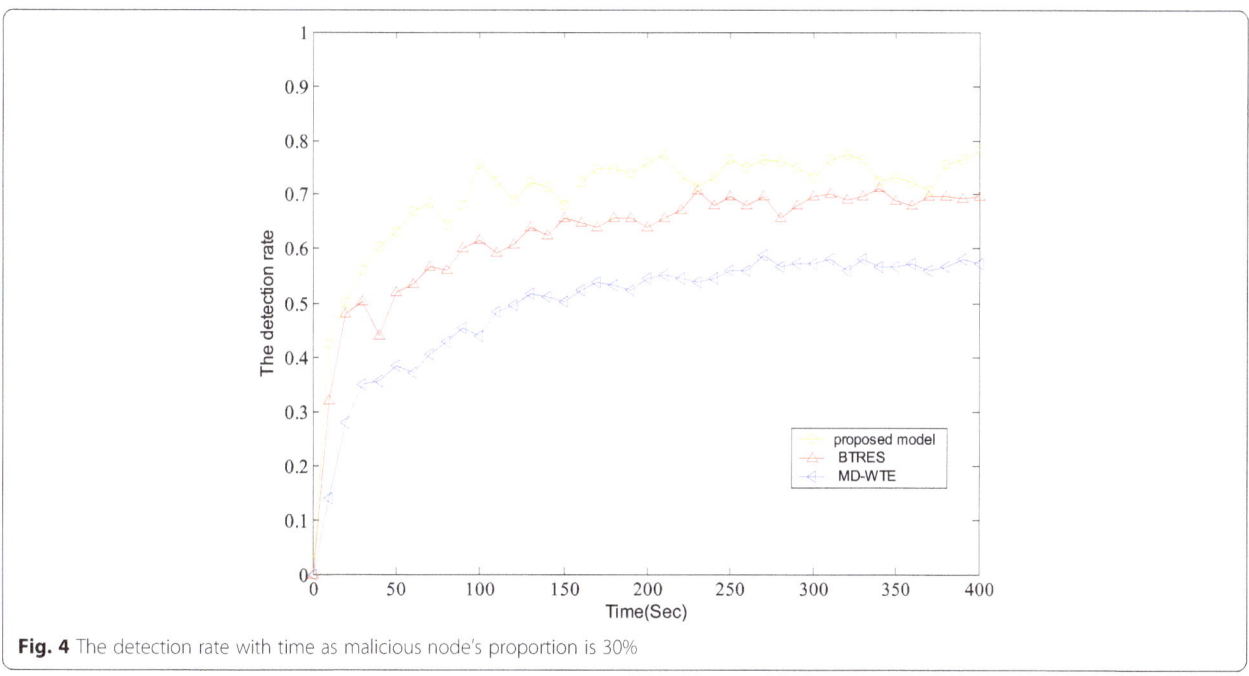

Fig. 4 The detection rate with time as malicious node's proportion is 30%

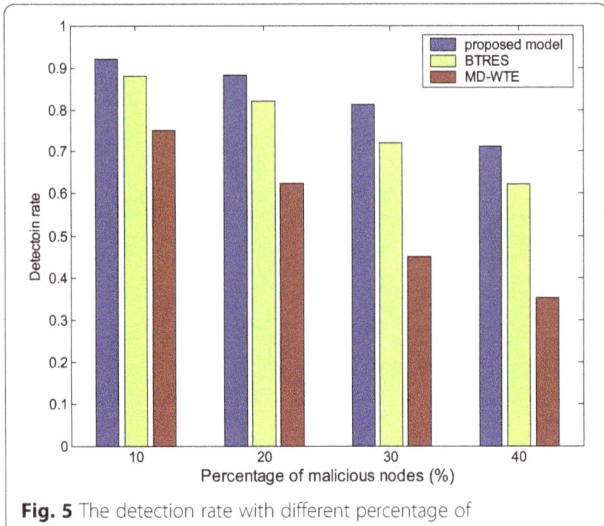

Fig. 5 The detection rate with different percentage of malicious nodes

conducive to the delivery of data by normal nodes in the network.

Furthermore, we evaluate the detection rate and false error rate of malicious nodes. Due to open environment and the performance of nodes being uncertain, not all nodes can be judged as trusted or untrustworthy state. Figures 3 and 4 show the differences in detection rate when malicious nodes are 15% and 30%, respectively. As can be seen from the results, our proposed model and BTRES have risen rapidly from the beginning of different scenarios and have maintained a high detection rate in the process of operation. The reason is that both of the methods can reduce the dependence on prior experience and the assumption of prior distribution, which improves the speed and accuracy of identifying the malicious nodes. Comparatively, after accumulating a certain amount of records, the detection rate in MD-WTE increases gradually. In our proposed model, entropy-based weight assignment improves the objectivity of trust evaluation and obtains fast convergence rate.

Next, we analyze and compare the detection rate and false alarm rate under different percentage of malicious nodes. As can be seen from Figs. 5 and 6, our proposed model and BTRES have higher detection rate and lower false alarm rate when the proportion of malicious nodes is small. As the proportion of malicious nodes increases, the detection rate of MD-WTE decreases rapidly, and the false alarm rate also increases sharply. The detection rate of our proposed model decreases slowly and gradually stabilizes, and the promotion of false alarm rate is relatively small. The reason is that the fuzziness and conflict of evidence increase as the proportion of

nodes decreases obviously; meanwhile, the trust value of malicious nodes increase. This trend also shows that when malicious nodes reach a certain proportion, the normal nodes in the network cannot effectively detect the abnormal behavior of malicious nodes. Hence, the average trust value of malicious nodes increases. Compared with the other methods, our proposed model can reduce the impact of malicious nodes more effectively. It illustrates that the direct trust and recommendation trust should be dynamically adjusted based on the proportion of malicious nodes. Once the percent of malicious nodes exceeds the extent, the normal behavior of nodes may be mistaken as abnormal. That will not be

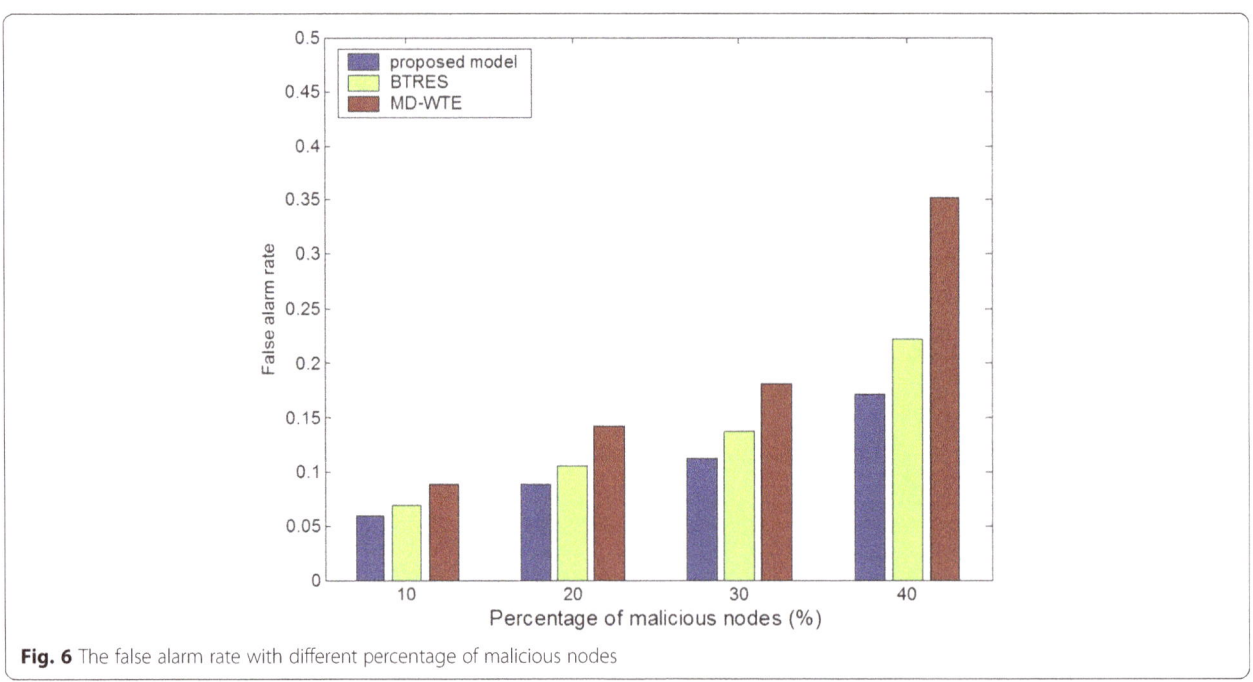

Fig. 6 The false alarm rate with different percentage of malicious nodes

malicious nodes increases. By means of selecting the proper value of the weight of direct trust and indirect trust, our proposed model improves the accuracy of the malicious node's detection and robustness of trust evaluation model.

6 Conclusions

In this paper, we proposed a trust evaluation model with entropy-based weight assignment for malicious node's detection in wireless sensor networks. Multidimensional trust indicators are derived from communication between adjacent sensor nodes, and direct and indirect trust values will be estimated based on the corresponding behaviors of those sensor nodes. To improve the validity of trust quantification and ensure the objectivity of evaluation, the entropy weight method is applied to determine the proper value of the weight. In our future work, we will devote to evaluate the trust level of sensor nodes in clustered WSNs and combine the trust model with secure data fusion. Besides, some professional techniques, e.g., fuzzy logic or pattern recognition, will be employed and discussed to reduce the fuzziness of behavior evidences.

Abbreviation
WSNs: Wireless sensor networks

Acknowledgements
The authors acknowledged the anonymous reviewers and editors for their efforts in valuable comments and suggestions.

Authors' contributions
XY proposed the innovation ideas and theoretical analysis, and he carried out the experiments and data analysis. SL conceived of the study, participated in its design and coordination, and helped to draft the manuscript. All authors read and approved the final manuscript.

Competing interests
The authors declare that they have no competing interests.

References
1. G. Han, J. Jiang, L. Shu, J. Niu, J. Comput. Syst. Sci. **80**(3), 602–617 (2014)
2. A. Ahmed, K.A. Bakar, M.I. Channa, A.W. Khan, K. Haseeb, Peer-Peer Netw. Appl. **10**(1), 216–237 (2017)
3. B. Wu, X. Yan, Y. Wang, C. Guedes Soares, Risk Anal. **37**(10), 1936–1957 (2017)
4. M.A. Simplício, P. Barreto, C.B. Margi, Comput. Netw. **54**(1), 2591–2612 (2010)
5. Y.M. Huang, M.Y. Hsieh, H.C. Chao, IEEE J. Sel. Areas Commun. **24**(7), 400–411 (2009)
6. J. Duan, D. Gao, D. Yang, C.H. Foh, IEEE Internet Things J. **1**(1), 58–69 (2014)
7. M.K. Denko, T. Sun, I. Woungang, Comput. Commun. **34**(3), 398–406 (2011)
8. W. Zhang, S. Zhu, J. Tang, J. Supercomput. **74**, 1779–1801 (2018)
9. B. Wu, L. Zong, X. Yan, C. Guedes Soares, Ocean Eng. **164**, 590–603 (2018)
10. D. He, S. Zeadally, N. Kumar, J.H. Lee, IEEE Syst. J. **11**(4), 2590–2601 (2016)
11. S. Ganeriwal, M. Srivastava, *Proceedings of the 2nd ACM Workshop on Security of Ad-Hoc and Sensor Networks Washington DC* (2004), pp. 66–77
12. M.J. Probst, S.K. Kasera, *Proceedings of 2007 International Conference on Parallel and Distributed Systems* (2007), pp. 1–8
13. R. Feng, S. Che, X. Wang, Int. J. Distrib. Sens. Netw. **9**(6), 1–9 (2013)
14. X. Anita, M.A. Bhagyaveni, J. Manickam, Sci. World J. **8**(1), 341–356 (2014)
15. E. Aivaloglou, S. Gritzalis, Wirel. Netw **16**(5), 1493–1510 (2010)
16. G. Wu, Z. Du, Y. Hu, et al., Soft. Comput. **18**(9), 1829–1840 (2014)
17. J. Jiang, G. Han, F. Wang, L. Shu, IEEE Trans. IEEE **26**(5), 1228–1237 (2014)
18. F. Bao, I.R. Chen, M.J. Chang, et al., IEEE Trans. Netw. Serv. Manag. **9**(2), 169–183 (2012)
19. T. Zhou, C. Wu, J. Zhang, D. Zhang, Saf. Sci. **96**, 183–191 (2017)
20. J. Tang, A. Liu, J. Zhang, Sensors **18**(3), 751 (2018)
21. H. Liao, S. Ding, Int. J. Distrib. Sensor Netw. **1**, 1–13 (2015)
22. R.A. Shaikh, H. Jameel, B.J. D'Auriol, et al., IEEE Trans. Parallel Distrib. Syst. **20**(11), 1698–1712 (2008)
23. J.M. Zhou, F. Liu, Q.Y. Lu, J. Sensors **68**(9), 907–913 (2014)
24. G.V. Crosby, N. Pissinou, J. Gadze, *Proceedings of Second IEEE Workshop on Dependability and Security in Sensor Networks and Systems* (IEEE, Columbia, 2006), pp. 13–22
25. J.S. Fu, Y. Liu, Sensors **15**(1), 2021–2040 (2015)
26. A. Boukerch, L. Xu, K. El-Khatib, Comput. Commun. **30**(12), 2413–2427 (2007)
27. M. Al Ameen, J. Liu, K. Kwak, J. Med. Syst. **36**(1), 93–101 (2012)
28. H. Marzi, A. Marzi, *Proc. of IEEE CIVEMSA* (2014), pp. 64–69
29. B.K. Kannan, S.N. Kramer, J. Mech. Des. **116**(2), 405–411 (1994)
30. H. Oh, C.T. Ngo, IEEE Sensors J. **18**(5), 2184–2194 (2018)
31. Y. Wang, E. Zio, X. Wei, D. Zhang, B. Wu, Int. J. Disaster Risk Reduct. **33**, 343–354 (2019)
32. C. Wan, X. Yan, D. Zhang, Z. Qu, Z. Yang, Transport. Res. E-Log **125**, 222–240 (2019)
33. M.Y. Zhang, D. Zhang, F. Goerlandt, X. Yan, P. Kujala, Saf. Sci. **111**, 128–143 (2018)
34. I.M. Atakli, H. Hu, Y. Chen, et al., *Proceedings of the International Symposium on Simulation of Systems Security* (2008), pp. 836–843
35. W. Fang, C. Zhang, Z. Shi, J. Netw. Comput. Appl. **59**, 88–94 (2016)

16

Wireless multimedia sensor network for rape disease detections

Zhou Libo[1,2], Huang Tian[2,3] and Guan Chunyun[1*]

Abstract

In order to reduce pesticides, it is also necessary to understand how plant diseases originate in the beginning and to be around them all around the world. One of the best ways to achieve this goal is to promote images that are not visible to the world (WMSN) in an agricultural country. However, the sending of a wireless device with little knowledge of the images will add more traffic to traffic, especially electricity. In this pamphlet, we create a diagnostic procedure that is aimed at driving WMSN's additional resources. From the study of the locally available nodes of the plants, this new technique can make the first choice on the vegetation of the plants, knowing how to send pictures to the control center to continue to explore, to move forward. Look at the Internet. The same method includes the distribution of images using color and shape, and its 2D profile is used as part of the stages. What happens on metal images and lack of food indicates that the correct route is 94.5%.

Keywords: Plant diseases, Object detection, Convolution neural networks, Filter banks, Rape diseases

1 Introduction

Agriculture is now changing from modern farming to modern agriculture. Agricultural marketing (IOT) will play a major role in promoting agricultural farming, including agricultural infrastructure development, the development of modern agricultural technology and agricultural use. The use of modern technology, computer technology, and computers is a computer-simultaneous work of coordinating the development of smart farming and productive farming [1].

Crop rotation as a major activity for agricultural material is of great importance [2]. Because of climate change, it is not only necessary to do a thorough investigation and research on the growth of the crop, as well as to understand the small crop yields. The growth of the crop and the growth of crop yields provides scientific and agricultural support. In order to maximize the profitability of farming, various species of biodiversity and biodiversity were established. The crop rotation system also confirms two types of information and re-engineering. The room for environmental creation produces heat and soil, such as heat, humidity, wind, wind,

rain, pH prices and so on [3]. Panel drawing photo gets a picture of the size of the plant. The size and size of the crop can be shown directly. Many numbers make up a moth-free area, and they can use the Internet.

The whole story is being prepared in this way. The second phase creates the design of the hardware platform. The third stage, the provision of scalar sensor node is asked. Then in the fourth stage, the formation of a photo snapshot picture is described. The fifth stage, the program is launched. The sixth stage produces test results. Finally, summarize this pamphlet.

1.1 Related works

Work offices use WSN node (or node) to learn more about the soil and send data to archives. Micro devices have sensors, systems and communication trips. The key points to consider are the need to use data, time and resources for communication, the size of the connection to node and sensors, crop requirements and so on. These articles are discussed in this section. Because the sequence of the list is divided into size groups, each part of the field is managed by one or more groups made up of one or more WSN additions. Soils will have more information about soil, climate, and / or crops. When airplane travels at each camp, the data is sent to a storage

* Correspondence: guancy188@163.com
[1]College of Agriculture, Hunan Agricultural University, Changsha 410128, China
Full list of author information is available at the end of the article

facility, which sends packages through the UAV airplane, which is represented as an airplane.

This natural body has been described as a comprehensive web site [4, 5]. It also allows for node sections to be explained in accordance with minimum requirements and requirements, regardless of communication. In this mode, the number and functions of node are upping as these factors are used to establish and grow in each field. In this sense, the area where the height of height is not too high (eg, temperature) will be monitored by a number. Instead, neighboring glasses should be used to monitor changes in amplitude (eg gas, shaking). Good bulletin and internet system are proven by trials: from the christmas movement, the number of natural tendencies is learned to describe big changes. With these data, the particles can be moved quickly, and the example used in the inner location is found [6].

However, communication skills should also take into account. The delivery of windows should re-do dust work, but to try to create a mesh bullet that permits full cooperation between the nodes nearby. It allows all the data to be stored by placing the flight number in the jungle, so avoid the need to find it correctly. Access to more information from any neighboring location also contributes to the best way to solve each phase. It seems to make the whole world run smoothly and causes problems related to communication [7]. This process works best if there is no bad grid that is meant to be originally meant for the environment, but without working properly. When node distribution really causes the spread of stories about communication becomes important only. Even at this time, construction can make sure that the work is more productive and increase the amount of time required. In addition, it is not only the way to prepare eggs for the eggs. As shown in these books, many of the methods used are described and described in a series of times that are sometimes outdated and time-consuming, not working and waiting for regular communication [8]. However, the house will be full of overflowing sequence where the main package is being sent out from each point to reduce the harvest time [9].

Convolutional neural networks are now considered to be a great way to explore something. With advanced technology, the most efficient machines are the same. Among them, we mentioned the best known methods and notes. We settled on three recent construction projects: high-speed construction in the areas (Fast R-CNN) [10] in the paper, Single shot multi-box detector (SSD) [11] and network-based convolution network (R- FCN) [12]. As explained in [13], although the meta-column was originally planned by another device (VGG, ResNet etc.), we now use unequal enhancements for infrastructure. Therefore, all construction must be connected with any drawings, depending on the work or needs.

Fast R-CNN, the test process is executed out in two steps. Firstly, the manufacturing area (RPN) takes the picture as a reflection and preparation by drawing [11]. Medium tools are used to define conflicts; everyone has a score. In order to teach RPN, the plan to consider whether it includes the ironic idea of the object in terms of the IOU, between what they want and world standards. In the second section, a box that is already made is designed to lift objects from the same graph. Therefore, these damage are fed into additional iron sheets to showcase the potential of the modern listings and boxes in all areas. All of this is done through a collaborative partnership, allowing the system to share all the tools related to the change and form of interface, so that it can achieve unique limits. Since the start of the R-CNN start-up, it has been linked to a number of jobs due to its good performance in known knowledge by the district.

SSD infrastructure [14] uses feedforward network convolution to solve problems. Websites make up a regular group of boxes and availability of items in each box. The nets can do different things by combining prophecies from different groups with different answers. In addition to that, the SSD includes the process to become a single internet, thus avoiding the process of disruption process as well as timing time [15].

The R-FCN program [16] provides the use of graphs that can use to solve the problem of transformation. This proposal is analogous to a Fast R-CNN, but not from the same previews that are invading the region, but from the last areas that are shaking sideways (a region with greater chance of including things or being part of it) before it can [17]. By using this method, the storage capacity used in the local population is reduced. In [16], they demonstrated that using the ResNet-101 if a transparent device can make competitions compared to the Faster R-CNN.

So as to combat the type of wireless weaknesses such as quick communication, power consumption, limited use and power conservation and low-density, a mixed product consisted by three different sections is made: 1. The search system. They are consisted from a machine with no electricity, which can manage the seed to make it known (for example, a good sample, storage, and drive). This is explained in detail in Section 2.1. 2. Mobile phone guide. The revised version is associated with the phone number that carries the vehicle and is used as the data collection section, described in section 2.2. 3. Long-term communication Finally, a computer-based network service provider that offers long-term services. This is explained in section 2.3. The activity on the package is displayed on Fig. 1.

In order to meet the demands of the project, for example, to look at the distance between the garden and

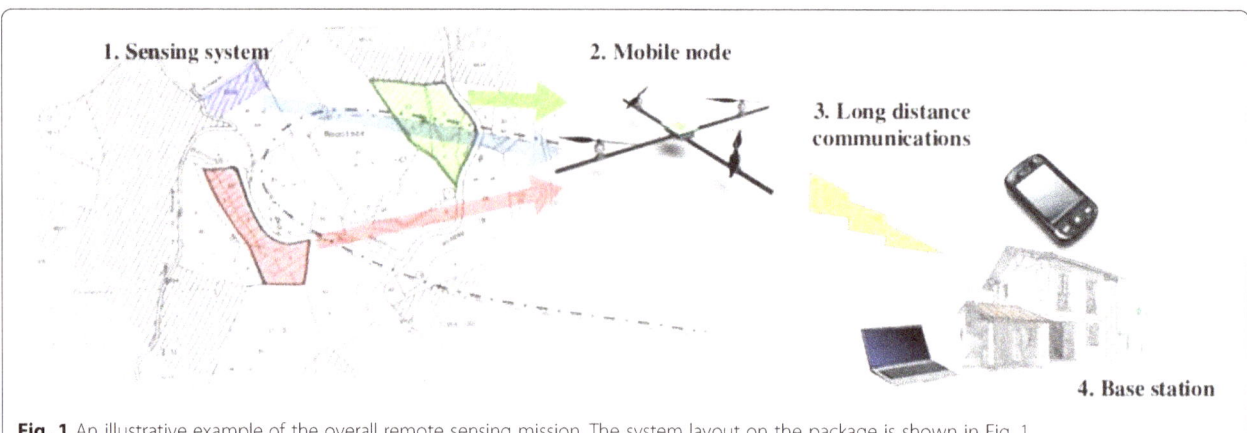

Fig. 1 An illustrative example of the overall remote sensing mission. The system layout on the package is shown in Fig. 1

the reserve, the package used for GPRS is allowed. Communication with an UAV database. GPRS connections allow data data to transmit from the dust of the vineyard until the train passes longer. The solution to this problem is just a few (QoS) of the catalyst machine, so it may seem like powerful and reliable. Unlike GPRS modules, modems placed on the UAVs allow data to be supplied to undermine the most advanced GPRS-outdoor areas outside of remote villages. The data is stored and then published as the UA arrives on the site with the interview. Another program is to use SMS (SMS), which allows regular messages to be sent to farm managers. To

increase their knowledge. The above-mentioned method is based on the Arduino Board board and the GPRS team, which includes HiLo SAGEM method (see fig. 2).

This node is a self-representative SoC THLK2405 and CMOS photo drawing OV7640. Photo preview is included in SoC THLK2405. The graphic section includes three stages: the image area of the image, data capture and stress area. OV7640 offers additional display or image enhancements 8 in different words; is monitored by the web camera interface. OV7640 has images that can run faster for 30 frames per second, and run the image on the character of the photography, publishing

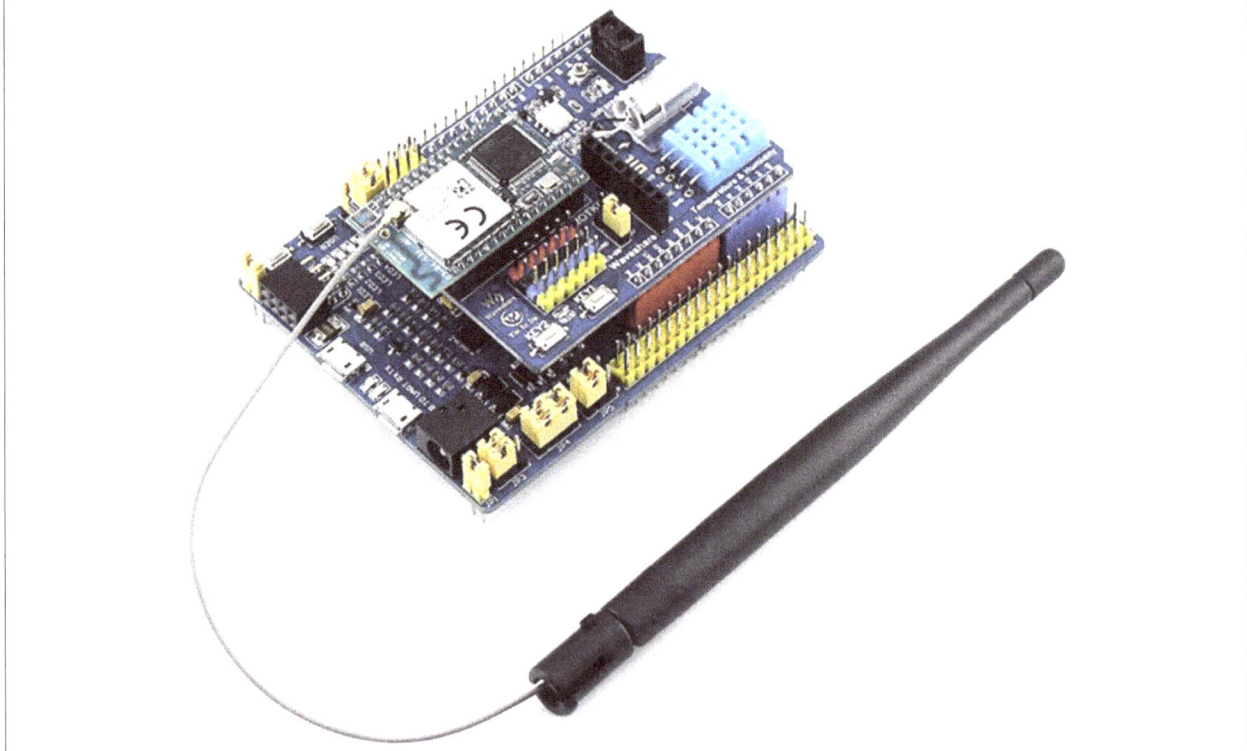

Fig. 2 Arduino system employed for communication with the base station, likewise for message delivery. The system mentioned above is based on the Arduino Uno board and the four-band GPRS module, which includes a HiLo SAGEM communication module (see Fig. 2)

and publishing data. All the basic artwork, including ensuring ensuring, gamma, white cleanliness, color display, loss and so on, can also be processed through SCCB form. Figure 3 shows the headings and images of pictures.

2 Methods

Tomato trees are protected against many diseases and damage that cause disease and pests. Several reasons may be due to the harvest: (I) abiotic complications due to nature, such as heat, humidity, nutrients, fertilizers, bright colors and plants; (2) pathogens from plants to plants such as whitefly, vegetable gardens, worms, and so on. (3) Many diseases, including bacteria, viruses, and fungi. The disease is a pest, as well as a plant that has a variety of shapes, such as shape, color, shape, etc. Therefore, because of similarities, these changes are difficult to distinguish, which makes their awareness difficult, recognition and preparation can prevent many of the damage. Based on the information we have discussed, we consider the following characteristics:

- infection: according to the life cycle of the disease, plants show different patterns and their infection status.
- Location of symptoms: disease affects not only the leaves, but also the other location of the plant, such as stems.
- Leaf pattern: symptoms of disease show a marked change in the front or back of the leaf.

- Type of fungus: determining the size of fungus can be a simple way to distinguish clearly certain diseases.
- Color and shape: according to the disease, the main plants body may be look like different size at different time of infection.

Figure 4 showed the characteristic of different conditions and varying diseases and pests identified in our method. A detailed research of the symptoms of each class and pest is described in [18].

At each step, the window will sliding to a new location, it will predict the current region proposal at the same time, so the size of the *reg* layer will be 4 times larger than the number of the windows, compute the positions number of k boxes. Based on the research results of others, we found that using a variety of different feet the feature map of degree is predicted, and better results will be obtained. Therefore, we choose to use the output of 6 convolutional layers as the prediction layer of our network, which are res3b3 [19], res5c/conv1–2 [20], res5c/conv2–2 [21], res5c. /conv3–2, pool6. There will be two branches after each layer, one branch is responsible for predicting candidate frames, and one convolutional layer prediction candidate frame can be directly followed by the feature map. The other branch is responsible for predicting which category the candidate box belongs to [22]. Of course, each The candidate box for a layer of prediction and the number of categories it belongs to are fixed. Therefore, we can get 6 sets of candidate boxes, then we only need to combine the 6 sets of

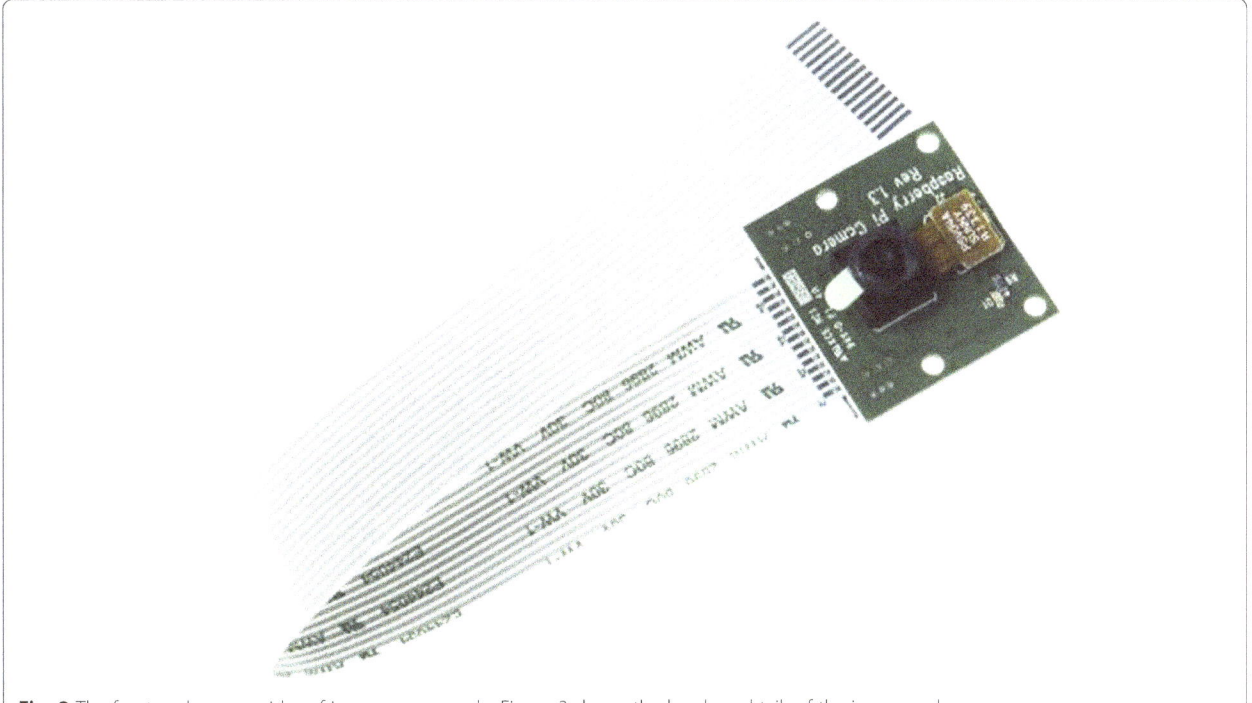

Fig. 3 The front and reverse sides of image sensor node. Figure 3 shows the heads and tails of the image nodes

Fig. 4 The representative of diseases and pests affecting tomato plants. In Fig. 4, we show representatives of different conditions and varying diseases and pests identified in our work. (**a**) gray mold, (**b**) canker, (**c**) leaf mildew, (**d**) plague Yersinia pestis, (**e**) leaf miner, (**f**) whitefly, (**g**) low temperature, (**h**) nutrition excess or lack of, (**i**) powdery mildew. Collect images under different changing and environmental conditions. These patterns help to distinguish between certain appropriate characteristics of each disease and pest

candidate boxes together and calculate the loss function, then update the parameters of the network. The detailed structure of the network is described below.

For a complete target process, the first thing we need to do is to build a data set for our algorithm to learn. This data set should include the original image, as well as the location of each object in the original image. The detailed information we have Introduced in 2.2. Next we need to do a simple processing of the data set. Due to the algorithm used in this paper, we need to adjust the image to a fixed size. The fixed size used in this paper is 300×300. Next, we need to make an enhancement to the data, mainly because the training data included in this article is relatively small, so we need to increase the amount of data. Our loss function for the image is defined as:

$$L = \frac{1}{N} \sum_i L_c\left(p_i, p_i'\right) + \lambda \frac{1}{N} \sum_i p_i' L_r\left(a_i, a_i'\right) \quad (1)$$

Here, i is the index of a small number of anchors, and p_i is the prediction probability of the anchor i as an object. If the anchor is positive, the ground live label p_i^* is 1, and if the anchor is negative, it is 0. t_i is the vector representing the four parameterized coordinates of the predictive boundary frame, and t_i^* is the vector of the ground live frame associated with the positive anchor. Classification loss L_{cls} is the logarithmic loss of two classes (object and non-object).

For regression loss, we use $L_{reg}\left(t_i, t_i^*\right) = R(t_i - t_i^*)$, where R is a robust loss function (smooth L1) defined in [23].

For one of the feature maps [24] in each of the above layers, a solid number of candidate boxes. For example, if the size of the feature map is m × n and the number of candidate frames generated by each position is k, then for a detection task of a class c object, $(c + 4) \times k \times m \times n$ data. We call him the prior box. Here's how k is generated.

It is necessary to generate k differences in the same position of the feature map. The candidate box is because it can handle the problem of objects of different scales. The aspect ratio used in this paper is {2,3}. In the feature map of res3b3, we only use the aspect ratio of 2, that is, in the layer of res3b3, each point on the feature map is generated. 3 candidate boxes, each of the other feature maps produces 4 candidate boxes. Thus we get m × n × k candidate boxes. The six candidate boxes consist of two squares, min_size×min_size, max_size×max_size, and the other two are calculated based on the aspect ratio and min_size. This way we have a lot of candidate boxes. Next we need to match the candidate box to the real box we have calibrated in our dataset to get the real box of our training. First we need to calculate the IOU [25] of the candidate box and the real box we generated in the current layer, and then select the positive sample with IOU greater than 0.5, and the negative sample with less than 0.5. However, there is a big problem in doing this. Because we generate a candidate box at every position, there will be a large number of samples that are negative samples, so that we will have sample imbalance when training. The final training results have a relatively large impact. Therefore, we need to deal with the phenomenon of sample imbalance here, using focal loss to deal with such a situation, mainly to balance the negative and positive samples on loss. The method used in this paper is to use the difficult sample mining method. In this paper, the value of the loss function of each candidate box is calculated in the forward calculation, and then several candidate boxes with the largest loss function value are selected to be retained. Other candidates The box is directly discarded to ensure that the ratio of positive and negative samples in the training process is 3:1. In this way, we have obtained all the training data.

3 Experiments/results

We tried the difficulties in our Cole and pathogens, for example [17]. It includes about 5000 images collected from many Cole fields in Korea. Pathogens and diseases can be carried out in different areas such as weather, space and moisture. Thus, the images are collected in different ways according to time (such as lighting), weather (such as heat, humidity), and sweet surfaces (using a greenhouse). Additionally, our dataset includes photographs and variations, samples from the beginning, middle and end endings, images contained with different pages of plants (eg, stems, leaves, fruits, etc.), different plants, green make-up plants, etc. components Examples for each class are illustrated in Table 1. After the additional supplementary techniques, the number of annotation figures is similar to the set limits for the image in the image: geometric change (change, crop, fluctuation, fluctuation) and radical change (contrast to lighting, color, sound). The previous class is a horizontal class

with a more detailed picture. When creating a CNN storage library, its box is used as an unpleasant example.

However, the alternate page view of Alternaria is a very similar type of rust, which determines their number. As shown in Fig. 5, an example of GoogLeNet Inception is used to remove the original statues of the image, making it more likely to deteriorate in different places. Therefore, the requirements used for CNN are a clear indication of the rape of the leaves.

In addition to the databases used [18], we have a new class with "yellow curl" images. As mentioned earlier, one of the major problems that we have found to reduce energy generated by the system is the balance between classrooms because of the limited resources and data available. This can be illustrated by the number of individual pictures, as shown in Table 1 and Fig. 6.

The size of the data cells affects the correctness of the apple's awareness. In this paper, two experimental teams are being conducted to show that the damage is real. The model was already taught and after data development. From the results shown in Fig. 7, it may be clear that the teaching method is a non-disclosure process, and eventually results in a total of 86.79%. However, the additional requirements and policies make up 97.62% accurately, making this total by 10,83% compared to those that were not lost. It can be seen from the results which are known as, as shown in Fig. 7, this phenomenon is due to the following reasons: (1) Systematic arrangements that are made up of various specialized photography techniques provide many opportunities for learning how to apply models on CNN; (2) the spread of different types of photographs to the teaching of weight gain in the example from CNN, but the photographic figure of the image is a wide range of variations and will cause serious problems; (3) The picture begins with the image of an apple photograph, so the CNN model can better appreciate the apple's character made from apple

Table 1 The list of the classes included in our Cole pest and disease dataset and its annotated samples

Class	Number of Images in the Dataset [1]	Number of Annotated Samples (Bounding Boxes) [2]	Percentage of Bounding Box Samples (%)
Leaf mold	1350	11,922	27.47
Gray mold	335	2768	6.37
Canker	309	2648	6.10
Plague	296	2570	5.92
Miner	339	2946	6.78
Low temperature	55	477	1.09
Powdery mildew	40	338	0.77
Whitefly	49	404	0.93
Nutritional excess	50	426	0.98
Background [3]	2177	18,899	43.54
Total	5000	43,398	100

[1] Number of images in the dataset; [2] Number of annotated samples after data augmentation; [3] Transversal category included in every image.

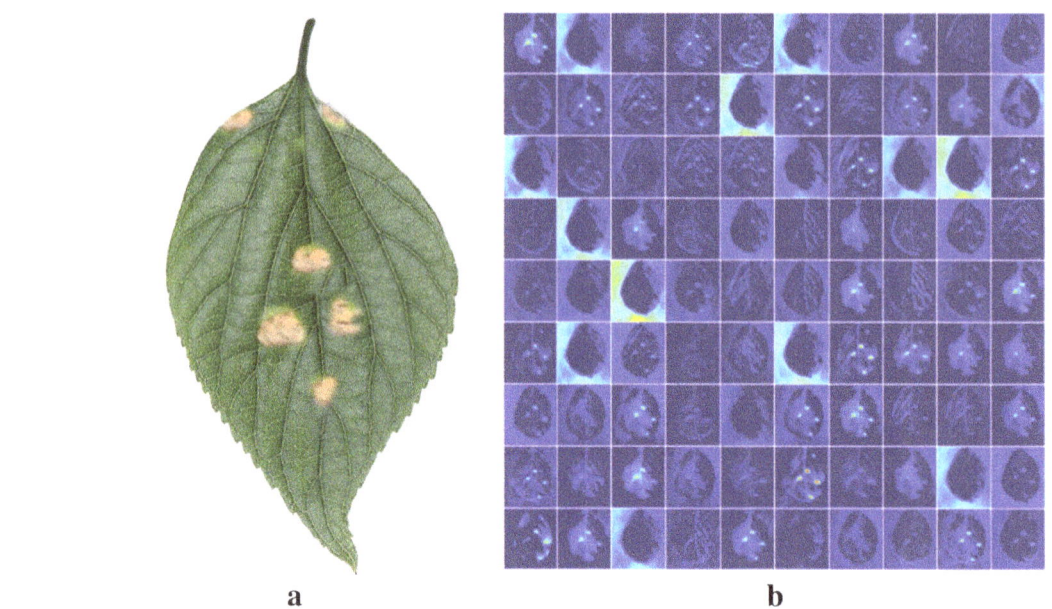

a   b

Fig. 5 Activation visualization: (**a**) original image; and (**b**) the learned weights by the first layer. As shown in Fig. 5, the model proposed by GoogLeNet Inception is used to extract the pathological features from the original image, which improves the automatic feature extraction in multidimensional space. Therefore, the proposed model based on CNN has better recognition ability for rape leaf disease

fruit. The display results show that data expansion may result in increased power levels.

Table 2 shows the final results of our refined system. The relative value proves that all categories have a satisfactory improvement compared to our previous results. The average accuracy has increased by about 13%. In fact, this is due to the implementation of a secondary diagnostic unit (CNN filter bank) that enables the system to filter misclassified samples, with a primary focus on reducing false positives.

Sample size and change are another key point influencing the end result. For example, in the case of gray mold, the number of samples is less than that of leaf mold. In addition, the gray model class shows high intra-class variability, which may confuse the system with other classes (see Fig. 7).

4 Discussion

On this paper, a new way to get an UAV with wireless clocks (WSN) is run to look different from the plots. The following approach can overcome the problems associated with the distribution of interest-oriented areas, such as how long it is not possible to use WSN, related and continue to look at different issues. These problems have been supported by using a basic WSN mobile device that can always remember the subnetworks that are available in all regions. Because the movement organs are the ones that exist, it is easy to deal with problems related to soil conditions. We have been asking for a strong - known knowledge of mysterious learning known by pathogens. This process creates a cure for disease and the location of tomato trees, which represents a significant difference in other ways of fighting disease. Our

Fig. 6 Some False Positives Produced In The Primary Diagnostic Unit. Figure 6 shows some false positives produced in the primary diagnostic unit: (**a**) the ulcer sample was detected as plague; (**b**) the gray mold sample was detected as ulcer disease; (**c**) the low temperature sample was detected as ulcer disease

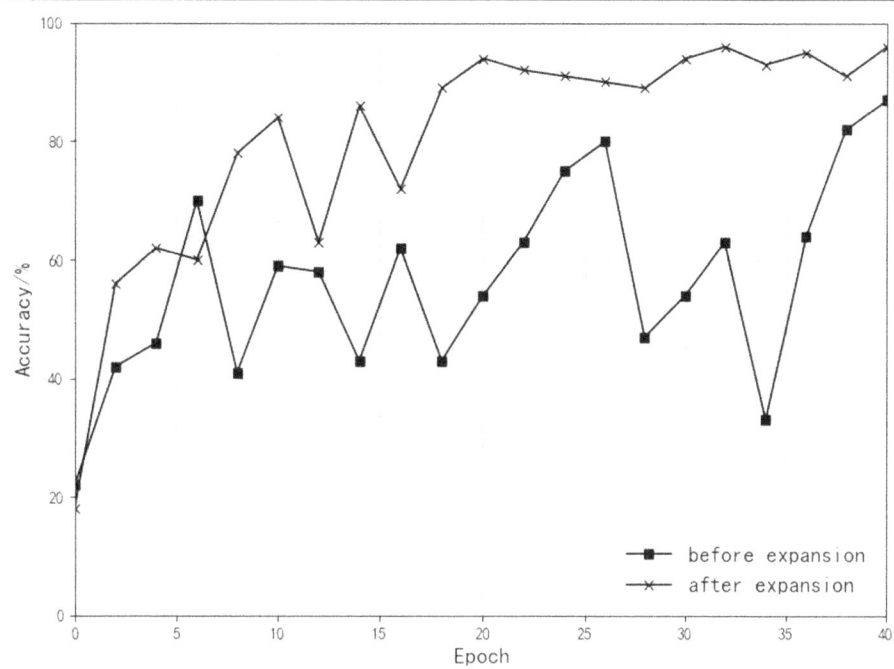

Fig. 7 The influence of expanded dataset. As shown in Fig. 7, this phenomenon is mainly due to the following reasons: (1) the extended datasets generated by various digital image processing techniques provide more opportunities for learning appropriate hierarchical features for models based on CNN; (2) the expansion of image diversity in image dataset is helpful to fully train the weight of learning in the model based on CNN, but the smaller image data set is lack of diversity and will lead to over-fitting problem; (3) the image preprocessing simulates the real collecting environment of apple pathological image, so the model based on CNN has better recognition ability to the natural apple pathological image obtained from apple orchard. The experimental results show that the extended data set can improve the generalization ability of the proposed model

review uses pictures that are carried out in situ with various camera tools and uses them using the latest technology and software programs using the GPU, rather than using a collection of natural resources (vegetation, vegetation) and research in the laboratory. In addition, our data on the diseases of tomato and insect pests have different functions, such as light, size, and so on. Our goal is to gain a deeper understanding of our work experience. Therefore, a comparison between the test results and the various depths of construction and the demonstration of how our detective detection system can find eight types of diseases and pests, including various and varied challenges. Additionally, we find that using advanced and more advanced technology can

Table 2 Comparison of our proposed method with other method

| Class/Feature Extractor | Meta-Architectures | | | | | | |
| | Faster R-CNN | | | | | R-FCN | SSD |
	VGG-16	ResNet-50	ResNet-101	ResNet-152	ResNeXt-50	ResNet-50	ResNet-50
Leaf mold	**0.9060**	0.8827	0.803	0.8273	0.840	**0.8820**	0.8510
Gray mold	**0.7968**	0.6684	0.449	0.4499	0.620	**0.7960**	0.7620
Canker	**0.8569**	0.7580	0.660	0.7154	0.738	**0.8638**	0.8326
Plague	**0.8762**	0.7588	0.613	0.6809	0.742	**0.8732**	0.8409
Miner	0.8046	0.7884	0.756	0.7793	0.767	**0.8812**	0.7963
Low temperature	0.7824	0.6733	0.468	0.5221	0.623	0.7545	0.7892
Powdery mildew	0.6556	0.5982	0.413	0.4928	0.505	**0.7950**	0.8014
Whitefly	0.8301	0.8125	0.637	0.7001	0.720	**0.9492**	0.8402
Nutritional excess	0.8971	0.7637	0.547	0.8109	0.814	**0.9290**	0.8553
Background	0.9005	0.8331	0.624	0.7049	0.745	0.8644	0.8841
Total mean AP	**0.8306**	0.7537	0.590	0.6683	0.711	**0.8598**	0.8253

* The bold numbers correspond the more challenging classes and best results among other meta-architectures.
AP: average precision.

achieve better performance. We hope that our prepared program will play a key role in the research process. Future work will improve the future results and further work will spread the knowledge of disease and insects to other plants.

The main purpose of the project is to use the UAV as a link to airline and to fix WSN change when the UAV is near your target area. Ad-hoc communications and communication systems have been used to solve these problems. In addition, the establishment of a user interface, can deal with data and monitor data. We need to emphasize that the cheap and reliable UAV platforms are now available and can be easily used for such purposes. The project provides an opportunity to try using four automobile vehicles in real-life situations. The UAV can stop the project and fly over a different distance.

Abbrevistions

ANN: Artificial Neural Network; BP: Back-Propagation; CNN: Convolutional Neural Network; LDA: Linear Discriminant Analysis; MLP: Multi-Layer Perceptron

Acknowledgements

Not applicable.

About the author's

Libo Zhou received his bachelor's degree in Physics from Hunan University of Science and Technology in 2001 and his master's degree in software engineering from Hunan University in 2012. He is currently pursuing a doctorate at Hunan Agricultural University in Changsha, China. His main research interests include image processing, image recognition algorithms, and high-speed information collection and processing.

Huang Tian received the B.S degree in aeronautical power plant control engineering from Nanjing University Of Aeronautics and Astronautics, Nanjing, China, in 1983, and the M.S. degrees in automatic control theory and application from National University of Defense Technology, Changsha, China, in 1988, and the Ph.D. degrees in mechanical manufacture and automation from Nanjing University Of Aeronautics and Astronautics, Nanjing, China, in 1992, he worked as a professor at the School of Information Science and Engineering of Central South University, Changsha, China. His research interests include artificial intelligence and application, bionic robot and intelligent bionic system, advanced control theory and advanced calculation, biomedical image processing.

Chunyun Guan Professor Guan, a mongolian ethnic, is an academician of Chinese Academy of Engineering, a leading doctoral mentor of the first-calss subject of crop science, and a doctoral and MPhil mentor of the subject of plant cultivation.

As a visiting senior scholar, he studied in Univesity of Alberta in Canada from 1993 to 1994. In the following five years (1995–2000), he was president of HUNAU. Currently, he is head of the Research Institute of Oil-bearing Crops of HUNAU; chief of the provincial key laboratory of crop genetic engineering in Hunan; and the person-in-charge of the Hunan Branch of the National Center for Improvement of Oil-bearing Crops. He also holds important posts such as: the councilman of Groupe Consultatif International de Recherche sur le Colza (GCIRC); vice-chairman of Hunan Scientific Institution; the convener of the Second Discipline Evaluation Subcommittee of the State Degree Committee; and an expert of the Discipline Evaluation Committee of the State Council.

Scope of Research: Modern Breeding of Rapeseeds and its Molecular Basis, ecophysiology and molecular biology of rapeseed, cultivation methods of high-yielding and quality rapeseed and utilization of such products.

Authors' contributions

LBZ designed the study, performed the experiments, and data analysis, and wrote the paper. LBZ and TH advised on the design of the system and analyzed to find the best method for efficient recognition of diseases and pests of Cole plants. GC provided the facilities for data collection and contributed with the information for the data annotation. All authors read and approved the final manuscript.

Competing interests

The authors declare that they have no competing interests.

Author details

[1]College of Agriculture, Hunan Agricultural University, Changsha 410128, China. [2]College of Information and Electronic Engineering, Hunan City University, YiYang 413002, Hunan Province, China. [3]Hunan Engineering Research Center for Internet of Animals, Changsha 410205, China.

References

1. G. Owomugisha, E. Mwebaze, in *Proceedings of the 2016 15th IEEE International Conference on Machine Learning and Applications (ICMLA), Anaheim, CA, USA, 18–20*. Machine learning for plant disease incidence and severity measurements from leaf images (2016)
2. M. Dutot, L.M. Nelson, R.C. Tyson, Predicting the spread of postharvest disease in stored fruit, with application to apples[J]. Postharvest Biol. Technol. **85**, 45–56 (2013)
3. P. Zhao, G. Liu, M.Z. Li, et al., Management information system for apple diseases and insect pests based on GIS[J]. Trans. Chin. Soc. Agric. Eng **22**, 150–154 (2006)
4. Y. Es-saady, I. El Massi, M. El Yassa, et al., in *2016 International Conference on Electrical and Information Technologies (ICEIT)*. IEEE. Automatic recognition of plant leaves diseases based on serial combination of two SVM classifiers[C] (2016), pp. 561–566
5. P.B. Padol, A.A. Yadav, SVM classifier based grape leaf disease detection[C]// Advances in Signal Processing. IEEE, 175–179 (2016)
6. S.S. Sannakki, V.S. Rajpurohit, V.B. Nargund, et al., "Diagnosis and classification of grape leaf diseases using neural networks"[C]// 2013 Fourth International Conference on Computing, Communications and Networking Technologies (ICCCNT). IEEE, 1–5 (2013)
7. T. Dean, M.A. Ruzon, M. Segal, et al., in *Proceedings of the IEEE Conference on Computer Vision and Pattern Recognition*. Fast, accurate detection of 100,000 object classes on a single machine[C] (2013), pp. 1814–1821
8. J. Donahue, Y. Jia, O. Vinyals, et al., in *International Conference on Machine Learning*. Decaf: A deep convolutional activation feature for generic visual recognition[C] (2014), pp. 647–655
9. Zeiler M D, Fergus R. Stochastic Pooling for Regularization of Deep Convolutional Neural Networks[J]. arXiv preprint arXiv:1301.3557, 2013
10. J. Dong, Q. Chen, S. Yan, et al., in *European Conference on Computer Vision*. Towards unified object detection and semantic segmentation[C] (Springer, Cham, 2014), pp. 299–314
11. D. Erhan, C. Szegedy, A. Toshev, et al., in *Proceedings of the IEEE Conference on Computer Vision and Pattern Recognition*. Scalable object detection using deep neural networks[C] (2014), pp. 2147–2154
12. M. Everingham, S.M.A. Eslami, L. Van Gool, et al., The pascal visual object classes challenge: A retrospective[J]. Int. J. Comput. Vis. **111**(1), 98–136 (2015)
13. P.F. Felzenszwalb, R.B. Girshick, D. McAllester, et al., Object detection with discriminatively trained part-based models[J]. IEEE Trans. Pattern Anal. Mach. Intell. **32**(9), 1627–1645 (2010)
14. S. Gidaris, N. Komodakis, in *Proceedings of the IEEE International Conference on Computer Vision*. Object detection via a multi-region and semantic segmentation-aware cnn model[C] (2015), pp. 1134–1142
15. S. Ruder, *An Overview of Gradient Descent Optimization Algorithms[J]. arXiv preprint arXiv:1609.04747* (2016)
16. A. Krizhevsky, I. Sutskever, G.E. Hinton, in *Advances in neural information processing systems*. Imagenet classification with deep convolutional neural networks[C] (2012), pp. 1097–1105
17. C. Szegedy, W. Liu, Y. Jia, et al., in *Proceedings of the IEEE Conference on Computer Vision and Pattern Recognition*. Going deeper with convolutions[C] (2015), pp. 1–9
18. A. Giusti, D.C. Cireşan, J. Masci, et al., in *2013 IEEE International Conference on Image Processing. IEEE*. Fast image scanning with deep max-pooling convolutional neural networks[C] (2013), pp. 4034–4038

19. P. Pawara, E. Okafor, O. Surinta, et al., *Comparing Local Descriptors and Bags of Visual Words to Deep Convolutional Neural Networks for Plant Recognition[C]//ICPRAM* (2017), pp. 479–486

20. M. Holmberg, F.A.M. Davide, C. Di Natale, et al., Drift counteraction in odour recognition applications: Lifelong calibration method[J]. Sensors Actuators B Chem. **42**(3), 185–194 (1997)

21. Cugu I, Sener E, Erciyes C, et al. A Novel Tree Classifier Utilizing Deep and Hand-Crafted Representations. arXiv 2017. arXiv preprint arXiv:1701.08291

22. J. Amara, B. Bouaziz, A. A Deep Learning-based Approach for Banana Leaf Diseases Classification[C]//BTW (Workshops). 79–88 (2017)

23. Bahrampour S, Ramakrishnan N, Schott L, et al. Comparative Study of Caffe, Neon, Theano, and Torch for Deep Learning[J]. 2016

24. Y. Kawasaki, H. Uga, S. Kagiwada, H. Iyatomi, in *Advances in Visual Computing, Proceedings of the 11th International Symposium, ISVC 2015, Las Vegas, NV, USA, 14–16 December 2015*, ed. by G. Bebis. Basic Study of Automated Diagnosis of Viral Plant Diseases Using Convolutional Neural Networks, vol 9475 (Lecture Notes in Computer Science; Springer, Cham, Switzerland, 2015), pp. 638–645

25. A. Johannes, A. Picon, A. Alvarez-Gila, et al., Automatic plant disease diagnosis using mobile capture devices, applied on a wheat use case[J]. Comput. Electron. Agric. **138**, 200–209 (2017)

17

Analysis of node deployment in wireless sensor networks in warehouse environment monitoring systems

Jia Mao[1*] (iD), Xiaoxi Jiang[1] and Xiuzhi Zhang[2]

Abstract

This paper mainly studies the deployment of wireless sensor network nodes in the warehouse environment monitoring system, discusses the deployment algorithm of wireless sensor network nodes in the warehouse environment, and finds out the node deployment scheme with better network performance through comparison. Wireless sensor network node deployment is the basis of wireless sensor network application in storage environment monitoring system. It affects the performance of the whole network and is the primary problem to be solved in network application. This paper discusses the advantages of wireless sensor network in the monitoring of storage environment, especially the deployment and simulation analysis of sensor nodes in the warehouse environment. Aiming at the influence of sensor perception model on the effectiveness of the node deployment plan, this paper proposes a node deployment collaborative perception model based on 0-1 perception model and exponential model. The sensor node deployment problem is transformed into a three-dimensional node deployment problem. Finally, the algorithm is applied to tobacco storage environment. In order to verify the effectiveness of the proposed algorithm, the scheme obtained by the proposed algorithm is compared with that obtained by the corresponding deployment algorithm in this paper. The comparison results show that the overall performance of the algorithm is better than that of the usual scheme.

Keywords: Warehousing environment, Wireless sensor networks, Node deployment, Detection model

1 Foreword

The biggest trend of modern logistics is networking and intelligence. Warehousing is an important part of modern logistics, linking up production and consumption, occupying an important position in the entire logistics system. Through warehouse environment monitoring systems, people get to understand warehousing status quo, make proper regulation of warehousing conditions, and guarantee warehousing quality and security of products. Traditional monitoring methods mostly take remedial measures only after the unforeseen circumstances, and cannot detect the potential safety hazards in the warehouse environment [1]. As people improve service quality requirements, businesses also pay more attention to the storage conditions of warehouse environment. They require that monitoring has strong flexibility, can accurately measure the environment parameters and make trend forecasting to reduce the possibility of disaster events; can ensure warehouse environment security, reduce losses; can ensure the product service quality and enhance customer satisfaction. In recent years, with the development of microelectronics and wireless communications technologies, wireless sensor networks (WSNs) have drawn the world's attention. A WSN contains wireless sensors that have the ability to sense, calculate, communicate, observe, and react to events that occur in a particular area. The application of WSNs technology is a development direction of warehouse environment monitoring systems. Warehousing intelligence is a key research issue of modern logistics [2]. The WSNs technology is in line with the trend of intelligent development. WSNs has the advantages of easy deployment and self-organization that can overcome the weaknesses of the traditional monitoring methods, it has a broad space for development in warehouse environment monitoring and is gradually replacing the traditional monitoring networks.

* Correspondence: jlmj1@163.com
[1]School of Transportation, Jilin University, Changchun, China
Full list of author information is available at the end of the article

2 Research status

Scholars have done a lot of research on the node deployment in WSNs. Younis M et al. [3] divided the node deployment in WSNs into two categories: static deployment and dynamic deployment. The classification criterion mainly depends on whether the entire sensor network needs to be operated or the sensor nodes need to be optimized during the deployment of sensor nodes [3]. Based on the deployment method and optimization object, deployment can be divided into two categories.

1. Deployment methods

WSNs nodes will be deployed in varied ways under different application scenarios. Deployment methods are closely related to specific applications. There are two types of deployment: deterministic deployment and random deployment [4]. When the application environment is known, the network operation status is relatively fixed and the sensor nodes are clearly located in space, it is suitable for the deterministic deployment. In order to find a solution, deterministic deployment can abstract the problem into a mathematical model and transform it into a linear programming problem or a static optimization problem. Coskun V (2008) [5] proposed to maximize the network life cycle algorithm, in order to ensure the maximum network coverage based on network connectivity, hexagonal grids are used for node deployment. The proposed algorithm is proved to be effective, scalable, and operable through simulation experiments. Although deterministic deployment can bring some convenience to the problem solution, the model is too idealistic and simplistic. However, it is obvious that random deployment has advantages for scenarios with harsh environment where man-made and large deployment is difficult.

Random deployment is an economical deployment method, but cannot guarantee full coverage. In order to achieve the desired coverage effect, many redundant nodes need to be deployed. This deployment method is applicable when coverage requirements are not strict. However, in some areas where deployment is difficult, it is necessary to increase the coverage effect by investing a large number of sensor nodes to achieve the coverage effect [6]. The traditional random node deployment will form coverage holes in the perceptual area. In response to this, Mohammed Abo-Zahhad et al. 2005 [7] proposed a WSNs deployment approach based on a multi-objective immune algorithm. This method redefines sensor nodes to reduce coverage holes and improve network coverage, saves energy consumption, and guarantees connectivity while limiting sensor mobile costs. By comparing with other deployment methods, it is found that this method can improve network coverage and reduce the mobile costs of WSNs [8].

2. Optimization object

According to the optimization object, node deployment can be divided into coverage-based deployment, network connectivity-based deployment, and energy efficiency-based deployment. The performance of the whole WSN depends on the network coverage. In the node deployment in WSNs, it is highly important to improve the network performance and coverage. Many scholars at home and abroad have done a lot of research on how to improve the network coverage. Hou Y et al. [8] proposed a node optimization algorithm by transforming the node deployment problem into a computational geometry problem]. Fan Zhigang [9] proposed a sensor node deployment algorithm based on cellular grid, which can accurately deploy the monitoring area, not only can achieve complete coverage but also can accurately deploy some redundant nodes to extend network life cycle and can be applied to areas with stringent control for the density of wireless sensor nodes. Fadi M. Al-Turjman et al. (2013) [10] proposed a general method to evaluate the average connectivity based on grid deployment strategy by analyzing the practical problems in deployment. Achieving complete coverage while guaranteeing connectivity is of practical significance in application. S.M. Nazrul Alam et al. (2015) [11] used Voronoi elements to divide the deployment space into several polyhedrons and calculated the furthest distance between any two points in the polyhedron so as to ensure that the wireless sensors communicate with each other and transmit information to the base station. After simulation, the deployment method proposed by S. M. Nazrul Alam et al. was proved to have good connectivity.

Network usage times are crucial for the operation of WSNs. Researchers at home and abroad have studied how to prolong the service life of WSNs. Hashim A. Hashim et al. (2016) [12] proposed an energy-efficient deployment strategy that uses artificial bee colony optimization of network parameters to limit the total number of redundant nodes. To avoid the NP-hard problem, the algorithm uses a two-tier structure based on cost constraints called ILDCC to improve the deployment life cycle. A cubic grid model is proposed in the SP 3D space. Although this deployment strategy is cost-effective and easy to implement, it has several drawbacks, such as the increase in the number of redundant nodes in the network will increase the possibility of collisions and conflicts. With the help of the data link layer protocol, the impact of collisions and collisions can be reduced. D Wang et al. (2008) [13] provided an analytical framework that obeys the coverage and usage cycles under two-dimensional Gaussian distribution. The paper proposes that the algorithm can better meet the coverage and usage requirements. The two algorithms proposed by the analysis framework prove that the network lifetime can be effectively increased, but the boundary problem existing in the deployment is not considered. Fadi M. AL-Turjman et al. (2015) [14] proposed a combined energy-efficient and k-tolerant heterogeneous node deployment WSNs strategy. In order to limit the unlimited search space to a controlled range of numbers, a 3D grid model is adopted. The main goal of the algorithm is to find the $Q_{SN} + Q_{RN}$ optimal node location at the Vth grid vertex location so as to maximize the network lifetime. Deployment strategy is mainly divided into two phases, the first phase is to find the optimal location of all nodes to minimize energy consumption, the second phase occurs after each round to improve the connection performance and release overload node pressure. After simulation, compared with several typical methods, the algorithm extends the use of time of node by 40%.

3 Problem description

Generally, the environmental parameters that need to be monitored in a warehouse mainly include temperature, humidity, illuminance, and oxygen content. To monitor the environment parameters in warehouse environment, a large number of sensor nodes need to be deployed in the warehouse environment to construct a reasonable node deployment plan to meet the requirements of warehouse parameter monitoring. A reasonable deployment of sensor nodes needs to be technical and economic, therefore, a reasonable node deployment WSN plan is essential for warehouse environment parameter monitoring.

By using the Voronoi diagram partitioning theory, the 3D space can be divided into several small spaces $V_i (i = 1, 2, \cdots n)$. These small spaces are polyhedral cells [15]. Each Voronoi region VR_{P_i} boundary is composed of multiple Voronoi edges, as shown in Fig. 1. Each Voronoi edge is the set of all the points closest to seed P_i.

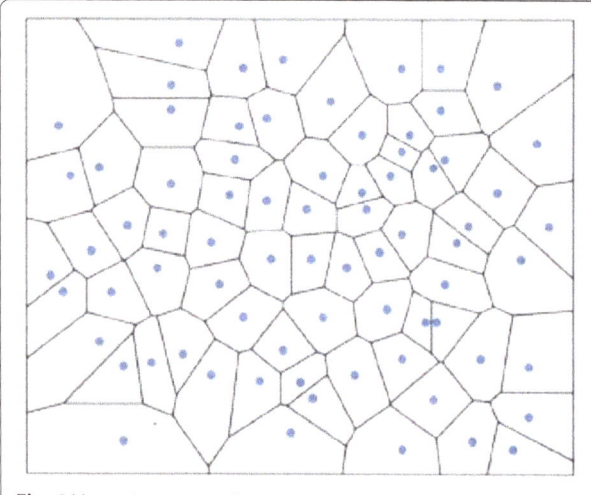

Fig. 1 Voronoi segmentation

This characteristic naturally divides the space into several regions, each of which is the influence area corresponding to the growing point [16].

Voronoi diagram partition theory can be used to divide the three-dimensional space into several small spaces, and these small spaces are mostly polyhedral elements, and for these polyhedra, can find an outside contact ball with a definite radius. In the space filling, the sensor node can be deployed in the center of the polyhedron, and the radius of the external sphere of the polyhedron is the radius of the unit perception model. Through the deployment of sensor nodes, the three-dimensional space can be fully covered and the whole storage environment can be monitored [17].

3.1 Research hypothesis

Through the description of the problem, to achieve node deployment in WSNs under warehouse environment, a suitable space division is the key. Although the best space division plan does not create exactly the same division cells, in order to obtain a suitable plan, it is assumed that all the division cells have the same shape for the following reasons:

(1). With the same shape as cells, one can be more concerned about the shape of the division cell, making the problem easier to handle.
(2). The same cell can provide certain rules. It is possible to deterministically establish the position of any cell by a simple set of equations, use the same cell for spatial division, and also make the deployment method implement distributed computation and make the deployment method have good scalability.
(3). In this paper, we choose the same type of wireless sensor, with the same sensing range and transmission range, so the same cell shape is the best choice.

If the shape of all cells is the same, then the number of cells can be reduced by maximizing the capacity of one cell.

The core of node deployment is to divide the 3D space into a number of small spaces, the sensor nodes are deployed in the center of each small space, so there are the following constraints:

(1). The number of cells divided in the 3D space must be fillable polyhedrons.
(2). The radius of the circumscribed sphere attached to the cell can not be larger than the sensing range R_s.
(3). The distance between two farthest points of adjacent cells can not be larger than the transmission distance R_c.

Among them, the first constraint limits the number of possible polyhedrons. Because when the 3D space of nodes to be deployed is fixed and when the sensor senses the same radius, the larger the volume of the polyhedron, the fewest number of the polyhedrons required. The second constraint ensures that the entire network covers all the areas to be monitored. The third constraint guarantees the connectivity between sensor nodes [18].

Figure 2 shows the basic structure of using a truncated octahedron to divide and fill a space. If the center $(0, 0, 0)$ of the sized $4 \times 4 \times 4m^3$ space is the starting point for truncated octahedral coverage, deploy a truncated octahedron at the geometric center $(0, 0, 0)$, then deploy 8 truncated octahedrons around the truncated octahedron, and the fillable polyhedra are all congruent shapes.

The position of the node can be calculated, that is, by calculating the relative displacement of the two types of nodes in these polyhedrons, the deployment position of the wireless sensor node can be obtained. For the convenience of representation, two types of nodes with different offsets are respectively recorded as class A nodes and class B nodes. Described by offsets $\lambda = (a_1, b_1, c_1)$ and $\gamma = (a_2, b_2, c_2)$, respectively. Therefore, as shown in Fig. 1, the node with λ is referred to as node A, and the coordinates of the point can be obtained by the formula (1) and represented by L_A. The node with γ is called node B, and the coordinates of the point can be obtained by formula (2) and represented by L_B.

$$L_A^{''} = L_A + \lambda \tag{1}$$

$$L_B = L_A + \gamma \tag{2}$$

Among them, the specific values of λ and γ for different space-fillable polyhedrons are shown in Table 1.

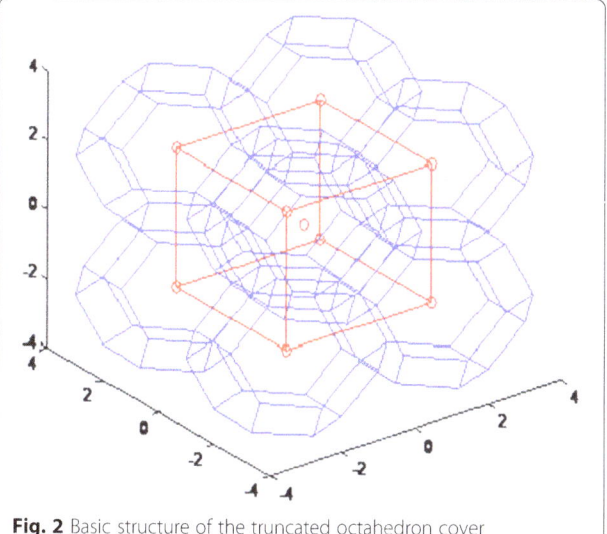

Fig. 2 Basic structure of the truncated octahedron cover

Table 1 Offset table

	Class A node offset	Class B node offset
Truncated octahedron	$(n \times \frac{4R}{\sqrt{5}}, m \times \frac{4R}{\sqrt{5}}, k \times \frac{4R}{\sqrt{5}})$	$(\frac{2R}{\sqrt{5}}, \frac{2R}{\sqrt{5}}, \frac{2R}{\sqrt{5}})$
Diamond dodecahedron	$(n \times 2R, m \times 2R, k \times 2R)$	$(2R, 2R, 0)$ $\{$ $(2R, 0, 2R)$ $(0, 2R, 2R)$
Optimal hexagonal prism	$(n \times \sqrt{6}R, m \times \sqrt{2}R, k \times \frac{2\sqrt{3}R}{3})$	$(\frac{\sqrt{6}R}{2}, \frac{\sqrt{2}R}{2}, 0)$

3.2 Algorithm analysis

The cell perception model is a sphere with an irregular probability range. If we only explore the cell perception model to achieve 100% coverage of 3D space, there must be overlapping coverage. In fact, the Voronoi network division provides a good suggestion to solve the problem of 3D space coverage. It divides the 3D space into multiple spaces and calls them Voronoi cells. If each Voronoi cell is the same, then each Voronoi cell should be a polyhedron that achieves a seamless spatial coverage called a spatially filled polyhedron such as truncated octahedron, rhombic dodecahedron, hexagonal prism, and cube. The geometric center of each refillable polyhedron corresponds to the location of one wireless sensor node in the WSN, and each refillable polyhedron in space should be within the sensing range of the wireless sensor node. In order to maximize the volume of the space-filled polyhedron, the radius of the circumscribed sphere of each refillable polyhedron needs to be equal to the radius R of cell perception model [19]. Therefore, the problem of node deployment is reduced to the issue of space-fillable polyhedron partitioning.

The known fillable polyhedra are a regular prism, a cube, a regular hexagonal prism, a truncated octahedron, and a diamond-shaped dodecahedron. Firstly, through problem description and combined with the relevant theories of Kepler's prediction and Kelvin's prediction, rhomboid dodecahedron and truncated octahedron are selected as spatial fillable polyhedra. Then, through further analysis, it is found that the method to solve the coverage problem in two-dimensional space is to use regular hexagon for two-dimensional coverage. However, a polyhedron with hexagon as its cross-section has no properties of space filling, and a polyhedron with space-filling properties and hexagon as its cross-section is a hexagonal prism. Therefore, the six-prism is included in the comparison. Since the cube is the only regular polyhedron subdivided in three-dimensional space, the cube and the truncated octahedron, the diamond-shaped dodecahedron and the hexagonal prism are selected as fillable polyhedra for spatial subdivision.

For any shape element, the radius of the catch is R_s, so the volume of the catch is $\frac{4}{3} * \pi R_s^3$, and the volume of the catch is also the maximum volume of the element. Here we introduce the concept of a V.Q., short for volume quotient, the ratio of the volume of a polyhedron unit to the volume of its external contacts, to compare units of different shapes. If the volume of a polyhedral unit is V, the volume quotient of the polyhedral unit is $\frac{3V}{4\pi R^3}$. V. Q. is always within the range [20]. Our goal is to find a space-filled polyhedron with a maximum volume quotient (i.e., close to 1) to determine a theoretically optimal partitioning solution that minimizes the number of deployed nodes and theoretically optimizes the performance of the deployed solution. During the actual investigation, we found that there are many practical factors affecting the deployment method of wireless sensor network nodes in the storage environment, such as signal conflicts between WSNs and transmitting power between wireless sensors. Therefore, it is very difficult to find an optimal deployment plan in both theory and practice. Therefore, this paper aims to find a deployment method that can achieve better results than common solutions.

Assume that the radius of the external circle of the fillable polyhedron is R_s, in the case of space filling with six-prism, the radius of the external contact cannot guarantee the unique six-prism. This is because it is possible to have a number of hexagonal prisms of different heights and sizes on the hexagonal surface and still have the same diameter for their external contacts. Here, the six-prism with the largest V. Q. is selected, V. Q. value is $\frac{3}{2\pi} = 0.477$, and it is called the best six-prism. Similarly, the volume quotient of the cube, the diamond-shaped dodecahedron, and the truncated octahedron can be obtained through calculation. The specific results are shown in Table 2.

By comparing the volume quotients of different fillable polyhedra, it is obvious that if the truncated octahedron is taken as the shape of the fillable polyhedron, the number of active nodes will be the minimum [21].

3.3 Algorithm process
Normally, the input of our algorithm is three parameters:

(1) Space restrictions to be monitored.
(2) The radius R of the unit sensing model that affects the offset is different. If the fillable polyhedron selected is different, the corresponding offset is also different.
(3) Coordinate of the initial point, which is directly related to the deployment location of later nodes.

Combined with the content described in Sections 3.1 and 3.2, the detailed algorithm process can be obtained as follows:

(1) Input units perceive the radius R of the model.
(2) Select the starting point, which can be understood as the location where the first wireless sensor node is deployed, and set the coordinates of the starting point to (x_s, y_s, z_s).
(3) To determine whether the coordinates of class A nodes exceed the limited range, which means go to step (8) to end; otherwise, go to step (4), and continue to obtain L_A.

The (x_a, y_a, z_a) is the general form of class A node coordinates, class A node coordinates calculation formula as shown in Eq. (1), the size of the offset vector is determined by the structure of space-filling polyhedron, due to the different space-filling polyhedron filling structure is different, the offset vector is different also, and the offset (a_1, b_1, c_1) can be calculated.

(4) By superposing the offset with the coordinates of the class A node obtained in step 3, the coordinates of the class A node are further obtained.
(5) To determine whether the coordinates of class B node exceed the limited range, go to step (8) to end, otherwise go to step (6), and continue to obtain L_B.

The (x_b, y_b, z_b) is the general form of class B node coordinates, class B node coordinates calculation formula as shown in Eq. (2), the size of the offset vector is determined by the structure of space-filling polyhedron; due to the different space-filling polyhedron, filling structure is different, the offset vector is different also, and the offset (a_2, b_2, c_2) can be calculated.

(6) By superposing the coordinates of the class B node obtained by offset γ and step (3), the coordinates of the class B node are obtained.
(7) Determine whether $L_A^{''}$ exceeds the limit range. If it exceeds the limit, go to step (8) to end; otherwise, go to step (4) and continue to get L_A. In which $L_B = L_A + \gamma$.
(8) The end. Output sensor node deployment location coordinates.

Table 2 Volumetric comparison of fillable polyhedra

Fillable polyhedra	The cube	Optimal six-prism	Truncated octahedron	Rhombohedral dodecahedron
Volume quotient (V. Q.)	$\frac{2}{\sqrt{3}\pi} \approx 0.36755$	$\frac{3}{2\pi} \approx 0.477$	$\frac{24}{5\sqrt{5}\pi} \approx 0.68329$	$\frac{3}{2\pi} \approx 0.477$

4 Design of node deployment scheme

4.1 Analysis of monitoring system requirements

Tobacco warehousing is an important part of tobacco production. The quality of tobacco warehousing affects product quality and production cost. High-quality warehouse environment occupies an important position in the production of enterprises and is conducive to ensure normal production. The application of WSNs to tobacco warehouse environment monitoring systems can make full use of the WSNs self-organizing, easy deployment and real-time advantages, and dispose wireless sensor nodes in the tobacco stacks to realize real-time, fine, and dynamic monitoring of the temperature, relative humidity, and gas concentration of tobacco warehouse environment. The application of WSNs in tobacco warehousing monitoring system is beneficial to promote work efficiency and improve the real-time performance and accuracy of tobacco warehouse environment monitoring, which is conducive to the security of warehousing and the quality of tobacco and tobacco products.

4.2 Overall structure of the monitoring system

Through the demand analysis of tobacco, warehouse environment monitoring systems, it is found that the systems need to deploy devices such as temperature and humidity wireless sensors, phosphide detection sensors, dehumidifiers, air conditioners, and cameras, giving full play to the self-organizing, easy deployment and multi-hop transmission advantages of WSNs, building the entire WSNs monitoring system. As shown in Fig. 3.

According to the above analysis of the characteristics of the tobacco warehouse environment requirements, the overall schematic of tobacco warehouse WSN monitoring system is designed, as shown in Fig. 4.

A number of wireless sensor nodes, including temperature sensors, humidity sensors, phosphide detection sensors, etc., are deployed at suitable locations within the tobacco stacks and warehouse. These nodes are utilized for data collection, then the collected data is transmitted to the network base station in real time via multi-hop transmission of WSNs, and the data is summarized, classified, and processed to ensure that monitoring personnel have more accurate and intuitive hold of the warehouse temperature and humidity, phosphine concentration and specific internal warehouse environment. A tobacco warehouse about $105 \times 77 \times 15$ m space size, in which multiple wireless sensor nodes are placed, periodically collecting environmental parameters to detect changes of the parameters of tobacco and warehouse that are to be monitored, so as to ensure warehouse environment and product quality of tobacco and tobacco products. The sensor nodes have a sensing radius of 5 m.

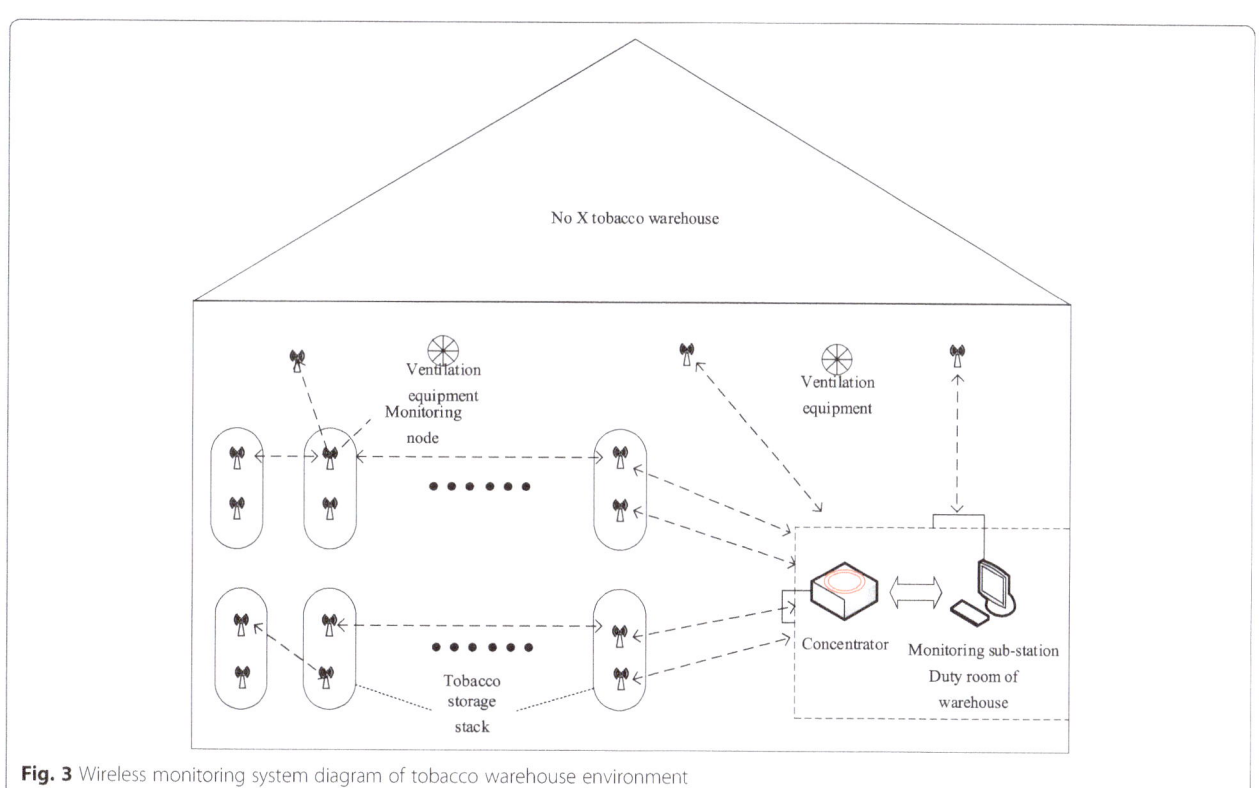

Fig. 3 Wireless monitoring system diagram of tobacco warehouse environment

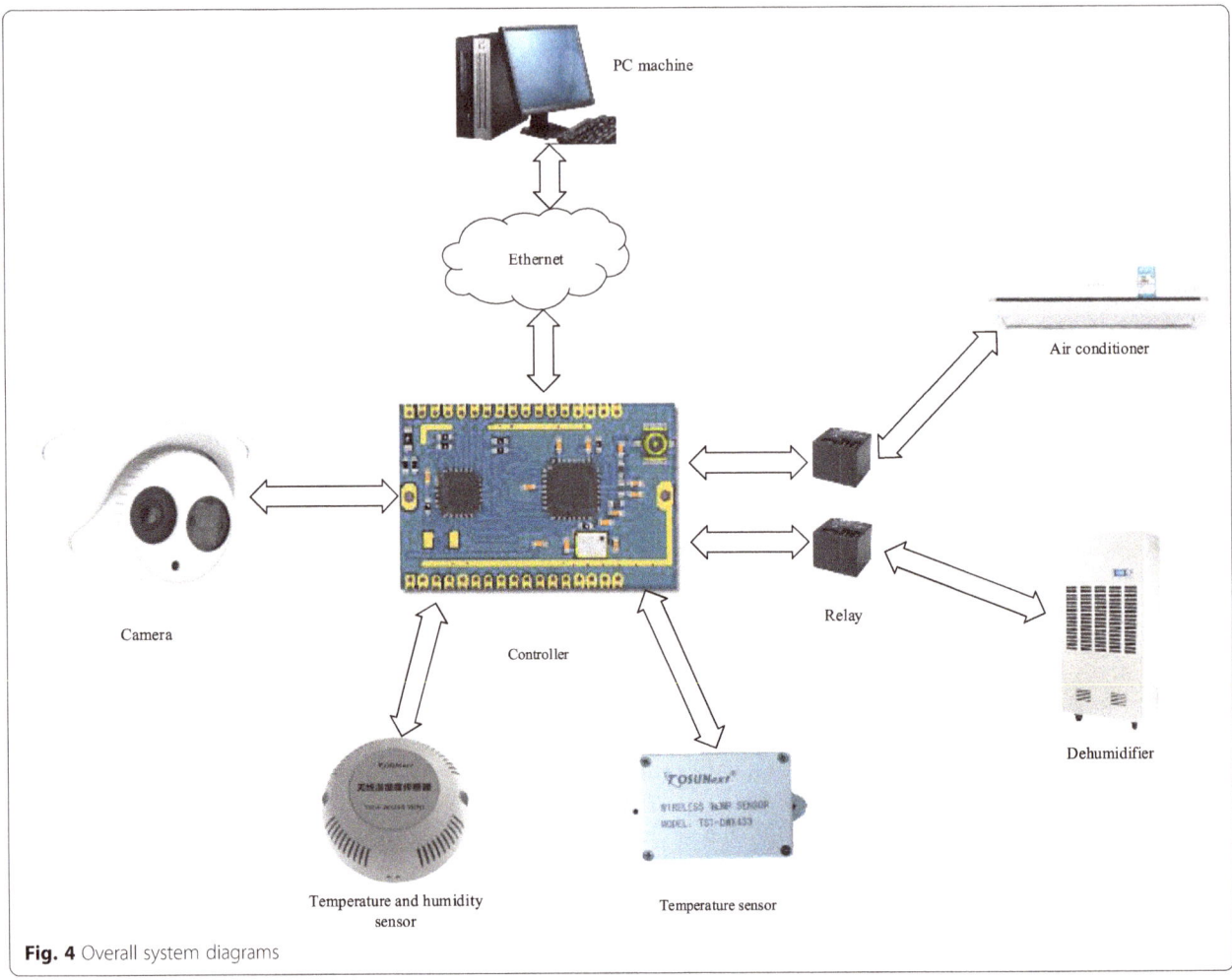

Fig. 4 Overall system diagrams

4.3 Design of node deployment plan
4.3.1 Model establishment and assumptions
A tobacco leaf warehouse is about 105 m × 77 m × 15 m in size. Multiple wireless sensor nodes are placed in the storage environment, and environmental parameters are periodically or periodically collected to detect changes in the parameters to be monitored in the tobacco leaves and warehouses to ensure the tobacco leaves and tobacco product storage environment and product quality.

In the stage of wireless sensor network node deployment in the tobacco warehouse, the following assumptions are made:

(1) The wireless sensor node can be placed at any position in the tobacco warehouse, and the wireless sensor network node is fixed in position after deployment, and the number of wireless sensor nodes that need to be deployed in the entire tobacco warehouse can be clearly understood through calculation.
Each wireless sensor location has its own identification code, which can clearly know that there is an abnormality in the parameters to be monitored at a certain location; each wireless sensor node has similar processing and communication capabilities and the transmission power can be adjusted.

(2) In the storage wireless sensor network monitoring system, the average distance between two adjacent wireless sensor nodes is smaller than the maximum communication radius between the wireless sensors, which can ensure the information transmission of the entire wireless sensor network, so that the base station can receive the environment. Parameters, and summary processing take appropriate improvement measures.

(3) In the wireless sensor network, each wireless sensor node has its own location without additional hardware such as GPS to help acquire. The transmit power of each wireless sensor node is limited. If the transmit power of the connected sensor node is known, the approximate distance between the wireless sensor nodes can be estimated from the received signal power.

(4) Each wireless sensor network node can guarantee a certain time of work, ensure long-term collection of information, and improve the storage environment.

4.3.2 Node deployment implementation

1. Parameter setting

(1) Determine the storage space to be monitored

The tobacco leaf warehouse to be monitored is about a rectangular parallelepiped space with length, width, and height of 105 m, 77 m, and 15 m respectively.

(2) Defining the perceived radius of the wireless sensor

Defining the perceived radius of the wireless sensor is the basis for solving the problem of node deployment. In the tobacco warehouse, a wireless sensor with a sensing radius of 5 m is selected.

(3) Choice of fillable polyhedron

According to the comprehensive analysis in Section 3.2, the truncated octahedron has the largest volume quotient. The spatial segmentation using the truncated octahedron can use the minimum number of wireless sensor nodes, save costs, and enable wireless under the premise of ensuring connectivity. The use time of the sensor network is maximized, so the warehouse space is three-dimensionally divided by the truncated octahedron in the tobacco warehouse.

When the tobacco warehouse space is three-dimensionally divided by the truncated octahedron, the offset of the coordinate of the class A node is $(n \times 4\sqrt{5}, m \times 4\sqrt{5}, k \times 4\sqrt{5})$, where n, m, k are integers, and the offset of the coordinates of the class B wireless sensor node is $(2\sqrt{5}, 2\sqrt{5}, 2\sqrt{5})$.

(4) Select the starting point

The first wireless sensor node is deployed starting from a corner of the warehouse, and the coordinates of the point are $(0, 0, 0)$. After the initial node deployment location is determined, the next wireless sensor node is determined until the wireless sensor network node covers the entire warehouse.

2. Algorithm implementation

Typically, the input to this method algorithm is three parameters:

(1) Space area limit to be monitored.
(2) The radius of the unit-aware model affecting the offset, if the selected fillable polyhedron is different, the corresponding offset is also different;
(3) The coordinates of the initial point, which is directly related to the deployment location of the node afterwards.

4.3.3 Simulation environment and simulation results

1. Simulation environment

Tobacco storage is an important link in the production process of tobacco enterprises. The quality of tobacco storage affects the quality of products and the production cost of enterprises. High-quality storage environment is conducive to ensure the normal production and plays an important role in the production of enterprises.

It is an important part of storage work to automatically collect storage environment parameters and transmit them to the control center, analyze timely storage environment, find hidden dangers, and deal with them in time. Humidity, temperature, illumination, gas concentration (O_2, CO_2, and other gas concentration, etc.), and dust are the main environmental parameters for monitoring.

(1) Air temperature

Air temperature is the air temperature in daily life, used to indicate the degree of cold and hot air. Air temperature has a negative correlation with the distance from the ground, which increases with the decrease of the distance from the ground, and decreases with the increase of the distance from the ground. In the actual storage process, the air temperature in the storage environment must be strictly controlled to ensure the quality of stored goods and reduce the consumption in logistics.

(2) Air humidity

Air humidity is a physical quantity that indicates the moisture content and humidity of the air. Under a certain temperature, air humidity is proportional to the moisture content in the air. If the moisture content in the air is higher, the air humidity will be higher; if the moisture content in the air is lower, the air humidity will be lower.

(3) The light

Illuminance, refers to the luminous flux received on the surface of the subject per unit area. Light has different effects on different items in storage. One is that light can adversely affect many things. The other is that light can protect objects under certain conditions.

The minimum relative humidity for most microbes is 80–90%. At 95%, microbes grow very vigorously, and below 75%, most items are not easily moldy. Most microbes die when the sun shines for 1–4 h and 3–5 m with an ultraviolet lamp. So a certain amount of light can protect stored goods.

(4) Gas concentration

There are various types of gases in the storage environment. When monitoring the concentration of gases, the concentration of gases including oxygen, carbon dioxide, and various harmful gases is mainly monitored.

Most moldy microorganisms, especially molds, grow fast in the environment with high-oxygen concentration, but grow slowly in the environment with high carbon dioxide concentration. It can be found that increasing or decreasing the concentration of carbon dioxide or oxygen can inhibit the life activities of microorganisms or even kill them by controlling the air composition in the storage environment. For example, during the storage process of grain, it is sensitive to the gases contained in the air. In order to ensure the quality of grain to the greatest extent and reduce the damage to grain by microorganisms and pests, it is necessary to control the oxygen concentration in the grain stack. Through a large number of experiments, the results show that the grain storage effect is best when the oxygen concentration in the storage environment is maintained between 2% and 5%.

(5) Dust illumination

Dust refers to the solid particles suspended in the air, and often refers to dust, dust, dust, powder, etc. The precision or sensitivity of precision instruments and electromechanical equipment will be affected by dust or sundries to different degrees. Similarly, corrosion-prone metal products stored in storage environments are prone to corrosion due to dust. For the storage environment where the vulnerable objects are stored, dust concentration detection files in the air should be established to record the results of regular detection. According to the records, appropriate measures should be taken to control dust in a reasonable way, such as installing partial exhaust hood and other improvement measures, so as to effectively control dust.

2. Simulation results

The specific parameters of the simulation scenario are shown in Table 3.

Using the algorithm of this paper in Section 3.2, using the java tool, according to the input parameter settings, the coordinates of the wireless sensor nodes can be obtained, and the coordinates of the deployment positions of some nodes are shown in Fig. 5. Figure 6 is a simplified stereogram of sensor node deployment, where the green area is the warehouse, the white point is the location of sensor node, and each cube vertex is the center point of the octahedron.

5 Comparison and analysis of node deployment before and after optimization
5.1 Evaluation index of node deployment
To verify the performance of the proposed deployment method based on space-fillable polyhedrons, the optimized deployment method proposed in this paper and the node deployment plan obtained from the pre-optimization deployment method are respectively studied in the same deployment environment, comparison is made from the aspects of coverage performance, connectivity performance, consumption performance, and economic performance.

(1) Coverage performance analysis

To further verify the effectiveness of the deployment method proposed in this paper, we compare the performance of the proposed deployment method with that of the pre-optimization method [22]. Since the truncated octahedron has the largest V.Q., and the required full coverage in the unit space is the minimum, the truncated octahedron is selected as the fillable polyhedron to be filled in the monitored space. Figure 7 shows the comparison of the coverage performance.

As seen from Fig. 7, the target coverage area ratio of WSNs has a positive correlation with the number of network nodes, and the area ratio of the WSNs target covering area increases as the number of WSNs nodes increases. The pre-deployment method simply sets all border nodes to be active, regardless of the target area boundaries, so that the number of active nodes is approximately 1.13 times that of the optimized deployment plan proposed in this paper. Therefore, the rational use of the boundary of the target area can effectively reduce the active nodes near the boundary.

Table 3 Specific parameters

Parameter	Value
Size of space	105 m ×7715 m × m
Wireless sensor sensing radius	5 m
Initial node coordinates	(0,0,0,)
MAC/Route	IEEE802.15.4/AODV

```
 Javadoc   Statement   Console
 stop  ykh2 [Java Application] C:\Program Files\Java\jre1.8.0_91\bin\javaw.exe (
        (87.207    33.541    06.708)Coordinates of Node deployment location
        (87.207    46.957    06.708)Coordinates of Node deployment location
        (87.207    60.374    06.708)Coordinates of Node deployment location
        (87.207    73.790    06.708)Coordinates of Node deployment location
        (100.623   20.125    06.708)Coordinates of Node deployment locati
        (100.623   33.541    06.708)Coordinates of Node deployment locati
        (100.623   46.957    06.708)Coordinates of Node deployment locati
        (100.623   60.374    06.708)Coordinates of Node deployment locati
        (100.623   73.790    06.708)Coordinates of Node deployment locati
        (06.708    20.125    06.708)Coordinates of Node deployment locatio
        (06.708    33.541    06.708)Coordinates of Node deployment locatio
        (06.708    46.957    06.708)Coordinates of Node deployment locatio
        (06.708    60.374    06.708)Coordinates of Node deployment locatio
        (06.708    73.790    06.708)Coordinates of Node deployment locatio
        (20.125    06.708    06.708)Coordinates of Node deployment locatio
        (33.541    06.708    06.708)Coordinates of Node deployment locatio
        (46.957    06.708    06.708)Coordinates of Node deployment locatio
        (60.374    06.708    06.708)Coordinates of Node deployment locatio
        (73.790    06.708    06.708)Coordinates of Node deployment location
        (87.207    06.708    06.708)Coordinates of Node deployment locatio
        (100.623   06.708    06.708)Coordinates of Node deployment locatio
    144 Number of nodes
```

Fig. 5 Wireless sensor node coordinates

(2) Connectivity performance analysis

To compare the impact of different deployment plans on the connectivity performance of WSNs, the transmission range of the optimized nodes is compared with the node transmission range before optimization [23]. By calculating the distance between the node locations in the adjacent subdivision cells, the comparison results of the minimum required transmission range are shown in Fig. 8.

As can be seen from the analysis of Fig. 8, the minimum transmission range required by the optimized deployment method is $1.7R_s$, and the minimum transmission range required before optimization is $1.9 R_s$. In warehouse environment monitoring systems, the deployment of WSNs nodes is limited by the transmission range of each wireless sensor node, and the transmission range required by different deployment plans is different. The greater the required WSNs node transmission range, the greater the probability that the network is not connected, the smaller the required transmission range, the better the connectivity of the entire WSNs. Therefore, the optimized

deployment method has slightly better connectivity than the pre-optimization deployment method.

(3) Consumption performance analysis

The consumption performance affects the performance of the entire WSNs in the warehouse environment, especially on the network usage time, so the energy consumption performance of the node deployment plan is comparatively analyzed. In the tobacco warehouse environment, supposing the number of packets transmitted and forwarded in different deployment methods is the same, then the use of time for deployment depends on the transmission range between the wireless sensor nodes and the number of wireless sensor nodes [24]. Compared with the network life cycle of the deployment method of this paper, the comparison results of different service life cycles are shown in Fig. 9.

It can be seen from Fig. 9 that the deployment of the optimized wireless sensor nodes has a longer life cycle and a better deployment effect.

(4) Analysis of energy consumption performance

Energy consumption performance affects the performance of the entire wireless sensor network in the storage environment, especially the impact of network usage time [25]. Therefore, the energy consumption performance of the node deployment scheme is compared and analyzed here. In order to verify the performance of the space-fillable polyhedron deployment algorithm proposed in this paper, in the same deployment environment, the algorithm proposed in this paper is simulated and compared with the algorithm in literature [26] and literature [27].

In the tobacco storage environment application, the number of data packets transmitted and forwarded in

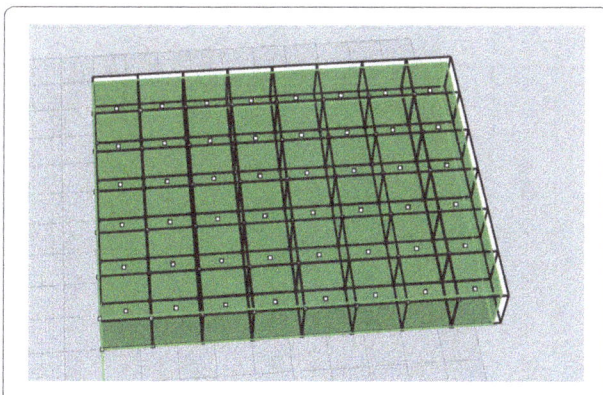

Fig. 6 Node deployment simplifies stereograms

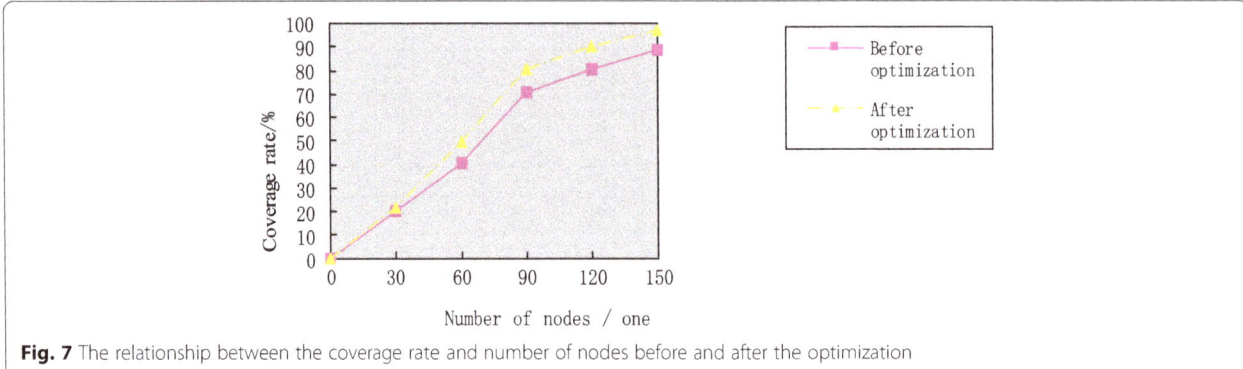

Fig. 7 The relationship between the coverage rate and number of nodes before and after the optimization

different algorithms is the same, then the deployment time of the deployment scheme depends on the transmission range between the wireless sensor nodes in the scheme and the number of wireless sensor nodes.

It is assumed that in a wireless network, the energy consumption of transmitting a data packet is proportional to the square of the transmission range, and the transmission ranges of the two split units A and B are γ_A and γ_B, respectively. In the split units A and B, the volumes of the individual units are V^A and V^B, respectively. If the life cycles of split units A and B are defined as L^A and L^B, respectively, then:

$$\frac{L^A}{L^B} = \frac{r_B^2}{r_A^2} \times \frac{V^A}{V^B} \tag{3}$$

It is known that the node deployment scheme in [26] mainly uses the optimal hexagonal prism to cover and fill the entire monitoring space. The literature [27] mainly uses the truncated octahedron to cover the entire monitoring space. In this paper, the tobacco warehouse deployment scheme mainly adopts the interception. The angular octahedron is spatially partitioned and filled. The life cycle of the proposed deployment scheme is used as the benchmark. Using

formula (3), the life cycle comparison results of the available schemes are as follows:

$$\frac{L^{[26]}}{L} = \frac{\left(r_s\sqrt{\frac{17}{5}}\right)^2}{\left(r_s\sqrt{\frac{7}{2}}\right)^2} \times \frac{\frac{r_s^3}{4}}{\frac{4r_s^3}{5\sqrt{5}}} = \frac{17\sqrt{5}}{56}$$

Compared with the network life cycle in the scheme obtained by the algorithm, the usage period of different schemes is shown in Fig. 10.

As can be seen from Fig. 10, the wireless sensor node deployment method using the truncated octahedron for spatial segmentation has the longest usage period, is superior to the optimal hexagonal prism and rhombohedral dodecahedron and the cube space splitting mode, and has the best deployment effect

5.2 Weight distribution of evaluation indexes

Through the analysis of the principle of AHP, the AHP can be used to determine the weight of the evaluation index of the wireless sensor network node deployment scheme in the storage environment monitoring system, so as to prove the optimal scheme [28]. The specific modeling steps are shown as follows:

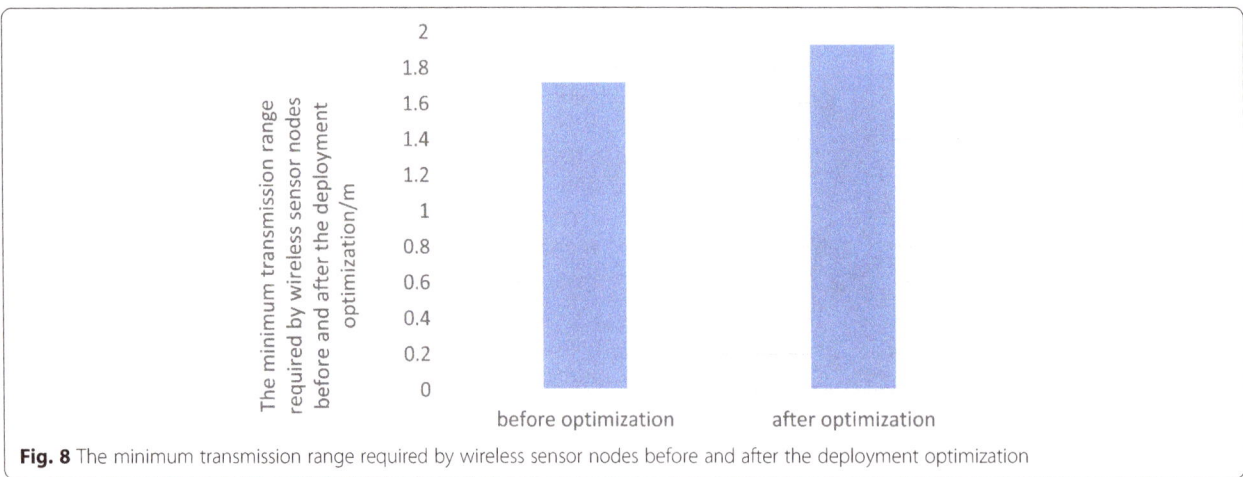

Fig. 8 The minimum transmission range required by wireless sensor nodes before and after the deployment optimization

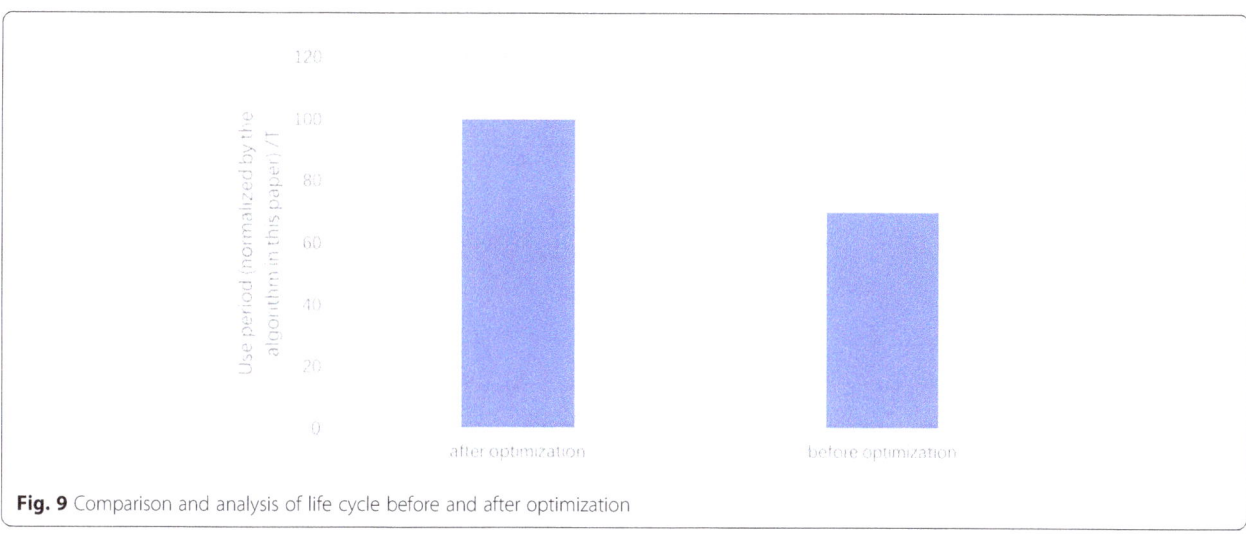

Fig. 9 Comparison and analysis of life cycle before and after optimization

(1) Establish evaluation index model

Refer to the evaluation index of wireless sensor network node deployment scheme preliminarily established in section 5.1 to draw the hierarchical structure model, as shown in Fig. 11.

The model is mainly composed of three parts, including target layer A, criterion layer B, and scheme layer C.

Target layer: effect of wireless sensor network node deployment scheme in warehouse monitoring system (A)

Criterion layer: covering performance index (B1), connectivity performance index (B2), energy consumption performance index (B3)

Scheme layer: overall network coverage (C1), coverage multiplicity (C2), coverage time (C3), number of active nodes (C4), blind area of connectivity (C5), communication distance of nodes (C6), energy consumption required for network coverage (C7), energy consumption required for connectivity (C8)

(2) Construct judgment matrix

For storage environment monitor system in wireless sensor network node deployment plan effect evaluation hierarchical structure model, based on the structure of judgment matrix, applying YAAHPO. 6.0 software can be calculated by analytic hierarchy process (AHP) judgment matrix the largest eigenvalue of $\lambda_{\max}$, the consistency ratio CR and the weight vector of W, RI, and combined with random consistency index value, calculate the consistency index of $CI = CR \times RI$, YAAHPO. 6.0 AHP software is a relatively advanced and mature analytical tool. Through a large number of empirical studies, it is proved that the results of this software have high reliability and validity. If $CR \leq 0.1$, it is considered that the

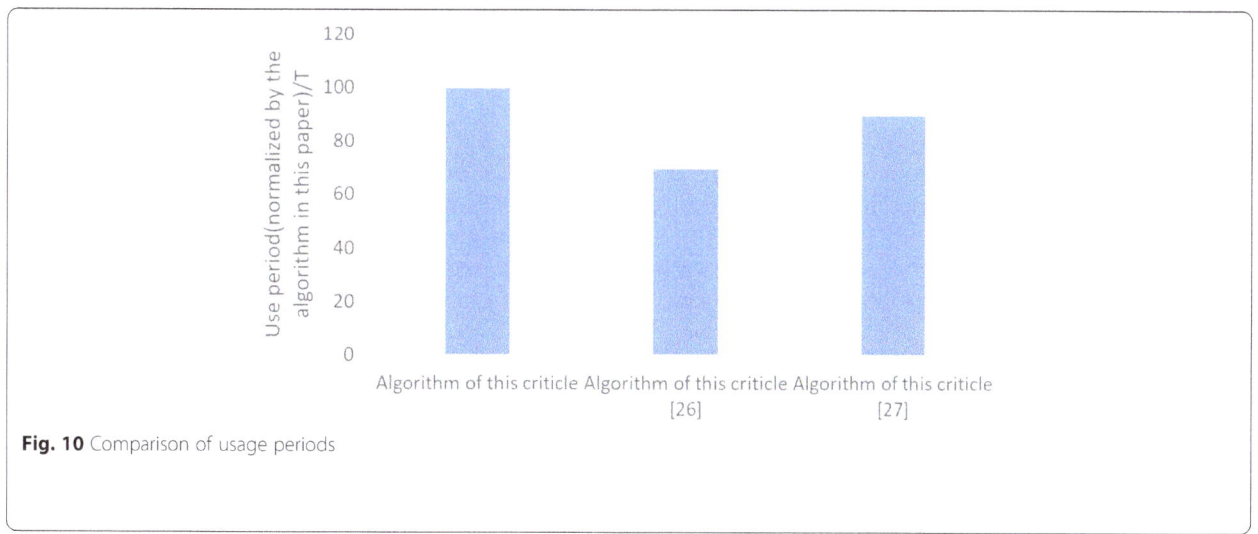

Fig. 10 Comparison of usage periods

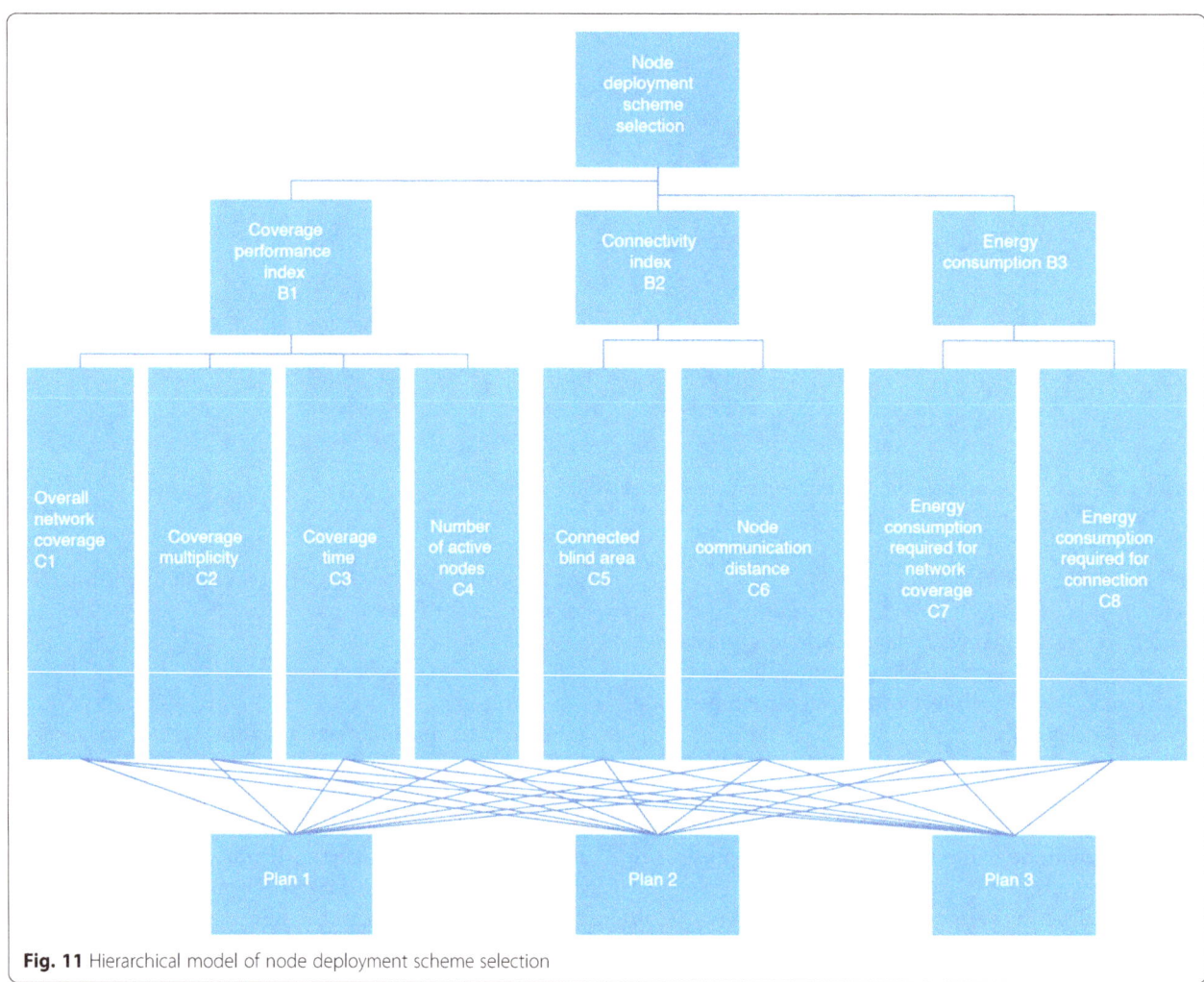

Fig. 11 Hierarchical model of node deployment scheme selection

judgment matrix conforms to the satisfactory consistency standard and can accept the result of hierarchical single ordering. If $CR > 0.1$, the judgment matrix needs to be reconstructed until it passes the consistency test. Finally, the weight of each index in the evaluation index system of wireless sensor network node deployment scheme in storage environment monitoring system is obtained.

Combined with the AHP, the comprehensive performance of the node deployment schemes obtained by different deployment algorithms is compared. The actual grade scores and comprehensive weights are shown in Table 4.

It can be seen from the Table 4 actual rating score table that the comprehensive weight of the deployment scheme obtained by the algorithm is 4.75, and the combined weights of the literature [26] and the literature [27] are 3.3253 and 3.8987 respectively. The comprehensive weight value of the optimized deployment plan is obviously larger than that of before optimization. The comprehensive weight value of the performance of the deployment scheme obtained in this paper is significantly greater than

the other. The comprehensive weight value of the performance of the two schemes, in summary, can be seen that when deploying wireless sensor network nodes in a specific space, the deployment method proposed in this paper has better performance, and the specific spatial partitioning method can be selected according to the actual situation, reducing the number of nodes required for the entire wireless sensor network sensor saves the deployment cost of the wireless sensor network node, improves the coverage of the entire wireless sensor network, and the connectivity between the nodes can be better, and the deployment method can ensure a longer time. The network has a long service life and better comprehensive performance, which satisfies the requirements of wireless sensor network node deployment in the warehouse environment monitoring system.

6 Conclusion and discussion

6.1 Conclusion

Wireless sensor network node deployment is the basis of wireless sensor network application in the warehouse

Table 4 The actual rating scale and comprehensive weights

Level 1 indicator	Level 2 indicator	Level 3 indicator		Actual rating			
		Index	The weight of the elements to the overall system	This article (after optimization)	Literature [26]	Literature [27]	Before optimization
	B1 (0.5714)	C1 (0.4763)	0.2722	5	4	4	4
		C2 (0.2536)	0.1449	5	3	4	3
		C3 (0.0934)	0.0534	4	4	4	4
		C4 (0.1766)	0.1009	4	3	3	3
	B2 (0.2857)	C5 (0.3333)	0.0952	4	3	4	4
		C6 (0.6667)	0.1905	5	3	4	3
	B3 (0.1429)	C7 (0.3333)	0.0476	5	3	4	3
		C8 (0.6667)	0.0952	5	3	4	3
				4.75	3.3253	3.8987	3.90

environment monitoring system, which affects the performance of the entire network and is the primary problem to be solved when performing network applications. In this paper, the deployment of wireless sensor network sensor nodes in the storage environment is taken as the main line. The theory and algorithm related to the deployment of wireless sensor nodes in the application of storage environment monitoring system are studied. The specific contents are summarized as follows:

(1) Explain the relevant theory based on WSN warehouse environment monitoring system

By introducing the importance of introducing monitoring system in the storage environment, the environmental parameters that need to be monitored in several storage environment monitoring systems such as temperature, humidity, illuminance, gas concentration, and dust are designed.

(2) Building a node deployment algorithm based on space-fillable polyhedron.

Based on the description and analysis of the deployment situation in the warehouse environment monitoring system, a node deployment algorithm based on the space partitioning of the fillable polyhedron is constructed, and the offset of the fillable polyhedron used in the deployment algorithm is calculated, and the algorithm is calculated. Finally, the performance of node deployment scheme is analyzed theoretically.

(3) Node deployment case study

Taking the tobacco storage environment as an example, the effectiveness and dominance of the algorithm based on spatially partitionable polyhedral node deployment is verified. Firstly, according to the characteristics

of the tobacco storage environment, the monitoring requirements of the tobacco storage environment are analyzed and the overall architecture of the monitoring system is designed. Secondly, the key design node deployment plan is adopted. Through the analysis of the storage environment structure model, the node deployment algorithm proposed in the paper is used. Deployment: Finally, the performance of the node deployment scheme in the warehouse environment is evaluated. By comparing the schemes obtained by the node deployment algorithm proposed in the literature, the validity and advantages of the proposed algorithm were further proved in the coverage performance analysis, connectivity performance analysis, energy consumption performance analysis and AHP method, and a relatively better scheme was obtained.

6.2 Discussion
Influenced by the complexity of warehouse environment, the limitation of the model and the limitation of personal research ability, the node deployment algorithm based on space fillable polyhedron partition proposed in this paper still has many shortcomings in the specific warehouse environment monitoring system. Future research work will focus on the following aspects:

1. In the design of the unit perception model, this paper constructs a unit probability perception model based on the error rate, which meets the coverage requirements of the current warehouse environment. However, the model is also a coverage model similar to the sphere when applied and less consideration is given to the irregular situation of the unit perception model. In the next work, it is necessary to study the unit perception which is more suitable for the warehouse environment application. Model.

2. The deployment strategy proposed in this paper only considers maximizing the space coverage of the storage environment on the premise of maximizing the coverage capacity of each sensor node. It does not consider the adjustable factors such as signal conflict between wireless sensor networks and transmission power between wireless sensors. In the next work, the deployment effectiveness of these factors will be considered comprehensively. The effect of fruit.

3. This paper only considers the deterministic deployment of wireless sensor nodes in the warehouse environment and does not take into account the random deployment of a considerable number of nodes in the actual warehouse environment applications. Therefore, in future research, the deployment of random nodes will be considered, which makes the model of node deployment in the warehouse environment monitoring system more perfect.

Abbreviations
AHP: Analytic hierarchy process; V.Q.: Volume quotient; WSNs: Wireless sensor networks

Acknowledgments
Not applicable.

Authors' contributions
JM is the main writer of this paper. He proposed the main idea. XJ and XZ gave some important suggestions for this paper. All authors read and approved the final manuscript.

Authors' information
Not applicable.

Funding
The authors acknowledge Science and Technology Research Program of Jilin Provincial Education Department (grant #:JJKH20190239SK) for supporting this study.

Competing interests
The authors declare that they have no competing interests.

Author details
[1]School of Transportation, Jilin University, Changchun, China. [2]School of Mechanical and Aerospace Engineering, Jilin University, Changchun, China.

References
1. S. Khan, *Wireless sensor networks [M]* (CRC Press, Boca Raton, 2016), pp. 10–12
2. O. Atsuyuki, *Spatial tessellations: concepts and applications of Voronoi diagrams* (Wiley, New York, 2010), pp. 32–45
3. M. Younis, K. Akkaya, Strategies and techniques for node placement in wireless sensor networks: A survey. Ad Hoc Networks 6, 621–655 (2008)
4. J. O'rourke, *Art Gallery. Theorems and algorithms. The International Series of Monographs on Computer Science* (Oxford University Press, New York, 1987)
5. V. Coskun, Relocating sensor nodes to maximize cumulative connected coverage in wireless sensor networks. Sensors 8, 2792–2817 (2008)
6. S. Tilak, N. Abu-Ghazaleh, W. B Heinzelman, *Proceedings of the First ACM International Workshop on Wireless Sensor Networks and Applications* (ACM, Atlanta, 2002)
7. M. Abo-Zahhad, S.M. Ahmed, N. Sabor, Rearrangement of mobile wireless sensor nodes for coverage maximization based on immune node deployment algorithm. Comput Electrical Eng 43, 76–89 (2015)
8. Y. Hou, C. Chen, B. Jeng, An optimal new-node placement to enhance the coverage of wireless sensor networks. Wireless Networks 16, 1033–1043 (2010)
9. F. Zhigang, *Wireless sensor network coverage and node deployment research* (University of Electronic Science and technology, Chengdu, 2008)
10. F.M. Al-Turjman, H.S. Hassanein, M. Ibnkahla, Quantifying connectivity in wireless sensor networks with grid-based deployments. J Network Comput Appl 36, 368–377 (2013)
11. A. SMN, Z.J. Haas, Coverage and connectivity in three-dimensional networks with random node deployment. Ad Hoc Networks 34, 157–169 (2015)
12. H.A. Hashim, B.O. Ayinde, M.A. Abido, Optimal placement of relay nodes in wireless sensor network using artificial bee colony algorithm. J Network Comput Appl 64, 239–248 (2016)
13. D. Wang, B. Xie, D.P. Agrawal, Coverage and lifetime optimization of wireless sensor networks with Gaussian distribution. IEEE Trans Mobile Comput 7, 1444–1458 (2008)
14. F.M. Al-Turjman, H.S. Hassanein, M. Ibnkahla, Towards prolonged lifetime for deployed WSNs in outdoor environment monitoring. Ad Hoc Networks 24, 172–185 (2015)
15. Y. Chack-shan, Fast 3D k-coverage determination algorithm for sensor networks based on cube partition. Comput Appl 27, 507–509 (2007)
16. X. Shunping et al., Maximum coverage space optimization based on Voronoi diagram heuristic and group intelligence. J Surveying Mapp V40, 778–883 (2011)
17. C. Jun, *Voronoi dynamic spatial data model* (Surveying and mapping press, Beijing, 2002)
18. Li H. Research on the service area division of the regional market town system based on weighted Voronoi diagram and gravitational model. Information Technology and Applications (IFITA), 2010 International Forum on. IEEE, 2010.
19. S.M.N. Alam, Z.J. Haas, Coverage and connectivity in three-dimensional underwater sensor networks. Wireless Commun Mobile Comput 8, 995–1009 (2008)
20. V.A. Kaseva, Wireless sensor network for hospital security: from user requirements to pilot deployment. EURASIP J Wireless Commun Netw 1, 1–15 (2011)
21. Z. Chang-Jian, On volume quotient functions. Indagationes Mathematicae 24, 57–67 (2013)
22. Oktug S, Khalilov A, Tezcan H. 3D coverage analysis under heterogeneous deployment strategies in wireless sensor networks. Fourth Advanced International Conference on. IEEE Computer Society, 2008, 199–204.
23. K.A.M. Almahorg, S. Naik, X. Shen, Efficient localized protocols to compute connected dominating sets for ad hoc networks. Proceedings of the Global Communications Conference, 2010 IEEE Global Telecommunications Conference GLOBECOM 2010 (IEEE, Miami, 2010)
24. N. Javid, M.B. Rasheed, M. Imran, An energy-efficient distributed clustering algorithm for heterogeneous WSNs. EURASIP J Wireless Commun Netw 151 (2015).
25. H. Na, C. Lee, A full-duplex SWIPT system with self-energy recycling to minimize energy consumption. EURASIP J Wireless Commun Netw 2018, 253 (2018)
26. L. Wang, Research on key technologies of warehousing environment monitoring system based on WSN (Beijing Materials College, Beijing, 2011)
27. A. SMN, Z.J. Haas, Coverage and connectivity in three-dimensional networks with random node deployment. Ad Hoc Networks 34, 157–169(2015)
28. S.A. Khan, F. Dweiri, Fuzzy-AHP approach for warehouse performance measurement. IEEE International Conference on Industrial Engineering & Engineering Management (2016)

Permissions

All chapters in this book were first published in EURASIP JWCN, by Springer; hereby published with permission under the Creative Commons Attribution License or equivalent. Every chapter published in this book has been scrutinized by our experts. Their significance has been extensively debated. The topics covered herein carry significant findings which will fuel the growth of the discipline. They may even be implemented as practical applications or may be referred to as a beginning point for another development.

The contributors of this book come from diverse backgrounds, making this book a truly international effort. This book will bring forth new frontiers with its revolutionizing research information and detailed analysis of the nascent developments around the world.

We would like to thank all the contributing authors for lending their expertise to make the book truly unique. They have played a crucial role in the development of this book. Without their invaluable contributions this book wouldn't have been possible. They have made vital efforts to compile up to date information on the varied aspects of this subject to make this book a valuable addition to the collection of many professionals and students.

This book was conceptualized with the vision of imparting up-to-date information and advanced data in this field. To ensure the same, a matchless editorial board was set up. Every individual on the board went through rigorous rounds of assessment to prove their worth. After which they invested a large part of their time researching and compiling the most relevant data for our readers.

The editorial board has been involved in producing this book since its inception. They have spent rigorous hours researching and exploring the diverse topics which have resulted in the successful publishing of this book. They have passed on their knowledge of decades through this book. To expedite this challenging task, the publisher supported the team at every step. A small team of assistant editors was also appointed to further simplify the editing procedure and attain best results for the readers.

Apart from the editorial board, the designing team has also invested a significant amount of their time in understanding the subject and creating the most relevant covers. They scrutinized every image to scout for the most suitable representation of the subject and create an appropriate cover for the book.

The publishing team has been an ardent support to the editorial, designing and production team. Their endless efforts to recruit the best for this project, has resulted in the accomplishment of this book. They are a veteran in the field of academics and their pool of knowledge is as vast as their experience in printing. Their expertise and guidance has proved useful at every step. Their uncompromising quality standards have made this book an exceptional effort. Their encouragement from time to time has been an inspiration for everyone.

The publisher and the editorial board hope that this book will prove to be a valuable piece of knowledge for researchers, students, practitioners and scholars across the globe.

List of Contributors

Maomao Zhang and Feng Zhai
China University of Mining and Technology, Xuzhou 221000, Jiangsu, China

Pan Feng, Danyang Qin, Ping Ji, Min Zhao and Ruolin Guo
Heilongjiang University, Harbin, China

Teklu Merhawit Berhane
Dire-dawa Institute of Technology, Dire Dawa, Ethiopia

BoSung Kim and JooSeok Song
Department of Computer Science, Yonsei University, 50 Yonsei-ro, Seodaemun-Gu, Seoul, Republic of Korea

Hui Li, Ming Lyu, Jie Zhang and Yuming Bo
School of Automation, Nanjing University of Science and Technology, Nanjing 210094, China

Baozhu Du
Institute for Automatic Control and Complex Systems, University of Duisburg-Essen, 47057 Duisburg, Germany

Yang Yu, Bo Xue, Zhuyang Chen and Zhiwen Qian
School of Electric Information Engineering, Jiangsu University of Technology, Changzhou 213001, China

Nattakarn Shutimarrungson and Pongpisit Wuttidittachotti
Faculty of Information Technology, King Mongkut's University of Technology North Bangkok, Bangkok, Thailand

Fuxiang Liu
Science College and Three Gorges Mathematical Research Center, China Three Gorges University, Yichang 443002, China

Yuanzhen Li, Yang Zhao and Yingyu Zhang
School of Computer Science, Liaocheng University, Liaocheng, Shandong 252059, People's Republic of China

Mengdi Cui and Yinghan Xu
School of Information Science and Engineering, Yanshan University, Qinhuangdao, China

Rongrong Yin and Xueliang Yin
School of Information Science and Engineering, Yanshan University, Qinhuangdao, China
The Key Laboratory of Special Fiber and Fiber Sensor of Hebei Province, Yanshan University, Qinhuangdao, China

Hong Zhang and Zhanming Li
College of Electrical and Information Engineering, Lanzhou University Technology, Lanzhou 730050, People's Republic of China

Wanneng Shu
College of Computer Science, South-Central University for Nationalities, Wuhan 430074, People's Republic of China

Jarong Chou
Deparment of Electrical and Computer Engineering, Michigan State University, East Lansing, MI 48824, USA

Runze Wan, Haijun Wang and Jun Shang
Hubei Co-Innovation Center of Information Technology Service for Elementary Education, Hubei University of Education, Wuhan, China

Naixue Xiong
Department of Mathematics and Computer Science, Northeastern State University, Tahlequah, OK, USA

Qinghui Hu
School of Computer Science & Engineering, Guilin University of Aerospace Technology, Guilin, China

Ye Chen and Anfeng Liu
School of Computer Science and Engineering, Central South University, Changsha 410083, China

Wei Liu
School of Informatics, Hunan University of Chinese Medicine, Changsha 410208, China

Tian Wang
Department of Computer Science and Technology, Huaqiao University, Xiamen 361021, China

Qingyong Deng
The Key Laboratory of Hunan Province for Internet of Things and Information Security, Xiangtan University, Xiangtan 411105, China
College of Information Engineering, Xiangtan University, Xiangtan 411105, China

Houbing Song
Department of Electrical, Computer, Software, and Systems Engineering, Embry-Riddle Aeronautical University, Daytona Beach, FL 32114, USA

Nengxian Liu
College of Mathematics and Computer Science, Fuzhou University, Fuzhou 350108, China

Jeng-Shyang Pan
College of Mathematics and Computer Science, Fuzhou University, Fuzhou 350108, China
Fujian Provincial Key Laboratory of Big Data Mining and Applications, Fujian University of Technology, Fuzhou 350118, China
College of Computer Science and Engineering, Shandong University of Science and Technology, Qingdao 266510, China

Trong-The Nguyen
Fujian Provincial Key Laboratory of Big Data Mining and Applications, Fujian University of Technology, Fuzhou 350118, China
Department of Information Technology, University of Manage and Technology, Haiphong 180000, Vietnam

Zhiyong Wei and Fengling Wang
Nanning College for Vocational Technology, Nanning 530008, Guangxi, China

Shining Li
School of Computer Science, Northwestern Polytechnical University, Xi'an 710029, People's Republic of China

Xueqiang Yin
School of Computer Science, Northwestern Polytechnical University, Xi'an 710029, People's Republic of China
The 15th Research Institute of China Electronic Technology Group Corporation, Beijing 100083, People's Republic of China

Guan Chunyun
College of Agriculture, Hunan Agricultural University, Changsha 410128, China

Zhou Libo
College of Agriculture, Hunan Agricultural University, Changsha 410128, China
College of Information and Electronic Engineering, Hunan City University, YiYang 413002, Hunan Province, China

Huang Tian
College of Information and Electronic Engineering, Hunan City University, YiYang 413002, Hunan Province, China
Hunan Engineering Research Center for Internet of Animals, Changsha 410205, China

Jia Mao and Xiaoxi Jiang
School of Transportation, Jilin University, Changchun, China

Xiuzhi Zhang
School of Mechanical and Aerospace Engineering, Jilin University, Changchun, China

Index